T5-CVD-938

CHICAGO PUBLIC LIBRARY
HAROLD WASHINGTON LIBRARY CENTER
R0006990889

Sulphide Catalysts, Their Properties and Applications

CZECHOSLOVAK
ACADEMY
OF
SCIENCES

Scientific Editor
Pavel Jírů

Scientific Reviewers
Václav Veselý
Ludvík Beránek

Translated by
Ota Sofr

Sulphide Catalysts, Their Properties and Applications

OTTO WEISSER
and
STANISLAV LANDA
Institute of Chemical Technology, Prague

PERGAMON PRESS
Oxford, New York
FRIEDR. VIEWEG + SOHN
Braunschweig

ISBN 0 08 017556 2 (Pergamon)
ISBN 3 528 08293 3 (Vieweg)

1973

Published in co-edition with Academia - Publishing House of the Czechoslovak Academy of Sciences, Prague

Library of Congress Catalog Card No. 72-93153

Printed in Czechoslovakia

List of Contents

B. *Special Section*

Preface

The volume of production using catalytic techniques is unusually high. It is estimated, that 75% of all chemical production is at present based on catalysed reactions and, that with newly built production facilities this proportion will rise to 90%.

Fuels are a major component of the overall industrial output and, of great importance are high-quality fuels for combustion in power plants and in other industrial equipment as well as in households. These fuels which must not contaminate the atmosphere with toxic chemicals when burnt are difficult to produce without the use of catalysis. It is in the processing of low-grade natural raw materials (high-sulphur petroleum, coal and products of coal carbonization, shale oils etc.) that highly resistant sulphide catalysts are of great importance, operating efficiently under reaction conditions where other types of contact catalysts fail completely.

Production processes in which sulphide catalysts are used account for a major part of the fuel industry. It is surprising therefore that one of the industrially most significant groups of catalysts is still being neglected to a large extent by theoreticians. In consequence, an empirical approach continues to prevail in the problems of sulphide catalysis and research continues to be concentrated on technological application problems. A wider fundamental research into the physico-chemical character of these catalysts is becoming increasingly necessary.

The present monograph, which continues the work started by an earlier bibliographical review [2177] originated from the need for a comprehensive assessment of this important field of catalysts. It includes basic literature and patent data up to the year 1971. The authors do not claim that the list of literature included in this monograph is absolutely complete. In the region of technological applications, recording and analysis of all articles and, especially, of all patents would be outside the scope of this work. For the sake of completeness, on the other hand, some processes are mentioned particularly in Chapter 7., which mainly deals with technologic aspects of the topic, which are no more relevant to contemporary industry (this mainly applies to some processes of coal and tar treatment with the use of sulphide catalysts). In many cases, however, these processes are important because similar modern processes have been developed from them or, experience gained in them has been utilised in other processes.

This book has been written for, among other reasons, a wish to stimulate theoreticians to take a deeper interest in this significant region of heterogeneous catalysis. If this meets with at least partial success, and should this monograph become a source of essential basic information for technologists, it will have satisfied the main purpose for which it was written.

The authors wish to thank the following Publishing Houses for their kind permission to reproduce in this book the figures published in the cited literature: Academic Press, Inc., New York (Fig. 9, 11, 34, 35, 36, 37, 38, 43, 44, 50, 51, 58, 59, 62, 86); Litton Educational Publishing International, New York (Fig. 9, 10, 57, 105); American Chemical Society Publications, Washington (Fig. 2, 23, 24, 30, 32, 33, 39, 40, 52, 54, 63, 67, 75, 82, 97, 98); American Institute of Chemical Engineers, New York (Fig. 65); Gulf Publishing Co. Publications, Houston (Fig. 68, 74, 84, 85, 87, 88, 89, 90, 91, 96); The Institute of Petroleum, London (Fig. 66); Japan Petroleum Institute, Tokyo (Fig. 71); Verlag Chemie, G.m.b.H., Weinheim, Bergstr. (Fig. 18, 34, 35, 36, 38, 48); Verlag W. Girardet, Essen (Fig. 92); Verlag Johann Ambrosius Barth, Leipzig (Fig. 1, 3, 8); VEB Deutscher Verlag für Grundstoffindustrie, Leipzig (Fig. 14, 31, 41, 45, 72, 104); Izd. Nauka, Leningrad (Fig. 99, 100); Izd. Nauka, Moscow (Fig. 6, 7, 12, 13, 16, 17, 19, 25, 26, 27, 28, 42, 54, 83, 101, 102, 103); Izd. Chimija, Moscow (Fig. 55, 56, 60); Izd. Nedra, Moscow (Fig. 64); Izd. AN Turk. SSR, Ashkhabad (Fig. 61); Izd. Bashk. Fil. AN SSSR, Ufa (Fig. 60); Academia, Prague (Fig. 4, 5, 46, 47).

At this point, we feel it is our duty to thank all those who helped us in completing this monograph by advice and critical discussion. We wish particularly to thank the scientific editor, Dr. P. Jírů as well as Prof. V. Veselý and Dr. L. Beránek, who reviewed this monograph and aided us with many remarks concerning its scientific content and wording. We are further indebted to Prof. J. Mostecký and Dr. R. Kubička for valuable discussion. We wish to thank many of our collaborators at the Department of Synthetic Fuels and Petroleum of the Institute of Chemical Technology, Prague, for technical aid in writing this book.

Our most sincere thanks are due to the staff of the Publishing House of the Czechoslovak Academy of Sciences as well as all others who contributed towards the publication of this book.

Prague, February 1971

The Authors

A.

GENERAL SECTION

1.

HISTORICAL REVIEW OF THE DEVELOPMENT AND APPLICATION OF SULPHIDE CATALYSTS

Technical heterogeneous catalysis, which doubtless influenced the development of modern chemical technology in a decisive manner, grew out of the pioneering work of Sabatier, Ipatieff, Kuhlmann, Haber and a number of other scientists. The research of these workers became the basis of modern catalytical large-scale production techniques like, for example, the synthesis of ammonia, sulphuric acid and of a number of other products of the organo-chemical and fuel industries.

The development of these efficient catalysts for the most important catalytical processes, however, was related from the beginning with the problems of deactivation and catalyst poisoning, their regeneration and reactivation. Especially sulphur compounds, which are frequently present in technical raw materials, had a deteriorating effect on the majority of catalysts employed, making it practically impossible to carry out the required reactions with classical types of catalysts unless these sulphur compounds were first removed. On the other hand, satisfactory removal of these compounds before the reaction is very unfavourable from the economic aspect and, in the special case of hydroprocessing of basic raw materials in the fuel industry (coal, petroleum, tar and petroleum distillation residues) preliminary sulphur removal is economically impossible. A satisfactory solution to this problem was only made possible by the suggestion, revolutionary at the time, of using sulphur compounds, either originally contained in the raw material or added intentionally, to catalyse the conversion of components of this raw material. Classical evidence of the successful application of this idea is the production of motor fuels by hydrogenation of coal and of its carbonisation products.

Berthelot's historical experiments [150], in which coal was successfully reduced at 270 °C by means of the effect of hydroiodic acid on distillable oil, were carried out some one hundred years ago. They became the starting point of wide-spread research in the field of coal hydrogenation, leading to the origin of a new field of technology. In hydrogenating different types of coal, Fischer [1101] replaced hydroiodic acid by other hydrogen sources, e.g. mixtures of sodium formate and water, carbon monoxide and steam, etc.

Considerable progress was made in 1910, when on the basis of Landsberg's suggestions, Bergius started to experiment with hydrogenation of heavy petroleum distillates using molecular hydrogen [144]. In this process, cleavage reactions were found to take place at temperatures above 400 °C, gasoline fractions being formed in the process.

In 1913, Bergius and his collaborator Billwiller achieved a decisive break-through by hydrogenating coal at 400 to 450 °C and a hydrogen pressure of 150 atm, obtaining an 80% yield of gaseous, liquid and benzene-soluble fractions [141, 142, 144, 145]. The authors named utilized the experience gained in hydrogenating coal to hydrogenate tars and oils, too [146].

Although the yield of liquid products in the Bergius' process of coal hydrogenation was large, the quality of the product of hydrogenation was inferior in terms of motor fuel application. The hydrogenation product mainly contained high-molecular weight hydrocarbons, as well as a large amount of unwanted oxygen, sulphur and nitrogen compounds. The reason for these difficulties was a very low rate of hydrogenation and hydrocracking, since the original Bergius process was non-catalytic [141, 1091, 1362, 2031].

Bergius ascribed the formation of high-molecular weight substance to the polymerising effect of sulphur contained in the raw material, since coal hydrogenation proceeds more readily when some oxides, mainly iron oxides, are added. With the use of these oxides, even bituminous coal containing around 85% carbon could be liquefied. Bergius believed that the oxide added only binds sulphur in the original material, thus eliminating its inhibitive effect on hydrogenation of the coal mass [143, 144, 2031].

It was only later that scientists of the I. G. Farbenindustrie A. G. [1091, 1092] as well as Varga [2104] showed, that the main effect of ferrous oxide was catalysis of the hydrogenation and coal mass cleavage processes as well as of desulphurisation of the raw material. Obviously, ferrous sulphide formed in the process also acted as a catalyst.

When, in 1924 [925] the research staff of the Badische Anilin u. Sodafabrik in Ludwigshafen continued the research started by Bergius, they were able to look back on the outstanding success achieved in the application of heterogeneous catalysis to ammonia and methanol syntheses. It was clear from the beginning, however, that new catalyst types, especially resistant to the effect of catalyst poisons, would have to be found for hydrogenating the highly complicated mixtures of organic substances which form the coal mass or tar. After systematic research led by Krauch and especially by Pier, metal sulphides, especially molybdenum and tungsten sulphides were found to be the catalysts required [1101].

Varga, who reported the positive influence of sulphur and sulphur compounds on the activity of molybdenum and tungsten catalysts in destructive tar hydrogenation [2106, 2115, 2120, 2122], arrived at similar conclusions a little later. In 1933, an interesting discussion of priority ensued in respect of this finding and of the application of sulphide catalysts between M. Pier [1625] and I. G. Farbenindustrie [822, 823] on the one hand and Makray [1276, 1277], Mestern [1362, 1363] and Szeszich and Hupe [2031, 2032] on the other; involved, also, was the problem of priority in the application of heterogeneous catalysts for the processing of fuels in general.

It must be stressed in this respect, that the use of catalysts in hydroprocessing of fuels was known to be advantageous even before Bergius. For example, Day [377] demonstrated the advantage of using platinum and palladium catalysts in hydrogenating petroleum distillates as early as 1906. In 1912, Theodorin reported in his thesis [1362] the results of hydrocracking of Romanian petroleum, using iron and nickel as catalysts. The most important results of these initial experiments were reported by Klever [1045], who used some metals and their compounds (Ag, Zn, Cd, Pb, Cu, etc.) as catalysts in the hydrogenation of petroleum distillation residues, tar oils and bituminous as well as brown coal bitumen. It is interesting, that Klever likewise mentions the possibility of simultaneous hydrodesulphurisation of the raw material, which takes place during the high- pressure hydrogenation process. McComb [1343] later used molybdenum and nickel to hydrogenate oils. Howard [801] used molybdenum and chromium oxide as well as cadmium sulphide to hydrogenate coal paste. Noting finally the application of sulphide catalysts in the field of gas production, where Carpenter [270, 271] used nickel subsulphide [503, 659] for hydrodesulphurisation of carbonisation gases, we have before us the state of the art of heterogeneous catalysis application in the field of fuel raw material hydrogenation. Using this information the staff of the BASF in Ludwigshafen set out in their systematic research of the catalytic activity of a wide range of inorganic substances with the purpose of finding the optimum catalysts for fuel hydrogenation.

The results of this intensive work, carried out by the Badische Anilin und Soda Fabrik and later by I. G. F. [1629] are laid down in a large number of patents and technical reports [416], while there are only a few experimental studies published in the technical literature of that time. The reasons for this secrecy were commercial at first, while later strategical reasons became important. The paper by Pier [1626], regarded as classical nowadays, proves that applied research into sulphide catalysts was accompanied by research into the basic properties of these catalysts.

After 1930, fundamental research into the properties and application of sulphide catalysts developed in some laboratories outside of Germany, too. The work of the Soviet scientists (Moldavskii, Karzhev, Nemtsov, Prokopets, Lozovoi, Eru, Orlov, Ŕapoport and others) dealt mainly with the mechanism of hydrogenation of pure substances, including hydrocarbons, phenols, and heterocycles [1722]. Furthermore, research into the hydrogenation of coal and of the coal mass was done by the Bureau of Mines (Storch et al.). Other studies in this field were published by British authors (Gordon, King, Cawley), Japanese (Ando, Kurokawa), Italian (Roberti) and other scientists.

In the period just before and during the Second World War, sulphide catalysts began to be applied to petroleum processing, too. This was mainly caused by the rapid development of catalytic hydrodesulphurisation, although sulphide catalysts also started to play a role in dehydrogenation of cyclanes to aromatics and especially research into the applicability of these catalysts to hydrocracking processes developed rapidly.

In the post-war period, a decline was recorded in chemical coal processing in Europe, while on a world-wide scale and especially in the USA, catalytic processes began to be applied intensively to petroleum processing. The application of sulphide catalysts especially in hydrorefining and hydrocracking of petroleum products developed rapidly. In this field, outstanding success was achieved by the staff of a number of American companies (Hydrocarbon Research Corp., Gulf Research Dev., Esso Research and Eng., Universal Oil Prod. Co., etc.). European catalytic procedures, e.g. those worked out by BASF, British Petroleum, I.F.P., the Leuna-Werke (GDR), Chemical Works in Most (Czechoslovakia), and in a number of Soviet research institutions achieved practical success, too.

The research and technical application of sulphide catalysts is developing at a rapid rate, today, evidenced not only by the rapidly increasing number of studies published in the literature, but also by the increasing number of technological processes in which these catalysts are employed.

2.

THE SIGNIFICANCE, TECHNICAL APPLICATION AND SPECIFIC POSITION OF SULPHIDE CATALYSTS

2.1.

The significance of sulphide catalysts in chemical technology

It will be clear from the brief historical survey presented in the preceding chapter, that the development of this type of catalyst is immediately related to the development of chemical technology of fuels and, recently, of petrochemistry. The volume of catalytic processes in the fuel industry is enormous nowadays [1619], exceeding by far the application of heterogeneous catalysis in other fields of chemical technology. Comparing the production of sulphuric acid (the world production was 65 million tons in 1965) and of ammonia (the world production was 17.5 million tons in 1965) [1784], as the two major industries which use heterogeneous catalysis techniques outside the fuel industry, with the annual volume of catalytic processes in the refining industry which, in 1966, was nearly three quarters of a billion tons (excluding the socialist countries) [314], we see that the petroleum industry processes, by means of catalytic techniques, amounts of raw materials greater by one order of magnitude, and that its annual consumption of catalysts will be at least proportionately greater.

On a world-wide scale, catalytic processes formed more than 50% of the overall processing capacity of refineries in 1966 [314]. In technologically highly developed countries (USA and Canada) the absolute majority of all fuel raw materials (85%) is being processed by means of catalytic processes.

From the point of view of sulphide catalysts, the proportion of hydroprocessing (i. e. hydrotreating and hydrocracking) in catalytic processing of petroleum materials is very significant. Considering the constant growth in the proportion of sulphur containing petroleum in the overall volume produced, desulphurisation of petroleum fractions becomes a major role of refining processes. Refining catalysts in the "steady state" therefore contain metal sulphides as significant active components, even though the original form of the catalyst is a different one (metals or oxides). Similarly, in the case of hydrocracking, increasing amounts of low-quality distillates and especially of residual fractions with a high content of sulphur and of other contaminants are being processed, so that here also the use of sulphide catalysts is achieving prime importance.

The development of catalytic hydroprocessing in the refineries in individual regions is summarised in Table 1. It is to be seen from this table, that in 1966 hydroprocessing already formed nearly 1/3 of the processing capacity of refineries in North America (reforming processes formed 19%, catalytic cracking 36.4%). In the next few years, considerable growth in the proportion of these processes is expected all over the world

Table 1.

Development of Catalytic Hydrogenation Processes (Hydrocracking and Hydrorefining) in Different Regions (Excluding the socialist countries, data in 10^3 m^3/day) [314].

Region	Year					
	1956	1958	1960	1962	1964	1966
USA — Canada						
Capacity	70.0	222.7	337.7	429.6	506.1	589.4
% of processing capacity	4.6	13.3	18.8	23.2	26.9	31.2
Western Europe						
Capacity	11.6	13.6	42.1	69.0	134.0	179.7
% of processing capacity	3.0	3.0	6.7	8.7	13.5	13.3
Middle East						
Capacity	0.6	4.6	4.6	5.4	10.8	14.6
% of processing capacity	0.3	2.3	2.0	2.2	3.8	4.8
Africa						
Capacity	—	—	0.9	1.6	7.0	12.7
% of processing capacity	—	—	4.9	7.1	11.9	13.0
Latin America						
Capacity	—	3.8	6.3	9.2	16.2	29.7
% of processing capacity	—	0.9	1.4	1.7	2.7	5.0
Asia — Pacific Ocean						
Capacity	1.2	0.9	8.1	15.2	33.7	72.0
% of processing capacity	1.1	0.5	3.7	5.0	8.3	12.9

[314]. In the USA, the major part of hydrorefining is hydrotreating of gasoline and middle petroleum fractions (over 85%) [314]. In Europe, the hydrorefining capacities for medium distillates are relatively greater because of the greater consumption of these fractions. Recently, a rising trend has been recorded in hydrotreating of raw materials for catalytic cracking as well as in hydrorefining lubricants. Production volumes of modern hydrorefining units are nowadays so large that they are comparable with the entire production capacity of former refineries: for example, the capacity

of a modern unit for hydrorefining of the raw material for catalytic cracking is nearly 2.3 million tons per year [314, 706].

A very rapid development is likewise recorded in hydrocracking, the speed of this development in the period 1963–1968 is similar to the rate of development of reforming in the 1949 to 1957 period, or that of catalytic cracking between 1938 and 1941 [1450]. While in 1963, the world-wide capacity of hydrocracking units was around 10,500 m^3/day (data excluding the socialist countries) the production volume of this process rose in the following years very rapidly to 43,000 m^3/day in 1966 [1838] and to 122,000 m^3/day (distillates hydrocracking) in June 1970 [110 A], In this field also, enormous units are being constructed at present, e.g. the Isomax unit in the Standard Oil of California refinery at Richmond has a capacity of about 3.7 million tons/year (80,000 BSD) [1768]. The hydrocracking process is becoming one of the main technological stages even now, when processing of residual fractions or production of petrochemical raw materials is involved [1622, 1838].

Thus, the significance of sulphide catalysts has shifted recently from the original field of coal and tar processing into the field of petroleum processing. Of the original 12 German plants hydroprocessing coal and tar, the production capacity of which was about 4 million tons/year in 1945 [119, 1101], only two remain in operation on the basis of the original or slightly modified processes (one continues in part to operate in Czechoslovakia). In 1958, hydrogenation of tar materials in the British plant at Billingham was stopped [320, 2161]. The wide experience, gained in catalytic hydrogenation of coal and heavy tar materials, is however being used nowadays in the development of new processes for the treatment of heavy petroleum residues and coal.

Especially in Europe, where radical technological change has taken place due to the conversion from coal to petroleum, a close continuity is seen in the field of catalytic hydrogenation processes and application of sulphide catalysts with the earlier results and experience. It is, moreover, clear with respect to limited petroleum reserves, that the return to the use of coal as a basic chemical raw material will become essential in future. Thus the large investments into modern coal hydrogenation research, especially in the USA [19, 73], and the related research into new types of hydrogenation catalysts are not surprising.

However, sulphide catalysts are beginning to be used outside the fuel industry, too. Their application is being considered in all those cases, where the raw materials processed contain substances which are catalytic poisons to the other, frequently more active and selective catalysts. For example, this is the case in hydrogenation of technical nitrocompounds to amines as well as in hydrogenation of sulphur-containing hydroformylation products. Practically all catalytic syntheses of sulphur-containing substances (mercaptans, sulphides, sulphur heterocyclics etc.) are catalysed by metal sulphides or by catalysts in which the sulphide component plays a decisive catalytic role.

2.2.

General survey of the properties of sulphide catalysts

In a number of properties, sulphide catalysts differ from the classical, mainly metallic catalysts. As in the case of oxide catalysts, metal sulphides are intermediate between metallic catalysts and insulators due to their semiconductive properties. Hence their polyfunctional catalytic character, so that besides oxidation-reduction properties, sulphide catalysts likewise have the properties of acid catalysts, i.e. they speed up isomerisation, cleavage and dehydration reactions.

Table 2.

Values of Apparent Activation Energy and Temperature Coefficients of Hydrogenation Reactions for Various Catalysts [21].

Catalyst	Temperature range °C	Substance hydrogenated	Apparent activation energy E (kcal/mole)	Temperature coefficient[a]) k_t
Ni	80—100	Benzene	12.7	1.12
Cu	160—200	Benzene	10.0	1.10
Pd on asbestos	140—225	Benzene	10.3	1.10
Pd on asbestos	140—225	Toluene	12.1	1.12
MoS_2	400—460	Toluene	23.1	1.25
MoS_2	330—370	Phenol	37.0	1.42
MoS_2	330—370	*o*-Cresol	33.0	1.37
MoS_2	210—270	Thiophene	34.0	1.33
NiS	210—270	Thiophene	32.0	1.36
MnS	210—270	Thiophene	28.0	1.30
CdS	210—270	Thiophene	28.0	1.30
CuS	210—270	Thiophene	17.0	1.18

a) The temperature coefficient k_t denotes the ratio of rate constants with a temperature interval of 10°C. The following approximate relationship holds between the temperature coefficient and activation energy [1520, 2137]: $k_t = \exp(6.9315/\gamma)$, where $\gamma = (1.37 \,.\, T^2/E - 1.37 \,.\, T)$ is the difference of temperatures which results in the reaction rate being doubled.

Metal sulphides are typical high-temperature catalysts. Differing from metallic catalysts, a number of which are active in hydrogenation at only room temperature, sulphide catalysts mostly become active at high temperatures only (from 200 to 250 °C). Comparison of the values of apparent activation energies and temperature coefficients of some hydrogenation reactions taking place in the presence of metals and sulphides, shown in Table 2, indicates that sulphide catalysts are considerably more sensitive to temperature variations. Altman and Nemtsov [21] demonstrate this fact by comparing the hydrogenation of toluene in the presence of palladium on asbestos and MoS_2 (Table 2). Assuming constant activation energies in the entire temperature interval considered, and assuming also that the reaction rate at 100 °C is 100 times greater on the palladium catalyst, it may be shown that at 450 °C MoS_2 would be thirteen times more active than the palladium catalyst.

A high adsorption capacity for hydrogen and unsaturated hydrocarbons is a significant property of the most important sulphide catalysts [416]. Compared with metal catalysts, MoS_2 for example is distinguished by a lower heat of hydrogen adsorption. The variation of the hydrogen surface concentration with temperature is considerably less on this catalyst compared with metallic ones, so that a sufficient hydrogen concentration is maintained on the catalyst surface in hydrogenation reactions even at high reaction temperatures [21].

The demands on the texture of sulphide catalysts are very great. These catalysts are only rarely used in the pure sulphide form. In most cases, they are supported on carriers, which, in practically all cases, have a significant influence on the overall catalytic efficiency of the supported catalyst, modifying considerably its selectivity for a given reaction. The activity of such a catalyst, e.g. for hydrodesulphurisation or hydrocracking depends on the fundamental physical properties, i.e. on the surface area, mean pore size and total pore volume. Increasing these properties to the optimum value increases the activity of the applied catalyst [157] (cf. p. 345).

The mean pore diameter is of special importance to the quality of the catalyst. The large size of molecules of hydrocarbon as well as of non-hydrocarbon components of technical mixtures makes it essential, that the catalyst has large inlet and outlet channels, as this is the only means of perfect utilisation of the inner catalyst surface, thus securing a high reaction rate [1730]. These demands are especially important when sulphide catalysts are to be used for hydrorefining of high-boiling raw materials in the mixed phase [2137].

Compared to other types, sulphide catalysts have a special position due to their outstanding resistance to the effect of catalyst poisons. Conventional catalyst poisons, especially sulphur compounds, which completely eliminate the function of metallic and of some other types of catalysts, not only do not decrease the catalytic activity of sulphides, their presence in fact is needed to maintain a high activity and long lifetime of the catalyst. Sulphide catalysts are very resistant to carbonaceous deposits on the catalyst surface, provided the partial pressure of hydrogen in the reaction is sufficiently high to keep up the process of "purification" of active catalyst centres. Thus, sulphide catalysts may be used in hydrocracking and hydrotreating even of easily adsorbing resinous and asphaltic substances, which coke easily on other catalysts, deactivating them totally.

There are only a limited number of substances which cause efficient deactivation of sulphide catalysts or, which act as true catalyst poisons. One of the worst types are

Table 3.
Examples of Typical Sulphide Catalysts Practically Used in the Industry.

Catalyst type		Catalyst notation	Catalyst composition in the fresh (nonworking) state	Origin of the catalyst	Application	Reference
Hydrocracking catalysts	For destructive hydrogenation of the heaviest fuels and petroleum residues	10927	$FeSO_4 + Na_2CO_3$ on dust from Winkler generator (5% Fe)	I. G. Farbenindustrie	Destructive hydrogenation of the heaviest fuels (I. G. F., Varga-process)	[317] [1101] [1172]
		—	$CoO + MoO_3$ on aluminosilicate (?)	Gulf Research Dev. Co.	Hydrocracking of residual petroleum fractions in the Gulf-HDS process	[155] [2137]
		—	$CoO + MoO_3$ on Al_2O_3	Hydrocarbon Research	Hydrocracking of residual petroleum fractions in the H-Oil process	[1618] [1621] [2137] [2148]
	For hydrocracking tar and petroleum distillates	6434	10% WS_2 on clay activated with HF	I. G. F.	Hydrocracking of tar distillates (vapour phase)	[1101]
		—	NiS on aluminosilicate (?)	California Research Corp.	Hydrocracking of medium petroleum distillates by the Isocracking process	[2148]
Refining and destructive-refining catalysts		3510	53.5% MoO_3 + 30% ZnO + 16.5% MgO	BASF	Hydrotreating and hydrocracking of medium oils from brown coal tar	[793] [1101]
		5058	WS_2	I. G. F.	Hydrotreating and hydrocracking of medium oils	[793] [1101]
		9062	7.5% MoO_3 + 7.5% WO_3 + 35% Al_2O_3 + + 50% SiO_2	VEB "Leuna Werke"	Hydrotreating and hydrocracking of medium oils	[1429]
		3076	NiS : WS_2 = 2 : 1 (unsupported)	I. G. F.	Hydrogenation catalyst of high efficiency (hydrogenation of olefins)	[317]
		—	NiS-WS_2 (40% W and 25% Ni, non-supported)	Shell Oil Co.	Unsupported desulphuration catalyst	[1743]

Refining and destructive-refining catalysts	8376	27% WS_2 + 3% NiS on γ-Al_2O_3	I. G. F.	Hydrotreating of tar-oils in vapour-phase, MTH-process	[1101]
	8376 ox.	26% WO_3 + 4.5% NiO + 69.5% Al_2O_3	VEB "Leuna Werke"	Hydrotreating of tars and petroleum distillates, MTH-process etc.	[1101] [1429]
	8376 ANTIAS	Type 8376 with increased Ni content	Chemical Works Záluží (ČSSR)	Hydrotreating of arsenic-containing tar	[2015]
	7846	10% MoO_3 + 3% NiO on Al_2O_3	I. G. F. resp. BASF	Hydrotreating of medium oils	[1429]
	8197	17% MoO_3 + 4.5% NiO + 78.5% Al_2O_3	VEB "Leuna Werke"	TTH-process	[1429]
	U 63	12% WO_3 + 8% MoO_3 + 4.5% NiO + 75.5% Al_2O_3	VEB "Leuna Werke"	Hydrotreating of tar and petroleum distillates	[1429]
	Co–Mo catalysts: BASF 0852, Nalco 471, Nalco 810, Aero HDS-2, Oronite, Gildler G-35 B, Houdry 200-A, Comox P. Spence, Cyanamid HDS 1, Co–Mo (USSR), 7362 (ČSSR)	For composition, see Table 78	Different American and European Companies	Hydrorefining and desulphuration catalysts for different purposes	[1743]
	BR 86	2.3% CoO + 0.8% NiO + 17% MoO_3 + 80% Al_2O_3	VEB "Leuna Werke"	Improved refining Co–Mo catalyst	[1429]
	—	1.0% Co + 0.5% Ni + + 8.3% Mo on Al_2O_3	Davison Chem. Co.	Improved hydrorefining Co–Mo catalyst	[1743]
Dehydrogenation catalysts	IG 5615	NiS : WS_2 ~1 : 2 (nonsupported	I. G. F.	Dehydrogenation of cyclanes to aromates	[1347]
	—	NiS : WS_2 ~2 : 1	Shell Dev. Co.	Dehydrogenation of cyclohexane and methylcyclohexane to aromatic substances	[1049]

arsenic compounds, which occur mostly in some tars and which are the cause of irreversible poisoning of hydrorefining sulphide catalysts [2001]. Similarly, the ash of some tars [703] or trace elements in petroleum fractions [1201] cause gradual deactivation and a change in the selectivity of sulphide catalysts. The relative ease of regeneration of supported sulphide catalysts, however, makes them economically feasible. Careless handling of sulphide catalysts in the air may, in some cases, also cause a considerable decrease in their activity [1429].

Thus, metal sulphides are highly resistant catalysts, the use of which especially in fuel processing, has enabled an intensive utilisation of heterogeneous catalysis in processing low-quality natural raw materials. They have therefore been one of the important factors in the origin of new, modern production processes.

The I. G. Farbenindustrie A. G., which was a pioneer of the application of sulphide catalysis in chemical technology, has introduced a numerical notation for the different catalysts produced by this company. In a number of cases this notation has been accepted generally, and continues to be used in some cases today. Table 3 summarises examples of the most important sulphide catalysts, which are either made immediately in the form of sulphides, or, oxides are converted *in situ* to sulphides. This may be done either directly with hydrogen sulphide or, sulphide formation may take place in the first stage of the process by means of hydrogen sulphide generated by the degradation of sulphur compounds in the raw material. Since details of the catalyst composition (especially in modern catalytic processes) usually are not available, only the most important catalysts are mentioned as examples, the composition of which is stated in the journals, patent specifications or catalogue literature.

3.

PREPARATION AND REGENERATION OF SULPHIDE CATALYSTS

3.1.

Preparation of sulphide catalysts

Natural sulphide minerals or, sulphides precipitated in the conventional way, are frequently quite inefficient as catalysts or, their activity may be very low (1394, 2175]. Therefore, catalytically-active sulphides must be prepared by means of special procedures.

Like other catalysts, sulphides may be used either in the pure form or supported on carriers; individual sulphides may be employed or their mixtures, in different combinations and ratios. To enhance the activity and lifetime of the catalyst and to achieve the required degree of selectivity, promoters are often added to the basic catalyst (cf. p. 91).

3.1.1.

Preparation of catalysts by the action of H_2S on suitable starting substances

Sulphide catalysts are most often prepared by the direct action of hydrogen sulphide or of some other sulphurising or sulphideforming agent on suitable starting substances. The preparation of individual sulphide catalysts starts in some cases from a solution of a suitable starting compound, which is precipitated with hydrogen sulphide under conditions which form a catalytically active sulphide. In this way, for example, active MoS_3 [946] and Re_2S_7 [226] catalysts were prepared; for details of the conditions of rhenium sulphide precipitation, see [1757a, 2065]. The preparation of an active molybdenum sulphide by precipitation of molybdic acid with hydrogen sulphide in the presence of H_3PO_4 was also reported [881]. Solutions of sulphides are sometimes used as precipitation agents: in this way, a molybdenum sulphide catalyst was prepared by precipitating an ammonium molybdate solution with sodium sulphide [1244, 1658], or an iron sulphide catalyst by precipitating a ferric chloride solution with ammonium sulphide [897, 1630]. Co and Ni polysulphide catalysts were precipitated from solutions of the respective salts by means of alkaline polysulphides [508, 509].

Precipitation from solutions is sometimes done in the presence of the carrier, so that the sulphide is supported on the carrier at the same time. In this way, for example, platinum sulphides supported on Al_2O_3 were prepared for gasoline reforming [466] and for hydrocarbon isomerisation [182]. Alternatively, there is a mixed sulphide catalyst Ni–Mo supported on magnesium silicate for use in hydrorefining [496] or W and Fe sulphides for pressure hydrogenation [896].

Sulphides, freshly precipitated from solutions of the respective salts react readily with oxygen. This partly oxidises adsorbed or occluded H_2S to elementary sulphur, which settles on the catalyst surface and to some extent influences the activity (especially initial activity) of the sulphide catalyst (cf. section 5.4., p. 107). Besides, partial oxidation of the surface layer takes place, again influencing the properties of the catalyst [1365].

In the majority of cases, active sulphides are prepared by sulphurising the starting compounds in the solid state. Sulphide formation is achieved by means of hydrogen sulphide or a sulphur compound which forms H_2S on hydrogenation (e.g. carbon disulphide, mercaptan, disulphides [4a, 303, 374b, 722c, 817a, 1446a, 2096, 2103f] etc.). Sulphide formation proceeds at elevated temperatures for several hours. In the process, the catalyst is usually converted to a sulphide completely, which is assured by introducing a large excess of the sulphide forming agent.

Sulphurisation is usually preceded by drying, calcination and catalyst forming. In order to assure complete conversion of the metal components to sulphides, the process is repeated twice [576, 2224]. In some cases, however, sulphide formation is intentionally carried out to a certain stage only; e.g. when preparing a catalyst for gasoline hydroforming, partly treated NiO was used [422].

Conversion of the starting substances to sulphides is not done by a unified procedure, since practically every sulphide catalyst demands special conditions of preparation and, in many cases, a special procedure for preparing the starting material. A typical example is the preparation of the active sulphide Ni–Mo refining catalyst, the purpose of which is to remove sulphur and nitrogen from light petroleum fractions [219]:

The initial catalyst mass is first pretreated with a steam-air mixture, by means of which the active components are converted to oxides. The catalyst is pre-heated in a stream of hot air so as to bring the temperature of the catalyst bed to a maximum of 400 °C. Steam is then added (the composition of the mixture should be 1 : 1 by wt.) and the pressure on the catalyst outlet is kept at 0.35 atm. The activation process is continued for 16 hours, after which the catalyst bed is cooled down to 230 °C and the catalyst is further cooled by a stream first of air and then of nitrogen.

The oxides formed are converted to sulphides at 150 to 200 °C, care being taken to keep the temperature from rising above 235 °C; the H_2S content in the sulphiding gas must therefore be controlled, usually at 8 to 12% by wt. When sulphide formation with H_2S is impracticable, a different source of sulphur may also be used, e. g. CS_2, mercaptans, etc., added to the raw materials in concentrations of 1 to 2% S, the temperature being kept below the critical limit.

Sulphurising is continued, until large amounts of H_2S start to penetrate into the effluent. The H_2S concentration in the sulphurising gas is then reduced to 3 to 5%, the temperature is raised to 315 °C, the H_2S inlet is stopped and the feed introduction is begun.

It is recommended to maintian the prescribed procedure accurately in order to obtain maximum activity of the catalyst. It is not recommended that the sulphurising stage be avoided by directly treating with the raw material containing sulphur compounds. In this case sulphurisation would be incomplete and the catalyst would be damaged by contact with hot hydrogen when in the nonsulphide state. Also, preliminary sulphurisation considerably decreases the initially very high rate of coke deposition on the catalyst surface. At the beginning of the sulphurisation process, a very low H_2S concentration must be maintained in order to avoid the temperature from rising in an uncontrolled manner by the influence of highly exothermic reactions. With greater H_2S concentrations in the sulphurising gas, higher sulphides are formed (in the case of Ni, Ni_6S_5 up to NiS), which however convert to stable modifications on reduction (Ni_3S_2). For perfect sulphurisation a minimum excess of 130% H_2S must be employed.

Oxidative pretreating with air and steam is advantageous in the case of a Ni–Mo refining catalyst, used for hydrotreating gasoline. For refining higher fractions it is not entirely essential, and the catalyst may be converted to sulphides directly [219].

Most often, active sulphide catalysts are prepared by converting the respective oxides to sulphides. The starting substances may be converted, if necessary, into oxides by means of calcination or oxidative preparation. Sulphurisation is usually preceded by reduction of the oxide material with hydrogen. Sulphurisation using oxides of metals of Group VIII of the periodic system is relatively easy and rapid [665, 690, 730, 1481]. Sulphurisation using Mo and W oxides [181, 1631] is more difficult, and does not always proceed in a regular manner. Romanowski [1765] showed, that in the reaction of gaseous H_2S with MoO_3 at 400 to 500 °C, reduction to MoO_2 and partial conversion to sulphide takes place, so that the final product is a mixture of MoS_2 and MoO_2. Instead of sulphurisation with gaseous H_2S, sulphur may be added to MoO_3, reduction and sulphide formation being carried out at the same time with hydrogen [404, 1320].

In some cases, conversion of the respective metal hydroxides (e.g. $Ni(OH)_2$ [95]) or carbonates [660, 839, 1511, 1854] to sulphides is employed to prepare active sulphides. Sulphides thus prepared are characterized by a great adsorptivity of oxygen, which oxidises H_2S to sulphur which is firmly chemisorbed to the catalyst surface [660]. Preparation of active Ni and Co sulphides by means of a direct reductive process involving the respective nitrates using a mixture of H_2 and H_2S [735], or, reduction of the respective sulphates or sulphites [662, 665, 666, 1043] has also been reported. To prepare an active hydrocracking catalyst, nickel fluoride chemisorbed on an alumosilicate carrier was converted directly to nickel sulphide; a highly active catalyst with long lifetime was obtained [1242].

Sulphides of metals of Group VI of the periodic system (Mo and W) were frequently prepared by sulphurising the respective acids [885] or ammonia salts of these acids to sulphides [832, 839, 849, 854, 856, 899] or, after supporting on a suitable carrier [96, 1368, 1369]. This procedure is employed especially for supported sulphide catalysts, the active component of which is a mixture of sulphides of Group VI and VIII of the periodic system. The modified preparation of the catalyst 8376 may serve as an example [576]:

γ-Al_2O_3 serving as carrier was prepared by precipitating a solution of $Al(NO_3)_3$ with a 15% aqueous ammonia solution at 95 °C, the pH of the mixture being kept at a value of 7. After filtration, the alumina precipitate was washed for several hours with hot water and formed into tablets. After calcination at 300—350 °C the tablets were impregnated with an ammonia solution of tungstic acid and nickel sulphate, dried at 140 °C and converted to sulphide at 400 to 450 °C. The tablets obtained were ground, pressed and sulphurised once more. The resulting catalyst contained 27% tungsten sulphide and 3% nickel sulphide on γ-Al_2O_3.

Sulphide catalysts have likewise been prepared by means of direct conversion of metals to sulphide. Signaigo [508, 1898] recommends sulphurisation with hydrogen sulphide of active (pyrophoric) metals at low temperatures (below 150 °C). Similarly, supported nickel and cobalt hydrogenation catalysts are prepared by converting metals to sulphides [734, 736]. Extensive reduction from the oxide state to the metal is recommended before the sulphurisation process in the preparation of the hydrorefining catalyst NiS–WS_2/SiO_2–MgO [2042].

Optimum conversion of the catalyst to sulphides secures high activity and required selectivity. This is especially important with catalysts which are prepared using hydrogen sulphide obtained by hydrogenating sulphur compounds contained in the raw material. In this case, the optimum conditions usually have to be determined experimentally for every catalyst type [930a] in order to achieve a sufficiently active catalyst. This is of outstanding importance in the case of desulphurising cobalt–molybdenum catalysts, the sulphurisation conditions of which will be discussed later (p. 85).

3.1.2.

Preparation of sulphide catalysts by decomposition of thiosalts

Another method of preparing sulphide catalysts, very often employed, is the decomposition of thiosalts, by means of which especially active molybdenum and tungsten catalysts are obtained. Preparation of thiosalts usually starts from the ammonium salt of the respective acid, which is precipitated in a medium of ammonia with hydrogen sulphide [1104]. The precipitation process is controlled in such a manner as to obtain an oxygen-free thiosalt (the tetrathiosalt in the case of Mo and W). When preparing ammonium tetrathiotungstate, precipitation of the ammonia solution of tungstic acid is carried out close to the boiling point of the solution [416] and, likewise, MoS_2 is prepared by precipitating the thiosalt from a hot solution (the heat liberated on precipitation of ammonium molybdate suffices [1163]). It was also found, that thiotungstate precipitation in an electromagnetic field of up to 1000 Oersted enables WS_2 with a superior activity for hydrorefining to be prepared [1135].

Decomposition of the thiosalt is performed either with the use of acids or by thermal means. The first method is less frequent and, in the case of Mo and W thiosalts, it leads to the formation of amorphous precipitates of the trisulphide which are further reduced by hydrogen to the respective disulphide [1958]. A mixed sulphide catalyst e.g. WS_2–NiS may be prepared by adding a solution of the second component to the Mo or W thiosalt solution and precipitating with the acid [83]. Precipitation of the trisulphide with an acid may also be carried out on a carrier which has first been impregnated with the respective thiosalt, e.g. on active carbon [1418].

Mostly, however, thermal decomposition of thiosalts prevails in the preparation of active Mo and W sulphides; this is usually done in an atmosphere of hydrogen [864, 878] so that the respective disulphide is formed at the same time according to the scheme:

$$(NH_4)_2MeS_4 + H_2 \rightarrow MeS_2 + (NH_4)_2S + H_2S \qquad (I)$$

where Me is W or Mo. The decomposition reactions will be discussed in detail in the following chapter, while here we shall give our attention to the practical application of this procedure in the preparation of Mo and W sulphides. On an industrial scale, this procedure was used to prepare pure WS_2 (catalyst 5058) [317, 1172].

The procedure was as follows: tungstic acid $WO_3.H_2O$ was dissolved in the mother liquor from the preceding preparation in a vessel fitted with heating mantle and agitator, the solution being saturated at the same time with ammonia at 60 °C. The clear solution was pumped into a precipitating tank, in which H_2S was introduced under agitation, the temperature rising during the first half-hour to about 60 °C. Hydrogen sulphide saturation continued for another 8 hours; after this time, the mixture was cooled to 20 °C and was agitated while being saturated with hydrogen sulphide. Yellow ammonium thiotungstate crystals were filtered in vacuo, dried and decomposed in a stream of hydrogen at about 420 °C. The WS_2 obtained was ground, formed into tablets and stored in well-sealing barrels in an atmosphere of N_2 or CO_2.

MoS_2, which is often used in laboratory-scale experiments, is easily prepared by reductive decomposition of ammonium thiomolybdate in an autoclave [1163].

The original ammonium sulphomolybdate was prepared by dissolving 60 g ammonium molybdate $3(NH_4)_2O.7MoO_3.4H_2O$ in a mixture of 180 ml distilled water and 100 ml ammonia (specific gravity 0.94) at a temperature of about 70 °C. The warm solution was filtered and 500 ml ammonia of the same concentration was added to the pure filtrate. After cooling, the warm solution was saturated with a rapid stream of H_2S. The yellow colour of the solution changed through orange to deep red, the mixture being heated by the reaction heat to 60 °C. After continued introduction of H_2S, red crystals of ammonia thiomolybdate started to precipitate. After complete precipitation, the crystals were filtered by suction, washed with a small amount of water and transferred into a 2.5 l autoclave, where 200 ml toluene was added. The autoclave was filled with hydrogen to 100 atm and heated to 300 °C, at which temperature it was kept for 30 min. After cooling the MoS_2 crystals were removed from the autoclave, the catalyst was washed with pure toluene and washed thoroughly with ether and finally with methanol. It is stored either under methanol or, it may be dried in a stream of nitrogen and transferred into well-sealed bottles with ground stoppers filled with an atmosphere of N_2 or CO_2. WS_2 is prepared on the laboratory scale in a similar manner [1159].

A method has also been described for the preparation of mixed MoS_2 and WS_2 catalysts with sulphides of iron group metals, a mixture of thiosalts and metals (obtained by decomposition of carbonyls) being decomposed and sulphurised at the same time [75]. Rather than in the pure form, Mo and W sulphides are more frequently used supported on suitable carriers which are impregnated with a solution of the respective thiosalt, thermal decomposition being carried out after drying. In this manner the hydrocracking catalyst 6434 (10% WS_2 on active clay impregnated with hydrofluoric acid) is prepared on the industrial scale [416, 1172], also WS_2 supported on iron silicate [1528]. The procedure of preparing MoS_2 on aluminosilicate is similar [959].

In some cases, the thiosalts were directly used as catalysts in hydrogenation reactions. In the course of the reaction, these thiosalts either form pure MoS_2 or WS_2 when ammonium salts are used [50, 1159, 1163] or, mixed sulphides are obtained when for example Co or Sn (catalysts for hydrogenation of phenols) [1336] or Cr [1206] thiomolybdates are used.

3.1.3.

Principles of preparing mixed and supported sulphide catalysts

The majority of the sulphide catalysts employed are mixtures of sulphides, usually supported on a suitable carrier. These mixed catalysts are prepared, similarly to other types of technical catalysts, either by mechanical mixing of the individual components (by mixing the finished sulphides or the initial substances) or, by impregnating finished sulphides or carriers or, finally, by precipitating solutions of substances, sometimes in the presence of a carrier suspension. The method employed to prepare a catalyst often determines its activity. For example, some cobalt–molybdenum desulphurisation catalysts prepared by impregnating the carriers are more active than when prepared by means of co-precipitation [815].

Another procedure involves a product which may be obtained by mixing, impregnating or precipitating, which is then dried or calcined, formed and finally sulphurised to convert the active components to the sulphide form. Each of these operations must be carried out under optimum conditions; calcination must be done with special care in order to avoid overheating the catalyst or recrystallisation of one of the components, which might have a serious effect on the activity of the catalyst [1431]. Some principles which apply to the very important sulphurisation stage have already been mentioned (p. 26); this operation will be discussed in detail in section 5.1. (p. 83, etc.).

The simplest method of preparing mixed or supported sulphide catalysts is mixing the components. In the process, either sulphides may be mixed together [460] or, sulphides may be mixed with the carrier [671, 713, 841, 1904]. The preparation of hydrogenation catalysts by evaporation of a solution of Mo or W thiosalt with salts (mainly carbonates) of an iron-group metal followed by additional sulphurisation of the evaporation residue was also reported [372]. The mechanical stability of sulphide catalysts for destructive hydrogenation was enhanced in one case by mixing them with a small amount of finely ground metals (Zn, Al, Be etc. [836]).

More frequently, mechanical mixing of suitable starting substances of the active components or, mixing with the carrier is employed. Components may be either mixed quite dry or in the moist state or in aqueous suspension. After perfect blending, the material is dried, formed and calcined (these operations are frequently carried out in a different order) and the preparation process is completed by sulphurisation of the catalyst. In this way, for example, nickel sulphide on diatomaceous earth is prepared (catalyst for selective hydrogenation of cracked gasolines [1347, 1803]), a nickel–molybdenum-sulphide catalyst for amination [1093], NiS–WS_2 catalyst for hydrogenating olefins [1511], Co–Mo refining catalyst [494] or a type 8376 catalyst [2157]. A typical example is the industrial preparation of the No. 3076 hydrogenation catalyst [317, 1431, 1854].

A mixture of $Ni(NO_3)_2$ and $NiSO_4$ is mixed and dissolved to make a solution containing 115 to 132 g/l Ni. While hot, this solution is precipitated with soda solution (the precipitation takes place at pH = 8 to 8.5). The precipitate formed is filtered and washed well with distilled water, dried at 100 to 120 °C and the dry product is ground. Dry nickel carbonate is mixed with tungstic acid and the mixture is homogenized in aqueous suspension, filtered, the paste is formed and dried at 100 to 120 °C and the dried mass is sulphurised. Sulphurisation is done at 420 to 460 °C for 5 to 6 hours with a mixture of H_2 and H_2S (1 : 14). The sulphurised catalyst containing 27 to 30% S is ground, 1–2% graphite is added and the mass is formed into tables 10 × 10 mm in size. The composition of the catalyst obtained is 1.7 to 2.2 NiS . WS_2, its bulk density being 2.3 to 2.6 kg/1. The catalyst 3076 supported on γ-Al_2O_3 is prepared in a similar manner.

One of the most frequently used methods of preparing mixed and supported sulphide catalysts is impregnation. Impregnation of ready sulphides with a suitable solution of a second component followed by sulphurisation of the mixture has been reported. For example, MoS_3 was impregnated with cobalt acetate solution and, after reduction and sulphurisation with a mixture of H_2 and H_2S, an active desulphurisating catalyst was obtained [493, 2132]. Impregnation of carriers with stable sulphide solutions was also used (e.g. platinum or tungsten) or, with solutions of sulphosalts [744, 1957, 1975].

In most cases, however, impregnation of carriers is carried out with solutions of suitable starting substances, which are converted to sulphides in the second stage. With multicomponent catalysts, the procedure may involve impregnation with a single solution containing all active components (so-called co-impregnation); impregnation is followed by drying, calcining and sulphurisation. This procedure is used for example to make type 8376 refining catalysts (p. 26) [576, 577, 1500, 2270], the NiS–MoS_2/Al_2O_3 catalyst [476], the iron sulphide catalyst [1771] and new types of sulphide Ni–W/Al_2O_3–SiO_2 hydrocracking catalysts [686, 688]. Some desulphurisating cobalt–molybdenum catalysts are prepared in the same way [1347].

It is important in the case of the co-impregnation method to select such a composition and acidity of the impregnation solutions as to avoid precipitation of the active components. To prepare stable solution of salts of Group VI and VIII metals, which are the most frequent components of supported hydrorefining and hydrodesulphurisating catalysts, Nahin and Huffman [1441, 1442] use in principle the following procedure: A solution of the salt of that metal which forms a basic oxide is added to an ammonia solution until the hydroxide is precipitated; this dissolves in the form of an ammonium complex, and the solution of the metal compound is added, which forms an acid oxide. In doing this, certain optimum concentration ranges must be maintained in all solutions. Warthen et al. [2162b] describe a different procedure for preparation of a stable Co–Mo solution for use in coimpregnating the carrier.

The Nahin-Huffman method is suitable for preparing stable cobalt–molybdenum impregnation solutions. For nickel–tungsten solutions it is better to set out from a mixture of aqueous ammonium metatungstate and nickel nitrate solutions, keeping the acidity of the final solution at a pH 2 to 5 [1347].

The stability of impregnation solutions may be enhanced by an addition of alkanolamines (usually added to the solution of that metal which forms an acid oxide) [470, 1489, 1506, 1887a, 1888]. Saturation of the carrier with a mixture of Ni and Mo acetylacetonates in isopropyl alcohol [592a] was also described.

The second impregnation method is double impregnation with starting solutions. Here, the order of the solutions is of some importance. For example, with cobalt–molybdenum desulphurisating catalysts, catalysts obtained were more active when the molybdenum component was applied first [755]. A sulphurised Ni–Mo refining catalyst was also prepared in this way [575, 1216, 1928]. Stage-wise impregnation was likewise employed for rhenium sulphide hydrocracking catalysts [433, 2219], a chromium–molybdenum sulphide refining catalyst [117], Co–Mo hydrocracking catalysts containing rare earth elements [1023a] and three-component Mo–W–Ni sulphide refining catalysts supported on Al_2O_3 [1219].

Uniform saturation of the carrier is important with impregnation methods. In this respect, catalysts prepared by co-impregnation are better than those prepared by double saturation. In spite of that, selective adsorption may take place on some parts of the carrier surface or, the components applied may migrate in the course of the calcination process. Addition of polyvalent alcohols, e.g. erythritol, to the co-impregnating solution was found useful [472]. Similarly, double saturation with intermediate calcination enabled the preparation of high-quality catalysts, e.g. Co–Mo catalyst [481].

The last of the group of important methods of preparing mixed or supported sulphide catalysts is precipitation of the original non-sulphide components (active sulphide catalyst being prepared in the same way as in the case of the impregnation method). The catalyst is prepared either by successive precipitation of the individual active components (or, precipitating the active components onto the carrier) or by means of co-precipitation, i.e. mutual precipitation of the active components (in some cases together with the carrier [1929]) when their solutions are mixed together.

Preparation of the WS_2–NiS hydrogenation catalyst may serve well as an example. A nickel nitrate solution is added to an ammonia solution of ammonium thiotungstate with excess ammonium sulphide, nickel sulphide precipitating in the process; the mixture is then acidified with dilute sulphuric acid to pH = 2 and tungsten trisulphide precipitates. The catalyst may also be obtained by co-precipitation, an acidified nickel nitrate solution being added to the thiotungstate solution. In both cases, the precipitate is reduced with hydrogen after drying, in order to convert higher sulphides to stable lower modifications [83, 327].

Catalytically active Co, Ni and Fe molybdates are obtained in a similar way [247, 1282].

The co-precipitation method was furthermore used to prepare a NiS–WS_2/Al_2O_3 refining catalyst [1905], a PtS catalyst for hydroforming hydrocarbons [2059] and a sulphide vanadium catalyst [526].

Precipitating procedures are likewise employed to prepare cobalt-molybdenum desulphurising catalysts. Unsupported cobalt molybdate is obtained by precipitating an ammonia solution of ammonium molybdate with cobalt nitrate solution. To prepare the supported catalyst, Al_2O_3 is suspended in cobalt nitrate solution, cobalt molybdate being then precipitated on the carrier surface by adding an ammonia solution of ammonium molybdate [1347]. In both cases, an active sulphide catalyst is obtained after pretreating of the material (washing, drying, grinding, tablet forming) by sulphurising in a reactor either with hydrogen sulphide (usually H_2S containing circulating gas) or directly with hydrogen sulphide originating in the sulphur-containing raw material. With these preparation methods, the final catalyst activity depends to a large extent on the precipitating procedure. It was found for example, that precipitation of cobalt molybdate on non-dried alumogel gives a catalyst, the activity of which is greater than that obtained with a dried carrier [245]. Similarly, a Mo–Ni/Al_2O_3 sulphide catalyst for hydrodenitrogenation prepared by a special procedure from a non-aqueous medium (methanol-propylene oxide) had an activity better than that of a catalyst prepared in the conventional manner [521].

3.1.4.

Special methods of preparation of sulphide catalysts

Besides the methods of preparation described, sulphide catalysts were prepared in some cases by means of special procedures. For example, an active cobalt sulphide catalyst for the synthesis of mercaptans was prepared from a Co–Al alloy by a procedure similar to the preparation of Raney catalysts. In this case, the alloy was etched with a sodium sulphide solution [508].

In another procedure, gaseous $MoCl_5$ was adsorbed on SiO_2 and an active hydrogenation catalyst was prepared by sulphide formation with a mixture of H_2 and H_2S [1496]. The preparation of sulphide catalysts in non-aqueous media was also tested; e.g. a $TiCl_4$ benzene solution was mixed with thiotungstate solution in cyclohexylamine, and the precipitate obtained was reduced (in the presence of H_2S) with hydrogen at 300 to 400 °C [842, 855]. From the same solvent, Mo and W sulphides were also precipitated [1638]. Highly efficient catalysts for hydrodesulphurisation of petroleum residues were obtained in the colloid form by means of thermal decomposition of a homogeneous mixture of a petroleum residue and an organometal complex in the presence of hydrogen and hydrogen sulphide [86, 592b, 611, 612, 613]. Other sulphur compounds may likewise be used to sulphurise the colloidal catalyst particles, e.g. CS_2, mercaptans etc. [2093, 2095]. Acetylacetonates [614], metal carbonyls [590, 612, 2095], xanthates [610], β-diketone complexes [591], phtalocyanine complexes [1966a] and others [86] were found to be suitable organic components for complexes of Groups V, VI and VIII metals, especially V, Mo, and Ni. Vanadium halides were also tested for this purpose [614a].

3.2.

Carriers for sulphide catalysts

Like the majority of technologically important catalysts, metal sulphides are mostly supported on suitable carriers. The only non-supported catalysts which are important for practical industrial use, are the catalysts 5058 (WS_2) and 3076 (WS_2.2 NiS). The first-named was soon replaced in fuel hydrogenation in the vapour phase by the supported catalyst 8376 and continues to be used in special processes only (e.g.

hydrorefining of lubricating oils [1477]). The use of the catalyst 3076 is likewise limited to special cases, in which the outstanding hydrogenation selectivity of this catalyst is utilised.

In industrial applications, sulphide catalysts are usually exposed to rather extreme conditions, involving high operating temperatures, large volumes of catalyst beds necessitating a high mechanical strength of the catalyst, and processing of low-quality materials which contaminate the active surface. To this is added the demand of a long lifetime or, of the possibility of repeated regeneration, so that it is essential for the carriers employed to be of high quality.

The carrier most often used is aluminium oxide, which is a part of the most important industrial sulphide catalysts: tungsten–nickel [416, 1101, 1626] and nickel–molybdenum [1429] hydrotreating catalysts and cobalt–molybdenum desulphurating catalysts [1347].

The active γ-modification of Al_2O_3 is suitable for catalytical purposes. This is a linear polymer with side chains, formed by dehydrated aluminium hydroxide groups. In preparing this highly active carrier, it is essential to maintain the optimum conditions of aluminium hydroxide precipitation, decantation, filtration and washing, gel ageing, drying and especially of calcining. In the precipitation process, much depends on the nature of the reagents used as well as their concentration, and furthermore on the acidity of the medium, temperature and rate of precipitation. Equally important is washing the precipitated hydroxide, where salts, the excess precipitant and impurities contained in the raw materials must be removed.

Calcining is usually carried out at 400 to 600 °C; at this temperature, a carrier with sufficiently large surface, 150 to 400 m^2/g and a suitable texture is obtained. Heating to temperatures higher than 750 °C is not recommended, as the active γ-form converts to the α-modification, which is non-active [1082, 1347, 1431].

Another active modification is η-Al_2O_3, prepared by calcining bayerite or gibbsite with rapid removal of the steam formed, or a calcined mixture of hydrargyllite and aluminium hydrate, prepared by precipitating aluminium salts [1080]. Another demand is high mechanical stability of the alumina employed. An active carrier of high mechanical strength is obtained by impregnating the powdered trihydrate with $AlCl_3$ solution and calcining [471]. Alumina is sometimes stabilized by a small addition of SiO_2, mainly securing better heat stability in the carrier [1347]; moreover, its acidity and, therefore, splitting properties are enhanced [1429]. On the other hand, virtually neutral alumina is sometimes required for the preparation of sulphide catalysts [1219d].

Another important type of carrier for sulphide catalysts are natural as well as synthetic aluminosilicates which serve as carriers in hydrocracking catalysts. The carrier of the splitting catalyst 6434 is active clay (Terrana) activated with HF [1626]. The carrier itself has a low splitting activity at pressures of up to 300 atm, which however increases considerably after supporting WS_2 on the carrier [416]. This hydrocracking catalyst has the disadvantage of being very sensitive to nitrogenous bases, and its operating pressure range is very high.

Besides natural aluminosilicates, synthetic aluminosilicates are being used, especially in new hydrocracking catalysts with active sulphide components [162, 532, 1198, 1203, 1651a, 1956]. With these carriers, the texture and acidity and, therefore, also selectivity may be controlled to some extent by varying the molar ratio SiO_2 : : Al_2O_3 [960]. The number of strongly acid centres in synthetic aluminosilicates may be increased further by the addition of halides.

Amorphous aluminosilicate carriers for hydrocracking catalysts are usually prepared by mixing solutions of water glass and an acidified aluminium salt; the activity of aluminosilicates is modified by adding acid components (hydrogen halides, halides). The carriers are produced in the form of tablets or globules.

The aluminosilicates prepared sometimes also contain other oxides (e.g. TiO_2, ZrO_2, ThO_2 or MgO); the SiO_2: Al_2O_3 ratio varies from 10 to 90% SiO_2.

Natural and synthetic zeolites, which act as molecular sieves with a single-size pore system may be expected to become important components of hydrocracking catalysts [1715]. Natural zeolites are used rarely; synthetic zeolites are sometimes subjected to special modification procedures [609]. An important property of molecular sieves is their considerably lower sensitivity to nitrogenous bases [2138a]. Active hydrogenating components are supported on acid carriers either by means of impregnation (co-impregnation or double impregnation) or, in some cases by co-precipitation or ion exchange [2138a].

Besides these most important sulphide catalyst carriers, other substances have also been described, for example silicagel [1496, 1528], SiO_2–MgO [214a], SiO_2–ZrO_2 [2237a] active carbon [337, 841, 1486, 1493, 1494, 1495, 1589] combinations of Al_2O_3 with zinc oxide [1948], magnesium oxide [96, 411], silica and titanium dioxide [942], and zirconium dioxide [244].

Means of influencing the activity of sulphide catalysts by way of the carrier will be discussed in greater detail in section 5.3. (p. 101).

3.3.

Regeneration of sulphide catalysts

In spite of their outstanding resistance to common catalyst poisons and a long lifetime, sulphide catalysts are subject to deactivation in the course of use, so that at the end of an operating cycle the catalyst must be either discarded or its activity must be renewed. The causes of deactivation will be discussed in detail in section 5.4. (p. 104). Here, we shall only analyse the processes employed in renewing the activity of sulphide catalysts, especially those which are of industrial significance.

Sulphide catalysts, the activity of which has decreased as a consequence of ageing or catalysts irreversibly poisoned (for example, by arsenic compounds) are either discarded or reworked. Reworking a catalyst involves mainly the isolation of valuable components and their re-use in preparing fresh catalysts. Especially with multi-component catalysts, this principle is mainly enforced by a shortage of the valuable components of the catalyst and, especially, by economical factors; an important factor is the concentration of the respective components in the deactivated catalyst. Before regeneration, the catalyst is usually made freee of carbon by calcining, whereupon it is converted to the oxide form; it is then ground and extracted with suitable agents. In re-working sulphide catalysts, main interest concentrates on the isolation of Mo, W, Ni and Co from refining and desulphurisation catalysts: the procedures employed are usually based on experience gained in extractive metallurgy [599]. In re-working MoS_2 catalysts, for example, extraction is done with 50% H_2SO_4 [946]; spent catalysts containing WS_2 are roasted and extracted with ammonium sulphide [845] etc.

Usually, sulphide catalysts, the surfaces of which have been deactivated by contamination, have to be renewed. Most often, carbonaceous deposits of a coke-like character are involved. Besides, other ballast components must also be removed, especially ash substances, residues of organo-metal compounds, etc.

Regeneration of catalysts contaminated in this manner usually involves preliminary purification, followed by oxidative removal of carbonaceous precipitates and removal of other contaminants.

During preliminary purification, residues of the material or products are removed from the catalyst by a strong stream of inert gases, hydrogen, steam or a hydrogen-steam mixture at an elevated temperature [321, 530, 753, 1475, 1568, 2236]. Care must be taken when using steam, however, because steam has a negative influence on the mechanical strength, structure and sometimes even activity of regenerated catalysts [1742].

In some cases, the catalyst is extracted with a suitable solvent, in order to remove residues of the raw material and of polymeric products [559, 1926, 2173]. Preliminary extraction of the sulphide catalyst, conversion of the active components (e.g. by means of acetic acid in the case of the NiS/Al_2O_3–SiO_2 catalyst) to salts easily decomposed by thermal means, and oxidative regeneration have also been described [344]. The hydrocracking catalyst NiS/Al_2O_3–SiO_2 has been regenerated with advantage by exposing it to the effect of sulphur halides before oxidative regeneration proper [1342].

Special extraction procedures are used in the case of those catalysts (especially desulphurisation ones), on the surface of which metal compounds, originally contained in the raw material processed, have been retained. For example, the major part of the contaminant may be removed by extraction with ammonium sulphide from a cobalt–molybdenum catalyst deactivated by vanadium [1736]. From catalysts in the oxide form, vanadium may be removed in part with the aid of some complex-forming acids or hydroxyacids [156].

In regeneration of a sulphide catalyst, the removal of carbonaceous deposits from the catalyst surface by means of careful combustion is involved. The basic demand is, that oxidation must take place under conditions which avoid overheating the catalyst and thus damaging its active structure. With the majority of regenerated catalysts, the maximum regenerating temperature is stated to be 700 to 750 °C [135, 326, 726, 1482, 2097, 2221]. A maximum operating temperature of 550 °C is recommended for hydrodesulphurisation Co–Mo catalysts [1479a, 2162a]. Controlled combustion is achieved by regulating the oxygen content of the regeneration gas, which usually is nitrogen or steam [274, 321, 326, 726, 753, 1560, 2097, 2221]; sometimes flue gas is also used. Regeneration may be done either at normal or at elevated pressure [208, 943, 1347, 2162a]. When oxidative regeneration is done very carefully and slowly at low temperatures, air may be used [208, 943]. The mechanical strength of the catalyst, which determines the possible number of regeneration cycles, is an important factor [274, 1327a]. Regeneration of sulphide catalysts used in dehydrogenation by means of SO_2 has also been reported [81, 82].

After combustion of the carbonaceous deposits, the active components are converted to the oxide form and, therefore, the catalyst must be reactivated. Cobalt–molybdenum catalysts are usually reduced with hydrogen after regeneration, after which they are sulphurised [737, 754, 1666], in spite of the fact that special sulphurisation is not essential since the catalyst is soon sulphurised by the material being processed [1347]. With other sulphide catalysts, reduction and sulphurisation are regular operations in the regeneration process [251, 274, 321, 326, 345, 588, 860, 928c, 943, 1482, 2097, 2103b, 2103c, 2228a]. With Group VI sulphides, the regenerated catalyst is converted to the thiosalt with the aid of ammonia and hydrogen sulphide. The thiosalt is then decomposed in an atmosphere of hydrogen to form the active sulphide [860, 943]. With catalysts for nitrobenzene hydrogenation, which contain sulphides of elements of Groups IB, IIB, IVB and VIIIA of the periodic system, the activity was considerably improved after oxidative regeneration by impregnating the catalyst with a sulphate solution and reducing in hydrogen at a maximum temperature of 300 °C [1830]. Schutt [1834d] describes the regeneration of Ni–Mo hydrocracking catalysts by means of impregnating with the Ni- or Mo-acetylacetonate complex.

Special regeneration procedures are required for sulphide hydrorefining catalysts deactivated by arsenic [930, 1927, 2016, 2269].

4.

PHYSICO-CHEMICAL PROPERTIES OF SULPHIDE CATALYSTS

The development of theoretical knowledge in the field of catalysis and the application of modern physico-chemical methods has enabled new, significant properties to be achieved in the development of sulphide catalysts; these properties have a substantial influence on the activity of the catalysts. For example, there exists a relationship between porosity, acidity and refining activity of multicomponent catalysts; also the identification of the true catalytically active components of cobalt–molybdenum desulphurisation catalysts by means of magnetic measurements, etc.

In spite of this partial success, the approach to the development of new types, especially of industrially important sulphide catalysts, continues to be rather empirical. It is only recently that increased attention has been given to determining the thermal stability of sulphide catalysts and their textural parameters. Their crystal structure and electrical and magnetic properties are also beginning to be studied in more detail.

4.1.

The crystal structure of sulphide catalysts

As with a number of other solid catalysts, the catalytic activity of metal sulphides is related to defects in their crystal lattice. The perfectly developed lattices of natural sulphides usually do not achieve the high activity of synthetic catalysts and, therefore, knowledge of crystal structure anomalies is important in terms of selection of the optimum conditions for preparing the catalysts.

The two most important sulphides, MoS_2 and WS_2, have a similar crystal structure. The crystallographic situation in molybdenite is illustrated in Fig. 1 [2210]. The symmetry of the crystal lattice is hexagonal [407, 618, 739], (the elementary cell dimensions are given in Table 4, for further data see [2242a]), each metal atom being surrounded by six sulphur atoms [407, 657, 660]. In the hexagonal molybdenite lattice (in which the individual crystallite layers are oriented antiparallel around the

Table 4.

Dimensions of the Elementary Crystal Lattice of MoS_2 and WS_2 [2210].

Sulphide	a (Å)	c (Å)	c/a	U (Å^3)
Hexagonal MoS_2	3.1602	12.294	2×1.9451	2×53.17
Rhombohedric MoS_2	3.163	18.37	3×1.936	3×53.1
Hexagonal WS_2	3.155	12.35	2×1.956	2×53.3
Rhombohedric WS_2	3.162	18.50	3×1.950	3×53.4

Note: a, c — Elementary lattice cell dimensions.
U — Elementary lattice cell volume.

c-axis – Fig. 1c) the metal layer is separated by two layers of sulphur atoms. The character of the bond between the two vicinal sulphur layers is not completely elucidated [207, 1575, 1763]; this bond, however, is rather weak, which explains the fact that MoS_2 as well as WS_2 are excellent solid lubricants.

Besides hexagonal MoS_2, a rhombohedric modification also exists [125, 1103]. The structure of this modification, which Semiletov [1856] studied in detail, is illustrated in Fig. 2 [935] and the elementary cell dimensions will be found in Table 4. A high yield of this modification is obtained by alkaline fusion of MoO_3 [2210] with sulphur or, by high-pressure and high-temperature synthesis from the elements [1900c]. The modification, however, is unstable, converting into hexagonal MoS_2 when heated [2210]. Zelikman et al. [2259] prepared, by means of a number of synthetic methods, MoS_2 with a defective structure, similar to that obtained on gradual heating of MoS_3 [2210] or on thermal decomposition of $(NH_4)_2MoS_4$ in hydrogen [1353].

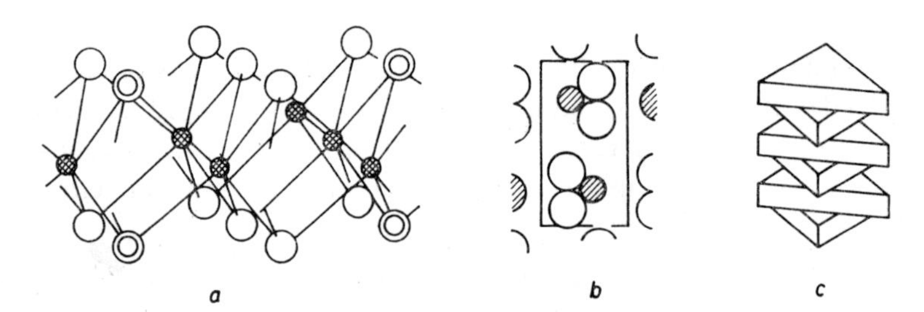

Fig. 1. Crystal structure of hexagonal MoS_2 (molybdenite) [2210]. *a* — Part of the crystal lattice (Mo ions are denoted by small shaded rings, sulphur ions by large rings; ions located in the same plane (100) are denoted by two rings). *b* — Section through the (110) plane (the *c*-axis is vertical, Mo ions are denoted by smaller shaded rings). *c* — Schematical model of hexagonal MoS_2 (neighbouring crystallite layers are oriented antiparallel).

Another molybdenum sulphide, used to prepare active MoS_2 catalysts, is the trisulphide MoS_3. This substance may be prepared in different ways, i.e. by acidifying an aqueous ammonium thiomolybdate solution, by the influence of H_2SO_4 on crystalline $(NH_4)_2MoS_4$, by the effect of HCl on piperidine and piperazine thiomolybdates [381, 615] or by their gentle thermal decomposition in vacuo at 200 °C. In all cases, an amorphous product is obtained, which is thermally unstable and converts into stable MoS_2 through a number of intermediate stages (section 4.2.).

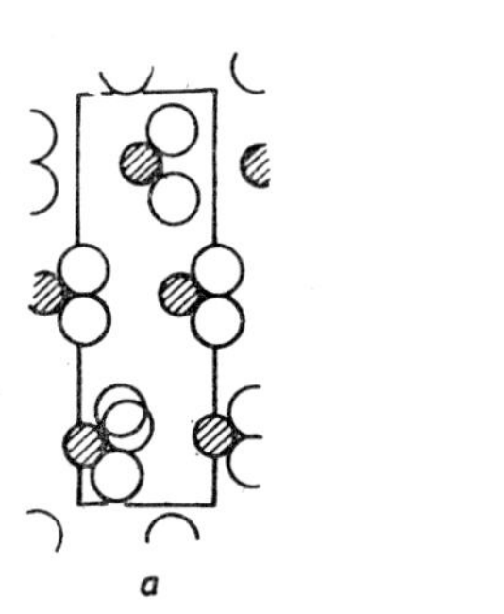

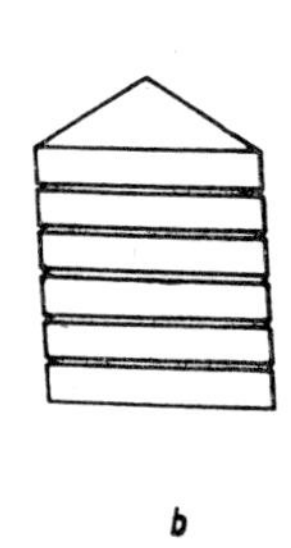

Fig. 2. Rhombohedric MoS_2 [125, 935]. *a* — Section through the (110) plane. *b* — Parallel orientation of crystallites around the *c*-axis.

Besides these modifications, Mo_2S_3 has also been described [683, 936, 1341, 1980]; this sulphide is formed by thermal decomposition of MoS_2 in an inert or hydrogen atmosphere [1540].

The series of tungsten sulphides is similar. WS_3 is even less stable than MoS_3, (especially in an oxidative atmosphere) and its preparation is similar to that of MoS_3 [381, 615]. The amorphous product converts to hexagonal WS_2 when heated to 1100 °C [88, 453, 615, 619]. The cell dimensions are given in Table 4; these values differ somewhat from those reported by the National Bureau of Standards [2019].

Wildervanck and Jellinek [2210] also described a rhombohedric WS_2 modification, the lattice constants of which are given in Table 4. High-pressure synthesis of this modification has been reported by Silverman [1900d].

Similarly to active MoS_2, the WS_2 catalyst likewise is most often prepared by thermal decomposition of the respective thiosalts, usually in a hydrogen atmosphere (for details, see section 4.2). The originally monoclinic ammonium thiotungstate crystals convert on thermal decomposition to WS_2 of monoclinic form, although a hexagonal structure may be proved by X-ray analysis [416, 1626]; the crystallite size is about 3×10^{-7} cm, corresponding to roughly five elementary WS_2 layers. Moreover, the diffractograms indicate that WS_2 is defective to some extent, due to the fact that the individual elementary crystallites are not oriented quite regularly [2210]. Hájek [716] sets out from Pier's original view (which is also interpreted by Kalechits [965]) and assumes that in the preparation of active WS_2 by thermal decomposition of ammonium thiotungstate, the hexagonal lattice formed is not quite precise, and catalytically active centres are formed by residues of the original monoclinic structure. Lozovoi et al. [1257] found by means of electron-microscopic investigation in WS_2

an agglomerate size of 300 to 20,000 Å. Samoilov and Rubinshtein [1778, 1797] found by means of a thorough phase analysis of the series of tungsten sulphides $WS_{2\cdot30}$–$WS_{1\cdot862}$ prepared by means of thermal decomposition of ammonium thiotungstate under different conditions, that irrespective of the W : S ratio, the catalyst only consists of the WS_2 phase with the same crystal lattice. Neither a WS_3 phase, nor other tungsten sulphides were found in the thiotungstate decomposition product, as was later confirmed by Wildervanck and Jellinek [2210]. Recently, Koestler [1063b] did some X-ray research on WS_2.

Another important group of catalysts are nickel sulphides. In the entire series of compounds described in the literature [177, 1134, 1151, 1152] (see also section 4.2.), NiS and Ni_3S_2 are the most important ones in terms of preparing active catalysts. NiS exists in three modifications: α-NiS is amorphous, β-NiS is trigonal (millerite) and γ-NiS has a hexagonal structure [20, 177, 728, 1067, 1976]. At medium temperatures, equilibrium is established between the β- and γ-forms; the hexagonal modification is stable above 396 °C [1039], the trigonal form (millerite) is stable below this temperature [1039] while at higher temperatures in a reducing atmosphere, NiS is unstable, converting quickly to the subsulphide Ni_3S_2 [1039]. Its crystal lattice is rhombohedric, very close to the cubic structure with the elementary cell parameters: $a = 4.04$ Å, $\alpha = 90.3°$. The atoms are located in the cell in such a manner, that every nickel atom is surrounded by four sulphur atoms while a sulphur atom is surrounded by four nickel atoms as the closest neighbours (Fig. 3) [2204]. The illustration also shows the distribution of atoms on the (100) and (111) planes, which is important from the point of view of desulphurisation reactions. (p. 213).

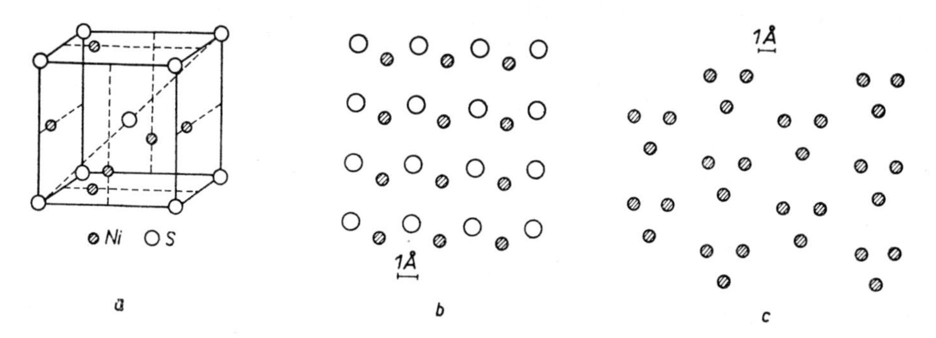

Fig. 3. Ni_3S_2 crystal lattice. *a* — Elementary lattice of the subsulphide [2204]. *b* — Distribution of atoms on the (100) plane [657]. *c* — Distribution of atoms on the (111) plane [657].

● Ni
○ S

The subsulphide Ni_3S_2 is likewise dimorphous. The low-temperature β′-phase is stable below 553 °C, the high-temperature β-phase is stable in the temperature range 620–818 °C. This β-phase forms a eutectic mixture with Ni (21.5% S, m. p. 645 °C) [1039].

The significance of the structural factor in nickel sulphide (or, in its mixture with ZnS), i.e. the relationship between interatomic distance in the reaction molecule and on the catalyst surface was studied by Rubinshtein et al. [1777] in the case of selective hydrogenation of butadiene and nitrobenzene under atmospheric pressure. These authors found, that sulphur atoms participate directly in the formation of active catalytic centres, taking part in the formation of the multiplet complex.

The greatest significance of nickel sulphide is its application as a component in mixed industrial catalysts of the 8376 type (27% WS_2 + 3% NiS supported on active γ-Al_2O_3) [1626] or the 3076 type (2 NiS.WS_2) [1431, 1854]. Although much attention was given to the crystal structure of the individual sulphides in these mixed catalysts, this problem has not yet been elucidated fully. With respect to the low concentration of the nickel component in the 8376 catalyst, its X-ray analysis is difficult. In the case of the WS_2–NiS catalyst (non-supported) Pier [1626] mentions an interesting variation of the splitting activity. While up to 400 °C the splitting properties of the WS_2 and WS_2–NiS catalysts are practically identical, the mixed sulphide loses this property at 425 °C. Without having lost the hydrogenation activity, its splitting activity decreased down to the level of the catalyst 8376. Whilst no substantial difference was found by X-ray analysis between the fresh and used catalyst, electron diffraction showed in the used catalyst a nickel component besides the WS_2 phase. On these grounds, Pier believes that the altered splitting activity is due to the disintegration of the original mixed phase in the surface layers of the catalyst.

Later research carried out by Samoilov and Rubinstein [1796] confirmed the difficulty of identifying the form of the nickel component in the 8376 catalyst. From X-ray studies of the fresh catalyst, the authors believe that the subsulphide Ni_3S_2 is present rather than the γ-NiS form, which was also confirmed by investigation of the sulphurised nickel catalyst [368] (cf. also p. 48).

Research into the crystal structure of newly developed mixed catalysts, for example the refining catalyst NiS–MoS_2/Al_2O_3 (catalyst No. 8167) [1433, 1764] as well as a study of the relationship between the crystal structure and catalytic activity of these mixed catalysts is only just beginning.

Cobalt sulphide was employed, for example, in hydrogenating some aromatic substances [1749, 1750] and in the synthesis of sulphur compounds [1211]. Caglioti and Roberti [248] believe the active catalyst, formed by reduction of CoS, to be the cubic subsulphide Co_4S_3. Later research, however, did not confirm the existence of this sulphide unambiguously [819,1243]. It was established by means of X-ray analysis, that the compound Co_9S_8 was probably involved in Roberti's study. However, cobalt forms a series of other sulphides with sulphur [462, 1266, 1347], so that the form of the cobalt sulphide in the catalyst will mainly depend on the reaction conditions under which this catalyst operates.

The most important field of application for cobalt sulphide is its use for activating as well as the active component of the cobalt–molybdenum refining catalyst, which converts in part to the sulphide form on sulphurisation. Richardson [1740] states, that under desulphurisating conditions, Co_9S_8 (which itself is catalytically poorly active) is present in the active catalyst. The remaining cobalt is bound in non-active spinel $CoAl_2O_4$ and in the highly active complex Co–MoS_2 which is the desulphurisating catalyst proper (p. 70).

Izmailzade [924] devoted a detailed study to zinc sulphide used as a dehydrogenating catalyst. X-ray analysis showed, besides ZnS (amorphous phase, crystal modification α and β), that ZnO exists in the catalyst used in dehydrogenation of ethanol and ethylbenzene (the oxide forms from oxygen adsorbed in the course of preparation of the catalyst). According to the author, ZnO is the most active component of the dehydrogenating catalyst; the activity of α-ZnS is less while β-ZnS does not contribute substantially to the overall activity. Similarly, Rubinhstein et al. [1776] believe that the sulphide is converted in part to ZnO in the decomposition of isopropylalcohol (dehydrogenation, dehydration) in the presence of ZnS.

4.2.

The behaviour of sulphide catalysts at elevated temperatures and some thermodynamic properties of sulphides

As with other catalyst types, freshly prepared sulphides need not necessarily be the catalytically most highly active form. In the majority of cases, the active catalyst surface is only formed under operating conditions. In this initial phase, the atomic ratio Me : S usually changes and at the same time that form of the active sulphide component forms which is relatively most stable under the respective operating conditions. Sulphide catalysts are typically used under very extreme operating conditions, i.e. usually at temperatures over 250 °C in an atmosphere of hydrogen under pressure.

Tungsten disulphide is a very stable substance [1623, 1868]. *in vacuo*, up to 1100 °C WS_2 does not decompose substantially; decomposition starts only above 1200 °C (about 60% S is split off within two hours) [1623]. Differing from tungsten disulphide, the fresh tungsten sulphide catalyst is characterized by a considerable content of non-stoichiometric sulphur up to $WS_{2.30}$ and therefore undergoes significant changes *in vacuo* at temperatures as low as 300 °C. The atomic ratio W : S, and with it the texture and catalytic activity of the catalyst alter [1794].

Differing from the disulphide, WS_3 is a very labile substance. Thermal degradation of WS_3 proceeds readily as low as 300 °C, being completed at 1100 °C and the result is hexagonal WS_2 [88, 453, 615, 2210].

From the point of view of activity of the tungsten sulphide catalyst, its behaviour in a hydrogen atmosphere is most important; recently this topic was studied by

Bartovská et al. [115]. The authors found an unusual course of the reduction reaction in the temperature range 500 to 800 °C,

$$\tfrac{1}{2}\,WS_2(s) + H_2(g) \rightleftarrows \tfrac{1}{2}\,W(s) + H_2S(g) \qquad \text{(II)}$$

which is to be seen in Fig. 4 (which includes all data reported to date [115, 282, 1014, 1574]). This anomalous behaviour may be explained on the basis of X-ray measurements and analogue results found in the case of MoS_2 [2210] which suggests that WS_2 crystallites are gradually oriented in the 500 to 800 °C range around the

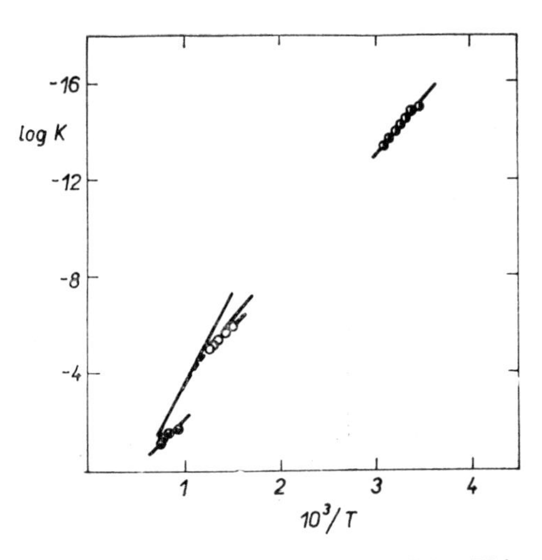

Fig. 4. Equilibrium diagram of reduction of WS_2 by hydrogen [115]. Comparison of experimental data reported by different authors (K — equilibrium constant of reaction II):

——— [115]
- - - - - transition region [115]
◑ ○ [282]
◒ [1574].

c-axis to form nearly a perfect hexagonal configuration of high-temperature WS_2. This transition to the stable form of the sulphide may be the main cause of the loss of catalytic activity on overheating the catalyst (usually above 600 °C) [958, 1586, 1794] which has already been mentioned by Pier [1626].

One of the reasons for loss of activity (which is mainly observed when sulphide catalysts are employed in industrial-scale hydrorefining of tar and petroleum materials) is a loss of sulphur from the catalyst. As Fig. 4 indicates, the equilibrium partial pressure needed to maintain the WS_2 modification is very low, e. g. at about 400 °C:

$$\log K \sim \log \frac{[H_2S]}{[H_2]} \sim \log \frac{p_{H_2S}}{p_{H_2}} \sim -6; \; \frac{p_{H_2S}}{p_{H_2}} \sim 10^{-6}.$$

In spite of this, feeds containing little or no sulphur must be sulphurised [1172].

It is clear however, on the other hand, that only a slight partial hydrogen sulphide pressure is required to convert a metal to the sulphide (in a metallic catalyst or a metal reduced, for example, from the oxide). When, therefore, substances containing even a small amount of sulphur are produced by hydroprocessing (hydrocracking, reforming etc.), there is always the possibility of sulphides being formed which therefore may influence the original catalytic system to a considerable degree. Comparison of equilibrium data for tungsten and molybdenum disulphide indicates, that WS_2 is slightly more resistant to a reductive loss of sulphur; obviously, older MoS_2 equilibrium data must be considered with care, cf. the differences in Fig. 4. This agrees with practical experience, according to which MoS_2 demand more sulphurisation than WS_2, but also more than would correspond to the equilibrium data [637].

Some important kinetic data concerning the formation of WS_2 by means of sulphurisation either of the metal [572a, 1156a] or of the trioxide [1327b, 1327c] were reported recently .

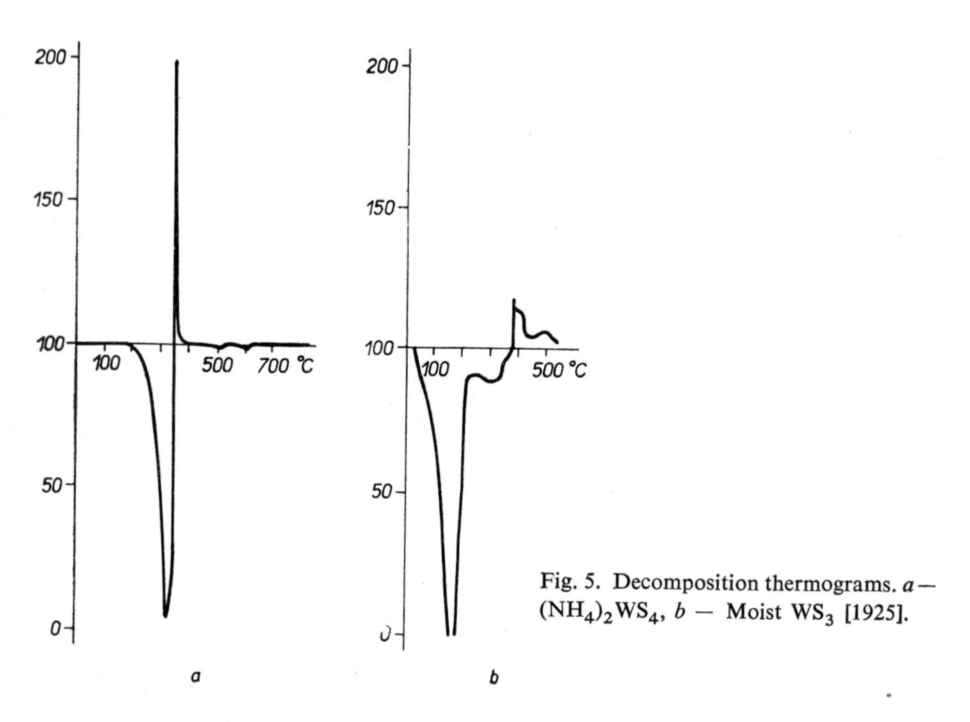

Fig. 5. Decomposition thermograms. *a* — $(NH_4)_2WS_4$, *b* — Moist WS_3 [1925].

The preparation of active tungsten as well as molybdenum sulphides is usually based on the respective ammonium thiosalts which are thermally split in a hydrogen atmosphere. Assuming that we set out from the tetrathiosalt, the respective sulphide is formed according to the reaction

$$(NH_4)_2WS_4 + H_2 \rightarrow WS_2 + (NH_4)_2S + H_2S \qquad \text{(III)}$$

and similarly, for molybdenum

$$(NH_4)_2MoS_4 + H_2 \rightarrow MoS_2 + (NH_4)_2S + H_2S. \qquad \text{(IV)}$$

In both cases, decomposition of the pure thiosalt was studied by means of thermal analysis. Fig. 5*a* shows the DTA diagram of ammonium tetrathiotungstate in an inert medium, from the course of which the decomposition mechanism may be

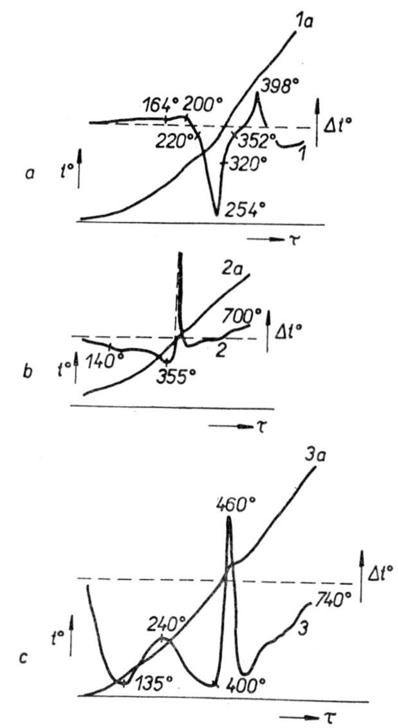

Fig. 6. Decomposition thermograms (in N_2 atmosphere, temperature °C) [1757]. *a* — $(NH_4)_2MoS_4$, *b* — $(NH_4)_2MoS_4$ heated preliminarily 4 hours in N_2 to 200 °C. *c* — MoS_3 (formed by decomposition of $(NH_4)_2MoS_4$ solution with hydrochloric acid). (1*a*, 1*b*, 1*c* — dependence of temperature on time; 1, 2, 3 — differential thermal analysis curves).

estimated. The first endothermic wave corresponds, according to Sokol [1925] to endothermic dissociation of the thiosalt

$$(NH_4)_2WS_4 \rightarrow WS_3 + (NH_4)_2S \qquad \text{(V)}$$

while the following strong exothermic effect, which may sometimes be the cause of explosive decomposition, involves several processes. Decomposition of the trisulphide

$$WS_3 \rightarrow WS_2 + S \qquad \text{(VI)}$$

is endothermic; the same applies to sulphur volatilisation. According to Sokol, the overall exothermic effect is most probably due to recrystallisation of the originally monoclinic lattice to a hexagonal one, and not to spontaneous recrystallisation of the WS_2 formed. This is likewise indicated by the thermogram of moist WS_3 (Fig. 5*b*), which has a distinct endothermic minimum as well as a small exothermic effect.

A similar curve was obtained [2175] on decomposing freshly prepared $(NH_4)_2MoS_4$ in a hydrogen atmosphere (the sample was prepared according to Krüss [1104]).

Differing from the above, Rode and Lebedev [1757] found a somewhat different behaviour in the molybdenum series (see Fig. 6). On thermal decomposition of ammonium tetrathiomolybdate in nitrogen (Fig. 6) a distinct endothermic peak starting at about 200 °C was found, corresponding to the decomposition

$$(NH_4)_2MoS_4 \rightarrow MoS_3 + 2\,NH_3 + H_2S. \tag{VII}$$

In the temperature range of 250 to 350 °C the trisulphide decomposes and sulphur distillates; the decomposition process is concluded with an exothermic peak at about 400 °C. Fig. 6 shows the decomposition curve of the tetrathiomolybdate which had previously been kept at 200 °C in nitrogen atmosphere. The thermogram shows an indistinct endothermic peak at 355 °C, corresponding to the overall effect of MoS_3 decomposition and sulphur volatilisation, followed by a distinct exothermic peak. This exothermic peak was also recorded on decomposition of wet MoS_3 (Fig. 6c): the first endothermic peak corresponds to dehydration, the second one to the overall effect of MoS_3 decomposition and sulphur volatilisation. Rode and Lebedev ascribe the exothermic peaks in the 400 to 460 °C range to the crystallisation effect, agreeing with Kingman [1038]. In this process, the crystalline disulphide is formed from amorphous MoS_2 containing a certain amount of non-stoichiometric sulphur formed by MoS_3 decomposition. The authors named believe, that the rhombohedric sulphide is formed first, converting later to the hexagonal MoS_2 modification.

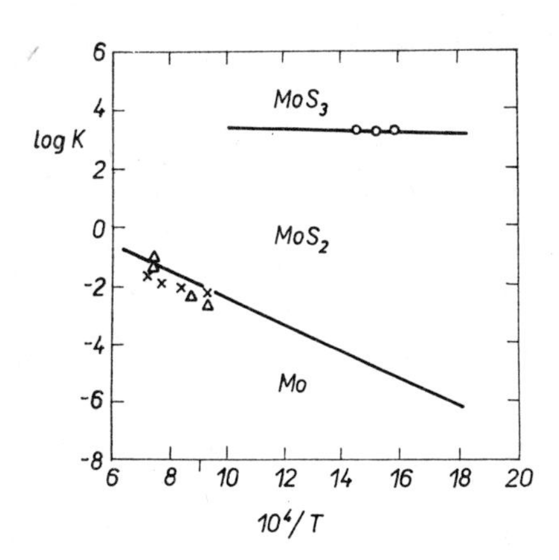

Fig. 7. Thermodynamic equilibrium of reduction of molybdenum sulphides by hydrogen [1347]. Equilibrium constant $K = ([H_2S]/[H_2])_r$

Experimental data: x, ○ [1573], △ [2258].

From the equilibrium diagram it follows (Fig. 7) [1014, 1015, 1572, 1573, 1769, 2258] that with the operating conditions under which MoS_2 is typically employed, MoS_2 only is stable [178, 581].

Thermal decomposition of MoS_3 (which may be prepared by different means, e.g. precipitating acid molybdate solutions with hydrogen sulphide, acidifying an aque-

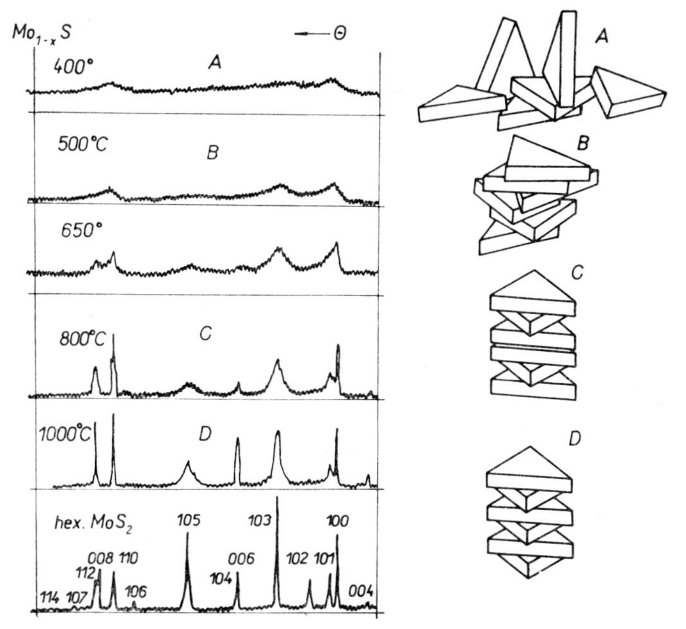

Fig. 8. Diffractograms of MoS_3 thermal decomposition products at different temperatures (Θ — glancing angle, heating time 24 to 36 hours) and models of the gradual build-up of the hexagonal MoS_2 lattice (*A*, *B*, *C*, *D*) [2210].

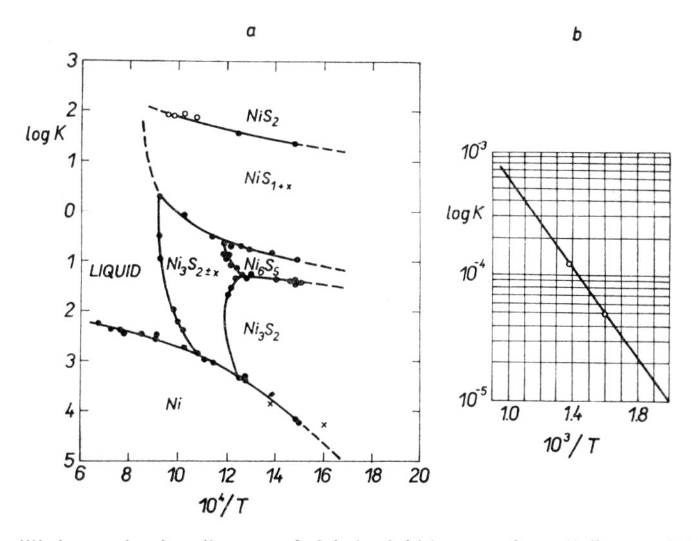

Fig. 9. Equilibrium reduction diagram of nickel sulphides. *a* — Overall diagram [1347, 1806, 1984]. *b* — Reduction $Ni_3S_2 + 2\,H_2 \rightleftarrows 3\,Ni + 2\,H_2S$ [1039].
Equilibrium constant $K = ([H_2S]/[H_2])_r$

Experimental data: ● [1766],
○ [177],
× [95].

ous thiomolybdate solution, the effect of dilute H_2SO_4 on crystalline thiomolybdate, reaction of hydrochloric acid with solid piperidine and piperazine thiomolybdates or their thermal decomposition) which was studied by Rode and Lebedev [1757] by means of thermal analysis, was amplified by an X-ray study reported by Wildervanck and Jellinek [2210].

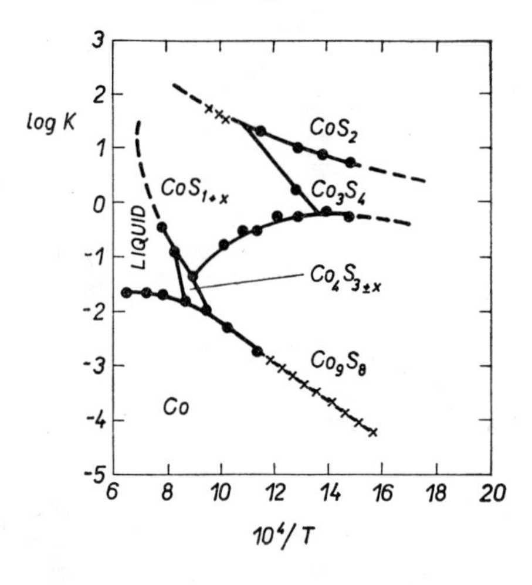

Fig. 10. Equilibrium reduction diagram of cobalt sulphides [1347, 1983].

Equilibrium constant $K = ([H_2S]/]H_2])_r$

Experimental data: ● [1766], × [819], + [1983].

Dissociation of MoS_3 and the formation of the hexagonal MoS_2 modification on heating to different temperatures in the 400 to 1000 °C range is illustrated in Fig. 8. On heating to 350 °C, MoS_3 loses some sulphur while the product continues to be amorphous. The crystallisation effect proper sets in at 350 to 400°C, when totally non-arranged crystallite aggregates begin to form (*A*). On further heating to 500 °C, aggregation takes place around the *c*-axis; the crystallite arrangement continues to be random, while the antiparallel orientation of the crystallites begins to prevail (*B*). At 800 °C the hexagonal structure is complete, but it has a considerable number of defects (*C*). At 1000 to 1100 °C the formation of the hexagonal lattice is completed (D). Differing from Rode and Lebedev [1757], Wildervanck and Jellinek [2210] did not observe any intermediate formation of a rhombohedrical modification; they assume, that neither in the case of the Russian authors was a rhombohedric modification involved, and that the erroneous conclusion was due to imprecise interpretation of the X-ray diagrams. Opalovskii and Fedorov [1540a] studied the formation of lower sulphides during thermal decomposition of MoS_2.

The equilibrium reduction diagrams of nickel and cobalt sulphides are shown in Figs. 9 and 10. The most stable nickel modification is the subsulphide Ni_3S_2 (in graph 9*b*, extrapolation is in accordance with the van't Hoff isobar [1039]) while under the reaction conditions under which NiS is employed (mainly in hydrogenation reactions) this catalyst converts to the subsulphide. This also applies e.g. to the mixed catalyst 8376, in which Ni_3S_2 was determined by X-ray means as the active component [2084]. The rate of reduction of NiS to the subsulphide is especially great at temperatures above 300 °C [654]. In hydrodesulphurisation reactions, either sulphurisation or hydrogenation of sulphur from the active Ni surface will depend on the type of substance desulphurated as well as on the hydrogen pressure [1772a, 2039].

Similarly, Co_9S_8 is the most stable sulphide modfication in the series of cobalt compounds, which may be considered for use as hydrogenation or refining catalysts [1347]. As will be seen later (section 4.5.) a part of the cobalt present in the refining cobalt–molybdenum catalyst takes this form (under operating conditions in an atmosphere of hydrogen); recently, it was found that the presence of a compound even poorer in sulphur (Co_4S_3) cannot be excluded [477].

The most stable sulphide modification of iron is FeS, the formation of which requires a very low partial H_2S pressure (at 400 °C, $\log([H_2S]/[H_2])_r \sim -4.5$ [1347]). Therefore, the catalytic effect of FeS and corrosion of the reactor walls and other parts of the apparatus must be kept in mind when hydrogenating sulphur compounds or sulphur-containing raw materials.

Sulphurisation kinetics of metals or of their oxides at different temperatures is likewise important for the formation of catalytically efficient, supported as well as unsupported sulphides. Such data were recently reported (besides tungsten – cf. p. 44) for example for Ni, Co [334a, 419a] and Cr [1327d].

Among other sulphides which are important in catalysis, mention must be made of sulphides of the noble metals and of some less common metals. In this group, especially platinum, palladium and rhenium sulphides were tested as reforming catalysts [226a, 466, 1372, 1376, 1377] (see section 9, p. 381). Besides, partial sulphurisation and contamination of the catalyst takes place in reforming sulphur-containing raw materials on technical platforming catalysts (Pt on Al_2O_3, temperatures around 500 °C); strong H_2S adsorption takes place on the surface, partial formation of platinum sulphides is possible, high-molecular sulphur compounds deposit and the catalyst is contaminated with FeS which is formed as a product of apparatus corrosion. These influences lead to a considerable activity decrease [190, 620, 769, 1374]. Although the existence of platinum sulphides is problematic under reforming conditions [2202] (for some thermodynamic data relating to different Pt sulphide forms, see the reference [176, 673, 1014] and their presence in the platforming catalyst was not directly proved by X-ray analysis [1373, 1430, 1432], yet their presence in the used catalyst cannot be excluded. On the other hand some recent results [743B, 743e, 1614a, 1656A, 1656b] showed that under certain circumstances sulphur compounds may considerably improve the efficiency of the platforming catalyst (see also section 9.4.1.).

Equilibrium data for the system palladium sulphides–hydrogen are reported by Japanese authors [1467]. Other thermodynamic data for catalytically important sulphides will be found in the following references: for Mo sulphides [1014,1274, 1275, 1921], for tungsten sulphides [334], for Ni, Co and Fe sulphides [1766], for Pt sulphides [1739, 1973a, 2266] and Re sulphide [1923a].

4.3.

Texture and adsorption properties of sulphide catalysts

Increased attention is being given recently to the texture properties of sulphide catalysts, particularly W, Mo and Ni sulphides which are of great importance in industrial applications. Measurements of adsorption and of other surface properties, however, are rendered considerably more complicated in this case by the difficulty of obtaining an absolutely pure sulphide surface, free of excess sulphur, adsorbed and occluded gases, sulphuric acid etc.

Most data available in the literature relate to the catalysts which are most important in industry, i.e. catalysts based on tungsten sulphide. The WS_2 catalyst (i.e. the

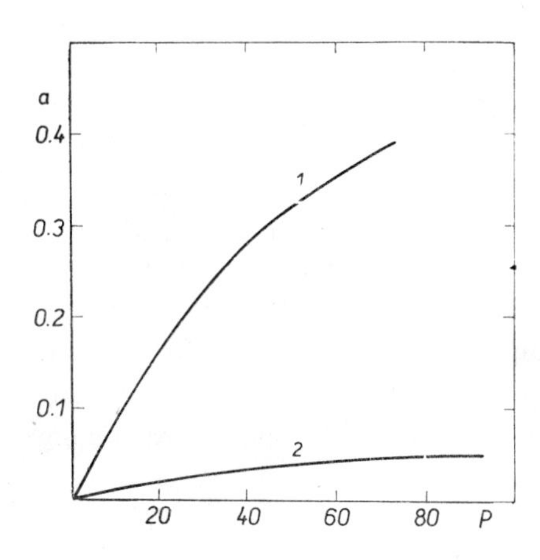

Fig. 11. Comparison of the adsorption capacity (a ml C_2H_6/g WS_2) of fresh (1) and used (2) WS_2; dependence on the overall pressure P (Torr) at 100 °C [416].

industrial catalyst 5058) is prepared practically exclusively by means of hydrothermal decomposition of the ammonium tetrathiosalt. Pier [1626] states that the surface area of a catalyst prepared in this way (on the basis of Ar adsorption) is 20 m^2/g (50 m^2/cm^3). Friz [564] obtained values which compare well with data published by Pier. He measured the adsorption of CO_2 and sulphur hexafluoride on the same catalyst. However, the surface of the catalyst varies considerably, depending on the conditions of preparation.

Compared to other adsorbents, WS_2 is characterized by a high degree of selectivity, especially for hydrogen and olefines [416, 1626]. Gonikberg and Levitskii [628] measured nitrogen and hydrogen adsorption on WS_2 at elevated pressures. For hydrogen adsorption, of which 3 to 4 times more is adsorbed under the same conditions compared to nitrogen, the following empirical relationship is satisfactory in the pressure range of 10 to 100 atm and between 300 to 400 °C:

$$X_H = \beta + \alpha P_H^{0,5} . \qquad \text{(VIII)}$$

Here X_H is the amount of hydrogen adsorbed and α, β are constants, P_H is the hydrogen pressure. Bernardini et al. [149a] likewise reported data for nitrogen adsorption on this catalyst.

The adsorption capacity of the deactivated catalyst is considerably lower, as shown by Fig. 11 (ethane adsorption at 100 °C was compared) [416]. Similarly, Kalechits et al. [953] found, that the surface area and pore volume (especially that of pores in the size range 20 to 80 Å) were decreased substantially in WS_2 used in hydrogenation of benzene. Partial recrystallisation to the less active hexagonal modification and a considerable decrease of the hydrogenation activity also took place. A sudden change in the texture and in the catalytic properties of WS_2 takes place when the catalyst is heated to 600 °C even for a short time. As was shown earlier, crystallites are oriented into a more or less regular lattice in the transition region (the temperature of 600 °C lies roughly in the middle of this region [115]). Kalechits et al. [958] actually observed that diffraction lines were substantially more distinct after the catalyst has been heated. The result of this change is a sharp decrease of the adsorption capacity, the pore volume (the greatest decrease is observed with pores of 10 to 20 Å) and the specific surface area (Table 5), together with a loss of activity, mainly in respect of hydrogenation. These observations are directly linked to the significance of non-stoichiometric sulphur in the WS_2 catalyst, as will be shown later (p. 78).

Table 5.

Influence of Heating on the Adsorption Characteristics of the WS_2 Catalyst [958].

WS_2 catalyst	Amount of adsorbed N_2 (g/g)	Pore volume (cm^3/g)	Surface area (m^2/g)
Original catalyst	0.046	0.057	66
Heated to 600 °C	0.041	0.051	54
Heated to 750 °C	0.029	0.036	23
Heated to 1000 °C	0.002	0.003	3

Detailed data on hydrogen adsorption on the WS_2 catalyst (prepared by decomposition of tetrathiotungstate) were reported by Decrue and Susz [382].

The work of Rubinshtein et al. [463, 1794, 1795, 1796, 1798, 1799], is of fundamental significance for the texture characteristics of catalysts based on WS_2. These authors likewise investigated the very significant influence of nonstoichiometric sulphur on the surface and catalytic properties of WS_2 catalysts. It was found by adsorption of nitrogen and benzene on fresh as well as spent WS_2 (used in the hydrogenation of high molecular weight oils), that pores of a diameter of about 30 Å persist [463] regardless of the type of adsorbate employed. Similarly, with fresh, non-used catalysts, similar values of the specific surface area were found: 66 and 72 m^2/g when N_2 and C_6H_6 were used, resp., i.e. a difference of about 10%. On the other hand, a difference of about 100% was found in the surface area value with used WS_2 containing 3.4% C (28 m^2/g with N_2, 55 m^2/g with C_6H_6) as well as in the total pore volume (0.03 and 0.070 cm^3/g). These differences were caused by adsorption of benzene vapour on the high molecular weight deposits present on the coke-covered

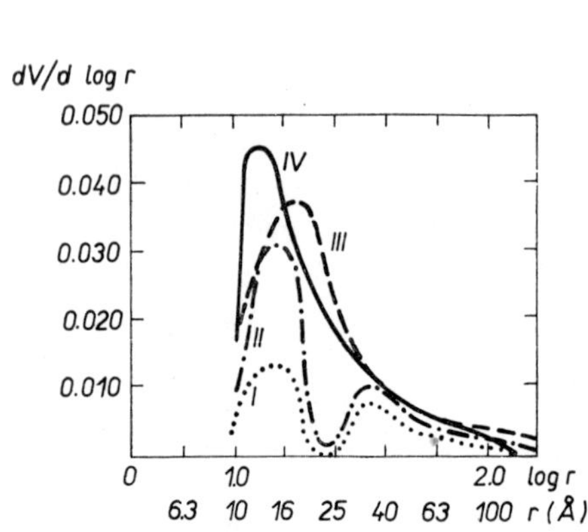

Fig. 12. Pore volume distribution curves for WS_2 — catalysts I to IV (for the notation see Table 6) [1794].

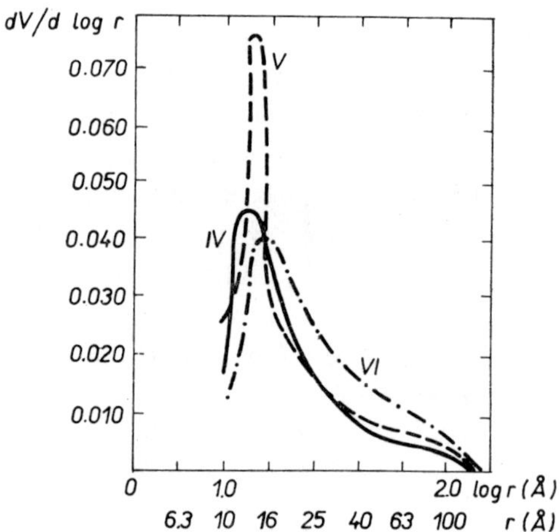

Fig. 13. Pore volume distribution curves for WS_2 catalysts IV—VI (for the notation, see Table 6) [1794].

used catalyst. On complete purification of the catalyst surface, e.g. by extraction with benzene, these differences were eliminated [1799]. It was found by comparing the values obtained with the use of N_2, that the surface area (pore volume) decreased in used WS_2 to about one-half the original values.

The problem of non-stoichiometric sulphur is significant for all sulphide catalysts, since the presence of this sulphur has a considerable influence on the physico-chemical and catalytic properties (p. 74) of these catalysts. On thermal reductive decomposition of the ammonium thiosalt, a disulphide with excess non-stoichiometric sulphur is always formed, its formula being MeS_{2+x}, where $x = 0$ to 0.5 depending on the conditions of preparation. There follows from the results of differential thermal analysis (p. 45), that the reaction intermediate in the formation of the disulphide, is the trisulphide. The degree to which sulphur is removed and, at the same time, the degree of organisation of the crystal lattice into the stable hexagonal form depends on the temperature at which the catalyst is processed: with increase in temperature,

Table 6.

Texture Properties of WS_2 Catalysts, Their Composition and Catalytic Activity [1794].

Catalyst notation	Characteristic of the catalyst	Total pore volume (cm^3/g)[d]	Volume of pores with a radius of $r_{ef} > 10$ Å (cm^3/g)[d]	Surface area BET (m^2/g)[d]	Catalyst activity (%)[b]	Catalyst composition % by wt.			Catalyst composition Atomic ratio
						W	S	W+S[c]	(S : W)
I	Initial WS_2[a]	0.0088	0.0075	3	—	64.02	25.65	89.67	2.30
II	I heated in vacuo to 200 °C	0.0229	0.0116	19	—	—	—	—	—
III	I heated in vacuo to 300 °C	0.0544	0.0231	59	—	—	—	—	—
IV	I heated in vacuo to 400 °C	0.0568	0.0199	66	90.8	72.28	25.28	97.56	2.006
V	I heated in N_2 to 600 °C	0.0510	0.0222	54	84.4	69.91	26.19	96.10	2.16
VI	I heated in N_2 to 750 °C	0.0364	0.0239	23	51.2	72.36	25.98	98.34	2.08
VII	I heated in N_2 to 1000 °C	0.0030	0.0017	3	14.0	71.22	25.62	96.84	2.05

[a]) The catalyst was prepared by means of reductive decomposition of the thiotungstate: $(NH_4)_2WS_4 + H_2 \rightarrow WS_2 + (NH_4)_2S + H_2S$.

[b]) The activity was measured by the degree of benzene conversion on hydrogenation in an autoclave ($t = 400$ °C, initial pressure $H_2 = 140$ atm [1586]).

[c]) The remainder up to 100% includes NH_3, C, H, and 0.

[d]) The surface area was measured by nitrogen adsorption using the BET method; the pore volume was calculated from adsorption at the pressure of saturated N_2 vapours. The pore volume distribution curve was calculated on the basis of the Kelvin equation from the desorption branch of the nitrogen isotherm.

loss of non-stoichiometric sulphur [1038, 1353], increase of the size of catalyst crystals [1038, 1626] and a decrease of the catalytic activity [1038, 1586, 1794] takes place.

Samoilov and Rubinshtein [1794] found, that decomposition of ammonium tetrathiotungstate results in a catalyst of the atomic ratio W : S ~ 1 : 2.3. On heating *in vacuo*, the excess sulphur is split off and, at temperatures of about 400 °C, the stoichiometric composition of the disulphide is nearly achieved. Similarly, at 600 °C in a stream of nitrogen, 50% of the non-stoichiometric sulphur is removed, and this process is practically complete at about 1000 °C. Non-stoichiometric sulphur (at least the major part – cf. the conclusions drawn by Kalechits et al. [974, 1589] in Chapter 5) is bound only weakly in the catalyst, so that it is easily removed, especially on heating *in vacuo*. Removal of non-stoichiometric sulphur on heating to an elevated temperature, however, alters the texture and, therefore, also the catalytic properties of the catalyst to a substantial degree. Table 6 summarises the results for WS_2 catalysts processed in different ways. Pore size distribution curves for these catalysts are shown in Figs. 12 and 13. These results indicate the relationship between the degree to which non-stoichiometric sulphur is removed, and the texture properties. In the fresh catalyst I, the surface of which is poorly developed (3 m^2/g) and which is only slightly porous, pores of radii of about 15 and 35 Å prevail. On heating *in vacuo* over 200 °C, non-stoichiometric sulphur is split off and the active surface starts to form; the specific surface area grows rapidly (increase by a factor of 6), and likewise the pore volume increases considerably (2.6 times), especially in the case of pores of about 15 Å in radius (Fig. 12). At the same time small pores of <10 Å radius start to form. On further heating, the volume of pores of $r_{ef} \sim 15-20$ Å increases, while the volume of pores with radii of up to 10 Å rises even more quickly: therefore, a surface with very fine pores forms at higher temperatures. Hence, the catalyst IV compared with the catalyst II has a three times larger surface, while the pore volume increases with the radius $r_{ef} > 10$ Å by a factor of 1.7 only. At temperatures of about 600 °C the formation of the catalyst surface is almost complete and with a further increase in temperature, the porosity and specific surface area decrease rapidly (see also [958]), so that at 1000 °C, when an almost perfect hexagonal lattice has been formed, WS_2 loses its adsorbent character and at the same time that of a catalyst (cf. the rapid decrease in the hydrogenation activity) (Table 6).

As well as a catalyst with stoichiometric composition or excess sulphur, a very poor catalyst with excess tungsten may be formed on thermal reductive decomposition of the tetrathiosalt [1799]. The fresh sample ($WS_{1 \cdot 92}$) has a small surface area (8 m^2/g) and low porosity (pore volume 0.004 cm^3/g). On heating to 400 °C *in vacuo*, more sulphur is eliminated and a sulphide having an atomic ratio $WS_{1 \cdot 53}$ is formed. This sulphide had a far larger surface area (49 m^2/g) and pore volume (0.059 cm^3/g) than it had before heating. At the same time, the acidity of the catalyst and, therefore, also its catalytic activity in phenol hydrogenation decreases substantially (cf. section 5., p. 76) [1798]. Thus in this catalyst sulphur is located in the pores, so that it can be removed easily *in vacuo*. X-ray analysis, however, indicated that non-stoichiometric tungsten does not form a separate phase; instead, according to Samoilov [1799], tungsten converts to some form of a "rigid skeleton" which is catalytically less active than the catalyst with an almost stoichiometric composition. Samoilov [1799] likewise investigated the texture of the industrially deactivated WS_2 catalyst, and found that high molecular weight coke-like deposits contaminate all types of pores, thus gradually lowering the specific surface area as well as the pore volume.

Table 7.

The Texture Characteristics of Type 6434 Catalysts and of Their Carriers [1795].

Adsorbent	Surface area (BET) (m^2/g)	Pore volume (cm^3/g)
Clay I (Terrana)	183	0.1966
Clay II	36	0.0635
WS_2-clay I	87	0.1278
WS_2-clay II	21	0.0344

When WS_2 (8 to 10% b.w.) is supported on active clay, the important catalyst 6434 is obtained [1626]. The carriers are usually natural aluminosilicates of different origin with different specific surface areas. When WS_2 is supported on the carrier surface, its surface area as well as pore volume is decreased considerably (about 2 times), as seen from Table 7 [1795]. At the same time, the pore volume distribution changes substantially: while the carrier had a majority of pores of a specific size, an adsorbent with a wide spectrum of pore sizes is obtained after supporting of WS_2. The catalytic activity (specially with respect to splitting and isomerisation reactions) is greater, the greater the specific surface area and pore volume [978, 1795].

Another important WS_2-based catalyst is the mixed refining catalyst WS_2–NiS–Al_2O_3 (catalyst 8376). Compared to WS_2, its most important property is a substantially lower splitting activity with either unaltered [1626] or even enhanced [1256] hydrogenation activity. This increased hydrogenation activity is sometimes explained by a considerable rise in the dispersion of the supported sulphides, resulting in far better utilisation of the inner surface of the active catalyst components [1257, 1730].

The refining properties of the catalyst 8376 are closely related to its texture. Greber [644], who compared two catalysts of the 8376 type with widely differing refining activity found, that the system with macropores having a radius of >1000 Å determines the kinetics of the refining process and, therefore, the quality of the catalyst. In this case, the term refining is used to mean a summary effect, measured by hydrogenation of aromatic substances, hydrogenolysis of phenols and nitrogenous bases and improvement in the colour of the tar material.

In the course of use, the texture of the catalyst 8376 alters considerably [2084] as will be seen from Fig. 14. The mean radius of micropores is about 57 Å, the optimum macropore radius, which ensures optimum diffusion conditions in the pores of the refining catalyst, being 4.8 to 5.4 10^{-5} cm [2084]. The catalyst surface grows slightly in the first 4 hours of operation; this is caused by the fact, that the micropore volume decreases only slightly during this forming period, while the macro-

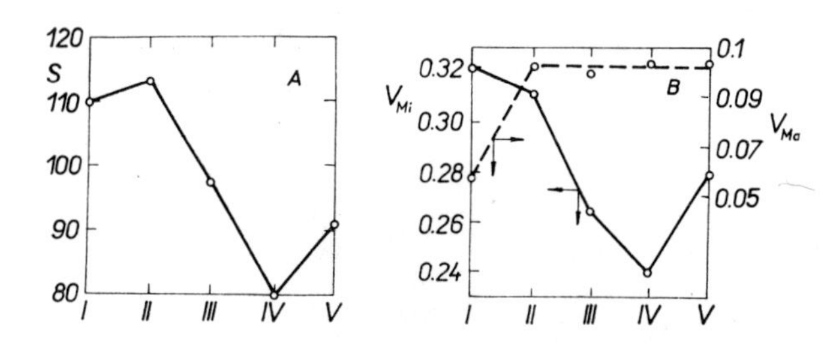

Fig. 14. Change of texture of type 8376 catalysts during hydrorefining [2084]. (The feed was brown coal tar, operating conditions: $t = 350$ °C, $P = 200$ atm H_2). *A* — Dependence of the specific surface variation S (m^2/g). *B* — Variation of the volume of micropores (V_{Mi}) and macropores (V_{Ma}) (both in cm^3/g). Fresh catalysts (I); the catalysts after 4 hours of operation (II), 24 hours (III), 9 days (IV), the catalysts after 9 days of operation, regenerated for 4 hours in hydrogen at 400 °C (V).

pores form very quickly. In the next few hours, the surface area decreases very rapidly as does the micropore volume, the macropore volume remaining unaltered. After 9 days of operation, the surface as well as the micropore volume are diminished by about 1/4. The micropores are blocked by high molecular weight deposits which can be removed in part by pure hydrogen. This relatively rapid decrease of the surface area and pore volume, however, is not followed by a decrease in the refining activity (with feeds which contain no ash or when refining in the vapour phase, the life-time of the catalyst is several months). These experiments therefore confirm the significance of macropores in maintaining the high refining activity of the 8376 catalyst. As Günther [696] has shown, texture is not the only factor that determines the refining activity of the catalyst 8376. Equally important is thermal activation of the fresh catalyst, which enables specific reactions to take place in the solid phase between the oxide components of the catalyst which, after sulphurisation lead to the formation of the most highly active system WS_2–NiS or, WS_2–NiS–Al_2O_3 on the wall of a pore of suitable size.

When the 8376 catalyst is heated, the surface area decreases systematically between 400 and 900 °C to nearly 10% of the original value, at the same time, the radius of the most frequent micropores rises slightly. At about 900 °C, γ-Al_2O_3 converts to the α-form (corundum), resulting in a sudden change in the catalyst activity, as will be seen from Fig. 15 [1431].

A closer study of the micropores revealed [1796] that the catalyst 8376 is an adsorbent having heterogenous pores, 10 to 150 Å in size, with a surface 1.5 to 2 times larger and porosity 3 times higher than pure WS_2. The unsupported mixed catalyst

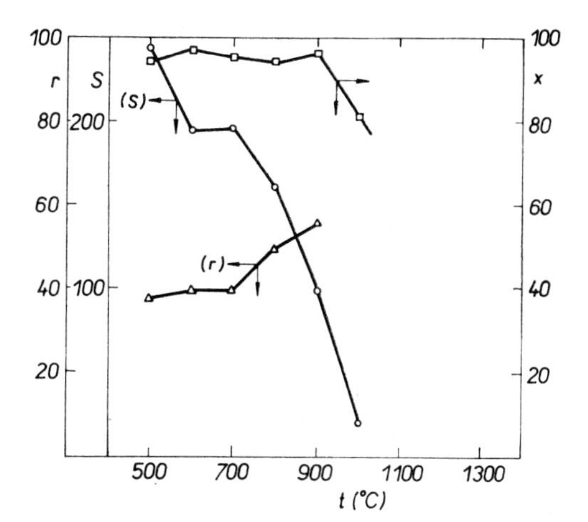

Fig. 15. Influence of heating the catalysts 8376 on its texture and catalytic properties [1431]. S — Surface area (m^2/g), r — Radius of pores occuring most frequently (Å). x — Degree of degradation of phenols (%) in hydrorefining of medium tar oil.

$WS_2.2\,NiS$ (catalyst 3076) is used less often in the industry compared to the catalyst 8376. Semenova studied its texture properties in connection with its optimum method of preparation [1854].

It is important in the case of supported sulphide catalysts to know what part of the surface is covered with the sulphide component. Lukens et al. [1265a] describe a very advantageous method of measuring this part of the surface, based on an exchange reaction between the sulphur of the catalyst and hydrogen sulphide containing $H_2{}^{35}S$.

Some adsorption data concern molybdenum disulphide [948a, 2216a]. Griffith et al. [95, 660] studied the adsorption of a number of substances (H_2, furane, tetrahydrofurane, thiophene, tetrahydrothiophene) on MoS_2 and $MoS_2.MoO_2$ catalysts; the surface of both catalysts was roughly 17 m^2/g. Activated adsorption starts on both catalysts at temperatures to 150 °C [95, 660]. Differing from hydrogen, 4 to 5 times more thiophene is adsorbed on the $MoS_2.MoO_2$ catalyst with a chemisorption maximum of about 150 °C. Compared to thiophene, tetrahydrothiophene is adsorbed less, decomposing at about 400 °C. Similar conditions were observed with adsorption of furane and tetrahydrofurane [95].

The surface area of MoS_2 formed by decomposition of MoS_3 in a hydrogen or inert atmosphere depends to a considerable extent on the rate of heating to the decomposition temperature [452]. When heated rapidly to 450 °C, the MoS_2 formed had a surface area of up to 158 m^2/g; on slow heating it was only 2 m^2/g. A maximum surface is obtained on decomposing at about 450 °C.

A number of authors have studied the adsorption of rare gases (Ar, Kr) [263, 264], methane [201], n-butane [1246a, 1246b], ammonia and water [265, 265a], cyclohexane [205], high-molecular weight n-paraffins [678, 679], benzene and heptane [107, 108] on MoS_2. On the other hand, relatively little data is found in the literature concerning the adsorption of hydrogen sulphide and other sulphur compounds on MoS_2. The work of Kemball et al. [2216], showed, however, that H_2S is absorbed far more strongly than H_2 on MoS_2.

Infrared spectroscopic methods were used to study thiophene adsorption on MoS_2 and on a sulphurised cobalt–molybdenum catalyst. Nicholson [1456, 1457] determined the adsorption bands in the IR spectrum, corresponding to one-, two- (through C atoms or a C and an S atom) and four-point adsorption of thiophene on the catalyst surface:

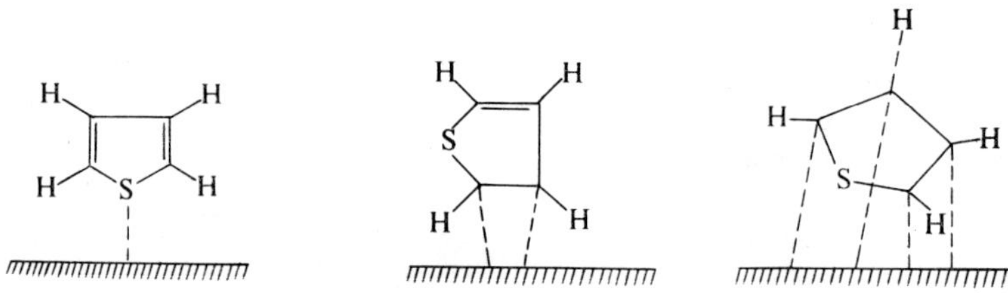

An analysis of the relevant IR spectra indicated, that the surface of the sulphurised cobalt–molybdenum catalyst mainly contains thiophene adsorbed in the four-point manner whereas pure MoS_2 contains this substance adsorbed in the four- as well as the two-point form, the former predominating. This also applies to MoS_2 supported on γ-Al_2O_3 where, however, the two-point form predominates. An addition of cobalt to MoS_2 causes the two-point adsorption form to predominate on this catalyst. Under hydrodesulphurisation conditions, one- and two-point adsorption is most important [1385a]. For the cobalt-molybdenum catalyst, Lipsch and Schuit [1243d] assume one-point adsorption through a Mo-S bond.

Recently, the development of new catalyst types has initiated a more detailed study of the relationship between catalytic activity and catalyst texture. Kalechits et al. [960] investigated this relationship in the case of MoS_2 supported on an aluminosilicate carrier with different molar ratios of $SiO_2 : Al_2O_3$. The selectivity of these catalysts can be controlled to a large degree by selecting the overall composition and the carrier composition. The maximum amount of MoS_2 which the carrier adsorbs on its surface with a given mode of preparation depends on its composition. When MoS_2 is supported on aluminosilicate, the specific surface area as well as the pore volume of the material are diminished; the mean micropore radius increases, especially in the case of low-acidity catalysts. For example, the distribution curve illustrated in Fig. 16, shows that the peak in the region of 10 Å (which is characteristic of the carrier) disappears, and the second peak shifts towards the narrower pores, the size of which is equal in order of magnitude to the thickness of a monomolecular layer of supported MoS_2. The effect of the carrier composition on the catalyst texture is even more marked; with a content of 10 to 30% Al_2O_3 the pore volume as well as surface area decreases very markedly. Parallel with this change, the splitting activity of the catalyst changes too (this is also related to the catalyst acidity, see Fig. 17) [960]

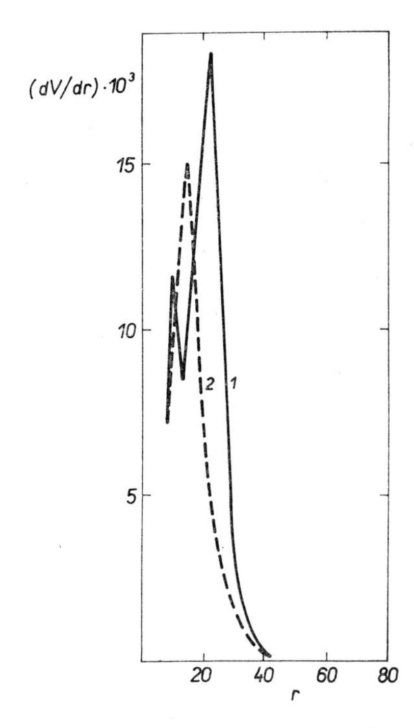

Fig. 16. Comparison of pore volume distribution of supported MoS_2 (about 10% MoS_2 on alumino-silicate, $Al_2O_3 : SiO_2 = 49 : 51$) and carrier [960]. 1 — Carrier. 2 — Catalyst. r — Pore radius (Å). $(dV/dr) . 10^6$ — (ml/gÅ).

as well as the hydrogenation efficiency (related to the decrease in the amount of MoS_2 in the catalyst) [961]. Some other relationships which are important for this catalyst, as well as an explanation of the changes in texture and catalytic properties will be discussed later (p. 103). Ganguli et al. [583a] studied the adsorption of hydrogen and cyclohexane on the hydrocracking catalyst MoS_2/activated clay.

Much attention is devoted in the literature to nickel sulphides. Griffith et al. [95, 654] measured the adsorption isobars of H_2, SO_2, CS_2, COS, CH_3SH, thiophene, tetrahydrothiophene, furane and tetrahydrofurane on Ni_3S_2 (on NiS also in the case of CS_2 and SO_2). The behaviour of CS_2, which is chemisorbed very strongly, is quite outstanding: its chemisorption is, for example, several times greater than that of thiophene. Besides the geometrical arrangement of the surface, this also explains the high catalytic activity of Ni_3S_2 in desulphurisating simple sulphur compounds, especially CS_2 [660].

Oxygen adsorption proceeds very quickly on NiS, even at very low temperature (−78 °C); on oxygen desorption at elevated temperatures, surface oxidation takes place followed by desorption of SO_2 [660]. Roginskii et al. [807, 809] studied the influence of Li and In sulphide admixtures on the adsorption of oxygen on NiS in the temperature range 25 to 65 °C. They found that additions of the two sulphides (the maximum content tested was 1% by wt. Li or In) decrease the rate of oxygen adsorption. A similar result was obtained with acetylene adsorption. On synthetic NiS_2, chemisorption of O_2, H_2S, H_2, H_2O vapour, SO_2 and C_2H_4 was measured in the tempera-

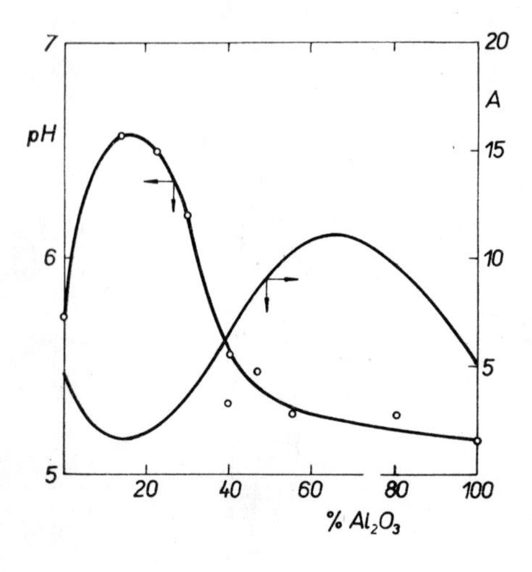

Fig. 17. Relationship between the composition of the carrier of the MoS_2–SiO_2–Al_2O_3 catalyst, the acidity of its surface (pH) and splitting activity in cracking of cumene (A, ml/g.s) [960].

ture range of 20 to 230 °C and the results related to the electrical conductivity of the sulphide [389], It was found that this catalyst is an n-type semiconductor with metallic conductivity [388].

Carbon monoxide is strongly adsorbed on NiS at −183 °C; at temperatures of 20 to 200 °C the amount of CO adsorbed is small. At 200 °C, hydrogen reduces NiS strongly [1968]. Adsorption of sulphur vapours on the NiS catalyst [2057] as well as chemisorption of CS_2 and H_2S on sulphurised NiO/Al_2O_3 [2057a] was investigated in connection with the synthesis of CS_2 from sulphur and methane, which is of technical importance.

The mechanism of catalytic hydrodesulphurisation cannot be studied without basic adsorption measurements on the catalyst used. The adsorption of H_2S, H_2, thiophene and butenes on sulphurised Co–Mo and Cr_2O_3 catalysts was studied [1565, 1567]. Although the two catalysts differ chemically they behave similarly during the hydrodesulphurisation of thiophene (300 to 400 °C) when both hydrogen and H_2S adsorb very strongly, while adsorption of butenes and thiophene is considerably weaker. This also agrees with the findings of Griffith et al. [658], who observed similar differences with sulphurised MoO_3 and Ni and Mo sulphides in the case of H_2 and thiophene.

In a study dealing with the catalytic effect of the cobalt–molybdenum catalyst, Lipsch and Schuit [1243d] observed very strong adsorption of water and hydrogen sulphide. The two substances decrease adsorption of thiophene and, particularly, of butenes. Pyridine is very strongly adsorbed, but does not influence the adsorption of thiophene and butenes as long as the catalyst surface is not saturated with pyridine. It would appear therefore, that thiophene and butenes are adsorbed on the same type of sites (sites I), while pyridine is adsorbed on another type (II). Pyridine is only adsorbed on sites I when all sites II are occupied. Hydrogen is chemisorbed on the cobalt–molybdenum catalyst from 220 °C on, the adsorption maximum being at 550 °C. Lipsch and Schuit found that hydrogen does not decrease the adsorption of thiophene.

Among the other sulphides, adsorption of steam on zinc sulphides was studied [1791].

4.4.

Electrical properties of sulphide catalysts

Relatively little data on the electrical properties of sulphides is to be found in the literature, and moreover, much of this data is contradictory. For example, with

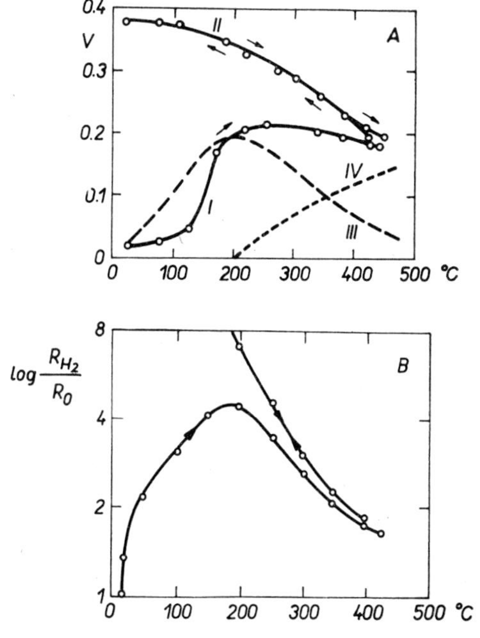

Fig. 18. Relationship between the electrical conductivity (*B*) of the WS_2 catalyst and the amount of absorbed H_2 (*A*) [564]. *V* — Volume of adsorbed H_2 ($cm^3/g\ WS_2$). R_0 — Specific resistance of the catalyst *in vacuo*. R_{H_2} — Specific resistance in an atmosphere of hydrogen (the experiments were carried out at $P_{2H} = 100$ Torr). I — Adsorption isobar with rising temperature. II — Adsorption isobar on cooling and heating. III — Desorption branch at temperatures above 200 °C. IV — Isobar of H_2 dissolution in the WS_2 crystal.

WS_2 which is the most important catalytic sulphide, metallic conductivity [562] as well as p-type [564] and n-type semiconductivity [383, 745] were found.

The significance of the relationship between adsorption and electrical properties of the WS_2 catalyst was discussed by Fritz [564], who measured the dependence of the electrical conductivity of this catalyst on the amount of adsorbed H_2 (Fig. 18). Up to 200 °C (Fig. 18A), rapid chemisorption of hydrogen takes place; at higher temperatures intensive desorption commences (curve III). At temperatures of 200 °C and above Friz states that hydrogen dissolves rapidly in the crystal (curve IV) in the spaces between the individual layers, so that the decrease in the overall amount of bound hydrogen is only slight above 200 °C (curve I). On cooling, this portion (IV) continues to be bound firmly in the crystal, and only the desorbed hydrogen is replenished. The process corresponding to curve II is reversible. The electrical resistance, however, only varies with the amount of hydrogen adsorbed. Friz explains the process by the fact, that WS_2 is a semiconductor because of the non-stoichiometric sulphur present (according to Friz it is a p-type semiconductor), having defects of acceptor character in the sites of excess sulphur. The adsorbed hydrogen, being a donor, transfers its electrons to these acceptor centres, the mobility of inner electrons is diminished and hence the conductivity decreases. When the temperature rises above 200 °C the amount of adsorbed hydrogen decreases and thus the conductivity increases. Hydrogen bound in the interionic layers does not influence these conditions. More recently Rohländer has explained this behaviour of WS_2 [1763].

The relationship between electrical conductivity and adsorption of ethanol, acetone and benzene vapours was likewise studied in the case of MoS_2. Experiments were carried out with single crystals of the sulphide [1236].

Roginskii et al. elucidated a number of important findings concerning the relationship between the electrical properties of semiconductors and their catalytic activity. These authors studied the degree of decomposition of isopropylalcohol in the presence of zinc compounds (ZnO, ZnS, ZnSe, ZnTe) [1106, 1107, 1761]. The activation energy of the dehydrogenation of isopropylalcohol (a typical oxidation-reduction reaction) decreases with decreasing width of the forbidden energy gap, or, with decreasing electronegativity difference Δx (according to Pauling) of the anion and cation. With increase in the Δx value, i.e. increasing ionic character of the bond, dehydration takes place more easily (a reaction of the acid-base type). This statement, however, must not be taken too literally considering the possible participation of acid admixtures, which may be present in the catalysts studied.

Roginskii et al. in their later work [807, 808] started from the well-known and clearly understood influence of additions of oxides of lower or higher valency (Li_2O, La_2O_3) on the semiconductor character of NiO [600, 742], extending this research to sulphides, too. It was found in a study of the decomposition of hydrazine in the presence of NiS (which, according to Haufe [741] is characterized by metallic conductivity with strongly developed covalent bonds [1046], similarly with NiS_2 [388] and its solid solutions with Li_2S and In_2S_3 which, differing from their respective oxides influence the catalytic activity of NiS in the decomposition of hydrazine considerably less. The relationship is illustrated in Fig. 19, exhibiting a weak maximum with a Li content of about 0.2% in the form of Li_2S. The same characteristic is also found in the relationship between the work function and the addition of Li and In sulphides. The character of the metallic conductivity of NiS likewise changes little on addition of Li and In.

Wilmot [2215a] considers the selective effect of nickel sulphide (supported on Al_2O_3) in hydrogenation of polyolefins to monoolefins to be a consequence of the semiconductor character of this catalyst. Sulphidation of the nickel compound with which the alumina is impregnated, gives virtually stoichiometric NiS, which

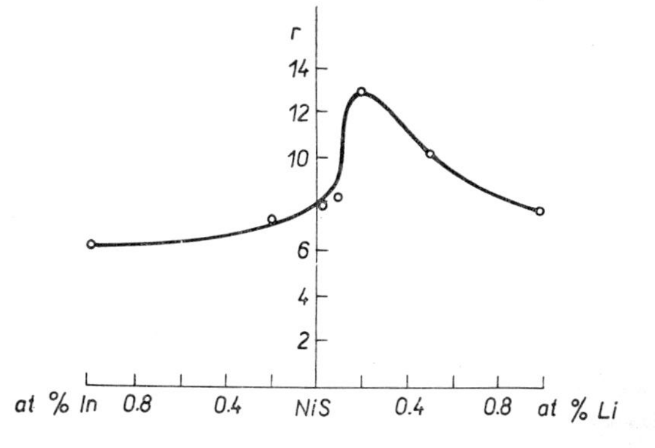

Fig. 19. Influence of the addition of Li and In sulphides on the catalytic activity of NiS in hydrazine decomposition: $3\,N_2H_4 \rightarrow 4\,NH_3 + N_2$ [808]. r — Rate of decomposition (mmoles/m^2 min).

forms the subsulphide Ni_3S_2 when reduced with hydrogen. Electrons remaining in lattice vacancies (in positions formerly occupied by sulphur) are very mobile, and the system behaves like an n-type semiconductor.

A study of the preparation of the desulphurisation catalyst V_2O_3–MoS_2 [2038] based on conductometric measurements revealed that partly reduced MoS_2, formed by sulphurising MoO_2, behaves as a p-type semiconductor. V_2O_3 (n-type semiconductor) forms a complex when present in sixfold excess to MoO_2, this complex is especially easy to sulphurise, forming an active desulphurisation catalyst.

Furthermore, the electrical conductivity and catalytic activity of α-HgS in ethanol dehydrogenation (200 °C) was studied in relation to the Cu content, added as an activator. In the concentration range 0.01 to 0.1% Cu, catalytic activity increases rapidly and, at the same time, conductivity passes through a minimum in the region of 0.01% [918].

From the point of view of the practical application of sulphide catalysts, however, studies of the relationship between the electrical properties of the catalyst and its activity in technically significant reactions is more important. This research is only just beginning, and the results obtained up to the present are rather qualitative. Rohländer [1763] measured the electrical conductivity *in vacuo*, in air and in argon in the temperature interval 20 to 450 °C and determined the type of semiconductivity in the case of the catalyst 5058 (WS_2). This highly active catalyst contains non-stoichiometric sulphur which is chemisorbed on the n-type semiconductive WS_2. The major part of this excess sulphur can be extracted from the catalyst, which thus loses its activity in a similar manner to that in which sulphur-free feeds are hydrogenated. The relative position of the Fermi level in the forbidden gap was determined from the experimental measurements. At room temperature up to 150 °C, active WS_2 is a p-type semiconductor, while at 450 °C the catalyst is an n-type semiconductor. The change of type of semiconductor occurs between 150 and 450 °C, the inversion point shifting towards the higher temperatures with increasing sulphur excess in the catalyst. The Fermi level shifts with rising temperature from the vicinity of the valence band towards the conduction band, and at the point of inversion it is located in the centre of the forbidden energy gap.

Based on these results, Rohländer [1763] explains the behaviour of WS_2 on hydrogen adsorption, as observed by Friz [564] (p. 61 and Fig. 18). Up to 250 °C hydrogen adsorption takes place on acceptor centres of the surface (S-ions); in the process, the Fermi level moves away from the valence band, arriving eventually in the centre of the forbidden gap. Above 250 °C sorption at acceptor centres stops and desorption follows, but at the same time, adsorption on donor centres (W-ions) of the surface increases with the temperature (the measurement was continued up to 450 °C). In this process, the Fermi level approaches the conduction band. On cooling, the Fermi level again falls and, on passage through the inversion point, sorption on acceptor centres is renewed. In this process, electrical conductivity alters continuously. According to Rohländer, therefore, WS_2 is a bifunctional catalyst with p- and n-type sorption centres on sulphur and tungsten ions.

Rohländer states that the semiconductor properties of WS_2 are related to its different catalytic properties under different reaction conditions, as characterized by Pier [1626]. Refining and hydrogenation reactions with no substantial change in the carbon skeleton take place up to 380 °C. Above 400 °C, the splitting and isomerisation properties of the catalyst begin to play an important role, the original carbon skeleton and mean molecular weight of the feed changing considerably. According to Rohländer, hydrogenation and refining reactions take place on the p-centres of the catalyst, while splitting and isomerisation take place in the region of the n-type semiconductor character of the catalyst.

In the p-type semiconductive region, i.e. at low and medium temperatures (150 to 200 °C), hydrogenation and refining reactions proceed at such a low rate, that it is impossible to operate under these conditions. A catalyst is therefore advantageous, the p-type semiconductive region of which shifts under the influence of non-stoichiometric sulphur towards higher temperatures, at which the rate of hydrogenation and refining reactions increases rapidly. A high-quality catalyst with good hydrogenation and refining properties, therefore, must have a certain sulphur excess. With these elevated temperatures, however, splitting and isomerisation reactions start to participate. The course of these reactions, which are typical ionic reactions, is governed by the acid character of the catalyst. According to Kalechits [1589] (p. 80 and following), however, it is certain to have this property in the presence of non-stoichiometric sulphur in an atmosphere of hydrogen.

The temperature at the same time determines the position of the Fermi level in the n-type semiconductive WS_2. At high temperatures (above 400 °C) this level shifts from the vicinity of the conduction band to a position over the centre of the forbidden band (at medium temperatures, inversion of the semiconductor character depends on the content of non-stoichiometric sulphur), up to the valence band at low temperatures. The catalytic properties of WS_2, i. e. activity and selectivity, are therefore determined by the position of the Fermi level in the forbidden energy gap.

The position of the Fermi level and, therefore, the catalytic activity of the sulphide is not influenced solely by the temperature and non-stoichiometric sulphur content, the type of carrier and presence of other components being also important. Especially a carrier of suitable acidity (e.g. γ-Al_2O_3) [1429] or addition of NiS to WS_2 decrease the original splitting activity of WS_2 considerably, the refining efficiency being maintained or even improved at the same time [416, 1626]. According to Rohländer [1763] the carrier or, added nickel sulphide influences the n-type semiconductivity of WS_2 by "sucking away" electrons in such a way, that even inversion of the type of semiconductivity may be the consequence. In this case, like in the presence of non-stoichiometric sulphur, the Fermi level shifts to a position close to the valence band. Mutual influence between WS_2 and NiS, considered by Mc Kinley [1347] to be a mutual promotion effect, is manifested in a marked manner especially with the unsupported WS_2.NiS catalyst. With this the splitting activity decreases sharply at

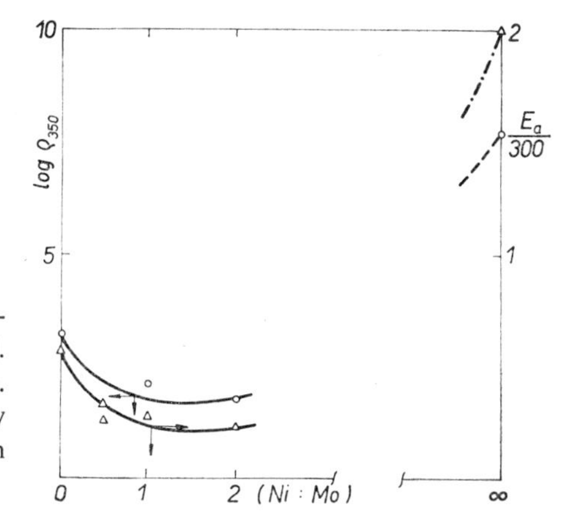

Fig. 20. Electrical properties of sulphide catalysts Ni-Mo/Al_2O_3 (70% carrier content) [1764]. ○ — Specific resistance at 350 °C (ϱ_{350}). △ — Activation energy (E_a) of conductivity (in eV). (Ni : Mo) — Atomic ratio of metals in the catalysts.

temperatures above 425 °C (compared to the catalyst 5058) until it is nearly equal to that of the refining catalyst 8376 [1626]. According to Rohländer, therefore, this important finding may be explained by a shift of the Fermi level to the vicinity of the valence band, causing a considerable decrease in the splitting efficiency of the catalyst. On the other hand, Pier [1626] considers this sudden change of activity to be the result of decomposition of the originally unified crystalline phase. Diffraction studies show the presence of a nickel component in the used catalyst (cf. p. 41), which was not present in the fresh catalyst.

Starting from the fact that WS_2 and MoS_2 have similar properties, Rohländer [1764] studied the electrical properties of the MoS_2–NiS–Al_2O_3 system, which is the basis of the catalyst 8197 [191]. The following fundamental conclusions may be drawn from the results obtained (cf. Figs. 20 and 21):

1. An addition of NiS (p-type semiconductor) [1968] to MoS_2 or, supporting of the two sulphides on a carrier, influences the electrical properties of the resulting catalyst to a substantial extent.

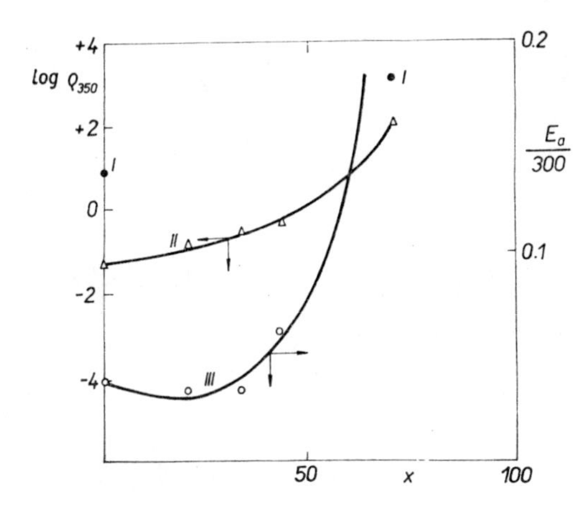

Fig. 21. Influence of the carrier content (x) on the electrical properties of MoS_2 and Ni–Mo/Al_2O_3 sulphide catalysts [1764]. E_a — Activation energy of conductivity (eV). x — % by wt. of the carrier in the catalyst, I — ϱ_{350} for pure MoS_2, II — ϱ_{350} for NiS : MoS_2 = 1 : 1, III — E_a dependence for NiS : MoS_2 = = 1 : 1 catalysts.

2. With a suitable carrier content, an addition of NiS to MoS_2 increases the conductivity of the catalyst to the maximum.
3. The activation energy of conductivity varies considerably with the changing catalyst composition.
4. Minimum activation energy of conductivity may be achieved by catalysts having different atomic ratio Ni/Mo and different carrier content.

It is clear, that Rohländer's views concerning the mechanism of the effect of sulphide catalysts, based on the semiconductor theory, cannot fully explain the complicated behaviour of these catalysts. However, they indicate the direction of further possible research in this field.

Kalechits et al. [974, 976, 1589] investigated the semiconductor properties of the WS_2 catalyst, especially with respect to their relationship to the non-stoichiometric sulphur content and catalytic activity. Three catalyst samples with atomic ratios of S : W = 2.3, 2.0 and 1.9 were studied. In the region of extrinsic conductivity, non-stoichiometric sulphur causes a decrease of conductivity and, therefore, increase of the activation energy of conductivity. On the contrary, a lack of sulphur has the opposite effect. The same also applies to the region of intrinsic conductivity, the transition from extrinsic conductivity to intrinsic conductivity taking place between 180 °C and 220 °C. As with sulphur, oxygen also diminishes the conductivity of the catalyst, but conductivity is also decreased on adsorption of hydrogen. The WS_2

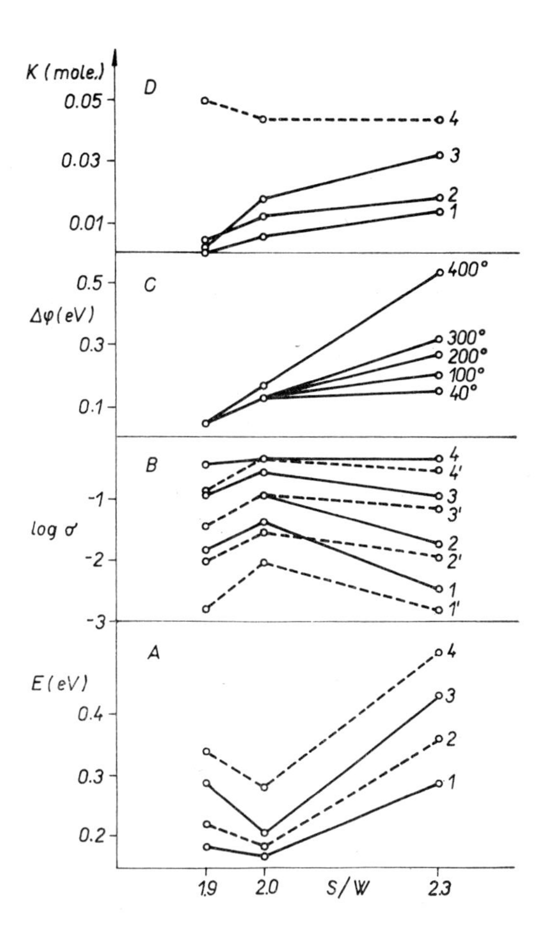

Fig. 22. Relationship between electrical and catalytic properties of WS_2 catalysts with different atomic ratios of S : W [974]. *A* — Relationship of the activation energy of conductivity (*E*, eV). 1, 2 *in vacuo*; 3, 4 in hydrogen (the dotted line corresponds to intrinsic conductivity, the fulldrawn line denotes admixture conductivity). *B* — Relationship of electrical conductivity (σ). 1—4 *in vacuo* at 100, 200, 300 and 400 °C. 1′—4′ in hydrogen. *C* — Relationship of the work function ($\Delta\varphi$, eV) at different temperatures. *D* — Conversion (*K*, mole) in benzene hydrogenation (cf. also p. 82 and Table 12). 1 — Isomerisation. 2 — Hydroisomerisation, 3 — Sum of isomerisation and hydroisomerisation, 4 — Sum of hydrogenation and hydroisomerisation.

catalyst appears to be a p-type semiconductor; non-stoichiometric sulphur in the catalyst acting as an acceptor admixture, decreasing the number of free electrons. A change in the magnitude of the work function and, a change in the catalytic activity of hydrogenation, isomerisation and hydrocracking reaction are related to this fact. These relationships are illustrated schematically in Fig. 22, where a change in some electrical properties of the WS_2 catalysts with different sulphur content is compared with catalytic activity changes with respect to benzene hydrogenation. The overall hydrogenation activity varies only slightly with decreasing sulphur content (Fig. 22D, curve 4), while on the other hand the isomerisation activity of the catalyst decreases (curve 2 in Fig. 22D), the work function also diminishing in this process (Fig. 22C).

The dependence of the catalytic activity of WS_2 may be explained by the fact, that non-stoichiometric sulphur (especially its firmly bound part) is an acceptor admixture, lowering the Fermi level and increasing the amount of positively charged particles on the catalyst surface [1589]. When the reaction is carried out in an atmosphere of hydrogen, chemisorbed hydrogen reacts with the positive holes of the lattice, the proton concentration rises and thus also the acidity of the catalyst surface. In this way, the course of ionic reactions is made easier (isomerisation, splitting).

For the sake of completion, let us mention some work dealing with the electrical properties of those metal sulphides which are used as important active components or specific admixtures in industrial catalysts. The data mainly relate to electrical conductivity of Mo sulphides [501, 524a, 524b, 900, 1284, 1530, 1531, 1531a, 1540b, 1800a, 1855, 1936, 2077], W sulphides [383, 1153, 1154, 1939a], Ni and Co sulphides [388, 436, 1146, 1740a], Pd sulphides [522], Cd [1436a] and Cr sulphides [981]. More data on the electrical properties of catalytically important sulphides will be found in the references [669, 670, 724, 1418a, 1982, 2036].

From the above analysis it is evident, that research into the relationships between the electrical and catalytic properties of sulphide catalysts is only just beginning. It is obvious, however, that in the selection of the optimum sulphide catalyst for a specified industrial process, an important factor to be considered will be the electronic state of the catalyst [742a].

4.5.

Magnetic properties of sulphide catalysts

The study of the magnetic properties of catalytically important solids (metals and semiconductors) enables their structure to be better understood, complementing the methods of X-ray and electron diffraction analysis in many cases, The magnetic properties of elements are related to the unpaired electrons in the d-orbitals of their atoms and depend, in semiconductive crystals of oxides and sulphides, on the valency of the individual ions. Systematic attention has mainly been given to the transition metals and oxides which are important in catalysis [1851, 1852]. Research into catalytically significant sulphides is only just beginning. The magnetic susceptibility of Ni and Co sulphides [177, 766, 1044, 1151], sulphides of the pyrites structure [765] and the ZnS. Cu luminophor [1109] has been studied. Opalovskii and Fedorov [1540b] and Dutta [437, 438] investigated the magnetic properties of MoS_2 and Loginov [1245] measured the susceptibility of vanadium sulphides. Some sulphides (e.g. FeS, NiS) exhibit ferromagnetic properties, which, however, are not even qualitatively understood nowadays. The cause of the difficulties is the unusually complicated preparation of their single crystals [1046].

Magnetic susceptibility measurements were utilized with success to investigate the preparation of active desulphurisation catalysts. Griffith et al. [95] prepared a nickel catalyst by sulphurising nickel hydroxide at 100 °C. When the sulphurisation product was reduced in hydrogen, the loss of weight was greater than that which would correspond to the formation of the subsulphide Ni_3S_2, and the product exhibited ferromagnetic properties, which disappeared when the catalyst was exposed to an atmosphere of CO at 110 to 130 °C. The cause of the susceptibility rise is the presence of metallic Ni in the reduced sample. A substantial rise of susceptibility only takes place with reduction at temperatures >350 °C (see Table 8, giving the reduction conditions and properties of the products).

Table 8.

The Magnetic Properties of Nickel Sulphide Reduced in Different Ways (the sulphide was prepared by sulphurisation of $Ni(OH)_2$ at 100 °C) [95].

Sample	Influence of gas (flow-rate 3.5 l/hour)			Magnetic susceptibility $\chi \cdot 10^6$
	Reduction gas	Time (h)	Temperature (°C)	
NiS prepared in the absence of air	—	—	—	2.1
NiS with oxygen-saturated surface	—	—	—	6.0
Ni_3S_2	—	—	—	4.3
NiS	H_2	5.3	300	7.1
NiS	H_2	3	370	171
NiS	H_2	6.5	370	341
NiS	CO	6.75	115	9.1

Magnetic measurements were utilized in studies of the equilibrium in the Ni_3S_2 + Ni system. A stream of hydrogen with different H_2S content (from 70 to 200 ppm) was passed over a layer of reduced Ni sulphide and the H_2S concentration was found for which, at different temperatures, the rate of change of magnetic susceptibility was zero. This H_2S concentration is the equilibrium one resulting in no more reduction of Ni_3S_2 at the respective temperature. For example, at 450 °C this H_2S concentration is 128 ppm, at 350 °C it is 50 ppm. For comparison, the equilibrium gas used to reduce NiS to the subsulphide at atmospheric pressure and 36 °C contains 44.5% H_2S, which shows that only Ni_3S_2 or its mixture with metallic nickel are stable under hydrogenation conditions.

Magnetic measurements also played an important role in the research of catalytically active components of the cobalt–molybdenum desulphurisation catalysts [1740]. In the original oxide form, the catalyst contains components which behave as follows on reduction at 400 °C and sulphurisation:

Al_2O_3 – a stable carrier
CoO – is reduced quantitatively to Co and sulphurised to Co_9S_8

$CoAl_2O_4$ (spinel) – is neither reduced nor sulphurised
$CoMoO_4$ – is reduced to form a mixture of Co + $MoO_{2.32}$, sulphurised to a mixture of Co_9S_8 + MoS_2 + MoO_2 [419]
MoO_3 – is reduced to $MoO_{2.32}$ and sulphurised to MoS_2.

On sulphurisation of the original oxide catalyst or, after activity stabilisation under hydrodesulphurisation conditions (cf. p. 86) an active mixed sulphide–oxide catalyst is obtained, containing components with the following activity for the desulphurisation reaction:

Al_2O_3, $CoAl_2O_4$ – non-active
MoS_2 – active, requiring however promotion by cobalt
Co_9S_8 – poor activity

The true desulphurisating component of the catalyst therefore is MoS_2 promoted by "active cobalt" supported on γ-Al_2O_3 of a suitable acidity. The composition and structure of the active desulphurisating complex is not yet known exactly: "active cobalt" is very stable, and is neither reduced nor sulphurised under the catalyst-forming conditions. Richardson [1740], who believes that the active Co–Mo complex is a highly efficient mixed cobalt–molybdenum sulphide formed from $CoMoO_4$, carried out a number of magnetic measurements in order to find out the amount of active cobalt in the fresh catalyst. The measurement was made possible by the

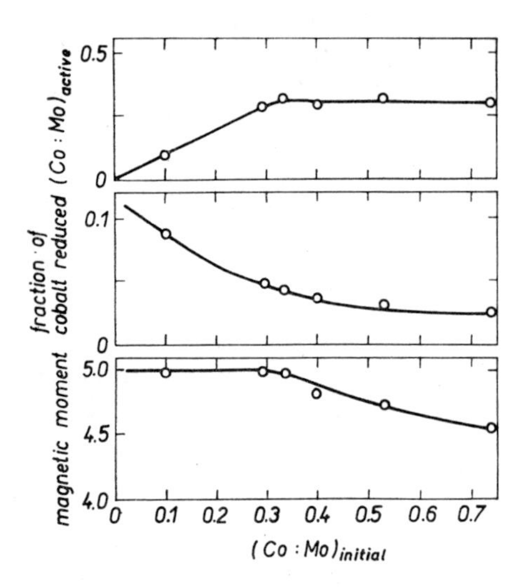

Fig. 23. The magnetic moment, proportion of reduced cobalt and $(Co:Mo)_{act}$ ratio as a function of $(Co:Mo)_{initial}$ (the catalyst was calcined at 650 °C, reduction was carried out with hydrogen at 400 °C; the initial system contained about 10% MoO_3 on Al_2O_3) [1740].

different behaviours of the individual cobalt components. The cobalt fraction bound in spinel was determined separately, the portion bound in the form of CoO was determined after reduction with hydrogen and "active cobalt" was calculated by difference. Richardson believes this active fraction to have the normal magnetic moment of the Co^{2+} ion in the octahedric arrangement, as for example in CoO, so that it is possible to determine the molar ratio of cobalt and molybdenum [$(Co:Mo)_{act}$] in the active desulphurisating complex. A typical result obtained after calcination of the original catalyst at 650 °C is illustrated in Fig. 23. No spinel $CoAl_2O_4$ is formed at first on addition of cobalt, up to a ratio of $(Co:Mo)_{initial} \sim$ ~ 0.3. Below this limit, the major part of the cobalt present forms the active cobalt–

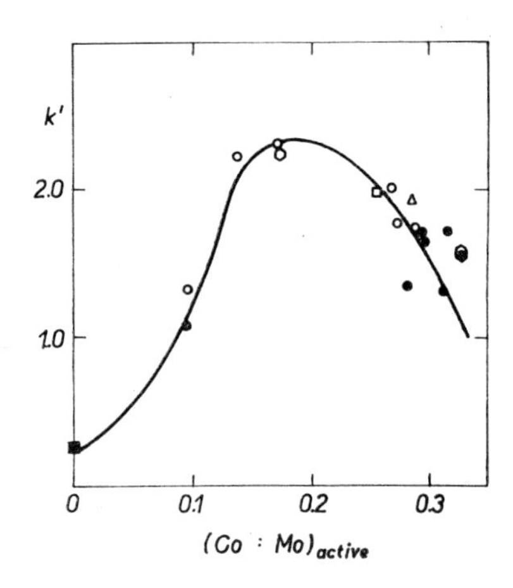

Fig. 24. Dependence of the desulphurisation activity (expressed by the intrinsic rate constant $k' = k \cdot 10^5$ cm/hour) of the cobalt–molybdenum catalyst (basic catalyst — 10% MoO_3 on γ-Al_2O_3) on the composition of the active complex $(Co:Mo)_{act}$. (Desulphurisation conditions: t = 400 °C, P = 56 atm, LHSV 2 l/l/hour, the raw material was gas oil with 1.92% S) [1740].

○ Catalyst calcined at 650 °C.

● Catalyst calcined at 580 °C.

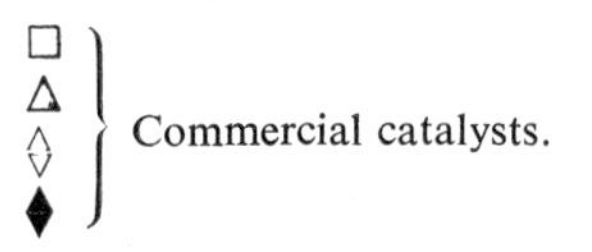
Commercial catalysts.

molybdenum complex, while only a small residue remains in the form of CoO. The catalytic desulphurisating activity (Fig. 24) which has a distinct maximum at $(Co:$ $:Mo)_{act} \sim 0.18$, changes when the catalyst structure changes. The activity of four samples of commercial cobalt–molybdenum catalysts was found to agree with the activity curve of catalysts prepared on the laboratory scale.

Besides other methods, like X-ray diffraction, infrared and reflectance spectroscopy, magnetic measurements were employed in some more recent work dealing with the composition and catalytic activity of a hydrodesulphurisation cobalt–molybdenum catalyst. Ashley and Mitchell [93a] consider, that the fresh (non-sulphurised) catalyst contains roughly 50% of the Co^{2+} present in the octahedral and 50% in the tetrahedral coordination, and Mo^{6+} in the tetrahedral coordination by oxide. Furthermore the authors found [93b] that there is a certain interaction between cobalt and molybdenum on their incorporation into alumina. The amount of one ion incorporated depends on the presence of the other ion. The stereochemistry of cobalt is likewise influenced by the amount of cobalt and molybdenum incorporated.

In another series of studies, Lipsch and Schuit [1243a, 1243b, 1243c, 1243d] first studied the unsupported catalyst, proving the presence of $CoMoO_4$. This compound exists in two modifications, molybdenum being octahedrally surrounded by oxygen in both modifications. Magnetic susceptibility measurements show that the major part of cobalt is surrounded octahedrally; it is partly surrounded tetrahedrally on the surface only. The authors state, that when supported on Al_2O_3 cobalt is distributed through the catalyst bulk in the form of $CoAl_2O_4$; only a small part, if any, is bound on the surface, but according to Lipsch and Schuit this cobalt does not take part in catalytic reactions. Molybdenum occurs in the catalyst in the form of MoO_3, which is probably spread on the alumina surface in the form of an unimolecular layer, covering about 20% of its surface. When freshly prepared, the catalyst is inactive, and must be activated by reduction. Lipsch and Schuit [1243d] state, that conversion of the catalyst to the sulphide form is not essential, although other authors, e.g. Hendricks et al. [754] or Blue and Spurlock [189a] accent the need of pre-sulphurising the catalyst (see also section 5.1, p. 85). It is not certain whether sulphidation of the cobalt–molybdenum catalyst leads to complete or only partial sulphidation of the metal components [1385a].

According to Lipsch and Schuit (1243d], the major part of the desulphurisation activity of the cobalt–molybdenum catalyst is due to MoO_3, active centres being located on sites of the "independent" Mo---O bond. On reduction, anion vacancies are formed on these sites, the sulphur-containing substance being adsorbed on them. Reduction of the catalyst should not be too extensive, as otherwise too many anion vacancies would be formed and too few oxygen atoms would remain for adsorption of hydrogen, which is needed for hydrogenolysis of C—S bonds. The authors believe, that no "overreduction" takes place when H_2S or mixtures of H_2 with H_2S are employed as milder reducing agents. When MoO_3 is converted to the sulphide by sulphurisation, hydrogen is adsorbed on the sulphur atoms, which fact may, under certain circumstances, increase the activity of the hydrodesulphurisation catalyst [1243d].

One disadvantage involved in the Lipsch and Schuit theory is the fact that it does not explain the influence of cobalt, which doubtless plays the role of a promoter for the overall hydrodesulphurisation reaction [1385a, 1832a]. Literature data reported up to the present likewise do not show clearly, whether the structural features discussed actually have a fundamental significance for the high activity of the cobalt–molybdenum catalyst used in practical operations [1832a].

Recently, the EPR method has come into use for research of the properties and character of active centres of sulphides [1065, 1932a, 2035a].

4.6.

Optical properties of sulphide catalysts

The relationship between optical and catalytic properties was studied in the case of the classical ZnS luminophor. Pure ZnS is a luminescent material, due to the presence of non-stoichiometric Zn in the sulphide [233, 238]. Luminescence is initiated by heating the sulphide to 950 °C, first in an inert atmosphere (2 hours in N_2), later in a reducing atmosphere. Heating the sulphide in the presence of sulphur vapours decreases the luminescent activity of ZnS rapidly. The activity is regenerated in an incompletely reversible manner on reduction of the sulphurised material [1869]. The catalytic activity of the sulphide varies accordingly (the model reaction employed was the decomposition of CH_3OH at 258 to 345 °C and pressures of 120 to 225 torr), so that clearly, active centres in ZnS occupy sites with non-stoichiometric Zn.

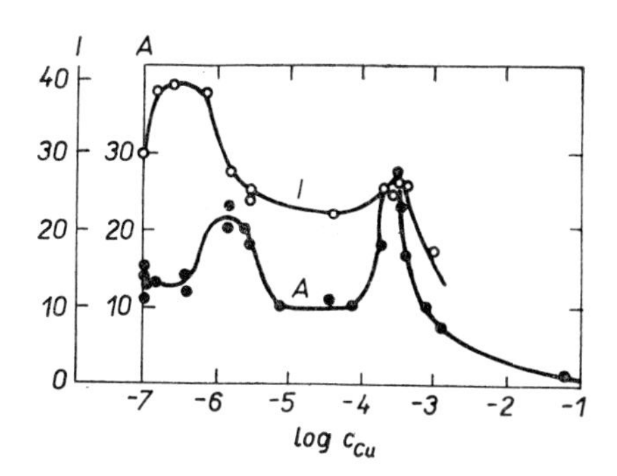

Fig. 25. Dependence of the luminiscence intensity of ZnS–Cu and of its catalytic activity (decomposition of CH_3OH at 315 to 375 °C in a flow reactor at normal pressure) on the activator content (Cu) [1108]. *I* — Relative luminescence intensity — *A* — Catalytic activity (measured by the decomposition of CH_3OH).

Kobozev observed a similar agreement of luminescent and catalytic activity in the case of ZnS, the luminescent activity of which was enhanced by the addition of an activator (Cu) [1061, 1062, 1108, 1111]. A typical relationship is shown in Fig. 25, from which agreement between the two relationships can be seen. Kobozev considers this agreement to be a corroboration of one of the main points in his theory of active sites, i.e. of the atomic character of catalytically active centres. A similar relationship was also observed between the catalytic activity of the ZnS.Cu luminophor with different Cu concentrations in the case of hydrogen peroxide decomposition, and the activity of this catalyst in exoelectronic emission, caused by the influence of X-rays [1063, 1110, 1864]. The relationship between chemisorption and luminiscence was likewise studied in the case of ZnS [745e].

The industrial catalysts WS_2, WS_2–NiS/Al_2O_3, WS_2 on alumosilicate and MoS_2 were subjected to a thorough electron-microscopic investigation [738, 1257]. Reflection spectra [2254a] and absorption spectra in the visible region of the spectrum and photoconductivity were also determined [563, 1540].

5.

THE ACTIVITY AND SELECTIVITY OF SULPHIDE CATALYSTS AND METHODS OF INFLUENCING THESE PROPERTIES

The basic factors which determine the activity and selectivity of sulphide catalysts include the type of sulphur-to-metal bond and more especially the non-stoichiometric sulphur content or, method of sulphurising the catalyst. The relationship between the activity and thermal processing of the catalyst during preparation, activation or operation in the reactor, is closely related to the above factors. Some important relationships have already been mentioned in connection with the preparation and texture properties of the catalysts. Finally, the catalytic activity of sulphides is influenced considerably by the type of carrier, the different activators and promoters employed and, also, by the deactivation reagents or, in exceptional cases, by catalyst poisons.

5.1.

The influence of non-stoichiometric sulphur, sulphur compounds and methods of sulphurisation on the activity of sulphide catalysts

After Pier et al. [1101, 1625, 1626] have succeeded in finding industrially useful catalysts, resistant to the influence of sulphur compounds and after Varga's discovery, that the activity of molybdenum and tungsten catalysts increases in the presence of hydrogen sulphide when used in the hydroprocessing of tar materials [2114, 2115, 2116, 2120, 2121, 2122], it was found that iron, nickel and cobalt catalysts behave in a similar manner [2031]. These results were later confirmed by other workers [402, 403, 571, 838, 1031, 1032, 1632, 1680, 1684, 2225, 2268] and notably during industrial applications. In these it was found that a continuous supply of sulphur (either by addition of H_2S to hydrogen or in the form of sulphur compounds in the feed) is essential to maintain a sufficient activity of the sulphur catalyst during the hydrogenation process, especially of acid oils [1382a]. A sulphur content of 0.2% in the feed is considered sufficient; an increase of this up to 2% brings

Table 9.

The Physico-Chemical Properties of WS_2 Catalysts and Their Activity in Pressure Hydrogenation of Phenol (for the hydrogenation conditions, see Fig. 27) [1798].

Sample	Atomic ratio (S : W)	Specific surface area (m^2/g)	Acidity (eqv. H^+/mole) $\times 10^{-3}$	Amount of reacted phenol (g)	H_2 pressure (atm)		$\Delta H_2/\Delta \varnothing$ $\left(\frac{\text{moles } H_2}{\text{moles phenol}}\right)$
					Initial	Final	
1	2.27	—	0.79	25.3	108	83	1.9
1_{400}[a])	—	65	0.1	23.8	113	95	1.5
3	2.30	3	1.46	27.1	110	81	2.1
3_{400}	2.006	66	0.29	22.7	104	84	1.7
4	1.92	—	1.24	21.0	110	91	1.8
4_{400}	1.53	49	0.0	17.0	112	99	1.5
5	2.25	—	0.81	25.0	109	88	1.7
5_{400}	—	50	—	22.5	113	96	1.5
9_{20}[b])	1.96	64	—	28.4	117	88	2.0
10	2.14	—	1.56	25.1	109	86	1.8
10_{400}	—	46	—	21.1	112	97	1.4
3 + 0.046 g S	—	—	—	29.8	115	80	2.3
3_{400} + 0.045 g S	—	—	—	28.7	113	82	2.1
4 + 0.043 g S	—	—	—	28.3	110	82	2.0
4_{400} + 0.043 g S	—	—	—	17.7	113	89	2.7

[a]) The index 400 by the catalyst number denotes that the catalyst was evacuated at a pressure of 5×10^{-5} Torr at 400 °C.
[b]) The catalyst 9_{20} was formed by extracting catalyst 3 with carbon disulphide at 20 °C.

about an increase in activity of 10 to 20%, but further increase has practically no effect [416]. The marked change in activity of sulphide catalysts was mentioned several times in Chapter 4 in connection with changes in the structure, texture and other physico-chemical properties. Analysis of these findings in greater detail shows that the change in activity is due to a change in physico-chemical properties, which in most cases are related to a change in the content or bonding of sulphur in the cata-

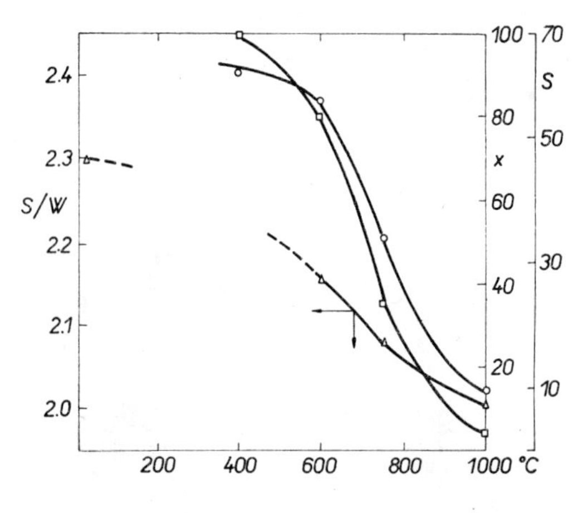

Fig. 26. Dependence of activity, sulphur content and surface area on the thermal treatment of the WS_2 catalyst [1794]. ○ — Catalytic activity (x — % of conversion in autoclave hydrogenation of benzene, t = 400 °C, initial hydrogen pressure = 140 atm) [1586].
△ — Atomic ratio S : W.
□ — Surface area S (m^3/g).
(WS_2 was prepared by reductive decomposition of ammonium tetrathiotungstate; heating at 600 to 1000 °C was done in N_2, heating at 400 °C was carried out *in vacuo*.)

lyst. Fig. 26 gives a summary of the relationship between the sulphur content of the catalyst, of its surface area and the catalytic activity in benzene hydrogenation in the case of WS_2. The decrease of the non-stoichiometric sulphur content, and the related changes of texture and structure which take place on heating the catalyst [958] result in a considerable decrease of catalytic activity in benzene hydrogenation. The critical temperature is about 600 °C, above which activity decreases rapidly [1794].

In destructive hydrogenation, a complex series of reactions is always involved, amongst which is hydrogenation of double bonds. Also hydrogenolysis (splitting) of the carbon chain (C–C bonds) as well as between carbon and other atoms (O, S, N and metals). Furthermore, isomerisation, condensation, etc. reactions are also involved. The sulphur contained in the catalyst does not influence all these reactions to the same extent. A change in the sulphur content of the catalyst or, a change in its mode of bonding, determines the selectivity of the catalyst.

Samoilov and Rubinshtein [1778, 1798, 1800] studied the relationship between activity and selectivity and the sulphur content in WS_2 in the case of hydrogenation of phenol to hydrocarbons. The most important results are summarised in Table 9. In semiquantitative terms, the overall activity of the catalyst may be measured by the amount of reacted phenol since, as shown in the studies [1798] and [1800], the course of reactions other than hydrogenation to hydrocarbons is negligible. Hydrogenation selectivity may be assessed from the specific hydrogen consumption, i.e. the number of H_2 moles consumed by one mole of reacted phenol. The following conclusions may be drawn from the data summarised in Table 9:

1. The high catalytic activity of WS_2 is entirely due to sulphur atoms bound on the surface. This follows from a comparison of the activity of the catalysts 3 and 9_{20}; in the latter case, the major part of the sulphur which was freely located in pores was removed by extraction with CS_2. This resulted in practically no change in the activity of the catalyst. The importance of surface sulphur atoms is also indicated by the fact, that a considerable increase in activity results when a small amount of "promoting" S is added to the catalyst (cf. the results for the catalysts 3, 3 + S, 4,

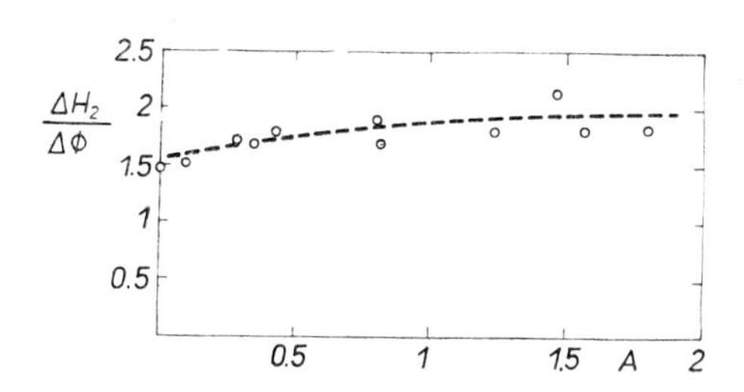

Fig. 27. Relationship between hydrogenation selectivity and acidity of WS_2 catalysts [1798]. (Hydrogenation of 50 g phenol with 1 g catalyst in a 0.5 l autoclave at 380 °C; initial pressure 110 atm, reaction time 1 hour).

ΔH_2 — Moles H_2 consumed in hydrogenation.
$\Delta \Phi$ — Moles of reacted phenol.
A — Acidity, expressed in eqv. H^+/mole.

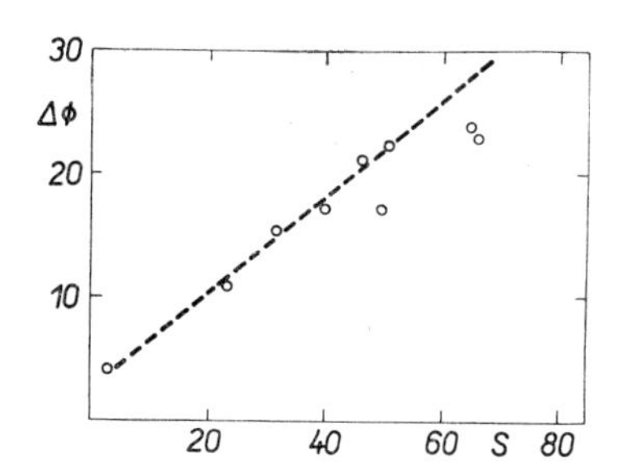

Fig. 28. Relationship between the activity of evacuated WS_2 catalysts and their surface area [1798]. (The catalysts were prepared by reductive decomposition of ammonium tetrathiotungstate at 400 °C and a vacuum of $5 \cdot 10^{-5}$ Torr.)

$\Delta \Phi$ — g phenol reacted out of the original 30 g. (For the reaction conditions see Fig. 27.)
S — Surface area (m^2/g).

3_{400}, 3_{400} + S). This promoting effect is more efficient, the smaller is the atomic ratio S : W in the original catalyst (cf. the results obtained with the catalysts 3, 3_{400}, 3 + S, 3_{400} + S). This may also explain why the catalyst activity increases only slightly when the sulphur content of the material processed rises above 2%.

2. With sufficiently prolonged evacuation, excess sulphur is removed from the pores as well as, in part, from the surface, resulting in a general decrease in the activity of the evacuated samples.

3. Samoilov and Rubinshtein also mention the importance of acidity for the activity of the catalyst in phenol hydrogenation. These authors compared, for example, results obtained with the catalysts 1 and 3, where the greater acidity of the catalyst 3 was the reason for a higher overall activity as well as hydrogenation selectiv-

ity (see cf. Table 9). In evacuated catalysts, acidity decreases to 0–20% of the original value, so that the significant activity decrease of these catalysts is obviously caused by the considerable decrease in acidity. Acidity, as a specific property, however, is not solely a function of the sulphur centent; it is due to a set of factors arising in the course of preparation of the catalyst. The dependence of overall activity and hydrogenation selectivity on acidity, therefore, is ambiguous (cf. for example results obtained with the catalysts 1, 3, 5 and 10). When the hydrogenation selectivity of the catalyst is considered (without reference to the other physico-chemical properties, e.g. the atomic ratio S : W, specific surface area etc.), it is found to increase only slightly with increase in acidity (Fig. 27). Samoilov and Rubinshtein used the molar hydrogen consumption per mole of reacted phenol ($\Delta H_2/\Delta \Phi$) as the criterion of hydrogenation selectivity. Similarly, the overall activity measured by the amount of phenol reacted is not necessarily a function of acidity (cf. Table 9, experiments with the catalysts 1, 3, 5 and 10).

4. The decrease of the sulphur content in the catalyst mainly influences the selectivity of the catalyst in hydrogenation reactions, as will be seen on comparing the individual data in Table 9. This effect is especially marked in the experiments with the catalyst 4, 4_{400} and 4_{400} + S.

5. The relationship between the overall activity of the catalyst (expressed in terms of the amount of reacted phenol or, in terms of the absolute value of the hydrogen consumption) and the atomic ratio S : W or, the specific surface area of the individual catalysts is not entirely satisfactory, since the acidity of the catalyst plays a large role. When, however, the acidity factor is eliminated, i.e. only evacuated catalysts are considered (the acidity of these are very low), the specific activity (i.e. activity related to a surface unit) is practically constant for the majority of catalysts studied (see Fig. 28, where the amount of reacted phenol is used as a measure of activity). The only exceptions are catalysts with a large number of ultramicropores (catalysts with surfaces of about 65 m^2/g), since these pores do not take part in the hydrogenation and hydrogenolysis reactions, and catalysts with a great lack of sulphur (catalysts with surfaces of about 50 m^2/g).

It is clear from these results, that sulphur atoms are significant components of the active centres in sulphide catalysts and that the quality of a catalyst is influenced decisively by sulphur bound on the surface. This obviously does not exclude a simultaneous activity influence of the metal component of the catalyst, especially in hydrogen chemisorption. According to Samoilov and Rubinshtein [1798] however, hydrogen adsorbed on sulphur atoms is considerably more active. Also, chemisorbed hydrogen is able to form a large number of the hydrosulphide groups –SH, which make a large contribution to the overall acidity of the catalyst.

The influence of additions of elementary sulphur was studied in the above-mentioned autoclave experiments. Under the respective reaction conditions (temperature 380 °C, operating hydrogen pressure 200 atm) and under the catalytic influence of sulphides [1171], elementary sulphur is easily hydrogenated to form hydrogen sulphide, provided it does not react with the catalyst sooner. Therefore, the promoting effect should rather be ascribed to the H_2S formed or, to that part of the H_2S which has reacted with the catalyst. Also, the water formed during the phenol hydrogenolysis may influence the selectivity of the catalyst employed to a marked extent. Under

Table 10.

Influence of the Carrier on the Activity and Selectivity of the MoS_2/Act. Carbon Catalyst in Benzene Hydrogenation (hydrogenation temperature 420 °C, pressure 300 to 320 atm, reaction time 3 hours) [974, 1589].

% MoS_2 in the catalyst (% by wt.)[a]	Atomic ratio[b] S : Mo	Specific activity[c]	
		Hydrogenation	Isomerisation
100	2.2–2.3 (fresh)	11.6	9.1
10	2.17	12.4	1.0
5	2.10	23.3	1.1
1	2.00	30.2	—

[a]) After impregnation of active carbon with thiomolybdate, the catalyst was reduced in hydrogen for 6 hours at 450 °C.

[b]) The analysis was carried out after reduction of the catalyst.

[c]) The specific activity in hydrogenation and isomerisation is expressed in terms of % conversion of the respective substance (benzene to hydrogenation products; isomerisation of the cyclohexane formed to methylcyclopentane) per 1% by wt. Mo on the catalyst surface.

the experimental conditions stated in Table 9, the amount of water formed is 100 times greater than the molar addition of sulphur in the case of 50% hydrogenolysis of the C—O bond.

For a more detailed study of the influence of non-stoichiometric sulphur on the activity of sulphide catalysts, Kalechits selected the model reaction of benzene hydrogenation, studying the selectivity of the catalyst using gas-chromatography to analyse the reaction products and labelling of atoms to find the type of sulphur bonds in the catalyst.

It was first found [397, 974, 1589] that supporting of an active sulphide, MoS_2 in this case, on an inert carrier with a large surface area (active carbon) considerably decreases the isomerisation activity of the catalyst. The specific hydrogenation activity of the supported catalyst increases with increasing dilution of the active component (Table 10), while the isomerisation properties of the catalyst do not alter markedly. For a semiquantitative comparison, activity data were related to 1% Mo in the supported catalyst. Supporting the sulphide on a carrier facilitates the degradation of the major part of the non-stoichiometric sulphur during the preparation of the catalyst, as shown by the decrease of the atomic ratio S : Mo with an increase of MoS_2 dilution in the resulting catalyst. The effect of this is a distinct change in the selectivity of the catalyst.

A similar change may also be caused by different degrees of reduction or sulphurisation of one and the same catalyst, as follows from results presented in Table 11

Table 11.

Influence of H_2 and H_2S on the Activity and Selectivity of the MoS_2/Active Carbon[a]) Catalyst in Benzene Hydrogenation (for the reaction conditions, see Table 10) [1589].

Gas employed[b])	Duration of contact of the gas with MoS_2 catalyst (hours)	Overall benzene conversion (% by wt.)	Cyclohexane conversion (% by wt.)	Specific activity[e])	
				Hydrogenation	Isomerisation
H_2	4[c])	72.2	2.4	12.0	0.4
H_2	1	76.0	2.4	12.7	0.4
H_2	0.5	78.2	2.7	13.0	0.5
H_2S	8[d])	70.0	10.0	11.7	1.7
H_2S	4	72.9	6.3	12.1	1.0
H_2S	1	76.6	4.5	12.8	0.8

[a]) The MoS_2 content was 10% by wt. in all catalyst samples.
[b]) The catalysts were exposed to the influence of the gas at 450 °C.
[c]) The atomic ratio was S : Mo = 2.03 after reduction.
[d]) The atomic ratio was S : Mo = 2.28 after exposition to H_2S.
[e]) For the specific activity, see Table 10.

[1589]. While the hydrogenation activity of MoS_2 in this case is only slightly influenced by the effect of H_2 or H_2S, sulphurisation of the catalyst with H_2S enhances the isomerisation properties of the catalyst considerably.

The sulphur content of the catalyst, especially of the non-stoichiometric sulphur, has therefore a decisive influence on the quality of sulphide catalysts. In agreement with the results obtained by Rubinshtein et al. (cf. p. 76), Kalechits also mentions the different types of sulphur bonding in the catalyst and, in consequence, the different effects of the individual sulphur groups on the overall activity of the catalyst. The different strength of sulphur bonds in the catalyst is related to the different rates of reduction, by means of which non-stoichiometric sulphur is removed on hydrogenation or, to the rate of exchange for sulphur contained in the material processed (e.g. radioactive isotope).

Kalechits et al. [397, 974, 976, 1589] performed hydrogenation experiments with benzene which, in some cases, contained carbon disulphide with labelled sulphur.

It was found when summarising the experimental results, that the majority of non-stoichiometric sulphur is rapidly removed from the catalyst on hydrogenation after the first experiment, only 6% sulphur capable of being removed by hydrogenation remaining in the catalyst. The last part of the non-stoichiometric sulphur is very difficult to remove or, cannot be removed by hydrogenation at all. On addition of a substance containing sulphur (which forms H_2S on hydrogenation) to the material to be hydrogenated, the sulphur content of the catalyst returns to the original value, sulphur in the catalyst being exchanged for sulphur in the raw material. The amount of this exchangeable sulphur is not more than 10% of the total sulphur content of

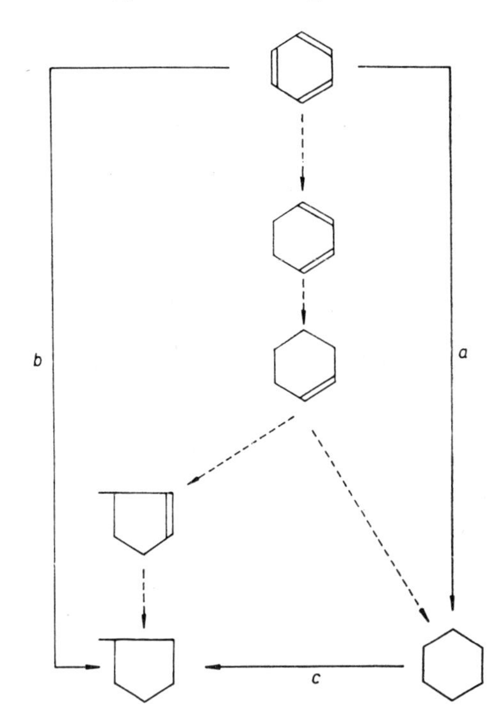

Fig. 29. Scheme of benzene hydrogenation.

the catalyst [1589]. On the basis of these results, Kalechits et al. [397, 974, 976, 1589] divides total sulphur in a fresh catalyst into the following groups:

1. Most of the sulphur (90 to 95%) is firmly-bound stoichiometric sulphur, which cannot be removed by hydrogenation under the respective conditions of preparation, activation or application of the catalyst; this sulphur is not exchanged or at least only with great difficulty.

2. The remaining part of sulphur is removable ("mobile"), but this property differs. The majority of the mobile portion is very easily removed on hydrogenation, the rest is removed with great difficulty or not at all. This second portion, however, is exchanged for sulphur contained in the raw material (after hydrogenation to H_2S) provided sulphur is present in the material processed.

Sulphur which is easy to remove is only very weakly bound to the catalyst mass. This is mainly mechanically bound molecular sulphur which can not only be hydrogenated but, also, rather easily evaporated *in vacuo* [1798]. Sulphur which is exchangeable but difficult to remove by hydrogenation is bound to the catalyst surface by a strong bond (chemisorbed sulphur) [976]. Both portions of this "mobile" sulphur are important to the catalyst's properties. Results obtained with a WS_2 catalyst [1589] confirm the observation made with the MoS_2 catalyst (Table 11) that, while the isomerisation and splitting activity of the catalyst rapidly decrease with decreasing non-stoichiometric sulphur content, the hydrogenation properties depend very little on this sulphur content. Non-stoichiometric sulphur therefore mainly influences the acid properties of the catalysts which are responsible for the course of ionic reactions (cf. p. 68).

Comparing this observation with the results obtained in phenol hydrogenation (Table 9) [1798] one finds, in the case of phenol hydrogenation, a greater sensitivity of hydrogenation selectivity to the sulphur content (either on its addition to the catalyst or comparing evacuated and non-evacuated catalyst samples).

To verify the influence of non-stoichiometric sulphur on the selectivity of the sulphide catalyst, Kalechits et al. studied the process of benzene hydrogenation in the presence of WS_2 with different sulphur content, using labelled sulphur [397, 974]. In the experiments, a mixture of benzene and cyclohexane was hydrogenated, one of the compounds containing radioactive carbon. The possible course of reactions in this process is shown in the diagram in Fig. 29.

Relative conversion values, present in Table 12, were obtained by evaluating the mass balance of the compounds hydrogenated. It again follows, that non-stoichiometric

Table 12.

Relative Conversion Values of Different Reactions in Benzene Hydrogenation (see reaction scheme in Fig. 29, mean values are stated) [974].

Reaction	Notation of the reaction in the scheme (Fig. 29)	Atomic ratio (S : W) in the used catalyst		
		2.3	2.0	1.9
Hydrogenation to cyclohexane	a	25	31.5	46.2
Hydroisomerisation to methyl-cyclopentane	b	18.5	12.5	3.8
Cyclohexane isomerisation	c	14.0	6.0	0.8
Sum of hydrogenation and hydroisomerisation	a + b	43.5	44.0	50.0
Sum of hydroisomerisation and isomerisation	b + c	32.5	18.5	4.2

sulphur very strongly enhances the isomerisation activity of the catalyst, while the overall hydrogenation efficiency (hydrogenation and hydroisomerisation) varies only slightly.

Similarly, with the sulphurised Fe-Cr catalyst, a significant influence of non-stoichiometric sulphur on activity was observed in the case of hydrodesulphurisation. It was found by means of X-ray analysis, that a solid solution of S in FeS is involved and, that the catalytically active centres are located in places where dissolved sulphur was removed by hydrogenation. This dissolved sulphur is highly reactive, far more than sulphide sulphur, and it likewise easily undergoes isotopic exchange with hydrogen sulphide [919, 920, 921] (cf. p. 214).

The problem of the origin of non-stoichiometric sulphur is a very important one. With W and Mo, active sulphides of which are prepared by reductive decomposition of thiosalts, there are two possibilities. Either it comes from sulphur formed by thermal decomposition of tetrathiosalts or of the trisulphide, in which case it could not be removed by hydrogenation or would be more difficult to hydrogenate:

$$(NH_4)_2WS_4 \rightarrow [WS_3 + (NH_4)_2S] \rightarrow WS_2 + S + (NH_4)_2S.$$

Alternatively, non-stoichiometric sulphur may be formed by reaction of the disulphide with hydrogen sulphide (either formed in the course of the reaction or added in the case where reductive decomposition is carried out with an $H_2 + H_2S$ mixture).

Kalechits and Deryagina [397, 974, 975] prepared and decomposed ammonium tetrathiotungstate in the presence of $H_2{}^{35}S$. They first prepared a fresh catalyst by decomposing $(NH_4)_2W^{35}S_4$ with a mixture of H_2 and H_2S, in a second experiment they decomposed $(NH_4)_2WS_4$ with a mixture of H_2 and $H_2{}^{35}S$. When WS_2 prepared in this way was used to hydrogenate benzene, the H_2S formed by degradation of non-stoichiometric sulphur contained very little $H_2{}^{35}S$ in the first case, while the reverse was true in the second case. Hence it follows, that mobile non-stoichiometric sulphur (called hydrogen-sulphide sulphur by the authors), originates in hydrogen sulphide which was present during preparation of the catalyst or, which was used in the catalytic reaction. The mechanism of the reaction of H_2S with the catalyst is not quite clear, but it seems that sulphur enters in this reaction into the surface layers of the lattice where it can be removed rather easily by hydrogenation [975].

The amount of non-stoichiometric sulphur and its ratio to sulphide sulphur can be measured directly by spectral means [589].

Sulphurisation carried out in a suitable manner and to a sufficient degree, is of the utmost importance for the optimum functioning of the catalyst. When sulphur-containing feeds are being processed, the greater part of the metal components of the sulphide catalyst are converted to sulphides. This was observed by Varga (p. 74); moreover, the presence of molybdenum and tungsten sulphides in the used catalyst were also confirmed in the case of the hydrorefining catalyst 3510

(MoO_3–ZnO–MgO), which was the first technically significant catalyst of its kind, and also the tungsten analogue of this catalyst [416, 1429]. It is also important to know that the oxide catalyst when sulphurised by the sulphur contained in the raw material has a refining activity which is equal or only slightly higher than that of the original oxide catalyst. On the other hand, it was observed in the case of the WO_3–ZnO–MgO catalyst that preceding sulphurisation of the catalyst with hydrogen

Table 13.

Influence of Preliminary Sulphurisation on the Activity of the Refining Catalyst 8376 [698]. (Hydrogenation of a tar feed for the TTH process, hydrogenated with circulation gas, temperature 340 °C, pressure 300 atm, LHSV 0.37; the initial catalyst was oxidic 8376).

Experiment[a])	Duration of experiment (days)	Properties of middle oil (180—350 °C) [c])		
		Aniline point (°C)	Content of phenols (g/l)	Content of basic nitrogen compounds (mg NH_3/l)
A	1	45.0	47.2	597
B	1	56.8	3.5	21
A	6—7	45.1	45.6	804
B	6—7	52.3	8.2	94
A	9—11[b])	44.5	49.8	724
B	9—11	50.3	10.7	120
Original material	—	33.0	130.0	1700

[a]) In both cases, the following mixtures were passed over the catalyst at 340 °C and 300 atm for 17 hours before the hydrorefining experiment, in the case A: fresh H_2 (97% H_2, the rest $CH_4 + N_2 + CO$, no H_2S); in the case B: circulating gas (80% H_2, 19% $CH_4 + N_2$, 0.7% H_2S). In the case B, the amount of H_2S introduced to the catalyst was greater than the amount needed to convert WO_3 and NiO to sulphides.

[b]) After 8 days, the catalyst was sulphurised in this experiment in the same way as in the experiment B.

[c]) Here, similar to a number of other cases described in this monograph, the refining activity was assessed by the degree of hydrogenation of the phenol content (expressed in g/l) and of nitrogenous substances (expressed in mg NH_3/l). Moreover, the specific gravity and colour of the hydrogenate were noted frequently.

sulphide at elevated pressures, in the absence of the material to be processed, leads to a considerable enhancement of the refining activity and, at the same time, decreases its sensitivity to nitrogenous bases [416, 1429]. These observations were documented recently in the case of other catalysts, too. This follows from experiments carried out with the catalyst 8376 [698], when the properties of the hydrogenation products were compared, which were obtained with a catalyst which was first reduced by hydrogen (series A experiments) or with hydrogen and H_2S (series B) (Table 13). Prior sulphurisation at 340°C in the absence of oil vapour leads to the formation of a highly active catalyst. Prolonged sulphurisation with sulphur from the material to be processed (after conversion to H_2S under refining conditions), does not improve the activity of the catalyst. Additional sulphurisation with hydrogen sulphide (for 8 days) also does not lead to improved activity. The same was found to apply to sulphurisation of the catalyst 7362 ($CoO–MoO_3–Al_2O_3$) when its desulphurising activity was studied by desulphurisation of diphenylsulphide dissolved in paraffin oil [1056, 2186].

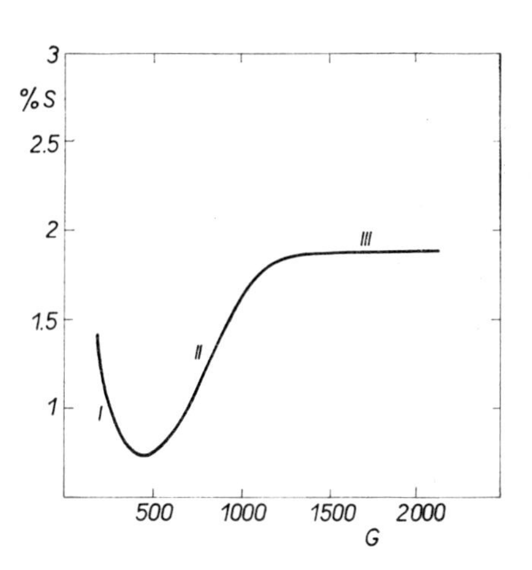

Fig. 30. Desulphurisation activity of the $CoO–MoO_3$ on bentonite at 343 °C and atmospheric pressure [243]. (100 ml catalyst, the feed was petroleum distillate with 3% S; the catalyst was heated to 345 °C during 3 hours in a stream of H_2 at atmospheric pressure.)

% S — Sulphur content in the product.
G — ml feed passed over the catalyst.

The problem of presulphurisation of oxide catalysts is a very important one [478] though it has not yet been solved completely. In the preparation of multicomponent oxide desulphurisation catalysts, sulphurisation is usually carried out after reduction, in order to secure a good desulphurisation activity and adequate lifetime of the catalyst (cf. p. 29) [754]. According to Mc Kinley [1347], however, presulphurisation is not essential for conventional desulphurisation catalysts, since the oxide catalyst is converted to the sulphide form rather quickly during the reaction. Presulphurisation, however, improves the refining activity of the catalyst in many cases [9, 137a, 189a, 339, 754, 1346a] and to some degree it also influences the selectivity of the catalyst. For example, a presulphurised Co–Mo catalyst has a greater hydrogenation activity, which is utilized with advantage in desulphurising the gasoline fraction in the Autofining process [1666]. On the other hand, the low dehydrogenation activity of the presulphurised catalyst is advantageous when the space velocity of the feed is low or, when heavy materials are being processed [1347, 1661]. Sulphurisation of the oxide catalyst or a catalyst after oxidative regeneration with sulphur originating in the material processed follows a characteristic mechanism [243], which is evident from the desulphurisation experiment carried out at atmospheric pressure (Fig. 30). In the first period (I) the desulphurising effect of the catalyst increases and the reaction between H_2S formed and the catalyst components is practically complete. In this period, the temperature varies and the catalyst surface may be overheated, so that the increase in activity of the catalyst may be deceptive. The H_2S partial pressure in the tail gas is low. When the curve has reached its minimum (i.e. maximum desulphurisation activity) the partial pressure of H_2S in the tail gas starts to rise, the sulphur content in the product rises slowly, too, and the metal components of the catalyst continue to be sulphurised (period II), until practically a stable state is achieved (period III). This latter period is the operating stage, with practically a constant catalyst activity. The activity change in this period is mainly caused by coke-like deposits which gradually occupy the active surface. In experiments under pressure, the character of the activity curve is similar, but the activity differences are less pronounced.

Referring back to the results obtained with the catalyst 8376 (Table 13), it will be seen that under specific conditions the activity of the catalyst may be influenced to a large extent during the preparation of the catalyst, and that sulphurisation leading to a maximum catalyst activity will necessitate suitably selected conditions. Günther

Table 14.

Influence of the Reduction and Sulphurisation Conditions on the Hydrorefining Activity of the Catalyst 8376 [698] (for the hydrogenation conditions, see Table 13).

Experiment No.	Sulphurisation temperature (°C)	Operating time of the catalyst (days)	Product properties			
			Density (d_4^{20})	Aniline point (°C)	Content of phenols (g/l)	Content of bases (mg NH_3/l)
Raw material	—	—	0.950	33.0	130.0	1700.0
I[a])	0–340	4–7	0.864	52.0	9.7	129.0
II[b])	0–340	4–5	0.873	44.0	46.6	533.0
III[c])	25	4–7	0.869	49.9	15.4	212.0
IV	113	4–5	0.865	50.2	16.2	188.0
V	184	4–8	0.864	52.7	11.8	163.0
VI	253	4–8	0.863	52.0	10.8	124.0
VII	296	3–4	0.874	46.1	31.9	383.0
VIII	340	2–3	0.879	44.1	45.2	683.0

[a]) The catalyst was heated within 17 hours to 340 °C in circulating gas (0.7% H_2S).
[b]) The catalyst was heated within 1 hour to 340 °C in circulating gas.
[c]) In the experiments III to VIII the catalyst was heated at a rate of 17 °C/hour in a medium of fresh hydrogen (free of H_2S) to the sulphurisation temperature, after which it was sulphurised for 24 hours with circulating gas at space velocity of 1000 at the respective sulphurising temperature; finally it was heated to the hydrogenation temperature in a medium of fresh H_2 at a rate of 17 °C/hour.

[698] therefore carried out a detailed investigation of the optimum sulphurisation conditions for the catalyst 8376. Initially he found, that there is a saturation limit for every temperature up to about 300 °C, i.e., all active centres are not sulphurised with equal ease. Above 300 °C, however, sulphurisation is powerful to such a degree, that all centres are saturated at a high rate (Fig. 31). Since in practice, e.g. in the Zeitz Works, GDR, the catalyst 8376 was sulphurised at a maximum temperature of 290 °C, the catalyst was not sulphurised completely when refining was started, although this was not detrimental to the activity of the catalyst.

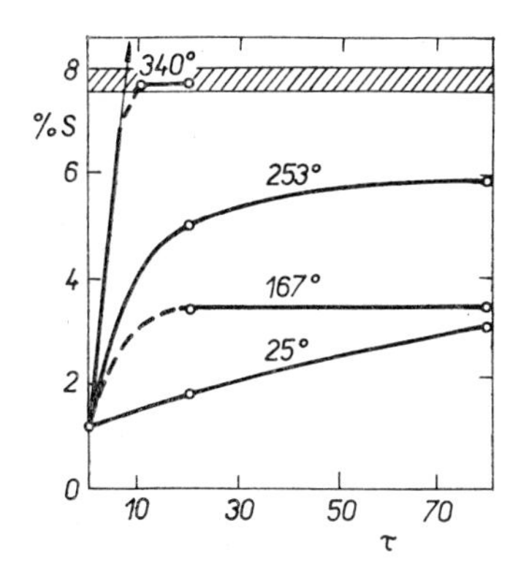

Fig. 31. Dependence of the degree of sulphurisation (% S) of the 8376 catalyst on the temperature and duration τ (hours) of the sulphurisation process [698]. (The oxide catalyst was heated at 300 atm to the operating temperature and sulphurised for 80 hours with circulating gas containing 0.7% H_2S, the space velocity of the circulating gas was 1000; analyses were carried out after 20 and 80 hours; the completely sulphurised catalyst should have 7.7. % S —shaded portion of the graph.)

From Table 14 it follows, that the optimum catalyst is obtained with a sulphurisation temperature of 253 °C. At lower sulphurisation temperatures the catalyst quality is inferior, but it is better than with temperatures of 300 °C and above. The quality of rapidly sulphurised catalysts (experiment II) is likewise low, while on the other hand low sulphurisation with a gradually increasing temperature (the normal method of catalyst forming and activation in the Zeitz Works, GDR [698]) results in a catalyst, the quality of which is equal to that of a catalyst obtained at the optimum sulphurisation temperature of 253 °C (cf. experiment VI). From the practical point of view, these results are important because a sufficient degree of sulphurisation is essential for achieving a high catalyst activity and, damage to the catalyst resulting from reduction with hydrogen deficient in sulphur at temperatures of 300 °C and more must be avoided. Günther [698] mentions that the circulating gas employed in his work (in the Zeitz Works, GDR) has rather a high H_2S content (0.7%). When the circulating gas is less rich in hydrogen sulphide or, where the petroleum or other materials processed are poor in sulphur, great attention should be given to sufficient sulphurisation of the refining catalyst.

The influence of sulphurisation or, of the presence of sulphur compounds in the feed is also clearly illustrated in the case of hydrocracking catalysts. In hydrocracking

Table 15.

Influence of Addition of Elementary Sulphur on the Hydrogenation of Olefins in the Gasoline Fraction b. p. 80 to 120 °C [2271]. (Catalyst WS_2–NiS–Al_2O_3, pressure 25 atm, temperature 200 °C.)

Sulphur content in the feed (%)	LHSV (l/l hour)	Olefine conversion (%)
0.04	1.5	87
0.30	1.5	44
0.02	1.5	90
0.20	1.5	65

of n-decane on the Ni–SiO_2–Al_2O_3 catalyst [1199] the overall reaction rate rises after sulphurisation and, moreover, there is a substantial increase in the hydrorefining activity. In the original un-sulphurised catalyst, a large number of the acid centres are blocked by Ni-salts; after sulphurisation, nickel is converted to the sulphide and thus new active acid centres are produced for splitting reactions. When the palladium sulphurised hydrocracking catalyst was used for the splitting of fuel oils, the iso/ /n-hydrocarbons ratio was considerably improved by the addition of a suitable sulphur-containing compound to the feed (butyl mercaptan, CS_2, H_2S or S) [384].

In some cases, the activity of the sulphurised catalyst also depends on the source of sulphur [374a]. In the case of dehydration of isopropylalcohol on nickel sulphide (prepared by sulphurisation of $Ni(OH)_2$), only when sulphurisation was carried out with a mixture of $H_2S + H_2$ (1 : 10) was an active catalyst produced. When H_2S was prepared by using a $CS_2 + H_2$ mixture, activities were very low, probably due to the blocking of active centres by carbon during the experiment (catalysts thus prepared had a considerably greater C content after the experiment) [1280]. Similarly, a more active catalyst was obtained by sulphurisation of a reactivated Ni–Mo–Al_2O_3 catalyst with a mixture of hydrogen and dimethyl disulphide [2096].

Sulphur compounds contained in the raw material sometimes have a substantial influence on the selectivity of sulphide catalysts. This fact is of considerable technological importance especially in the refining and desulphurisation of olefinic and aromatic feeds.

For example, methyl mercaptan and carbon disulphide strongly inhibited the selective hydrogenation of butadiene in mixtures containing C_5-olefines using a Co–Mo sulphide catalyst, while H_2S did not retard the reaction at all [1041]. Cole and Davidson [321] observed an inhibitive effect of sulphur compounds (mainly of thiophenic character) on the hydrogenation of olefins in the desulphurisation of cracked gasoline in the presence of WS_2–NiS. This was confirmed later for the same catalyst by Meerbott and Hinds [1347, 1348] who described the effect mentioned, not only for cracked gasoline, but also for synthetic heptane + 1-octene (1 : 1) mixture (the sulphur compounds added were a mixture of butanethiol and thiophene). From the results shown in Fig. 32 it follows, that with a sulphur content of 0.3 to 0.4% in the feed, the conversion of olefins is decreased by some 20 to 35%, depending on the space velocity of the feed.

Zyryanov et al. [2271] arrived at a similar result, finding a strong inhibitive effect of added elementary sulphur in the hydrogenation of olefins in gasoline fractions (see Table 15). In the case of hydrogenation of the pyrolysis gasoline in the presence of the WS_2–NiS–Al_2O_3 catalyst, H_2S in the circulating gas likewise inhibited the reduction of olefins [632].

Hydrogen sulphide contained in the reaction mixture also decreases the rate of hydrogenolysis of sulphur compounds [1385a]. For example in the case of thiophene

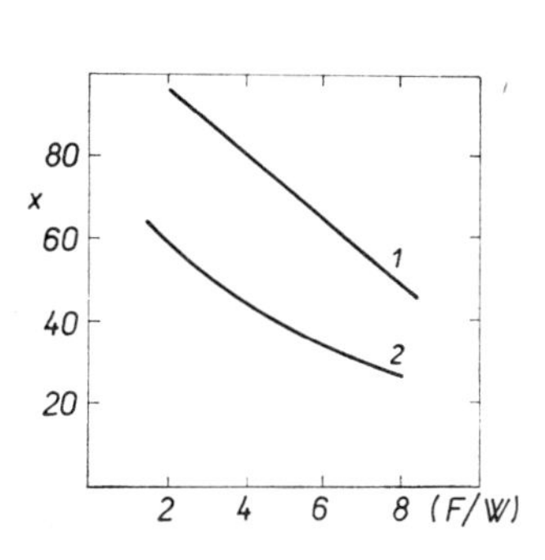

Fig. 32. The influence of sulphur compounds on hydrogenation of 1-octene [1348]. (Raw material: mixture of 1-octene and heptane (1:1); sulphur compounds: mixture of butanethiol and thiophene; catalyst WS_2.NiS, temperature 315°C, pressure 52 atm.)
1 — Raw material with no addition of sulphur. 2 — Raw material with addition of 0·3 to 0·4 % S. F/W — LHVS (l/l/hour). x — Degree of reduction of olefins (%).

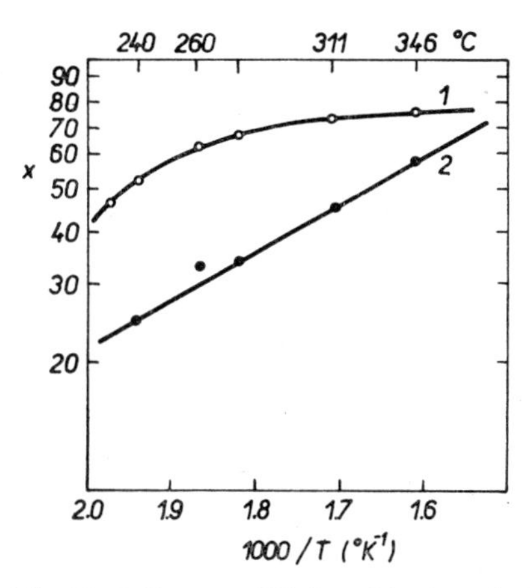

Fig. 33. Influence of H_2S on thiophene hydrogenolysis [1565].
1 — Result with no H_2S. 2 — After preliminary addition of H_2S (microreactor technique, 3.6 g Co–Mo/Al_2O_3 catalyst, 6 l H_2/hour, 5 ml doses of H_2S 3 to 5 sec. before thiophene feed, 4 µl thiophene). x — Degree of thiophene conversion in %.

hydrogenation on the MoO_2–MoS_2 catalyst ($T = 300$ °C, atmospheric pressure, 250 ppm thiophene in H_2) the degree of desulphuration decreased in the presence of H_2S (500 ppm) by 28% [658]. The same effect was observed in the case of hydrogenation of thiophene on the MoS_2 and Co–Mo catalyst [1066, 1565] (using the microreactor technique, cf. Fig. 33 [1565]), and equally in the hydrodesulphurisation of petroleum fractions [317a, 1363a], e.g. heavy gas oil obtained in coking [1347].

Hammar [721] explained this relationship by using reaction kinetics in the case of the Co–Mo catalyst, as he found that the following form of the kinetic equation is satisfactory for the rate of thiophene hydrogenation:

$$r = k' \frac{a_S a_H}{(1 + K_{H_2S} a_{H_2S})^2} \tag{IX}$$

here k' is the rate constant, a_S, a_H, a_{H_2S} are the activities of thiophene, hydrogen and hydrogen sulphide, resp. K_{H_2S} is the equilibrium adsorption constant of H_2S (cf. also p. 227). Besides H_2S, the amount of sulphur compounds in the feed material influences the extent of their hydrogenolysis, e.g. in the case of desulphurisation on WS_2–NiS [1348]. Satterfield and Roberts [1803a] recently found, that thiophene hydrogenolysis is at the same time retarded by thiophene itself. They therefore amplify the adsorption term in relation (IX) by the term $K_s . a_s$ which corresponds to thiophene.

The retarding effect of hydrogen sulphide was also recorded, though to a lesser degree, in the hydrogenolysis of dibenzothiophene on a sulphurised Co–Mo catalyst [1518]. For the other sulphide catalysts a similar kinetic analysis has not yet been carried out.

The relationships mentioned are important from the technological point of view, especially in the case of desulphurisation of cracked gasoline, gasoline obtained by pyrolysis etc., where operating conditions must be suitably selected in order to obtain maximum desulphurisation with a minimum decrease of the octane number. The mechanism of the effect of H_2S and of sulphur compounds on the hydrogenating activity of sulphide catalysts is not yet quite clear [1147], since the amount of experimental data available is very limited.

The main reason is most probably the high adsorption of H_2S and of other sulphur compounds on the catalyst surface, which blocks active hydrogenation centres, with the result that the rate of hydrogenation of olefins and olefinic intermediate products formed by hydrodesulphurisation of sulphur compounds is lowered. This is also indicated by low pressure experiments, in which conversion of olefines (hydrogenation of butenes on a sulphurised Co–Mo catalyst [1041, 1565] and of ethylene on WS_2 and MoS_2 [2182]) decreased considerably after adsorption of H_2S or, in its presence. This effect, however, depends on the catalyst type employed; e.g. it is substantially lower on sulphurised Cr_2O_3 [1567].

Švajgl and Smrž [2011, 2012a], who found that H_2S inhibited the high pressure hydrogenation of aromatic substances (but not of olefins) believe, that the cause of the high catalyst activity is the existence of a metallic nickel phase in the catalyst, the amount of which decreases due to the influence of H_2S. Thus the hydrogenation activity of the catalyst diminishes.

5.2.

The influence of promoters

A promoter is a substance which, when added to the catalyst in a certain (generally small) amount favourably influences one of its properties. There are numerous examples for this effect in sulphide catalysts; the majority of these are very important from a technical aspect. Also, the majority of carriers have a promoting effect besides their main function of supporting the catalyst. The next section will be devoted to the influence of carriers while, in this section, attention will be concentrated on promoters other than carriers.

The majority of sulphides or other compounds, added to the main sulphide component, act as chemico-structural (electron) promoters (p. 65) which create new catalytically active centres. As a result of this, the activity of the catalyst is considerably improved or, its selectivity is enhanced.

The development of technically important sulphide catalysts is closely connected with the promotion effect. The mixed catalyst 3510 which was the first one used in hydrorefining (MoO_3–ZnO–MgO, the oxide components of which were converted to sulphides during operation) was satisfactory in destructive hydrogenation of middle tar oils. With respect to its marked sensitivity to nitrogenous bases, this catalyst was not suited to processing brown- and especially bituminous-coal tar [416, 1429, 1430] and it was soon replaced by the catalyst 5058 (pure WS_2). This catalyst has outstanding hydrogenating properties, a suitable lifetime and effective splitting activity. From

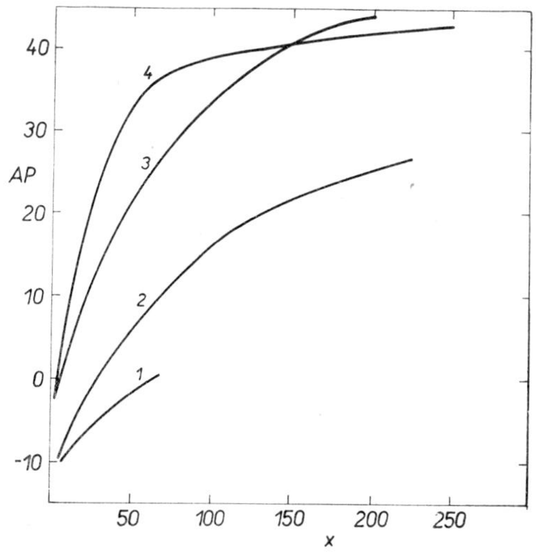

Fig. 34. Influence of NiS additions on hydrogenation activity (expressed by the aniline point AP °C of the tar middle oil hydrogenate) of the catalyst MoS_2 (or WS_2) + Al_2O_3 [416, 1626]. x — g active component/l catalyst. 1 — Ni_2O_3, no additive. 2 — MoO_3, no further additive. 3 — WO_3 varies, the catalyst has a constant content of 3% Ni_2O_3. 4 — MoO_3 varies, the catalyst has a constant content of 3% Ni_2O_3.
(Data concerning the composition of sulphide catalysts are expressed in terms of the respective oxides, experiments were done with middle tar oil with an aniline point of 12 °C. The phenols content was 17%, nitrogen content 0.75%. 1% CS_2 was added to the feed; the H_2 pressure in the experiments was 250 atm, temperature 450 °C, LHSV 0.8 kg/l hour).

the point of view of its technological application, however, the excessively high hydrogenating activity of this catalyst was a drawback. The gasoline obtained did not have enough aromatic character and the octane number was too low. Also the high price of this catalyst was a reason why methods of modifying the properties of WS_2 were investigated and other sulphide catalysts were tested [416, 1429, 1626]. Initially, the catalyst was "diluted" by supporting it on an active carrier, which limited the

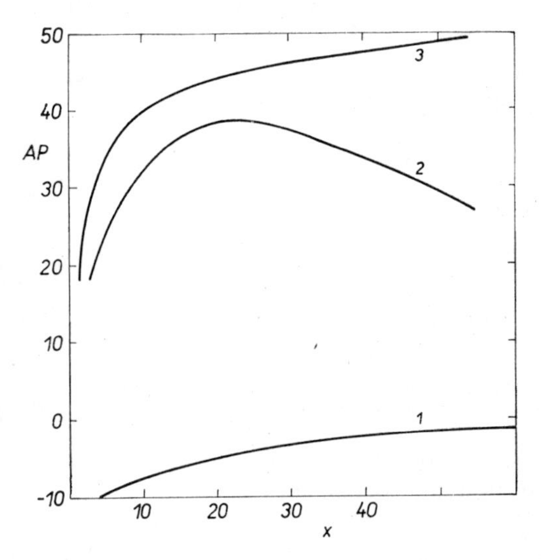

Fig. 35. Influence of NiS additions on the hydrogenation activity (expressed in the same way as in Fig. 34) of the catalysts MoS_2 (or WS_2) + Al_2O_3 [416, 1626]. x — g Ni_2O_3/l catalyst. 1 — Ni_2O_3, no other additive. 2 — 150 g MoO_3/l catalyst (constant), the Ni_2O_3 content varies. 3 — 250 g WO_3/l catalyst (constant), the Ni_2O_3 content varies. (Data concerning the composition of sulphide catalysts are expressed in terms of the respective oxides, the operating conditions are the same as in Fig. 34.)

hydrogenating activity of the catalyst. When aluminosilicate carriers were used (in some cases, with modified acidity) a hydrocracking catalyst was obtained (catalyst 6434, containing about 10% WS_2) with outstanding splitting activity and, at the same time, a very limited hydrogenating and refining activity, so that gasoline obtained using this catalyst was considerably more aromatic. However, the catalyst is unusually sensitive to the presence of nitrogenous bases, so that efficient pretreating is inevitable.

After separating the vapour-phase hydrogenation process into hydrocracking (Benzinierung) and hydrorefining (Vorhydrierung), the catalyst 6434 gave very good service. However, there remained the need for a suitable catalyst with a high refining and limited splitting activity for the hydrorefining step (in order to avoid the formation of low-quality non-aromatic gasoline). This catalyst had at the same time to be sufficiently resistant to the deactivating effect of nitrogen-containing and other harmful substances contained in the feed or formed in the course of the refining process [1101, 1429]. The development of these refining catalysts was especially intensive in Germany and in the USA. In Germany, attention concentrated on developing an optimum catalyst for the hydrorefining of coal tar and, for hydrorefining benzene from the coke industry; in the USA, where petroleum prevailed, catalysts were developed for hydrorefining kerosines, gas oils etc. [416].

Systematic testing of the catalytic properties of sulphides of different metals and their mixtures alone or supported on various carriers has shown, that the optimum combination is a mixture of sulphides of metals from the Groups VI and VIII of the Periodic Table and, that the most advantageous carrier is active γ-Al_2O_3. Here, achievement of the optimum properties for a specific process (refining activity with

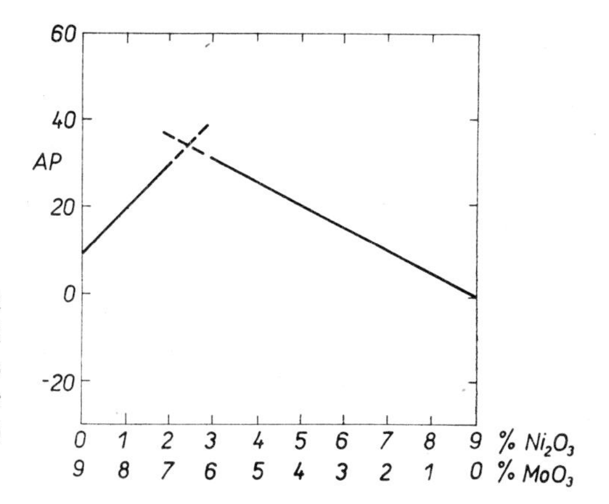

Fig. 36. The hydrogenation activity (expressed by the aniline point AP °C of the middle oil hydrogenate) of the MoS_2–NiS–Al_2O_3 catalysts [416, 1626]. (The sulphide content of the carrier is expressed in terms of the amount of oxides.)

respect to O and N compounds, desulphurisation activity, controlled hydrogenation and splitting activity) is a consequence of the mutual promoting effect [1347, 1626] of the active catalyst components between each other as well as the carrier. The results

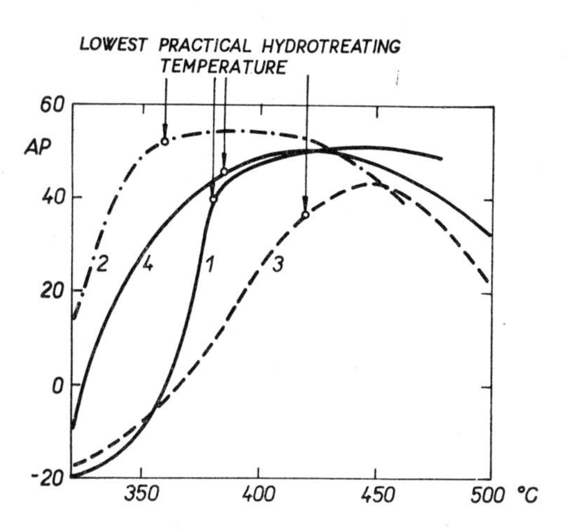

Fig. 37. Dependence of the hydrogenation activity of mixed sulphide catalysts on temperature [416]. *AP* — Aniline point of the middle oil hydrogenate (°C). 1 — WS_2. 2 — WS_2 + NiS. 3 — MoS_2 + NiS + Al_2O_3. 4 — WS_2 + + NiS + Al_2O_3. (The operating conditions are stated in Fig. 34, the lowest practical hydrotreating temperatures are indicated on the curves.)

for the sulphide systems W–Ni–Al_2O_3 and Mo–Ni–Al_2O_3 are reported in Figs. 34, 35 and 36.

Comparison of the activity of the individual catalysts was carried out with a typical hydrorefining reaction, i.e. hydrogenation of middle tar oil (cf. Fig. 34) [416]. The hydrogenation activity (measured by the aniline point of the hydrogenate, the initial feed had an AP value of 12 °C) of the supported catalysts can be controlled over a very wide range by varying the mutual ratio of active sulphides as shown in Figs. 34 and 35. With the combination Mo–Ni, optimum hydrogenating activities are achieved with an atomic ratio of about 3 : 2 (Mo : Ni) (Fig. 36). This mutual promoting effect is also evident in the case of non-supported sulphides when the hydro-

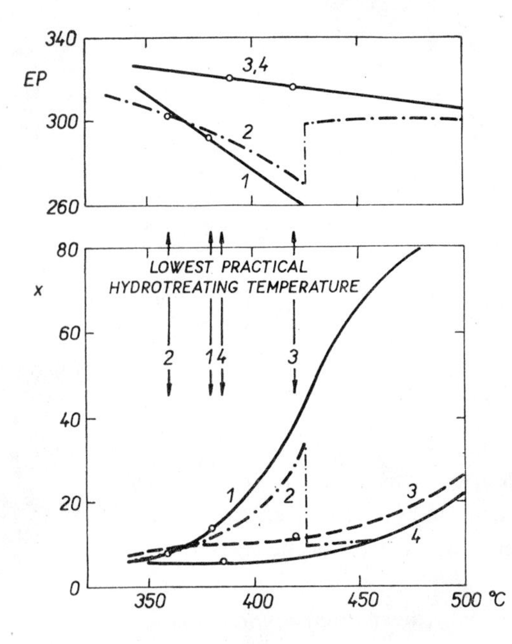

Fig. 38. Dependence of the splitting activity of different catalysts on the temperature t (°C) [416, 1626].

1 — 5058 (WS_2). 2 — WS_2 + NiS. 3 — MoS_2 + + NiS + Al_2O_3. 4 — WS_2 + NiS + Al_2O_3. x — % product up to 150°. *EP* — End of distillation (°C). (Middle tar oil was hydrogenated under the conditions stated in Fig. 34, except the temperature; with individual catalysts, the lowest practical hydrotreating temperatures are stated, i.e. temperatures at which almost complete degradation of phenols and nitrogenous bases is achieved; cf. Fig. 37.)

genation activity of the catalyst 5058 (WS_2) is compared with that of the catalyst 3076 (unsupported WS_2 + NiS) (Fig. 37).

The promoting effect of nickel sulphide is even more clearly shown by the splitting activity of W and Mo sulphides (Fig. 38). The decrease of the splitting activity of WS_2 is most marked after an addition of nickel sulphide at temperatures above 425 °C, when the splitting activity of the catalyst almost ceases (cf. also p. 41 and 65). When the two sulphides are supported on a carrier (γ-Al_2O_3) the splitting activity is limited considerably within the entire temperature range in which these catalysts can be practically employed.

The hydrorefining activity (i.e. activity for hydrogenolysis of oxygen- and nitrogen-containing substances, especially phenols and nitrogenous bases), is likewise improved by the promoting effect of nickel (cf. the difference between the catalysts WS_2 and WS_2 + NiS in Fig. 37, where the lowest practical hydrotreating temperatures are marked for the respective catalysts). When the two sulphides are supported on a carrier, the hydrorefining activity of the catalyst 8376 varies only slightly compared with the catalyst 5058 (WS_2). The promoting effect of Ni on the activity of the catalyst WS_2–Al_2O_3 is quite specific, as shown by Table 16.

From Pier's results it follows that although Ni promotes both W and Mo as refining catalysts, supported tungsten catalysts are better in all respects (their hydrogenation and refining activity is higher, their splitting activity is lower, cf. Figs. 34, 35, 37, 38). Earlier results (from the year 1942) obtained with the catalyst 7846 (MoS_2–NiS–Al_2O_3) also indicated [1429] that the catalyst 8376 is better.

Table 16.

The Specific Influence of the Promotion Effect of Nickel Compared with the Effect of Fe and Co in 8376-Type Refining Catalysts (27% WS_2, roughly 3% of a sulphide of a metal of group VIII, supported on γ-Al_2O_3) [416]. (Middle tar oil was hydrogenated under the conditions stated in Fig. 34.)

Metal of Group VIII	Product properties			
	% Fraction up to 180°	Middle Oil		
		Aniline point (°C)	Content of phenols (%)	Content of N bases (mg NH_3/l)
Ni	23	47	0.01	3
Co	23	11.5	0.03	40
Fe	20	−5.5	0.05	700

Recently, Münzing, Blume et al. [1429, 1433] found that the quality of the carrier employed, γ-Al_2O_3, has a marked influence on the overall refining activity of the catalyst. The 8197 catalyst the oxide form of which contains 17% MoO_3, 4.5% NiO and 78.5% Al_2O_3, exhibited the same hydrogenation activity as the catalyst 8376. The refining activity however was better (Table 17, cf. p. 103), so that the inferior quality of Ni–Mo catalysts observed earlier [416, 1429] was caused by the inferior

Table 17.

Comparison of the Overall Hydrorefining Efficiency of the Catalysts 8376 and 8197 in the TTH Process [1429].

Catalyst	Operating temperature (°C)	Properties of the hydrogenation product		
		Content of nitrogenous bases (mg NH_3/l)	Content of phenols (%)	Aniline point (°C)
8197 (Ni–Mo)	348	38	0.13	57.5
8376 (Ni–W)	358	61	0.40	57.5

quality of the carrier. According to Švajgl [2012c), however, these conclusions are not quite unambiguous.

Differing from its analogue, the cobalt-containing catalyst (which is the most important hydrodesulphurisating catalyst used in medium-pressure desulphurisation of petroleum products), the catalyst 8197 is considerably less sensitive to nitrogenous bases. In spite of the somewhat smaller desulphurisation activity, it can be used very well in hydrorefining tar products or their mixtures with petroleum fractions [1429]. The advantages of Ni–Mo refining catalysts will be discussed in greater detail in Chapter 7.

Table 18.

Composition and Activity of Hydrprefining Catalysts [1429].
(Raw material: fractions of diesel fuel from Romashkino petroleum, b. p. 240—360°C, 1.37% S, 0.19% phenols, 41 mg NH_3/l basic compounds, bromine number 11.9; hydrogenation at 357 °C, 50 atm, LHSV 2.3 l/l/hour.)

Catalyst notation[a])	Properties of the hydrogenation product				
	Sulphur content (%)	Content of phenols (%)	Content of basic compounds (mg NH_3/l)	Aniline point (°C)	Bromine number
8197	0.035	0.036	6.2	70.3	1.4
U 63	0.075	0.03	9.7	69.7	1.3
BR 86	0.024	0.02	4.5	70.3	3.4

[a]) The catalyst composition will be found in Table 3.

The fact, that petroleum is being increasingly used even in those countries which had formerly mainly processed tar-based materials, as well as the increasing application of hydrorefining processes, necessitated the development of new universal catalysts. These were required to have both high desulphurisation activity and reasonable activity in the hydrogenolysis of nitrogenous bases and phenols [191, 1429]. The hydrorefining properties of these catalysts are listed in Table 18. In connection with this, it is interesting to note the marked promoting effect of NiS by means of which the cobalt–molybdenum catalyst, originally highly sensitive to the presence of nitrogenous bases, achieved an excellent hydrorefining activity (especially for refining diesel fuel) [2191]. The same promotion effect of NiS was also observed in the case of a catalyst with aluminosilicate carrier [1535]. Here we should mention the work of Prokopets and Eru [1684] who, 35 years ago, observed a promoting effect of NiO in a MoS_2 catalyst. Medium-pressure refining catalysts may further be improved by modifying the carrier, i.e., by altering its composition as well as choosing a suitable method of preparation (cf. p. 102) [191].

From the point of view of the desulphurisating activity, the most highly active catalyst system is a mixture of molybdenum and cobalt sulphides. This combination was used for probably the first time in hydrorefining benzene. It was found that this mixed catalyst is active at both low and medium pressures [835, 867]. Later, a good hydrodesulphurisation activity was obtained with a mixture of cobalt sulphide and

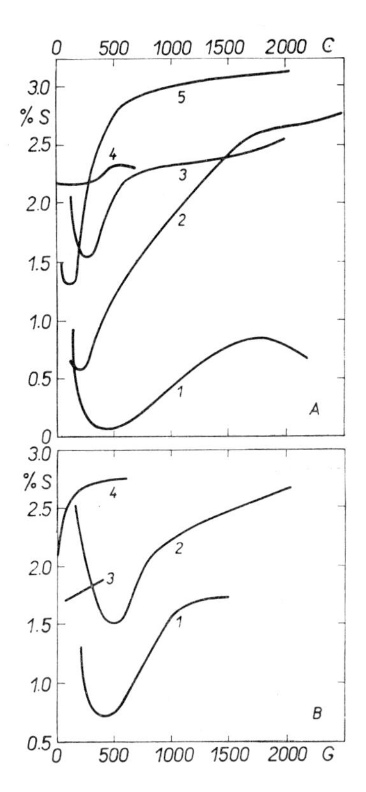

Fig. 39. The desulphurisation activity of some mixed catalysts (the feed in the hydrodesulphurisation process was a petroleum distillate containing 3.5 % S in the case *A* and 3% S in the case *B*). The reaction was carried out at atmospheric pressure and 350 °C in a flow reactor. *G* — ml feed passed over the catalyst [243]. *A* — Desulphurisation activity of molybdates: 1 — cobalt, 2 — copper, 3 — zinc, 4 — iron, 5 — aluminium. *B* — Desulphurisation activity of catalysts supported on bentonite: 1 — cobalt molybdate, 2 — mixture of cobalt and molybdenum oxides, 3 — cobalt oxide, 4 — molybdenum oxide.

molybdenum oxide [1590]. Byrns et al. [243] studied the mutual promoting influence of oxides and sulphides in detail. It is clear from the results of desulphurisation experiments (Fig. 39) that the promoting effect of cobalt on the desulphurisation activity of molybdenum oxide or sulphide (the majority of the active components are converted to the sulphide form in the course of desulphurisation) is much higher than the effect of other metals tested. This effect is expressed in terms of relative activities in Table 19.

Nickel is close to cobalt in terms of efficiency, since the desulphurisation activity of nickel–molybdenum catalysts is only slightly lower than the corresponding cobalt–molybdenum ones [154, 1347] (cf. also fig. 40). On the other hand, Ni–Mo catalysts are more resistant to nitrogenous bases, so that they are found rather more useful as refining catalysts for processing materials with a high content of nitrogen- and oxygen-containing substances (e.g. tar, low-quality hydrocracking feed etc.). The effect of cobalt molybdate is quite specific, as the activity of the pure components as well as of their mixtures is considerably lower (Fig. 39 B).

The most suitable carrier for cobalt–molybdenum catalysts is active γ-Al_2O_3, sometimes with modified acidity and porosity [154]. The desulphurisation activity rises rapidly at first with increasing concentration of the active components. When

Table 19.

The Relative Hydrodesulphuration Activity of Different Molybdenum Catalysts (the relative activity is based on the activity of cobalt molybdate = 100) [243, 1347].

Catalyst	Approximate relative activity[a])
Cobalt molybdate	100
Zinc molybdate	36
Ferrous molybdate	34
Cupric molybdate	21
Aluminium molybdate	6

[a]) The activity was compared for the case of the "equilibrium" catalyst, i. e. with a catalyst after sulphurisation by the sulphur contained in the feed, when the desulphurisation activity is constant (cf. Fig. 39).

the activity limit is achieved, however, further increase in the metal content is less effective. Usually, the overall content of cobalt and molybdenum oxides in the fresh catalyst varies from 7 to 20%, the optimum concentration being 13 to 16% [1347].

The atomic ratio of the active components is important for high desulphurisation activity [1385a]. As we see from Fig. 40, a maximum promotion effect is obtained with nickel in the atomic ratio of Ni : Mo $\sim$ 0.5–0.6, even in those cases where a high hydrocracking activity is desired at the same time [996a]. In the case of cobalt, the promotion maximum corresponds to a ratio of Co : Mo $\sim$ 0.3 to 0.4 and this maximum is rather sharp [154]. Similarly, Engel and Hoog [473] found that the optimum desulphurisating activity is achieved with an atomic ratio of about 0.2. Sulimov et al. [1985] state a rather wider range.

Recently, Ahuja et al. [11a] once more investigated the influence of the atomic ratio of Groups VIII and VI metals (combinations of Co–Mo, Co–W, Ni–W and Ni–Mo supported on different carriers) on the activity of hydrorefining catalysts (particular attention was paid to hydrogenation of olefins, and aromatic substances and hydrodesulphurisation). It was found that the activity optimum is usually close to the atomic ratio $Me_{VIII}/Me_{VI} \sim 0.25$. This fact could not be explained in detail neither by means of X-ray analysis nor by ESR spectrometry. In the case of the cobalt–molybdenum catalyst, however, the authors observed a maximum sulphur content in the catalyst in the region of the atomic ratio Co/Mo ~ 0.25 (although this was only 65% of the amount, theoretically required to convert Co and Mo to Co_9S_8 and MoS_2). Ahuja et al. consider MoS_2 to be the catalytic agent in the hydrodesulphurisation reaction but its formation and high activity requires intervention of cobalt.

In the case of the hydrocracking catalyst 6434, clay (Fuller's earth) activated with HF may be considered to be the splitting catalyst, while WS_2 is a promoter which modified the properties of the clay. Clay alone (Terrana) activated with HF

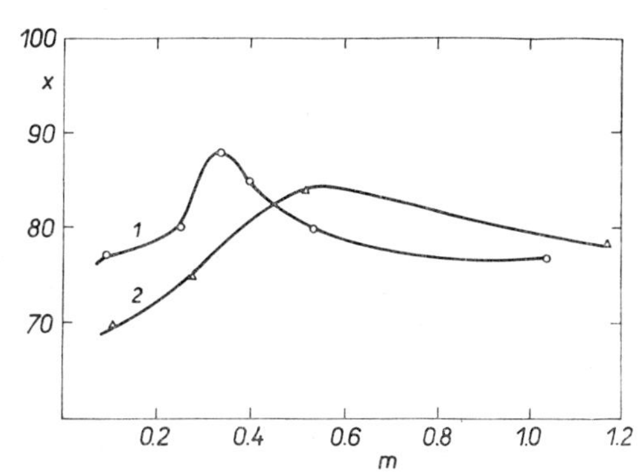

Fig. 40. The promotion effect of Co (1) and Ni (2) on the hydrodesulphurisation activity of the MoO_3–Al_2O_3 catalyst [154]. m — Atomic ratio Co (or Ni): Mo. x — Degree of desulphurisation (%). (Gas oil with 1.93% S was hydrogenated; the activity of the non-promoted catalyst was 50.2%.)

has rather a low splitting activity at pressures of around 200 atm, but increases with increasing hydrogen pressure. On addition of 10% WS_2 however the splitting activity is considerably increased, achieving at 200 atm the activity level of the clay with no sulphide component at 600 atm. Table 20 shows the data for hydrocracking pretreated medium tar oil and, for comparison, the result obtained with non-splitting active carbon is included. With the catalyst 6434, gasoline of a lower octane number was obtained due to the lower content of aromatic substances and higher content of n-paraffins (the hydrogenation activity of the catalyst 6434 is higher, and the isomerisation activity lower compared to clay) [416].

At present, new hydrocracking catalysts are being developed, in which other components, mainly iron-group metals, are added to the basic hydrogenation component (e. g. WS_2) [2018]. These additives increase the activity of the resulting catalyst considerably. Details will be discussed in Chapter 7 (p. 304].

Among other less conventional promoters, phosphorus (added in the form of ammonium phosphate, ammonium or cobalt phosphomolybdate) and fluorine or, the two together improve the desulphurisating activity of cobalt–molybdate catalysts [213, 729, 821, 1420, 1662]. This is also true of potassium which increases the desulphurisation selectivity of this catalyst in processing gasoline with a high olefine content [467, 473] (cf. also p. 329). The promoting effect of P, B and Li was observed in the case of the molybdenum sulphide catalyst used for the hydrogenation of heavy tar [661].

Cyanides of some metals were tested as promoters in the preparation of sulphide catalysts of hydrogenating oils and lignites [337, 2054]. The promotion effect of titanum oxide on the activity of the NiS/Al_2O_3 catalyst [1324] and the effect of Se, Te, and P on WS_2 in hydrogenating phenols [1800] were also investigated.

Table 20.

The Promotion Effect of WS_2 on the Hydrocracking Activity of Clay Activated with HF [416]. (The hydrocracking feed was prerefined middle tar oil.)

Product properties and operating conditions		Catalyst		
		6434	Clay activated with HF	Active carbon
Operating pressure (atm)		200	600	600
Operating temperature (°C)		407	425	475
Gasoline properties	Density	0.722	0.728	0.738
	% gasoline up to 100 °C	41	39	28
	End of distillation, °C	202	180	178
	% Paraffins	72	66	65.5
	% Naphthenes	19.5	21	21
	% Aromatic substances	8	12.5	13
	% Olefins	0.5	1.5	0.5
	Octane number	64	72	51

As stated earlier, in most cases the effect of chemico-structural promoters is involved in most of the examples mentioned. Data concerning the influence of texture promoters on sulphide catalysts are very rare as yet. In order to avoid sintering the active MoS_2 surface, the catalyst was prepared from the salt of a suitable heteropolyacid, the promoter being the central atom of the complex. Two basic types of heteropolyacids were used: $R_n[X(Mo_2O_7)_6]$ and $R_n[M(MoO_4)_6]$, where X is mainly P and Si, M = Cr, Ni, Co, Cu and R = H, NH_4 or K. Ammonium phosphomolybdate [1037, 1038] was found particularly useful. Similarly, Fleck [525] found that when molybdenum oxide is supported in the form of the heteropolyacid or its salts, e. g. in mixed Co–Mo catalysts, the influence of MoO_3 on sintering the surface of active γ-Al_2O_3 is considerably limited. In these cases, however, other effects are involved besides the texture effect, e.g. a change of the catalyst acidity [1429]. This texture-promoting effect is probably exerted by additions of ThO_2 and ZrO_2 to supported oxides or sulphides of Groups VI and VIII metals. Here, active and mechanically stable catalysts for steam reforming of hydrocarbons [411] or, hydrogenation of high-molecular weight petroleum hydrocarbons [1324] are involved. Additions of K_2O, BaO, Cr_2O_3 and Al_2O_3 to MoS_2 [1308], additions of K_2S to sulphurised Ni–Mo, Ni–W, Co–Mo and Fe–Cr catalysts for water gas conversion [12a], Group II metal oxides to the Co–Mo hydrodesulphurisation catalyst [937a] or of K_2O to the hydrocracking catalyst WS_2–Al_2O_3 [587] probably involve chemical effects.

5.3.

The influence of the carrier and acidity of the catalyst surface on the activity of sulphide catalysts

Sulphide catalyst carriers do not only act in the capacity of carriers. Their main effect is a promoting one helping to create the overall activity of the catalyst. A typical example is the catalyst 8376, in which active γ-Al_2O_3 acts in quite a specific manner (especially in hydrogenating phenols) and cannot be replaced by a different carrier, even by a substitute of equal texture [1626]. In the case of polyfunctional catalysts, e.g. sulphidic hydrocracking catalysts, the carrier takes a direct part in the catalytic reaction (in the case of hydrocracking catalyst it is responsible for the splitting properties) [80].

The composition of the carrier and the method of its preparation very often influences the overall activity and selectivity of the catalyst. Kalechits studied this in detail with the MoS_2–Al_2O_3–SiO_2 catalyst (cf. also p. 58). The method employed in preparing the carrier is very important for the hydrogenating activity of the catalyst. The preparation of an aluminosilicate carrier by saturating a silicic acid gel with aluminium sulphate followed by addition of ammonia and washing, then by coprecipitation of the two oxides and finally by prolonged mixing of the two moist gels was investigated. MoS_2 was supported on the carrier by impregnation with ammonium thiomolybdate solution and reduction with hydrogen. Catalysts with the carrier prepared by mechanical mixing always had a lower hydrogenation activity, while the two other methods gave catalysts of better hydrogenating efficiency. In the case of carriers prepared by impregnation and coprecipitation, however, the composition of the surface differs from that of the remaining catalyst mass (it is richer in Al_2O_3). This may be one of the reasons for the different hydrogenation activities of the catalysts mentioned [959, 966].

The amount of MoS_2 supported on the surface depends on the carrier composition; it is a minimum (2 to 3%) with an Al_2O_3 concentration in the carrier of about 20% [961]. This is related to the decrease in the hydrogenating activity of the catalyst with this carrier composition even when the relative hydrogenating activity only is considered (i.e. activity related to a certain amount of surface-bound MoS_2 molecules, cf. Fig. 42). Therefore, this decrease is not only caused by a lower content of the hydrogenating component in the catalyst (MoS_2) but also by a change in the physico-chemical nature of the catalyst surface (cf. the acidity change of the catalyst surface with the varying carrier composition in Fig. 17) [960, 967].

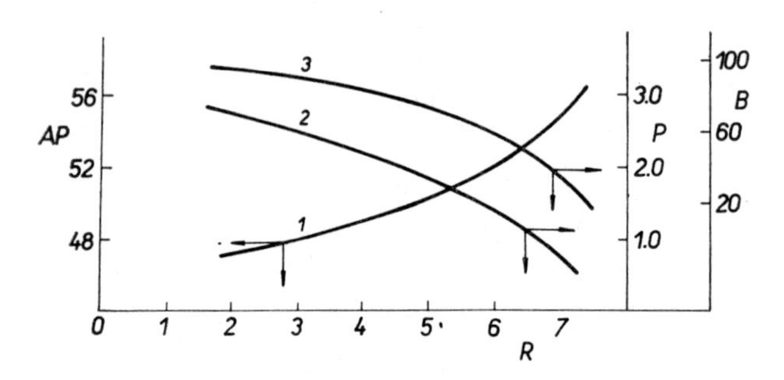

Fig. 41. Relationship between activity and residual acidity of different hydrorefining catalysts [1429]. *R* — Residual acidity (10^{-4} equiv. H^+/g). Activity was measured according to the following hydrogenate properties: *AP* — aniline point °C (curve 1), *P* — content of phenols (%) (curve 2), *B* — content of basic compounds (mg NH_3/l) (curve 3).

The acidity of the surface, which significantly alters the quality of the catalyst, is an important property of sulphide catalysts. The relationship between acidity and selectivity in refining and hydrocracking sulphide catalysts was already mentioned by Pier [1626] especially in connection with WS_2-based catalysts.

In later research these basic findings were developed in greater detail. As we have already mentioned, WS_2 itself is acid to a certain extent, depending on the method of preparation. This acidity influences the selectivity of the catalyst [1798] (p. 77). Moreover, γ-Al_2O_3 cannot be considered as a neutral carrier, since the residual acidity of alumina [1428] is very important for a high refining activity, as follows from Fig. 41.

The differing acidity of the carriers employed was probably the reason for discre-

Table 21.

The Influence of Conditions of Carrier Preparation on the Refining Activity of the Catalyst 8376 [1429].

Catalyst number	Conditions of precipitation in the preparation of Al_2O_3			Residual acidity (10^{-4} equiv. H^+/g)	Product properties (TTH process)		
	pH	Temperature °C	Structure		Content of the basic compounds (mg NH_3/l)	Content of phenols (%)	Aniline point (°C)
10	6	30	Böhmite	5.0	—	1.98	50.6
11	7	30	Böhmite	5.6	77	1.68	51.2
12	8	45	Böhmite	7.45	27	0.42	56.1
13	8	90	Böhmite	7.0	28	0.83	54.5
14	>12	30	Bayerite	3.0	87	2.55	49.2
15	Raw alumina		Hydrargillite	1.7	98	2.82	46.8

pancies in data reported by Pier [1626] and by Münzing et al. [1429] concerning the refining activity of the catalysts 7846, 8376 and 8197 (cf. also p. 95). The degree to which the catalyst activity is sensitive to preparation conditions is illustrated by data for the catalyst 8376 (Table 21 and Fig. 41). From these results it also follows, that the maximum of the refining activity agrees with the residual acidity maximum. At the same time, the catalyst acidity has a favourable influence on the activity in all basic refining reactions, i.e. hydrogenation of double bonds, hydrogenolysis of phenols and nitrogenous bases [1429]. Control of the splitting activity of the catalyst 6434 by a change in the carrier acidity has been discussed already. Similarly, modification of the carrier acidity by addition of HF caused an improvement of the activity of the catalyst MoS_2/Al_2O_3 in the hydrogenation of motor fuels [469].

Kalechits studied the details of the acidity/activity relationship again with the MoS_2 catalyst supported on aluminosilicate carriers [959, 960, 961]. In the case of these carriers, aluminosilicates with prevailing Brønsted acidity (SA_{BR}) were formed at low Al_2O_3 contents of about 10 to 20%, while aluminosilicates with Lewis acidity (SA_L) were obtained with 40 to 60% Al_2O_3 [960]. With 20% Al_2O_3 in the carrier, the MoS_2–Al_2O_3–SiO_2 catalyst has a minimum hydrogenation activity. Fig. 42 shows the pattern of the relationship between the relative hydrogenation activity and the carrier composition over the entire range. It is evident, that the relative activity of pure MoS_2 is always higher, i.e. that all components of the aluminosilicate carrier (SiO_2, Al_2O_3 and, therefore, the SA_{BR} and SA_L aluminosilicates) decrease

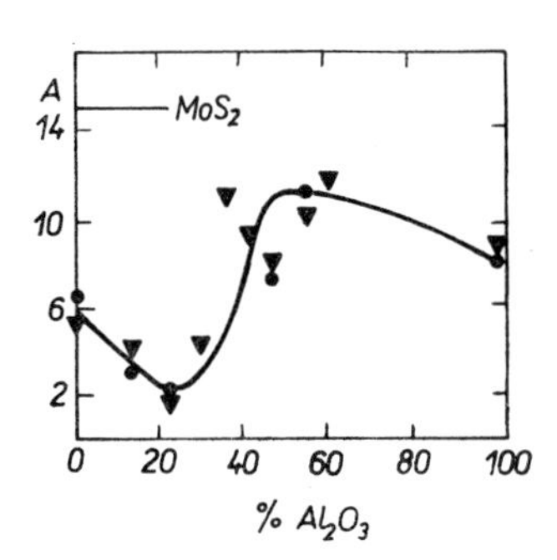

Fig. 42. Relationship between the hydrogenation activity (A) of MoS_2–Al_2O_3–SiO_2 catalysts on the composition of the carrier. (Benzene hydrogenation at 200 atm, 430 °C and LHSV = 3; A is the degree of benzene conversion related to 10^{21} molecules of MoS_2 on the catalyst surface [961].)

the hydrogenation activity of MoS_2, differing from neutral carriers (e.g. active carbon) which increase the relative hydrogenation activity [956]. The extents by which the hydrogenation activity is diminished are not equal, however, as is apparent from the shape of the relative activity curve. The greatest activity decrease is observed with a concentration of about 20% Al_2O_3 in the carrier (aluminosilicate with Brønsted acidity). The formation of strong complexes of MoS_2 with aluminosilicates SA_{BR}, causes the acidity and therefore the splitting activity to decrease (see Fig. 17). The maximum of the specific hydrogenating activity lies in the region of the maximum SA_L concentration. When MoS_2 is supported on these carriers, a substantial change takes place in the surface structure. Part of the acid aluminosilicate is destroyed, MoS_2 is adsorbed on Al_2O_3 and SiO_2 and a new phase is thus formed. The shape of the specific hydrogenating activity curve indicates, that the hydrogenation activity of MoS_2 is decreased most by SA_{BR} aluminosilicates and least by SA_L aluminosilicates. The carrier components SiO_2 and Al_2O_3 have an intermediate effect, MoS_2 bound to Al_2O_3 being more efficient in terms of hydrogenation.

These results are an example of the rather wide possibilities of influencing the selectivity of sulphide catalysts. In the case discussed [961], for example, the catalyst which has a maximum content of acid SA_{BR} aluminosilicates is most selective in the hydrogenation of polycyclic aromatic compounds in the presence of monocyclic ones.

5.4.

Deactivation of sulphide catalysts

One of the most valuable properties of sulphide catalysts, i.e. resistance to most catalyst poisons and long-term stability of activity even under quite severe operating conditions, has its limits with these catalysts, too. The activity of sulphide catalysts decreases in the course of use, the rate of this decrease depending on the quality of the catalyst, character of the feed processed, reaction conditions and, under practical conditions, on the care of the operators, quality of the control system, etc.

Ageing takes place slowly in high-quality sulphide catalysts. For example, the catalyst 5058 was used in a hydrorefining reactor for 1.5 to 2 years and some batches for up to five years, without any significant activity decrease. In appearance, the used catalyst was almost the same as the fresh one. X-ray analysis showed slightly more distinct diffraction lines due to an increase in the dimensions of crystallites [1626] or, a further increase in the symmetry of the lattice (p. 43). The situation is obviously different after local or overall overheating of the catalyst, where an almost perfectly symmetrical lattice is formed due to recrystallisation (e.g. with WS_2, the hexagonal modification is formed – cf. p. 39).

Chemical reactions may be another cause of ageing. For example, systematic loss of sulphur in the catalyst causes a change in the surface structure such that the catalyst loses a large part of its original activity. In the case of supported sulphide catalysts, overheating or long-term use may even lead to phase reactions between the active sulphide component and the carrier, the majority of active centres being destroyed in the process.

Loss of activity of sulphide catalysts may sometimes also be due to mechanical causes, mainly wear, disintegration or an overall loss of mechanical strength. These changes, which are not very frequent with fixed-bed catalysts, are mainly encountered in the case of supported catalysts due to damage or a poor carrier quality. In the case of the WS_2 catalyst, tablets disintegrate when the sulphur supply to the catalyst is poor or, when low-sulphur feeds are being processed.

5.4.1.

The influence of nitrogen- and oxygen-containing substances and of some hydrocarbons

The most frequent cause of deactivation of sulphide catalysts is, however, the effect of some substances which have a specific effect and which cannot be simply called poisons (since they usually are effective in concentration higher than the values which are characteristic of poisons), which however cause a distinct, reversible or irreversible deactivation.

The most important group of these deactivation agents are nitrogenous substances. In hydrogenation of ethylene, at atmospheric pressure at 425 °C, on MoS_2, a strong inhibitive effect of NH_3 and especially of pyridine was observed (decrease to 37 and 15% of the original conversion value, resp.) [55]. However, basic nitrogenous compounds have a deactivation effect even under pressure; at 200 atm and 300 °C the original activity of MoS_2 in the hydrogenation of the benzene ring is almost completely lost in the presence of pyridine or of basic compounds from shale tar. Hydrogenation of the benzene ring is satisfactory again only when the temperature is raised to 420 °C, when intensive hydrogenolysis of nitrogenous compounds takes place [294]. Particularly polycyclic nitrogen-containing heterocycles are strong poisons [155a].

Nitrogenous substances are strongly adsorbed on the surface of sulphide catalysts [305, 308] and especially, they decrease the acidity of the surface. Therefore the rate of ionic-type reactions (splitting, isomerisation) which need a certain acidity level in the catalyst, decreases far more in the presence of nitrogenous substances compared to the rate of hydrogenation. These conditions may be utilised with advantage, for example when hydrogenating diesel fuel. In the presence of NH_3 and nitrogenous bases, the splitting activity of WS_2 is considerably reduced, and a high-quality fuel with high cetane number is obtained [416]. Similarly, the splitting activity of sulphides of Groups VI and VIII metals was decreased intentionally in the case of hydrogenation of middle oils [884]. The degree to which the splitting activity of the WS_2 catalyst is decreased under the influence of nitrogenous bases will be seen from Table 75 [2196, 2197]. The sensitivity of sulphide catalysts to the presence of nitrogenous substances is decreased considerably in the presence of hydrogen sulphide. This is one of the reasons, why it is essential for H_2S to be contained in the refining gas or, why sulphur-containing or sulphurised feeds must be used in order to maintain a good refining activity [416].

Hydrocracking catalysts are especially sensitive to nitrogenous substances because of their higher acidity [319] (Fig. 43 and Table 22). Their inhibiting effect is very strong and its character is that of reversible poisoning [416]. Therefore, in practice,

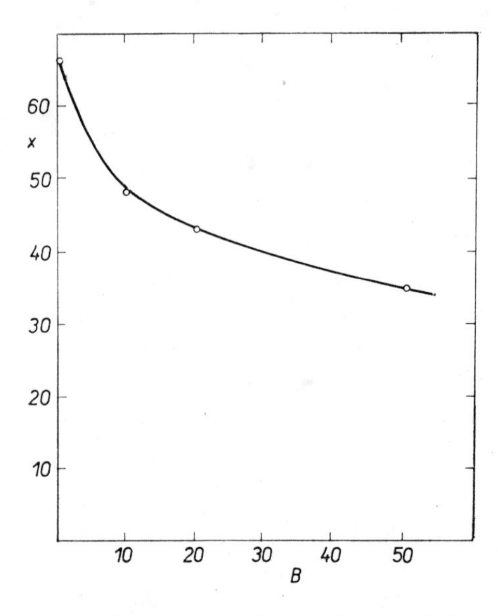

Fig. 43. The influence of nitrogenous bases on the splitting activity of the catalyst 6434 [416]. (Hydrocracking of gas oil with addition of the respective amount of nitrogenous bases). *B* — Content of basic compounds in the feed (mg NH_3/l). *x* — Conversion to gasoline (%).

e.g. the hydrocracking phase of the classic destructive hydrogenation of coal and tars (Benzinierung), circulating hydrogen must be freed of ammonia and prerefined feeds only must be processed [1101, 1172] or, high-activity hydrocracking catalysts of low sensitivity must be employed (p. 303).

Table 22.

The Influence of Aniline on the Hydrocracking Activity of Some Catalysts [416].

Catalyst	Relative splitting activity with aniline content (%)[a])				
	0.01	0.017	0.04	0.25	0.5
3510 (Mo–Zn–Mg)	—	—	—	—	70
5058 (WS_2)	—	96	60	44	—
6434 (WS_2/HF-clay):					
200 atm	—	44	—	—	—
600 atm	—	58	—	—	—
HF-clay 600 atm	65	34	25	—	—

[a]) The relative splitting activity is based on the activity of the respective catalyst in the absence of aniline (= 100); the aniline addition to the feed is expressed in % by wt.

Nitrogen compounds likewise influence the selectivity of catalysts in hydrodesulphurisation and refining reactions to a large extent. Kirsch et al. [1040] found in refining cracked gasoline and artificial mixtures of olefins and thiophene, that nitrogenous bases decrease particularly the hydrogenation activity of the catalysts employed (sulphurised Ni/Al_2O_3, sulphurised Co–Mo/Al_2O_3), while the desulphuration activity decreases to a lesser extent. Kolboe and Amberg [1066] likewise observed only a slight deactivating influence of NH_3 on MoS_2 activity in thiophene

Table 23.

Influence of Nitrogen-Containing Compounds on the Refining Activity of the NiS/Al_2O_3 Catalyst [1040].

(Feed: heptane-heptenes mixture, bromine number 31.4; the sulphur content in the form of thiophene was 0.46%; the catalyst was sulphurised Ni/Al_2O_3; temperature 371 °C, pressure 21 atm, LHSV 6 l/l/hour, molar ratio H_2: feed = 3.)

Content of nitrogenous compounds in the feed (%)	0	0.22
Desulphurisation (%)	63.9	45.0
Hydrogenation of olefins (%)	23.9	0.0
Selectivity[a])	62.6	100.0

[a]) $\text{Selectivity} = \frac{\%\ \text{desulphurisation} - \%\ \text{hydrogenation}}{\%\ \text{desulphurisation}} \cdot 100\ (\%)$.

hydrodesulphurisation. Thus the selectivity of the catalyst with respect to desulphurisation increases in the presence of nitrogen compounds, and consequently the desulphurised gasoline has a higher octane number due to the higher content of olefines. Typical results obtained with synthetic mixtures are listed in Table 23. Using a microreactor pulse technique, Lipsch and Schuit [1243d] observed a strong deactivating effect of pyridine on the catalyst Co–Mo/Al_2O_3 in thiophene hydrodesulphurisation as well as in hydrogenation of olefins (butenes).

The deactivating influence of oxygen and of oxygen-containing substances which was mentioned several times by Donath [416] was directly observed in the case of the MoS_2–ZnO (95 : 5) catalyst in low pressure ethylene hydrogenation. It was found, that the activity is decreased by oxygen as well as by steam. This deactivation is proportional to the concentration of the deactivating agent. It is reversible in the case of oxygen up to 300 °C, while irreversible poisoning sets in at 400 °C [1307].

The influence of these substances was studied in greater detail in the case of the WS_2 catalyst. The influence of atmospheric oxygen on WS_2 is manifested by a significant decrease of activity in hydrogenation of 2-octene at atmospheric pressure and 250 °C [149]. This deactivation is reversible, the catalyst recovered practically all the original activity on reduction in H_2 at 400 °C. In the hydrogenation of benzene and isomerisation of cyclohexane [1234] the catalyst selectivity was found to be influenced decisively by steam, which decreases the hydrogenation activity while increasing the isomerisation activity of the catalyst. Oxygen increases the isomerising effects of the catalyst [1235a], while it lowers the hydrogenation activity to a lesser degree than water. The hydrocracking activity of WS_2 increases in the presence of oxygen [1235a]. This effect may be explained either by the considerably stronger adsorption of water on the catalyst compared to benzene or, by a change of the semiconductive properties of the sulphide after adsorption of H_2O or O_2 on the surface [1234, 1235a]. The deactivating effect of water on the hydrodesulphurisation activity of Co–Mo catalysts is likewise worth mentioning [1243d].

This observation is of prime importance when evaluating the activity and selectivity of the catalyst, for which equal conditions must be maintained throughout and purified hydrogen must be used. The deactivating effect of H_2O and O_2 may likewise enable an explanation of the discrepancies found in the literature concerning the activity of different sulphide catalysts. The possibility of utilising the effect of H_2O to influence the selectivity of a catalyst in practical operations is given in those cases, where the expense of drying and oxygen removal would not be expensive (e.g. hydrogenation of benzene to cyclohexane etc.). In refining processes involving large amounts of raw materials this possibility cannot be used because, among other factors, water is always present as a reaction product of hydrogenolysis of oxygen-containing substances.

Münzing et al. [1429] reported the significant influence of oxygen on the quality of the refining catalysts 8376 and 8197. Oxygen is adsorbed very strongly on freshly sulphurised catalysts, the catalyst is heated (some pure sulphide catalysts may even be pyrophoric), partial surface oxidation sets in and the refining activity is lost. Moreover, additional sulphurisation cannot regenerate the original activity. Hence the necessity of carrying out high-pressure sulphurisation operations directly in the reactor and of avoiding any contact of the catalyst with the air when handling the sulphurised material. The authors observed similar results with the catalyst 8197 [1429].

Contrary to the above observations, Brewer and Cheavens [218] state, that the activity of Ni–Mo desulphurisation catalysts are improved when the catalyst is sulphurised after preliminary oxidation with an air–steam mixture at a maximum temperature of 400 °C. Young [2241] obtained a highly active palladium hydrocracking catalyst ($Pd–SiO_2–Al_2O_3$) by partial oxidation of the catalyst immediately after sulphurisation and reduction. Kolboe and Amberg [1066] found a distinct activating influence of oxygen in thiophene hydrodesulphurisation and, particularly, in hydrogenation of olefins in the presence of MoS_2 (at low total conversion). After a single pulse of oxygen, the rate of thiophene hydrogenolysis had doubled and the rate of hydrogenation of butenes rose 50 to 100 times. The authors assume, that the altered catalyst activity is related to changes in the semiconductor properties of MoS_2 after adsorption of oxygen, which is a strong electron acceptor.

The deactivating influence of some condensed aromatic hydrocarbons, especially with respect to the splitting activity of some sulphide catalysts, is important from the point of view of contamination of the active catalyst surface with carbon deposits. Polycyclic aromatic hydrocarbons have a tendency to undergo condensation reactions under hydrocracking conditions, high-molecular weight polymers being deposited

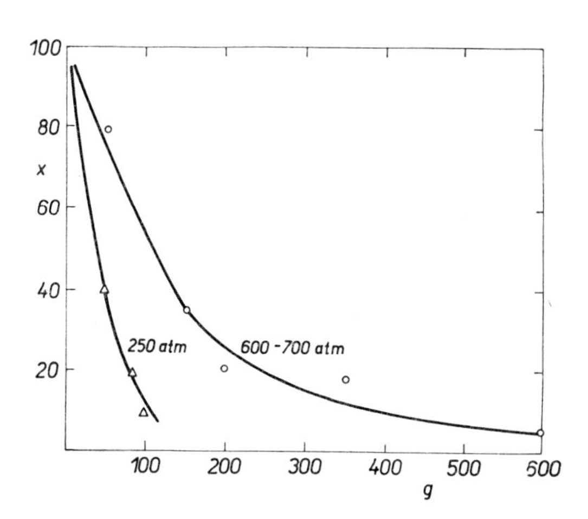

Fig. 44. Decaline hydrocracking on the WS_2 catalyst in the presence of coronene [416]. x — % decaline splitting. g — g coronene/1000 ml decaline. (The reaction was carried out in an autoclave at 425 °C and H_2 pressure of 600 to 700 atm; 300 ml (800 g) WS_2 were used, the batches were 1000 ml decaline with 0 to 600 g coronene.)

on the activated catalyst surface. The rate of hydrogenation decreases with increasing molecular weight of the aromatic substances, but the tendency towards condensation rises. Hydrocracking of these hydrocarbons takes place at high pressures only (around 700 to 800 atm), when polycyclic aromatic materials are hydrogenated to the respective perhydroaromatic substances. These do not undergo condensation reactions any more and their splitting is easier. Polycyclic aromatic compounds likewise decrease the rate of hydrocracking of other substances. Fig. 44 shows the inhibiting effect of coronene on decaline hydrocracking in the presence of WS_2 [416]; a considerably more rapid decrease of the splitting activity is to be seen from the graph at lower pressures. After these experiments, the catalysts contained a greater amount (about 5%) of carbon compounds which could not be extracted with benzene.

5.4.2.

The deactivating effect of carbonaceous deposits and ashes

Deposition of carbonaceous residues is not only a consequence of easily condensed polycyclic compounds present initially, as these deposits are also formed as a consequence of secondary reactions of products and intermediates. In a number of cases, more deposits are formed by these secondary processes than from the primary components of the feed [1064]. When natural materials of petroleum and tar origin are processed, contamination of the active catalyst surface by coke-like deposits is a regular phenomenon. The rate of formation of these deposits is especially higher when the content of asphaltic and resinous components in the feed is greater (i.e. also with a higher content of Conradson-type carbon), and it is influenced by the presence of sulphur and nitrogen compounds as well as ashes.

To understand the mechanism of coke deposition in the catalyst it is essential to know the distribution of carbonaceous deposits in catalyst particles. With non-supported WS_2 (from destructive hydrogenation of fuels) it was found, that the coke content is a maximum at depths of 0.05 to 0.1 mm from the tablet surface and, that the position of this maximum remains practically unaltered with increasing coke content. According to Zaidman et al. [2253] the formation of coke on a catalyst of given porosity is related to the hydrogen concentration in a certain location on the catalyst. The hydrogen concentration decreases quite rapidly with the distance from the tablet surface, so that diffusing molecules may undergo secondary condensation reactions at quite small distances from the surface. At this depth, therefore the carbon content increases to a maximum value and a kind of carbon barrier is formed, which hinders later diffusion of molecules deeper into the catalyst and, therefore, the carbon concentration again decreases towards the centre of the tablet.

Köhler [1064], however, found with catalysts supported on γ-Al_2O_3 (8376 and 8197, employed in the TTH process) practically the same carbon content in the surface layer as well as in the nucleus of the tablet (in a tablet of originally 10 × × 10 mm, the surface layer and a nucleus 5 mm in diameter were studied).

Deposition of carbonaceous substances causes slow deactivation of the catalyst. This activity decrease may be compensated for in most cases by an increase in the reaction temperature, the selectivity of the catalyst changing only slightly in the process. Ash contained in the feed is very important from the point of view of a long lifetime of sulphide catalysts. In most refining processes the catalyst bed acts as a filter on which the majority of the ash content is retained and causes a loss of activity. With regard to the large feed volumes which pass through the catalyst bed during an operational cycle, the amount of ash substances is often greater than the amount of active components originally contained in the catalyst after several regeneration cycles. For example, the catalyst 8376 on which brown coal tar was processed by the TTH process, contained nearly 7% of foreign oxides after the fifth regeneration cycle [703], and the desulphurisating and hydrocracking catalyst in the Gulf-HDS process contained up to 11.5% ash after use [1339]. Therefore the influence of alkali metals, alkaline earths, and iron, as well as of phosphorus and arsenic are specially important in processing tar from the point of view of the activity of refining and desulphurisation catalysts. The most important metallic components are vanadium, nickel and iron [703, 1121].

The harmful influence of ash substances on catalytic activity usually occurs later, especially in regeneration, where solid phase reactions take place between the carrier and active components of the original catalyst on the one hand and deposited ash substances on the other [701, 1064].

The gradual deactivation of the refining catalyst 8376 after individual regeneration cycles is illustrated in Table 24. The individual ash components differ in their deactivating effect. Günther [703] studied this problem in detail, and the following conclusions may be drawn from the results presented in Table 25:

Table 24.

The Deactivating Influence of Ash on the Refining Catalyst 8376 [703].
(A mixture of tar and light oil with 30 mg ash/kg was hydrogenated; 300 atm, 340 °C, LHSV 0.375 l/l/hour, catalyst particle size 3 to 8 mm.)

Catalyst	Ash content (%)[a]	Hydrogenate properties			
		Density	Middle oil quality		
			Aniline point (°C)	Content of phenols (g/l)	Content of bases (mg NH_3/l)
Fresh catalyst	1.1	0.866	53.5	8.0	147
Catalyst regenerated 1×	3.1	0.869	53.4	8.1	191
Catalyst regenerated 2×	4.6	0.869	52.1	13.7	332
Catalyst regenerated 3×	5.7	0.873	50.9	19.8	471
Catalyst regenerated 4×	6.4	0.875	48.0	26.5	615
Catalyst regenerated 5×	6.7	0.878	48.1	34.9	640

[a]) The ash content is expressed in terms of the total content of foreign oxides in the catalyst.

1. Some ash components have no deactivating influence, they only "dilute" the active catalyst components. These include graphite, alumina, aluminium sulphate, silicagel (the catalysts (4), (6), (8), (9)); aluminium sulphate actually acts as weak promoter.

2. Ash substances which are mechanically mixed with the catalyst mass have the same effect. When, however, ash substances are capable of reacting in the solid phase (at high temperatures, e.g. on regeneration) with the catalyst components, the decrease in activity is considerable (the catalysts (2) and (3)).

3. Admixtures of $CaSO_4$ and FeS cause no change of activity (the catalysts (10) and (16)). Admixtures of Na_2SO_4 and Fe_2O_3 cause a slight loss of activity (the catalysts (12) and (14)), but no distinct deactivation is involved. On tempering, however, the effect of these components is quite different,; with $CaSO_4$ (the catalyst (11)) there is no harmful reaction in the solid phase during tempering which would decrease the activity. With the other ash substances (the catalysts (13), (15) and (17)), however, the refining activity falls off, the influence of Fe_2O_3 being especially marked (the catalyst (15)).

Table 25.

Comparison of the Deactivating Effect of the Individual Ash Components on the Hydrorefining Activity of the Catalyst 8376 [703].

(Reaction conditions in measuring the catalyst activity are the same as in Table 24; the catalyst was formed into tablets 5 × 5 mm, heating of the catalyst, so-called tempering, was done for 120 hours at 480—500 °C.)

No.	Composition[a])	Analysis after tempering (%)	Properties of middle oil		
			Aniline point (°C)	Phenols (g/l)	Bases (mg NH_3/l)
(1)	Basic 8376, fresh, tempered	4.1 NiO 24.8 WO_3	54.0	5.8	130
(2)	85% (1) + 15% tar ash		52.4	8.4	127
(3)	(2) tempered		46.7	50.0	700
(4)	(1) + 4.5% graphite, tempered		53.3	7.3	199
(6)	(1) + 15% active Al_2O_3, tempered		53.5	6.3	143
(8)	(1) + 15% $Al_2(SO_4)_3$, tempered		54.9	5.0	117
(9)	(1) + 15% SiO_2, tempered		51.4	15.0	204
(10)	85% (1) + 15% $CaSO_4$		54.3	7.6	162
(11)	(10) tempered	5.20 CaO	53.4	7.6	162
(12)	85% (1) + 15% Na_2SO_4		51.9	10.6	244
(13)	(12) tempered	6.00 Na_2O	47.9	31.9	580
(14)	85% (1) + 15% Fe_2O_3 as hydrate		51.5	16.7	275
(15)	(14) tempered	14.9 Fe_2O_3	46.2	59.5	1300
(16)	89% (1) + 11% FeS	11.3 FeS	50.6	16.2	174
(17)	(16) tempered	10.4 Fe_2O_3	49.2	21.5	307
(18)	(1) + 5.3% CaO as $CaCO_3$, tempered		51.2	21.2	445
(19)	(1) + 5.8% Na_2O as Na_2SiO_3, tempered		39.2	73.7	2203
(22)	2.2% As_2O_3 as $(NH_4)_3AsO_3$, tempered		48.8	24.6	830
(24)	12.2% P_2O_5 as $(NH_4)_2HPO_4$, tempered		46.2	44.8	913

Table 25, continued

No.	Composition[a])	Analysis after tempering (%)	Properties of middle oil		
			Aniline point (°C)	Phenols (g/l)	Bases (mg NH_3/l)
(25)	(1) + 10% NiO, tempered	22.0 WO_3 14.0 NiO	46.6	30.7	724
(26)	(1) + 10% WO_3, tempered	34.4 WO_3 4.0 NiO	53.2	7.9	134
(27)	(1) + 10% WO_3 + 10% NiO, tempered	28.2 WO_3 12.0 NiO	47.2	23.1	426
(28)[b])	8376	4.5 NiO 24.8 WO_3	53.5	4.3	118
(29)[b])	8376 enriched with nickel	14.8 NiO 24.5 WO_3	51.7	11.9	341
(30)[b])	8376 enriched with nickel	19.0 NiO 24.2 WO_3	50.0	16.0	411

[a]) The basic catalyst and the added component were mixed either dry or wet, dried and pressed into tablets 5 × 5 mm.

[b]) These catalysts were prepared directly with the respective nickel content by co-impregnation of alumina with a mixture of $Ni(NO_3)_2$ and $(NH_4)_2WO_4$ ammonia solutions.

4. The influence of the anion on the degree of deactivation is likewise important. Comparison of the catalysts (11), (18) and (19) shows that carbonates are more dangerous than sulphates. The inert behaviour of sulphates was also confirmed with other sulphide catalysts employed for hydrodesulphurisation [1888e]. In the case of silicates, mechanical influences e.g. cementation may also participate.

5. Very efficient deactivating components are phosphorus (the catalyst (24)) and especially arsenic (the catalyst (23)).

6. Additions of the active components to the catalyst 8376 have different effects. A higher tungsten content (the catalyst (26)) had practically no effect. On the other hand, the effect of nickel (the catalysts (25), (29) and (30)) was surprising. Günther found, that a higher Ni content in the catalyst (achieved either by addditional saturation of the original catalyst 8376 or by increasing the Ni content during the preparation of the catalyst) had a distinct deactivating influence.

The influence of ash substances is especially important in coal hydrogenation, where even relatively slight amounts of a certain component which has a specific effect may influence the activity of the catalyst. For example, the catalytic or, promoting effect of FeS_2 is well known while, on the other hand, for example alkali metal compounds decrease the degree of conversion in coal hydrogenation in the presence of MoS_3 [1722].

Švajgl discussed the specific influence of arsenic on the activity of the refining catalyst 8376 in a number of studies [1991, 1992, 1993, 1996, 1997, 1998, 2000, 2001]. A high arsenic content (around 60 g/ton) is typical of North-Bohemian brown coal. On carbonisation, a part of the arsenic content passes into the tar and hence is present in catalytic processing. Most difficulties were encountered in vapour-phase hydrorefining of tars, where the catalyst 8376 ceased to be active after some 2 to 3 months, although otherwise it is very resistant [2001]. Fig. 45 shows the dependence of the rate of deactivation of the catalyst 8376 on the arsenic content in the feed (the catalyst activity was judged by the degradation of nitrogenous bases, which is the most sensitive reaction with respect to deactivation of refining catalysts). Švajgl showed, that deactivation of the catalyst 8376 is mainly caused by conversion of active NiS into catalytically far less active NiAs. The deactivation mechanism was proved by increasing the Ni content of the catalyst. It follows from Figs. 46 and 47, that the lifetime of the catalyst is directly proportional to the Ni content (in the range of 2 to 7% investigated) [1991, 1996, 1997, 2001]. The deactivation mechanism was confirmed by X-ray analysis (in the case of higher Ni and As concentrations) [2001] as well as by adding nickel in order to reactivate the

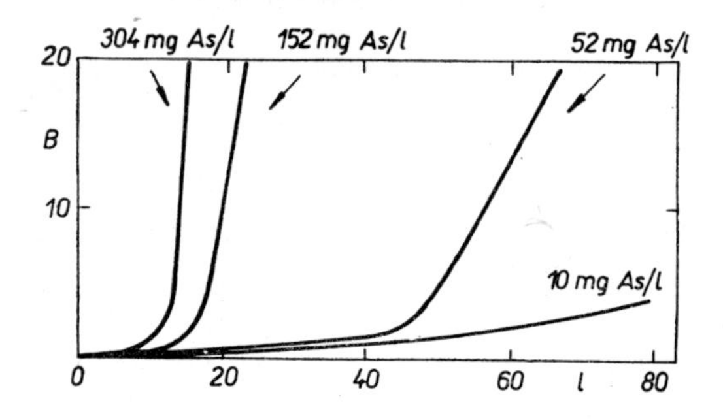

Fig. 45. Deactivation of the catalyst 8376 by arsenic compounds from tar (hydrogenation of a tar distillate with different As content; 420 °C, 320 atm, feed 2 kg/ /l/hour) [1997]. 1 — Volume of raw material processed. *B* — Content of bases in the hydrogenate (mg NH_3/l).

catalysts after calcination and partial removal of arsenic. On the basis of these results it was suggested that to increase the lifetime of the catalysts 8376 either preliminary catalytic distillation of the raw material in the presence of a nickel catalyst [2002] or, prerefining [2013] and partial degradation of arsenic compounds in the feed on deactivated 8376 catalyst beds [1999] should be carried out. A catalyst with higher nickel content (denoted 8376 ANTIAS) was successfully used industrially [2000, 2001]. Arsenic also deactivates the catalyst 5058 (pure WS_2) though it does not react chemically with this catalyst; it decreases both the hydrocracking and refining activity of the 5058 catalyst, the splitting activity decreasing more rapidly [2009, 2010].

Compared to data reported by Günther [703], Švajgl recorded no deactivating influence of Ni on the hydrorefining activity of the WS_2–NiS–Al_2O_3 catalyst. Contrary to this, as will be seen from Fig. 46, he found a slight increase in activity. Differing from the refining catalyst, Brennan and Den Herder [217, 1850] state that

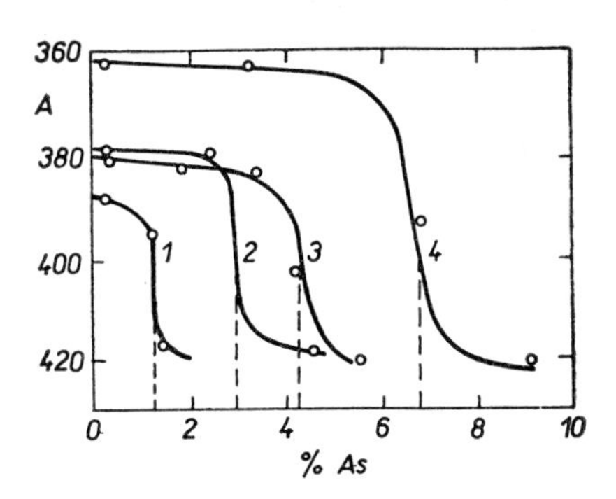

Fig. 46. Sensitivity of the NiS–WS_2–Al_2O_3 catalysts with different Ni content to deactivation by arsenic [1996]. *A* — Catalyst activity expressed by the temperature (°C) at which standard refining takes place, i.e. in the case of tar distillate hydrogenation this involves removal of phenols down to 250 mg/l, of organic compounds down to 5 mg NH_3/l and a rise of the aniline point to 50 °C. 1 — 1.76% Ni, 2 — 3.40% Ni, 3—5.14% Ni, 4 — 6.70% Ni.

the arsenic-containing nickel catalyst (on a SiO_2–Al_2O_3 carrier) has a good hydrocracking activity.

The increasing amounts of petroleum residues and high-boiling petroleum fractions which are processed by catalytic means made it essential to solve the problems of deposition of trace metals from petroleum on the catalysts and the influence of these deposits on the catalyst activity. In the case of catalytic cracking this problem has been studied for some time (especially with respect to Ni, V, Fe and alkali metals) and the deactivation mechanism is reasonably well understood [1449]. In the case of catalytic desulphurisation and hydrorefining, especially of petroleum residues, a very complicated process is involved, in which contamination of the catalyst with metals is accompanied by contamination with condensation and polymerisation pro-

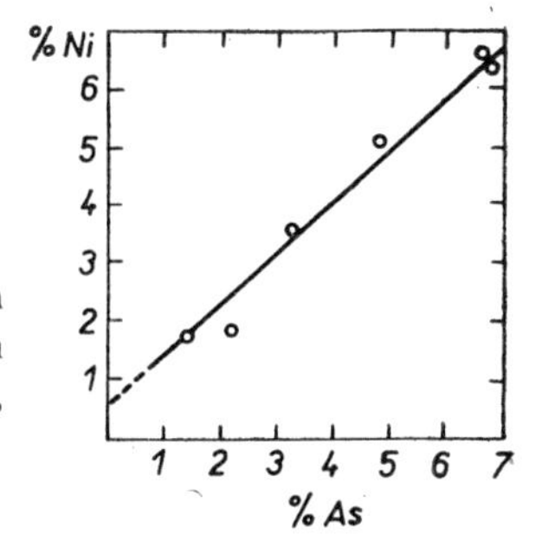

Fig. 47. Dependence of the Ni content on the As content in freshly deactivated WS_2–NiS–Al_2O_3 catalysts (i. e. catalysts in which a rapid activity decrease is observed, cf. Fig. 46) [1996, 2001].

ducts, so that it is difficult to separate the effects of the metals and the metal compounds. Therefore, the effect of metal deposits is mainly a secondary one, as they catalyse some polymerisation reactions and thus increase the amount of deposits [363].

There is not enough data in the literature, at present, dealing with the deactivation effect of trace metals. Some existing catalysts are highly resistant; e.g. the catalyst used in the Gulf-HDS process contained, as we have stated already, up to 11.5% V after a number of regeneration cycles, without any damage to the desulphurisating activity [1339]. Vanadium has a slight deactivating influence on conventional desulphurisation catalysts (cobalt–molybdenum and nickel–molybdenum) [1056, 2192]. The deactivating effect is increased at higher vanadium concentrations on the catalyst (over 5%) and at lower pressures [2186]. The deactivating effect is considerably higher when vanadium and alkali metals are present together [1480a].

B.

SPECIAL SECTION

In the second part of this monograph we shall deal with the most important applications of sulphide catalysts in technology and in a number of research projects which preceded the industrial application of sulphide catalysts.

The processes which take place on destructive hydrogenation, hydrocracking, cracking, hydrorefining as well as in dehydrogenation, reforming and isomerisation of petroleum and tar materials, are usually composed of a very complicated complex of simultaneous processes, each of which should be influenced by the presence of the sulphide catalyst in such a manner as to give a product of the required properties. Thus, the preparation of sulphide catalysts with the required selectivity necessitates a knowledge of the chemistry and kinetics of model reactions which take place in the presence of these catalysts and, therefore, much attention was given in the literature to studies of hydrogenation and of other reactions of individual organic compounds. The application of these basic findings enabled a deeper understanding of the processes which take place in the course of catalytic processing of natural raw materials and allowed rapid development of the most modern technological processes in the fuel and chemical industries.

A considerable part of this monograph will therefore be devoted to the influence of sulphide catalysts on the reactions of individual substances, namely in hydrogenation, dehydrogenation and isomerisation. The less important applications of sulphide catalysts in oxidation and polymerization reactions, and some special applications of these catalysts will be mentioned as examples of the wide range of application of this important group of catalysts.

6.

HYDROGENATION, SYNTHESIS AND SOME REACTIONS OF INDIVIDUAL ORGANIC COMPOUNDS IN THE PRESENCE OF SULPHIDE CATALYSTS

The following sections will deal with the application of sulphide catalysts in the hydrogenation of hydrocarbons, alcohols, ethers and phenols and furthermore of carbonyl compounds, acids and esters and oxygen containing heterocyclic compounds. Special attention will be devoted to the hydrogenation of sulphur and nitrogen compounds and to some important syntheses of sulphur, nitrogen and oxygen compounds which are catalysed by metal sulphides.

6.1.

Hydrogenation of unsaturated hydrocarbons

6.1.1.

Hydrogenation of monoolefins

Hydrogenation of olefins is very easily accomplished using sulphide catalysts. For example, hydrogenation of butylenes proceeds at a measurable rate in the presence of WS_2 at sub-atmospheric pressure at 0 °C, and far more quickly at 50 °C [416, 1626], Analysis of kinetic data of isobutylene hydrogenation at atmospheric pressure on the WS_2–NiS catalyst indicated [608], that a surface reaction is the rate-determining step. On the MoS_2–ZnO catalyst, low pressure hydrogenation of ethylene was studied in relation to the deactivating effect of oxygen (cf. section 5.4., p. 107) [1307]. Efficient catalysts of ethylene hydrogenation at atmospheric pressure are MoS_2 and WS_2 [149, 1167, 2182].

At lower reaction temperatures (60 – 110 °C) an efficient deactivating effect was observed in ethylene hydrogenation, caused by preferential adsorption of the unsaturated hydrocarbon [149]. The result of a hydrogenation experiment with the mixed MoS_2–WS_2/Al_2O_3–SiO_2 catalyst illustrated in Fig. 48 indicates, that deactivation takes place after saturation of the active catalyst surface. In this process, ethylene sorption does not take place preferentially on the centres which determine the hydrogenation rate – Fig. 48*A*. Deactivation is the more rapid, the higher is the partial pressure of ethylene in the initial mixture. The original catalytic activity is regenerated, however, on heating the catalyst to 400 °C even when the catalyst was deactivated by prolonged contact with air (see Fig. 48*B*). Bernardini and Brill [149] proved the deactivating effect of adsorbed ethylene by direct means too, i.e. by saturating the catalyst with ethylene before the hydrogenation experiment.

At normal pressure and very low space velocities, higher olefins can also be hydrogenated on sulphide catalysts. For example, 2-octene was 86% hydrogenated on WS_2 at 250 °C (LHVS = 0.12, molar ratio H_2 : olefin = 100) [149].

The use of sulphide catalysts in hydrogenating olefins at elevated pressures is quite conventional, as this reaction is one of the basic processes in hydrocracking as well as in hydrorefining. In practice, the sulphide catalyst which has the highest

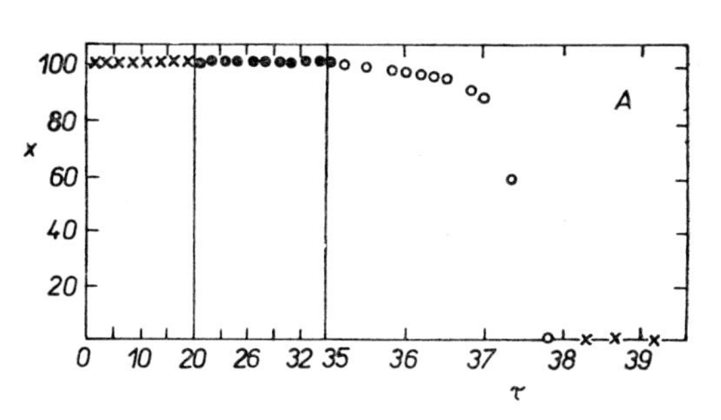

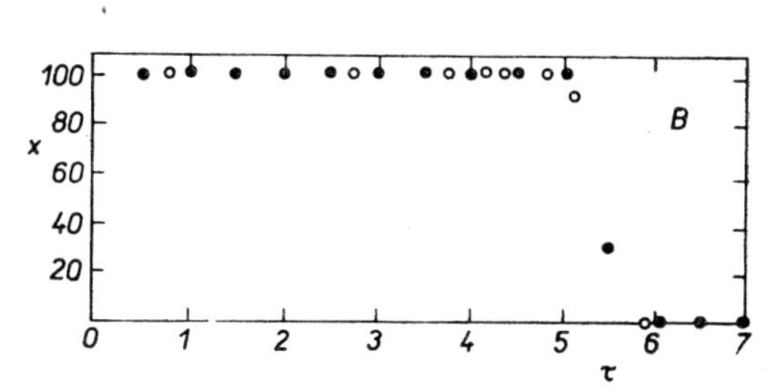

Fig. 48. Hydrogenation of ethylene on the MoS_2–WS_2/Al_2O_3–SiO_2 catalyst at atmospheric pressure [149]. τ — Reaction time (hours). x — Conversion to C_2H_6 (%). *A* — Hydrogenation at 110 °C and molar ratio of (H_2 : C_2H_4): × — 8.1, ● — 4.2, ○ — 3.0. *B* — Hydrogenation at 60 °C (eightfold molar excess of H_2). ● — fresh catalyst, ○ — the same catalyst, kept 10 days in air and regenerated with hydrogen at 400 °C.

hydrogenating efficiency, WS_2–NiS is used to hydrogenate double bonds e.g. in diisobutylene or triisobutylene [1101]. In these hydrogenation reactions, the catalyst 5058 [416, 1626] can also be used, but the hydrocracking activity of this catalyst must be diminished by decreasing the pressure to 50 atm (at the same time, lower space velocities are needed in order to achieve complete hydrogenation). Mixed sulphide catalysts on the basis of NiS were used to hydrogenate polyisobutylenes in other studies too [890, 898, 1945]. In the hydrogenation of 1-octene on the WS_2–NiS catalyst, the double bond was found to shift towards the centre of the molecule during the reaction, but skeletal isomerisation was not observed [1348]. Rhenium heptasulphide [226, 226a] is a very efficient catalyst of double bond reduction. Cyclohexene, 2-methyl-2-pentene and α-methylstyrene were quantitatively hydrogenated in a flow reactor at 150 to 200 °C and 20 atm pressure on sulphurised Pd/Al_2O_3 [134a].

Günther [697] studied some problems of the kinetics of olefin hydrogenation in the presence of the catalyst 8376. It was found in the hydrogenation of a mixture of olefins of b. p. 50 to 150 °C, that the reaction is pseudo-first-order with respect to olefins and that the reaction rate is practically independent of the hydrogen pressure in the 50 to 300 atm range, which agrees with observations by Heinemann et al. [746] relating to the hydrogenation of heptenes on a sulphurised Co–Mo catalyst.

Comparison of the hydrogenation of olefins and aromatic substances in the presence of sulphide catalysts shows a very marked difference between the reaction rates of reduction of the two types of hydrocarbons, which facilitates the selective hydrogenation of olefins in the presence of aromatic hydrocarbons. Günther [697] found a great difference in the hydrogenation rates of monoolefins and benzene in the presence of the catalyst 8376. It was similarly found with MoS_2, that the rate constant of benzene hydrogenation is some 2 orders of magnitude lower compared to hydrogenation of 4-methyl-cyclohexene [1258]. The structure of the olefin is important in this respect, since e.g. 1-methyl-1-cyclohexene is hydrogenated far more slowly [1253]. Besides these catalysts, hydrogenation of olefins mixed with aromatics is sufficiently selective when WS_2 and $Co–Mo/Al_2O_3$ are employed [420, 1924].

The hydrogenation kinetics of olefins in the presence of sulphur compounds is very important in technical respects (cf. pp. 89 and 328). Analysis of kinetic data in hydrogenation of heptenes in mixtures with isomeric heptanes and thiophene in the presence of a sulphurised $Co–Mo/Al_2O_3–SiO_2$ catalyst has shown, that activities of olefins as well as of thiophene occur in the rate equations of hydrogenation of heptenes [971]. This may indicate a mutual influence between the two reactions. At higher temperatures (about 375 °C) the retardation effect of thiophene on the hydrogenation of olefins is quite distinct; this is due to strong adsorption of the sulphur-containing heterocyclics on the catalyst surface. Therefore, the authors do not agree with Hammar [721], who believes that the two reactions take place on different active centres and do not influence each other. From some recent work by Amberg et al. [399, 1565], however, it follows that Hammar's original view is justified, since hydrogenation of olefins and hydrodesulphurisation take place on qualitatively different catalyst centres (details of this subject will be found on p. 222).

Kalechits et al. [971] also investigated the influence of internal diffusion on hydrogenation kinetics of olefins, finding that under the reaction conditions employed, internal diffusion has a considerable influence on thiophene hydrogenolysis in the presence of large-grained catalysts.

In some cases, hydrogenation of olefins is accompanied by skeletal isomerisation. Under the reaction conditions which are conventionally used for work in the presence of sulphide catalysts (temperatures up to 330 °C) structural isomerisation of n-olefins proceeds to a very limited extent or, not at all [1348, 2175]. At higher hydrogenation temperatures (around 400 °C) the polyfunctional character of sulphide catalysts starts to participate to a greater degree [2184] and, isomerisation takes place, besides hydrocracking [1459].

Cyclic olefins undergo skeletal isomerisation on hydrogenation far more easily. For example, isomerisation of a six-membered ring to a five-membered one was observed in the case of cyclohexene hydrogenation in the presence of MoS_2, WS_2, CoS_3 or NiS_2, especially at temperatures over 300 °C, in autoclaves as well as in the pulse microreactor [2187].

Very intensive isomerisation takes place on hydrogenation of some terpenes. For example, technical pinene (a mixture of 80% α- and 20% β-pinene) gave, besides pinane, *p*-menthane as the main product of hydrogenation in the presence of MoS_3 at 300 °C. Besides other isomerisation products, *p*-cymene was isolated from the reaction mixture as the product of aromatisation [1024]. It was proved by means of direct isomerisation of α-pinene in the presence of the WS_2 or MoS_2 catalysts [2185] that *p*-cymene is formed as one of a large number of reaction products, most probably after previous isomerisation of α-pinene to dipentene or terpinolene [697, 1024, 1027].

6.1.2.

Hydrogenation of polyolefins

Sulphide catalysts play a significant role in selective hydrogenation of polyolefins to monoolefins. This reaction is very important from the technological view-point since unstable polyolefins must be removed when stabilizing cracked gasolines, while monoolefins should remain (p. 328). The rate of hydrogenation of conjugated diolefins and monoolefins differ from each other sufficiently to allow good selectivity to be achieved when hydrogenating dienes. Hoffmann et al. [776, 776a] found, that the rate constants of hydrogenation of the two types of hydrocarbons (WS_2–NiS catalyst) differ by practically two orders of magnitude (Table 26). Maximum selectivity was achieved with a molar ratio of hydrogen to feed = 1 : 1. The process is characterised by the use of low operating pressures, from atmospheric [1340] to a maximum of 35 atm [777], the optimum being about 10 atm. Very high space velocities are also used (p. 328].

Mixtures of unsaturated C_4, C_5 and higher hydrocarbons [30, 1340], cyclopolyolefins [286c, 335, 593a] and furthermore, olefinic diene-containing materials used to make carboxylic acids by addition of $CO + H_2$ [2152] or for alkylation [30] are hydrogenated selectively in the presence of metal sulphides. Selective NiS-containing catalysts [439, 652, 1340, 1416, 1777] e.g. NiS on Al_2O_3 [30, 335, 1505 1900a, 2152] or the mixed catalyst WS_2–NiS [776, 777] are employed.

Low pressure butadiene hydrogenation in the presence of WS_2 (temperature 238 °C) proceeded at a high rate, the apparent reaction order being 0, indicating very strong adsorption of butadiene on the catalyst [1023]. In selective hydrogenation of butadiene in mixture with butenes, a sulphided Co–Mo catalyst was employed, the active component of which was the mixed sulphide $CoS.MoS_2$. This catalyst is especially selective in the presence of hydrogen sulphide which increases the acidity and decreases the hydrogenating activity of the catalyst, so that conversion of monoolefins is limited considerably. Also H_2S decreases the extent of side reactions by its competing adsorption, which is a consequence of the great adsorption of olefins [1041, 1042]. Piperylene was hydrogenated quantitatively to amylenes in the presence of the NiS/Al_2O_3 catalyst, the main products being *cis*- and *trans*-2-pentenes. The reaction was carried out at atmospheric pressure and 300 °C with a space velocity of hydrocarbon vapours of 400 to 800 l/l/hour and a molar ratio of piperylene: $H_2 = 1:4$. The pentane content of the product was not more than 1.5% [560].

Besides other sulphide catalysts, MoS_2 was used in selective hydrogenation of diolefins contained in monoolefins fractions of up to C_5 [527], in the olefinic fraction destined for oxosynthesis [234] or in hydrogenating 1,5-hexadiene and 1,3-cyclohexadiene [1490]. A recent development involves the use of rare metal sulphides (e.g. Ru, Pt or Pd) for selective hydrogenation of alkadienes [651a, 2086]. Rhenium

Table 26.

Comparison of the Rate of Hydrogenation of Olefins and Conjugated Diolefins in Cracked Gasoline [776]. (Gasoline from thermal cracking, WS_2–NiS catalyst, LHSV 3—18 l/l/hour, pressure 2.8—14 atm.)

Temperature (°C)	Rate constant[a]) (s^{-1})	
	for conjugated dienes, k_K	for total unsaturates, k_T
205	0.067	0.0006
230	0.17	0.0018
260	0.33	0.0045
285	0.63	0.0096

[a]) Kinetic data were evaluated using the following relationships:

$$\log \frac{1}{1 - x_T} = k_T \left(\frac{W}{F}\right) \frac{P_H}{P_0} \quad \text{(X)}$$

or

$$\log \frac{1}{1 - x_K} = k_K \left(\frac{W}{F}\right) y_H \quad \text{(XI)}$$

where: x_T, x_K are the degrees of conversion of all unsaturated hydrocarbons or diolefins resp., k_T and k_K are the rate constants of hydrogenation of all unsaturated hydrocarbons and diolefins, resp., (W/F) is the reciprocal of the space velocity, P_H and P_0 are the partial pressures of H_2 and of all unsaturated hydrocarbons, y_H is the molar fraction of H_2 at the reactor inlet.

heptasulphide was used for selective hydrogenation of the double bond in several natural substances [1653a, 1787d].

In a number of cases, terpenic hydrocarbons were hydrogenated in the presence of sulphide catalysts. Limonene was selectively hydrogenated in the liquid phase to the monoolefin (MoS_2/Al_2O_3–SiO_2, 150 °C, 105 atm) [1490]. Similarly, NiS/Al_2O_3 was used at 150 °C and 70 atm to convert limonene to a product containing only monoolefins, 90% of which was *p*-methylisopropylcyclohexene [649]. In the products of low-temperature (200 °C) hydrogenation of dipentene and d-limonene (catalyst MoS_3), *p*-3-menthene and α-terpinene were found besides *p*-cymene. At 300 °C the main product was *p*-menthane as well *p*-cymene [1027]. *p*-Menthane was likewise the hydrogenation product of pyrolytic dipentene (obtained by pyrolysis of waste rubber) in the presence of the catalysts 5058, 8376 and MoS_2 [1669, 1670, 1672]. *p*-Cymene is regularly found among the products on hydrogenation of dipentene or limonene or, in hydrogenation of terpenes which form dipentene on isomerisation (cf. p. 121) [697, 1024, 1027].

6.1.3.

Hydrogenation of acetylenes

Elimination of acetylenes from basic olefinic hydrocarbons (ethylene, propylene and the C_4-fraction) produced by pyrolysis of gaseous or liquid hydrocarbon mixtures, is a very important task. The content of acetylenes in pyrolysis products is relatively high (e.g. in the case of autothermal ethane dehydrogenation, the reaction gases contain 0.6% C_2H_2 [380]). This concentration must be diminished considerably for some important petrochemical syntheses, e.g. down to a level of 10 ppm.

Hydrogenation of acetylenes to olefins is very easy and, therefore, gentle conditions suffice to eliminate acetylenes from mixtures with olefins; temperatures generally in the 200 to 250 °C range, pressures of 1 to 20 atm and space velocities of the gaseous feed up to 3000 Nm^3/m^3/hour, depending on the acetylene content. Sulphide catalysts used most often for this purpose are those which contain nickel sulphide as the active component [12, 1149, 1726]. Sulphurised Ni (the catalyst is prepared by sulphurisation of the decomposition product of nickel formate on sepiolite [783]) or partially sulphurised NiO [114] are very efficient. Mixed sulphidic catalysts, containing Ni, Co and Cr sulphides [457, 528, 531] are also highly efficient. A nickel catalyst, sulphurised by a special procedure, was used for selective hydrogenation of acetylenes in dienes [783a].

Acetylenes are also removed on catalytic purification of gases. For example, acetylene was removed from coke oven gas together with oxides of nitrogen and carbon oxysulphide by hydrogenation at 150 to 200 °C on NiS, CuS or FeS catalysts [1076].

Molybdenum sulphide may be used for the same purpose [298]. Acetylenes were selectively hydrogenated in the ethylene fraction of coke-oven gas in the presence of Ni and Mo sulphides on Al_2O_3 [1418b].

6.2.

Hydrogenation of aromatic hydrocarbons

Hydrogenation of the aromatic ring, which is stabilised by mesomerism, is more difficult than that of substances which contain an isolated double bond in the carbon chain [479]. When metal sulphides are used as typical high-temperature catalysts [21] a sufficiently high reaction rate is achieved with hydrogenating aromatic substances at temperatures of at least 300 °C. At this temperature, sufficient conversion is achieved by a considerable increase of the partial pressure of hydrogen in the region of 50–100 atm.

The hydrogenation mechanism of aromatic substances in the presence of sulphide catalysts is little known at present [1920], and only its relationship to the geometrical factor has been partially clarified. According to Balandin's multiplet theory [103, 1570], adsorption of the benzene ring is assumed to take place on the surface of a suitable crystal lattice having the optimum distance between catalytic centres for the hydrogenation of the aromatic ring, and similarly for the dehydrogenation of the cyclohexane ring. Typical dehydrogenation and hydrogenation catalysts (cf. Chapter 9, p. 372) are metals which crystallise in face-centred cubic or hexagonal lattices [1047]. The necessary triangular arrangement is available on octahedral faces of the first type of lattice or, on basopinacoidal faces of the hexagonal crystals. This arrangement allows the sextet centre to be formed, which facilitates the course of the dehydrogenation and hydrogenation reaction.

After removal of the surface sulphide layer by hydrogenation, a triangular arrangement of active molybdenum atoms is available in the hexagonal lattice of sulphide catalysts e.g. MoS_2, although the distance between the centres is 3.15 Å [657] which is more than the critical distance of 2.77 Å above which the activity of metal catalysts in the dehydrogenation reaction of six-membered rings is considerably lower [1570]. A suitable combination of sulphides or, supporting the catalyst on a carrier probably also influences the arrangement of catalytic centres; this may explain, along with other factors, the improved activity of some sulphide catalysts (e.g. the catalyst 8376 in benzene hydrogenation or the WS_2–NiS catalyst in cyclohexane dehydrogenation, cf. p. 373).

In spite of that, hydrogenation of the aromatic ring and dehydrogenation of the cyclohexane ring takes place under suitable reaction conditions and in the presence of unsupported sulphide catalysts, i.e. pure WS_2 and MoS_2 sulphides. The relative difficulty and the lower selectivity of the two reactions which take place in the presence of suphide catalysts however are indications of the fact, that the true hydrogenation and dehydrogenation mechanism is more complicated than would appear from the simple scheme of the multiplet theory.

6.2.1.

Kinetics and chemistry of hydrogenation of aromatic hydrocarbons in the presence of sulphide catalysts

Lozovoi and Senyavin [1253, 1254, 1255, 1258] have carried out a detailed study of hydrogenation kinetics of aromatic hydrocarbons in the presence of sulphide catalysts. In autoclave experiments with MoS_2 and WS_2 catalysts at 150 to 220 atm and 380−475 °C, relative hydrogenation rates were measured with a number of alkyl-substituted aromatic and polycyclic substances. Under the experimental conditions employed, no hydrocracking was observed with the majority of the hydrocarbons tested, only under the most extreme experimental conditions, penta- and hexamethylbenzene were found to undergo weak hydrodealkylation. Table 27 shows the relative hydrogenation rates of individual aromatic hydrocarbons on WS_2 and MoS_2 catalysts (for comparison, results obtained with two low-temperature metal catalysts, Pt [1919] and Ni/Al_2O_3 [1249, 1250, 1251] are included, too).

Lozovoi and Senyavin derived the following relation for the relative rate of hydrogenation of polymethylbenzenes in the presence of WS_2 [1254]:

$$v_n = v_0(1.3 + n) \tag{XII}$$

where v_n is the relative rate for alkylbenzene and v_0 for benzene and n is the number of methyl substituents on the nucleus. We see from Table 27 that the relationship (XII) applies in the range of $n = 1$ to 5, since the rate of hydrogenation of hexamethylbenzene is far lower than would follow from the empirical relationship. This relationship does not apply to the MoS_2 catalyst, though its properties are similar to those of the WS_2 catalyst [1253]. In some earlier work by Altman and Nemtsov [21, 22] a higher relative hydrogenation rate is stated for alkylbenzenes in the presence of MoS_2.

The two sulphides, as well as the nickel catalyst, differ considerably in the hydrogenation rates of individual groups of hydrocarbons. Not only are monoolefins hydrogenated several times more quickly compared to aromatic substances (p. 120) [697], but moreover the rates of hydrogenation of condensed linearly annulated aromatic substances are considerably higher than with monoaromatic compounds (on WS_2, naphthalene is hydrogenated 23 times, anthracene 62 times more quickly than benzene). Angularly annulated chrysene, however, is hydrogenated more slowly, actually with greater difficulty than benzene. Another regular feature is a decrease

Table 27.

Relative Hydrogenation Rates of Aromatic Hydrocarbons in the Presence of MoS_2, WS_2, Pt (Hydrogenation rate of benzene in the presence of the respective catalyst under comparable

Hydrocarbon hydrogenated	Product
Benzene	Cyclohexane
Toluene	Methylcyclohexane
o-Xylene	1,2-Dimethylcyclohexane
m-Xylene	1,3-Dimethylcyclohexane
p-Xylene	1,4-Dimethylcyclohexane
Ethylbenzene	Ethylcyclohexane
1,2,3-Trimethylbenzene	1,2,3-Trimethylcyclohexane
1,2,4-Trimethylbenzene	1,2,4-Trimethylcyclohexane
1,3,5-Trimethylbenzene (mesitylene)	1,3,5-Trimethylcyclohexane
Isopropylbenzene	Isopropylcyclohexane
1,2,3,4-Tetramethylbenzene	1,2,3,4-Tetramethylcyclohexane
1,2,3,5-Tetramethylbenzene	1.2,3,5-Tetramethylcyclohexane
1,2,4,5-Tetramethylbenzene	1,2,4,5-Tetramethylcyclohexane
p-Cymene	*p*-Menthane
Pentamethylbenzene	Pentamethylcyclohexane
Hexamethylbenzene	Hexamethylcyclohexane
Naphthalene	Tetralin
Tetralin	Decalin
Anthracene	9,10-Dihydroanthracene
9,10-Dihydroanthracene	Tetrahydroanthracene
Tetrahydroanthracene	Octahydroanthracene
Octahydroanthracene	Perhydroanthracene
Chrysene	Tetrahydrochrysene
Tetrahydrochrysene	Octahydrochrysene
Octahydrochrysene	Dodecahydrochrysene
4-Methyl-1-cyclohexene	Methylcyclohexane

[a]) Hydrogenation at room temperature and atmospheric pressure, CH_3COOH as solvent [1919]

[b]) High-pressure hydrogenation, $t = 80-200$ °C [1249, 1250, 1251]; in the relative rate formula, n = number of substituents on the ring.

of the reaction rate with the degree of hydrogenation of the initial polycyclic hydrocarbon (cf. the decrease in the anthracene series).

By comparing kinetic hydrogenation data, Lozovoi and Senyavin [1258] found a higher hydrogenating activity in the case of the MoS_2 catalyst employed by them. For the apparent activation energy of benzene hydrogenation on MoS_2 in the temperature range of 410 to 480 °C they found a value of 19250 cal/mole compared to

Table 27, continued

and Ni/Al_2O_3.
conditions = 100.)

Relative hydrogenation rate in the presence of the catalyst					
$Pt^{a})$	Ni/Al_2O_3 [b])	MoS_2 [c])	WS_2 [d])	Calculated with the use of the formula	
				$v_n = v_0 \cdot 2^{-n}$ [b])	$v_n = v_0(1.3 + n)$ [d])
100	100	100	100	100	—
62	50	99	230	50	230
32	24	—	—	25	330
49	23	108	330	25	330
65	31	—	—	25	330
—	43	78	130	50	—
14	—	—	—	12.5	430
29	—	—	—	12.5	430
58	10	111	430	12.5	430
33	—	—	—	50	—
10	—	—	—	6.25	530
11	—	—	—	6.25	530
18	3.8	—	—	6.25	530
43	33	—	—	25	—
3.5	0.5	92	606	3.2	630
0.2	immeasurably small	—	150	1.6	730
—	314	1409	2300	—	—
—	24	287	250	—	—
—	326	—	6210	—	—
—	308	—	1380	—	—
—	147	—	460	—	—
—	4.4	—	299	—	—
—	—	—	80	—	—
—	—	—	75	—	—
—	—	—	95	—	—
—	13,400	18,000	—	—	—

[c]) High-pressure hydrogenation, $t = 420$ °C [1253, 1258].
[d]) High-pressure hydrogenation, $t = 400—420$ °C [1254, 1258]; in the relative rate formula, n denotes the number of methyl groups on the ring; the formula applies for $n = 1—5$.

24 500 cal/mole for the WS_2* catalyst. The two catalysts differ in their apparent reaction order, which is zero in the case of benzene and approximately 1 with respect to hydrogen with MoS_2 (Altman and Nemtsov [21, 22] arrived at the same conclusion

*) The apparent activation energies were calculated from the temperature coefficients of the respective reaction.

in toluene hydrogenation). With WS_2, the apparent reaction order is close to 0.5 with respect to benzene and close to 1.5 with respect to hydrogen. Therefore, the WS_2 catalyst is considerably more sensitive to temperature as well as to variations of the hydrocarbon partial pressure.

Lozovoi and Senyavin [1258] tried to explain the differences in the catalytic effect of MoS_2 and WS_2 as well as of low-temperature metal catalysts under the reaction conditions employed, starting from the different activity of the catalyst employed and the structure of the original substance. Obviously, these considerations must be amplified by a detailed study of the reaction mechanism on the individual catalysts.

The hydrogenation kinetics of mononalkylbenzenes with alkyl groups of different size was studied, too. Kupryanov and Dorogochinskii [1136, 1137] hydrogenated a number of aromatic compounds (benzene, toluene, cumene and technical alkyl benzenes: amyl-, heptyl-, nonyl-, decyl- and dodecylbenzene) in the presence of the catalyst 8376 in a flow reactor, with space velocities of 0.5 and 1 l/l/hour at 240 °C and 250 atm. They found that the rate of hydrogenation decreases with the increasing size of the alkyl group and, that the numerical value of the rate constant satisfies the empirical relationship

$$\log k_n = 0.1(k_0 - n) \qquad \text{(XIII)}$$

where k_n and k_0 are the apparent rate constants of the alkylaromatic compound and of benzene, and n is the number of carbon atoms in the side chain (the values of n are whole numbers with the exception of O).

On technical high-temperature catalysts, hydrogenation of aromatic compounds is relatively difficult and the kinetics of this reaction is influenced to a substantial degree by macrokinetic factors and by internal diffusion effects. Orochko and Shavolina [1554, 1865] found on hydrogenating benzene and toluene on the catalyst 8376, that with medium space velocities and medium conversion (below 40%), the hydrogenation kinetics can be described using the rate equation for a pseudomonomolecular reaction, which however becomes invalid at high feed rates. For practical engineering calculations, these authors proposed a modified kinetic equation in which the linear velocity of the reactants is taken into account. The authors named have likewise observed a strong retardation of the reaction rate of benzene and toluene hydrogenation on the catalyst 8376 due to the influence of internal diffusion effects. With catalyst grains 10 × 10 mm in size, the degree of utilisation of the inner catalyst surface was found to be about 60% in benzene hydrogenation, this value rising quickly with decreasing particle size. The retarding influence of internal diffusion was marked, especially when the molar hydrogen excess was increased, since the lower partial pressure of the hydrocarbon caused the rate of its diffusion into the inner space of the catalyst pores to be slowed down and, therefore, the degree of utilisation of the internal catalyst surface was lower, too. The retarding influence of internal diffusion on hydrogenation of aromatic substances in medium tar oil was already mentioned earlier by Reitz [1730] who also discussed the serious technological consequences of this phenomenon (cf. p. 312).

Although hydrogenation of aromatic compounds in the presence of sulphide catalysts is discussed in a large number of studies, the mechanism of this reaction has not yet been substantiated by a detailed kinetic analysis. In the presence of the catalyst 8376, which is one of the most efficient sulphide catalysts for hydrogenation, the reaction starts to proceed at hydrogen pressures of 150 to 180 atm at about 250 °C [1602]. Under these conditions, the corresponding naphthenic hydrocarbon is formed, while side reactions i.e. isomerisation to the five-membered ring and splitting of the ring, commence at temperatures above 300 °C [1857]. On hydrogenation in a flow reactor in the gas phase, benzene may be hydrogenated to cyclohexane on the catalyst 8376 at hydrogen pressures of 250 to 300 atm and up to 350 °C, the side reactions participating to a very limited extent only [1235, 1826, 1902, 2155]. According to Pier [1626] the catalyst 5058 can also be used to hydrogenate benzene to cyclohexane at very low feed rates (0.1 kg/l hour).

When the reaction is carried out in an autoclave, the degree of catalyst utilisation is considerably lower and hydrogenation is far slower compared to the flow system [1159, 1163, 1253, 2175]. To speed up the reaction, temperatures of 360 to 380 °C must be adopted: at these temperatures, however, the polyfunctional character of sulphide catalysts is more marked and the yield of the corresponding naphthene is

Table 28.

Benzene Hydrogenation (autoclave experiments, temperature 420 °C, pressure 300 atm, reaction time 3 hours, 30% of catalyst) [955].

Data relating to hydrogenation product		Catalyst			
		WS_2	MoS_3	WS_2–NiS–Al_2O_3	WS_2/aluminosilicate
Total benzene conversion (%)		69.4	45.9	71.7	35.9
Liquid hydrogenate yield (% by wt. of feed)		95	99	99	85.7
Naphthenic-paraffinic fraction (% by wt. in the hydrogenate)		67.5	45.4	71.4	25.2
Composition of the naphthenic-paraffinic fraction (% by wt.)	Pentanes	1.0	—	0.5	9.9
	Hexanes	5.6	6.0	1.1	8.5
	Cyclopentane	—	0.8	0.2	1.6
	Methylcyclopentane	55.6	29.1	13.7	40.4
	Cyclohexane	37.8	64.1	84.5	3.7[a])

[a]) The rest are paraffins, naphthenes and aromatic substances C_7 and higher.

decreased by isomerisation and cleavage products. For comparison, Table 28 shows the results of benzene hydrogenation on four basic sulphide catalysts under conditions when hydrocraking reactions start to play a role [955]. In spite of the incomplete conversion (36 to 72%) sulphide catalysts exhibit a low degree of selectivity in respect of hydrogenation to the corresponding naphthenes. The results obtained also indicate the order of activities of the individual sulphide catalysts in respect of hydrogenation, isomerisation and splitting reactions [952, 955, 965], which agrees with results reported by Lozovoi et al. [1256, 2248] and by Chinese authors [295].

The investigation of processes taking place in the presence of sulphide catalysts at temperatures higher than 360 to 380 °C is very important for understanding the mechanism of hydrocracking of aromatic hydrocarbons. As early as in the work of Pier et al. [416] and later in the work of Puchkov and Nikolaeva [1707, 1708, 1710] and of the Chinese authors [293] (catalyst MoS_2), a considerable isomerisation to compounds with a five-membered ring was observed at temperatures above 400 °C. Kalechits et al. devoted very much time to investigating the details of this reaction [968, 969, 972, 973, 976, 1587]. It was shown by a number of experiments using labelled atoms, that the simple consecutive scheme

$$\text{benzene} \longrightarrow \text{cyclohexane} \longrightarrow \text{methylcyclopentane} \longrightarrow \text{hexanes} \qquad \text{(XIV)}$$

is highly improbable. Kalechits showed on the basis of a number of hydrogenation experiments (temperature 420 °C, pressures around 300 atm, catalyst WS_2) with benzene–cyclohexane mixtures in which one or the other component was labelled with radiocarbon ^{14}C, that all three basic reactions, i.e. hydrogenation to cyclohexane, hydroisomerisation to methylcyclopentane and hydrocracking start from a common intermediate product, cyclohexene or the carbonium ion formed from it.

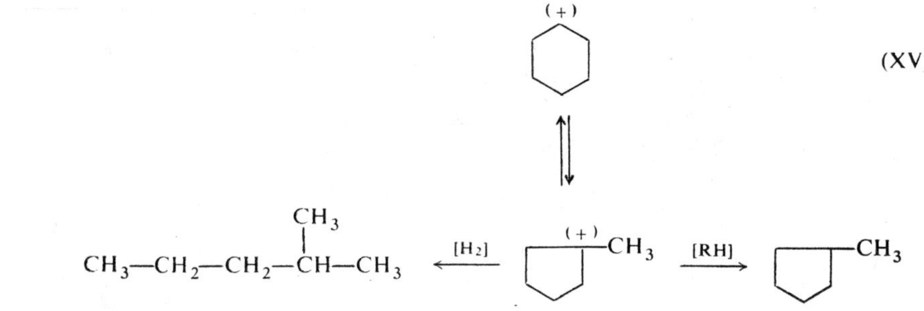

[H2] [H2] [H2]

[H+]

(+)

(XV)

$$CH_3{-}CH_2{-}CH_2{-}CH(CH_3){-}CH_3 \xleftarrow{[H_2]} \text{(+)}{-}CH_3 \xrightarrow{[RH]} {-}CH_3$$

It was found by comparing the experimental data that the most rapid reaction at high hydrogen pressures was benzene hydrogenation, followed by hydroisomerisation to methylcyclopentane, and isomerisation of cyclohexane to methylcyclopentane (in an equilibrium mixture at 420 °C, thermodynamic data show [1770] that 90% of the original cyclohexane should be converted to methylcyclopentane). The slowest reaction was hydrocracking to paraffinic hydrocarbons. Under the experimental conditions employed, the cyclohexene ⇄ cyclohexane equilibrium is shifted in the direction of cyclohexane and, therefore, the rate of isomerisation to methylcyclopentane is low. The hydrogen cation needed to form the carbonium ion is obtained from the acid catalyst surface, protonisation being the more intensive according to Kalechits (and, therefore hydroisomerisation and splitting are the more rapid), the greater the amount of non-stoichiometric sulphur contained in the catalyst (cf. p. 82).

The above conclusions are especially significant in studies of hydrocracking reactions of aromatic and naphthenic hydrocarbons (p. 138) and in determinations of the optimum conditions in technological hydrocracking of hydrocarbon fractions (cf. p. 299).

6.2.2.

Hydrogenation of non-condensed aromatic hydrocarbons

Hydrogenation of benzene to cyclohexane is nowadays a very important technological operation, due to the increasing demand for polyamide fibres. At present, units with capacities of 100,000 tons/year, producing very pure cyclohexane at medium pressures and temperatures of about 200 to 250 °C are quite common [71, 2220]. In these processes, highly efficient metallic catalysts are used, e.g. finely dispersed Ni on a carrier, or Pt on Al_2O_3. These are however very sensitive to sulphur so that materials to be processed must be highly purified. An advantage however is the fact that pure cyclohexane is obtained.

Investigations enabling the use of resistant sulphide catalysts in the hydrogenation of aromatic hydrocarbons and processing raw benzene directly to cyclohexane date back to quite an early period. Catalysts employed were WS_2, a mixed WS_2–NiS (catalyst 3076) [593, 901, 988a, 1101, 2101] and especially the catalyst 8376 [1235, 1826, 1902, 2155]. In the high-pressure process (250 atm) temperatures were 340 to 355 °C and a high molar excess of hydrogen was used. To improve heat dissipation, the catalyst was mixed with ground porcelain [1235]. At medium pressures (60 atm) space velocity was 0.35 kg/l hour at 370 °C, 99.2% conversion to cyclohexane being obtained [2155]. On the catalyst 8376, ethylbenzene may be hydrogenated with success to ethylcyclohexane with practically no splitting or isomerisation [2250]. The same applies to aromatic compounds with long side chains, e.g. isodecylbenzene and isodecyltoluene [1136]. MoS_2 supported on active carbon is very selective, too [956, 965].

Among other sulphide catalysts, MoS_2, CoS, NiS, FeS and SnS were tested for the hydrogenation of simple aromatic substances [904, 1256]; these however are far inferior to NiS–WS_2 catalysts in terms of selectivity as well as activity. Steffgren [1959b] assumed that platinum sulphide is formed and takes part in catalytic hydrogenation of sulphur- and nitrogen-containing aromatic hydrocarbons on the Pt/Al_2O_3 catalyst.

6.2.3.

Hydrogenation of condensed aromatic hydrocarbons

Hydrogenation of naphthalene to tetralin in the presence of sulphide catalysts proceeds far more easily than the hydrogenation of the simple benzene ring [1258, 1891] (cf. also Table 27). The reason for this higher reactivity is the fact that the resonance energy of the second ring is less than that of the basic benzene ring [813], so that naphthalene is less aromatic in character than benzene [172]. Taking the extreme view, the second six-membered ring may be regarded as a butadiene configuration [1047].

Hydrogenation of napthalene to decalin is a consecutive reaction [1562], the second stage being considerably slower than hydrogenation to tetralin (5 to 8 times slower with sulphide catalysts, on metal catalysts the difference is even more marked – cf. Table 27). Technical naphthalene containing sulphur compounds, especially thionaphthene, may be hydrogenated on the MoS_2 or MoS_3 catalyst. The main product is tetralin mixed with decalin and a certain amount of non-reacted naphthalene, depending on the reaction conditions used in hydrogenation [287, 719, 1452, 1485, 1679, 1681]. MoS_2 supported on active carbon [876, 880] or on aluminosilicate [1977] was used, as well as MoS_2 promoted with chromium [1206] or cobalt [1683]. In autoclave experiments (380 °C, initial hydrogen pressure 100 atm) with MoS_3 prepared by a special procedure, naphthalene was hydrogenated practically quantitatively to tetralin [1682]. In a number of cases, WS_2 [488, 870, 982, 2105, 2117] and the catalyst 8376 [3, 4, 871b, 1100b, 1867] were used in hydrogenation. Ovsyanikov and Orechkin [1562] suggested a continuous process of naphthalene hydrogenation on the catalyst 8376 with simultaneous production of tetralin and decalin (a 65% decalin fraction is obtained in the process which gives 99.4% pure decalin in the second stage at 310 to 320 °C, 300 atm and LHSV = 2).

Baker et al. [102] found, that in pressure hydrogenation of naphthalene in the presence of WS_3 at 385–410 °C, decahydroazulene is formed as well as decalin:

$$C_{10}H_8 + 5H_2 \longrightarrow C_{10}H_{18} \qquad \text{(XVI)}$$

The authors named also proved, that the azulene structure is formed by isomerisation of one of the hydrogenation intermediates and not by isomerisation of decalin itself.

Differing from other sulphide catalysts, rhenium heptasulphide is characterised by an unusually low activity in naphthalene hydrogenation. At 300 °C and 210 atm, no hydrogenation was observed in the presence of this catalyst, although benzene was partly hydrogenated under these conditions [226].

Anthracene is hydrogenated very easily to 9,10-dihydroanthracene. The reaction takes place with pressurised hydrogen at about 300 °C in the presence of MoS_2 [1163, 1687] or on NiS [355].

This reaction is about five times more rapid in the presence of WS_2 compared to hydrogenation to tetrahydroanthracene [1258] and the rate of further hydrogenation decreases gradually with the increasing content of hydrogen in the molecule (cf. Table 26).

On hydrogenation of anthracene in the presence of MoS_2 at 350 °C, symmetrical and asymmetrical octahydroanthracene was isolated, the yield of the asymmetrical isomer being greater with a lower reaction temperature [1688]. Among the products of anthracene perhydrogenation, Prokopets and Boguslavskaya [1691] also found perhydrophenanthrene.

In the presence of the Ni–W sulphide catalyst, up to 80% yields of 9,10-dihydroanthracene were obtained at operating pressures of 105 atm and 250 °C. At higher temperature, 1,2,3,4-tetrahydroanthracene is formed as another intermediate product, the major portion of which (70% in the reaction mixture) is formed at 435 °C. With further temperature increase, the reaction proceeds with hydrocracking to naphthalene derivatives and hydrogenation to 1,2,3,4,5,6,7,8-octahydroanthracene. Hydrogenation of anthracene on the catalyst mentioned proceeds as a pseudomonomolecular reaction, the rate-determining step being apparently orientation and adsorption of the reacting molecule on the catalyst surface [2225a].

Under the same conditions, phenanthrene is hydrogenated in the presence of MoS_2 to the tetrahydroderivative with greater difficulty compared to anthracene [1687]. On the WS_2–NiS/Al_2O_3 catalyst (space velocity 1 l/l/hour, 70 atm) practically complete conversion of phenanthrene to a mixture of tetra- and octahydrophenanthrenes was achieved [778]. This catalyst was also used to hydrogenate phenanthrene and other condensed aromatic compounds contained in creosote oil [288].

Higher polycondensed aromatic substances, e.g. chrysene, are very difficult to hydrogenate, the differences in the rates of hydrogenation of the individual hydroderivatives being practically negligible (cf. Table 27) [1258]. The hydrogenation product of coronene obtained at 300 to 600 atm and 270 °C on the WS_2 catalyst was not clearly identified [194].

6.3.

Hydrocracking, hydrodealkylation and cracking of hydrocarbons

6.3.1.

Hydrocracking of hydrocarbons

Hydrocracking is a complex of splitting, hydrogenation and isomerisation reactions, the kinetics of which may be influenced by altering the reaction conditions and character of the catalyst. The significance of isomerisation (of raw materials and intermediates) as the reaction step which influences the overall rate of hydrocracking, was recognized even in the first studies dealing with the chemistry of hydrogenative splitting of hydrocarbons [416, 1626]. Depending on the nature of the catalyst employed and the reaction conditions, either the ionic mechanism (when acid catalysts are used) or, the radical mechanism (on neutral or weakly acid catalysts and at higher temperatures) of splitting prevail in hydrocracking.

A number of authors have presented a modern interpretation of the ionic mechanism of hydrocracking [80, 349, 2194, 2196, 2197]. Welker [2194] applied Coonradt's and Carwood's explanation [349] of the hydrocracking mechanism to the hydrocracking of n-paraffins, showing that an efficient hydrocracking catalyst must have the necessary acidity (Brønsted and Lewis centres) and the necessary hydrogenation-dehydrogenation activity. On acid centres, carbonium ions are formed and isomerised (especially to tertiary ions which are the most thermodynamically stable; on hydrogenation-dehydrogenation centres, dehydrogenation of the C—C bond and formation of an initial olefinic intermediate product takes place, as well as hydrogenation of all olefinic products formed by the reaction. Here, the hydrogenation activity of the catalyst decides to a great extent the degree of isomerisation which takes place during the hydrocracking process [349, 2148].

The hydrocracking process is a complex of consecutive and parallel processes, so that (disregarding diffusion and sorption effects) the resulting rate of the process will be determined by the kinetics of the slowest partial reaction involved. The following scheme is obtained for a typical n-paraffin hydrocracking reaction (Fig. 49) [349, 2194]:

The primary process is dehydrogenation of the adsorbed paraffin with formation of the respective olefin; the secondary process is the addition of a proton from the acid centre of the catalyst and formation of a carbonium ion. The ion formed is either split or isomerised to the thermodynamically most stable tertiary ion. The isomerised ion is again either split or, it transfers the proton over to the catalyst. An isoolefin is formed which, on hydrogenation, forms the isomer of the initial substance.

The composition of the reaction product will therefore depend on the mutual ratio of reaction rates of the individual processes and, therefore, on the reaction conditions, activity and selectivity of the catalyst employed.

The olefin concentration on the catalyst depends on the ratio of rates of reaction r_1 and r_2. On the hydrocracking catalyst (which differs from the cracking catalyst by the presence of the hydrogenating component) $r_1 \gg r_2$ and, therefore, the hydro-

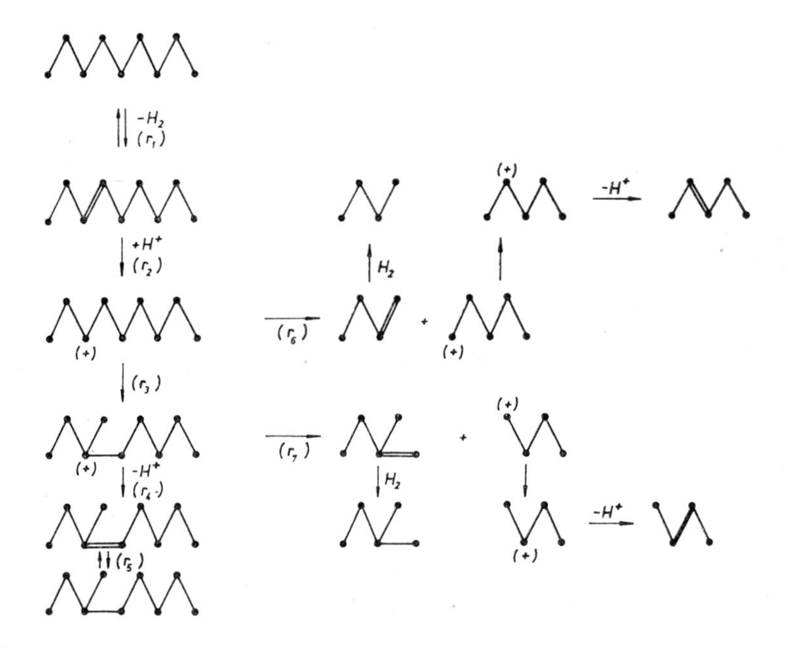

Fig. 49. Scheme of n-nonane hydrocracking on an acid-hydrocracking catalyst [349, 2194].

genation-dehydrogenation equilibrium is established. The magnitude of the hydrogenation activity will now decide the concentration of the isomerised raw material and the composition of the reaction product. When the hydrogenation rate r_5 is greater than the splitting rates r_6 and r_7, the product will contain a large amount of the isomerised feed. On the other hand, when the hydrogenating activity of the catalyst is low the splitting rates r_6 and r_7 will prevail and the amount of material isomerised will be a minimum (catalysts with WS_2 or NiS, i.e. components of a relatively low hydrogenation activity) [80, 532].

When a sufficiently high temperature is employed or, the hydrogenolytic activity of the catalyst is sufficient [2194] intensive splitting of ions takes place in the β-position of the carbonium ion and lower olefins are formed which, on hydrogenation, convert to lower paraffins (or isoparaffins). Alternatively, they are re-protonised and undergo secondary splitting in this form or after preliminary isomerisation.

Some alkylation reactions may also take part as side reactions [2194]. The hydrocracking rate of n-paraffins differs greatly with lower hydrocarbons. Fig. 50 shows the relationship between the splitting "half-time" of C_2 to C_{16} hydrocarbons and temperature, indicating the considerably higher reactivity of the branched hydro-

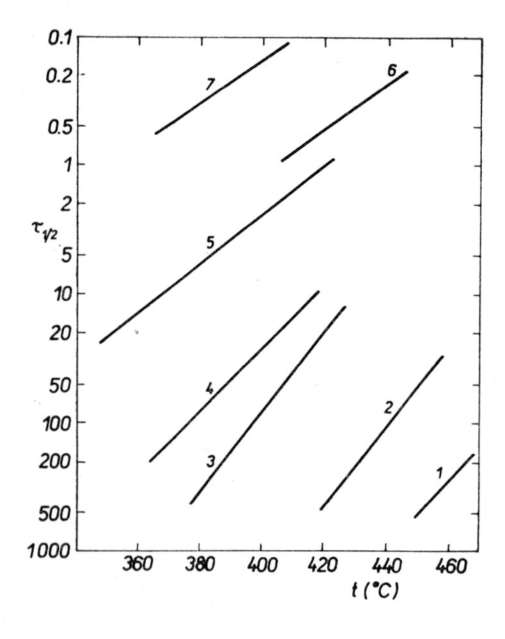

Fig. 50. Relationship between the "half-time" ($\tau_{1/2}$, hours) of hydrocracking of some hydrocarbons and temperature [416] (autoclave experiments, WS_2 catalyst, H_2 pressure 200 atm).

1 — ethane, 2 — propane, 3 — n-+iso-butane, 4 — isobutane, 5 — n-heptane, 6 — cetane, 7 — isooctane.

carbons; e.g. isooctane splitting is twenty times more rapid than n-heptane hydrocracking [416], isobutane being mainly formed [1259, 1261, 2247, 2249]. The hydrocracking rate of paraffinic hydrocarbons is lower in the presence of sulphide catalysts compared to metal catalysts of the Group VIII [1023b].

With a high hydrogenating activity of the catalyst (hydrocracking catalyst with a metallic component, e.g. Pt or Ni) secondary splitting takes place to a limited extent

Table 29.

Relationship Between the Ratio of (iso/n) Paraffins in the Splitting Products of n-Decane Hydrocracking with Catalysts Supported on Aluminosilicates (pressure 83 atm) [2148].

Hydrocracked component	Molar ratio of (iso/n) hydrocarbons							
	232 °C			288 °C				
	Equilibrium composition	Ni	NiS	Equilibrium composition	Pt	PtS	Co	CoS
Pentanes	3.4	1.6	10.5	2.9	0.09	2.0	0.4	8.2
Hexanes	3.1	3.0	10.8	2.7	0.07	2.5	0.5	8.5

only. This is a great advantage in technological respects due to the lower formation of gaseous hydrocarbons and lower hydrogen consumption [2148].

On the other hand, a lower hydrogenating activity of the catalyst (e.g. NiS/aluminosilicate, sulphurised platinum catalyst, sulphurised Co–Mo/Al_2O_3–SiO_2 etc.) is advantageous in terms of maintaining a high, above-equilibrium ratio of branched hydrocarbons to n-hydrocarbons in the splitting products [1729a, 2148].

The use of catalysts with a highly efficient hydrogenating component (Pt or Ni hydrocracking catalysts) leads to a low, even sub-equilibrium ratio of (iso : n) hydrocarbons in the splitting products [1839, 1840, 1841). The difference in the composition of n-decane hydrocracking products using catalysts with sulphide components is seen in Table 29. However, there is also a considerable difference in the molecular weight of the cleavage products; it is typical that demethanisation is easier on metallic Ni [1532]. On the other hand, the sulphide component has a higher hydroisomerisation effect [568, 2148].

Practically no substances with quarternary carbon atoms are formed in hydrocracking. The reason is, that a substance of this type would have to be formed from a tertiary carbonium ion:

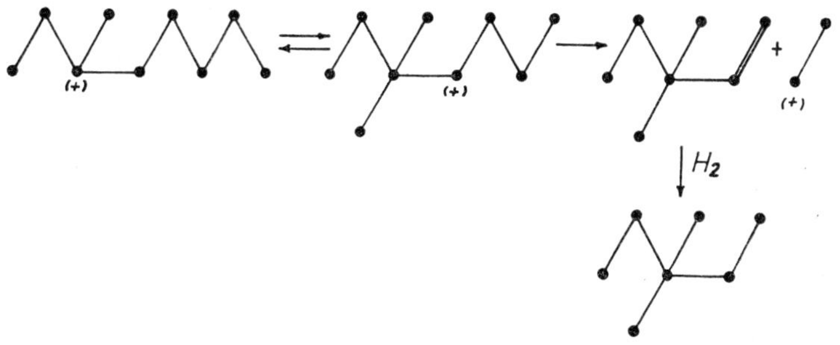

(+) (+) + (+) H_2

(XVII)

However, the isomerisation equilibrium is shifted more in the direction of the thermodynamically most stable tertiary ion, so that the mechanism indicated is very improbable.

The influence of temperature on the course of hydrocracking by the ion mechanism is very important. Practically all the simultaneous reactions are substantially speeded up with increasing temperature. It is also important to note, that the increase in the number of secondary carbonium ions which are split in the β-position is relatively largest (the reaction rate r_6 rises, cf. Fig. 49), so that the ratio of (iso : n) hydrocarbons in the splitting products decreases with increasing temperature. Günther [699] found that on hydrocracking spindle oil (with a mean number of 17 carbon atoms in the molecule) in the presence of the catalyst 5058, the splitting reaction as a whole is apparently a first-order reaction at 320 to 350 °C, taking place with an apparent activation energy of 64 kcal/mole.

At higher temperatures (above 400 to 450°) radical reactions of thermal splitting also start to participate [1425], proceeding by a different mechanism. Firstly, the yield of low-molecular weight hydrocarbons C_1 to C_3 grows, these being quite untypical for the ionic course of hydrocracking [2247, 2249, 2252].

The influence of the other reaction conditions (pressure, feed purity etc.) which are mainly of technological significance, will be discussed in section 7.2.2. (p. 300).

Olefins undergo hydrocracking more easily than paraffins. The reaction mechanism is most probably identical with the two types of hydrocarbons and it involves the same intermediates.

Naphthenes, especially cyclic ones (e.g. decalin) are hydrocracked more readily than the corresponding n-paraffins. Splitting takes place mostly after isomerisation of six-membered rings to five-membered ones [416, 795, 2247, 2249] and besides isoparaffins, monocyclic alkylnaphthenes are the main hydrocracking products of polycyclic naphthenes. The isomerisation stage with formation of five-membered rings was confirmed in the hydrocracking of cyclohexane on MoS_2 [950, 1462, 1463, 1709] and in the hydrocracking of some alkylcyclohexanes on the WS_2/aluminosilicate catalyst [979].

Higher alkylnaphthenes (C_{10} and higher) are hydrocracked on the NiS/Al_2O_3–SiO_2 catalyst by a mechanism which is an analogue of that of polyalkylaromatic compounds (the "paring" reaction, see later) [449]. The process is independent of the character of the alkylderivative, i.e. the same products are obtained with polymethylderivatives or naphthenes with higher alkylsubstituents (e.g. C_{10}-alkylcyclohexane differently substituted always formed cyclohexane and isobutane as the main products). An important difference from hydrocracking of alkylaromatic compounds on NiS/Al_2O_3–SiO_2, however, consists in the fact that the reaction takes place at a lower temperature (lower by 50 to 100 °C) and, furthermore, that transfer of methyl groups does not take place, so that an alkylnapthene with four less carbon atoms is always obtained after isobutane is split off. The preparation of the hydrocracking catalyst NiS/Al_2O_3–SiO_2 was described by White [2206].

Hydrocracking of aromatic hydrocarbons with alkyl groups on the ring proceeds either by cleavage of the ring or, in the form of hydrodealkylation [719, 1712, 2249]. A temperature rise mainly favours the splitting and isomerisation reactions, which is moreover supported by the higher catalyst selectivity. For example, the WS_2 catalyst speeds up the isomerisation process, while MoS_2 increases the rate of the hydrodealkylation or other cleavage reactions [949].

Hydrocracking of polyphenylalkanes takes place in the aliphatic chain. For example, diarylmethanes form alkylbenzenes as well as condensation products of the initial material at 400 °C and a hydrogen pressure of 14 atm on WS_2–NiS[1492]. Similarly dibenzyl is hydrocracked to toluene in the presence of sulphide catalysts [875]. A good yield of durene was obtained by hydrocracking the condensation product of the C_9-aromatic fraction (containing mainly pseudocumene) with formaldehyde. Sulphides, oxides or their mixtures supported on active carbon were used as catalysts [516].

Splitting of the aromatic ring must be preceded by hydrogenation, so that the rate of hydrocracking of the aromatic ring is controlled by the hydrogenation rate of aromatic hydrocarbons. As we have already shown (p. 132) hydrogenation rates of different types of aromatic compounds on sulphide catalysts differ by one or two orders of magnitude [1255, 1258] and, therefore, the rate of hydrocracking will depend on the type of initial aromatic compound.

Dicyclic and tricyclic aromatic compounds (naphthalenes, phenanthrenes) are hydrocracked most easily. When these hydrocarbons are partly hydrogenated, they are hydrocracked the more readily, the more hydrogen they contain [483, 484, 1252, 1258] (Table 30).

Table 30.

Relative Hydrocracking Rates of Di- and Tricyclic Aromatics and Naphthenes[a]) [1252, 1258]. (Autoclave experiments, 5% MoS_2, H_2 pressure 200—220 atm.)

Hydrocarbon	Summary formula	Relative hydrocracking rate[b])			
		380 °C	420 °C	475 °C	Mean value
Naphthalene	$C_{10}H_8$	1.00	0.11	0.43	0.27
Tetralin	$C_{10}H_{12}$	1.00	1.00	1.00	1.00
Decalin	$C_{10}H_{18}$	2.99	1.52	2.29	2.27
Anthracene	$C_{14}H_{10}$	3.13	1.71	1.70	2.18
9,10-Dihydroanthracene	$C_{14}H_{12}$	3.13	1.76	2.66	2.52
Octahydroanthracene	$C_{14}H_{18}$	3.83	3.10	5.23	4.05
Perhydroanthracene	$C_{14}H_{24}$	39.35	33.51	—	36.43
Phenanthrene	$C_{14}H_{10}$	—	3.75	2.46	3.11
1,2-Benzanthracene	$C_{18}H_{12}$	—	11.18	—	11.18
n-Dodecane	$C_{12}H_{26}$	66.10	47.45	—	56.77

[a]) Only hydrocarbons with a lower number of C—C bonds were considered as hydrocracking products.

[b]) The relative hydrocracking rate of tetralin at each temperature = 1.00.

Much attention was devoted to naphthalene hydrocracking in the presence of MoS_2 and WS_2 catalysts [197, 483, 484, 970, 977, 1052, 1252, 1258, 1588). Kalechits et al. [196, 197, 198, 977] carried out a thorough analysis of the product obtained on hydrocracking a mixture of naphthalene, tetralin and decalin in the presence of WS_2 (420 °C, 300 atm), identifying more than 100 different hydrocarbons and hydrocarbon groups. The composition of the paraffinic-naphthenic component of the hydrocracking product is shown in Table 31. The distribution of the reaction products by the number of carbon atoms in the molecule was, in molar %: 1.3% C_1, 4.4% C_2, 7.6% C_3, 16.5% C_4, 5.2% C_5, 8.2% C_6, 3.9% C_7, 3.4% C_8, 3.8% C_9 and 45.7% C_{10} [977]. From these results it follows, that ionic reactions (hydrocracking and isomerisation) are accompanied to a relatively small extent by radical

Table 31.

The Composition of the Naphthenic-Paraffinic Fraction in the Hydrocracking Product of a Naphthalene-Tetralin-Decalin (7 : 63 : 30% by wt.) Mixture on a WS_2 Catalyst (420 °C, 300 atm) [977].

Hydro-carbons	Composition of the naphthenic-paraffinic fraction of the hydrogenate (% by wt.)							Paraffins/cyclo-pentanes ratio
	Total	Paraffins	Cyclo-pentanes	Cyclo-hexanes	Bicyclo-[3,3,0]-octanes	Bicyclo-[4,3,0]-nonanes	Bicyclo-[4,4,0]-decanes	
C_5	1.47	1.02	0.45	—	—	—	—	2.27
C_6	8.27	3.30	4.52	0.45	—	—	—	0.73
C_7	4.59	1.41	2.11	1.07	—	—	—	0.67
C_8	4.59	1.00	2.06	1.53	0	—	—	0.49
C_9	5.82	0.61	4.30	0.91	0	0	—	0.14
C_{10}	75.26	8.08	10.55	7.89	24.27	7.69	16.78	0.76
Total	100.00	15.42	23.99	11.85	24.27	7.69	16.78	0.64

cleavage (demethylation and deethylation). The liquid naphthenic-paraffinic fraction mainly consists of C_{10} hydrocarbons (75%), a full one-half being composed of decalin and its isomers dimethylbicyclo-[3,3,0]-octanes and methylbicyclo-[4,3,0]-nonanes. In the group of monocyclic naphthenes, alkylcyclopentanes prevailed. The reaction mixture contained no non-alkylated bicyclic naphthenes, i.e. bicyclo-[3,3,0]-octane and hydrindane, which is explained by the authors [977] by the high rate of splitting of the isomerised five-membered ring, which is lowered considerably by alkylation of this ring.

A detailed analysis of the aromatic component of the hydrocracking products of a mixture of naphthalene, tetralin and decalin [197] showed, that the products are mainly n-butylbenzene and 1-methyl-2-n-propylbenzene besides lower alkylbenzenes and polyalkylbenzenes of C_{10}, differently substituted due to migration of methyl groups during the reaction. Other important products are methylindanes.

Flinn et al. found [532] that light alkanes with a high ratio of (iso : n) hydrocarbons are formed by tetralin hydrocracking on the NiS/Al_2O_3–SiO_2 catalyst at 370 °C.

The basic reaction was hydrogenolysis of the hydroaromatic ring and hydrodealkylation. Benzene, alkylbenzenes and a small amount of naphthalene were isolated. Neither formation of hydrocarbons higher than C_{10} nor cyclisation reactions were observed. On the other hand, Sullivan et al. [1987] observed with tetralin hydrocracking on the same catalyst (288 °C, 82 atm, LHSV = 1, total conversion 73.4%) that tricyclic perhydroaromatic hydrocarbons were formed, which is evidence of the transfer of alkyl groups followed by cyclisation.

Similarly interesting is the influence of the position of alkyl groups in naphthalene on the rate of the hydrocracking reaction. It was found [1588] that hydrogenation, isomerisation and splitting of β-methylnaphthalene is faster in the presence of WS_2, while the methyl group in the α-position has practically no influence.

Hydrocracking of phenanthrene on the NiS/Al_2O_3–SiO_2 catalyst (293 °C, 82 atm, LHSV = 16 and 2) is relatively easy (total phenanthrene conversion was 75% and 95% at LHSV 16 and 2, resp.; conversion to hydrocarbons other than tricyclic ones was 15.5 and 45.1%. resp.) [1987]. The definite prevailing reaction products are tetralin and methylcyclohexane. It is interesting to note, that with both space velocities tested, only very small amounts of alkanes are formed. It is clear from this, that butane is formed to a very limited extent when tetralin is being formed. The authors assume a set of subsequent reactions, among which the reaction (XVIII) takes place to a limited extent only:

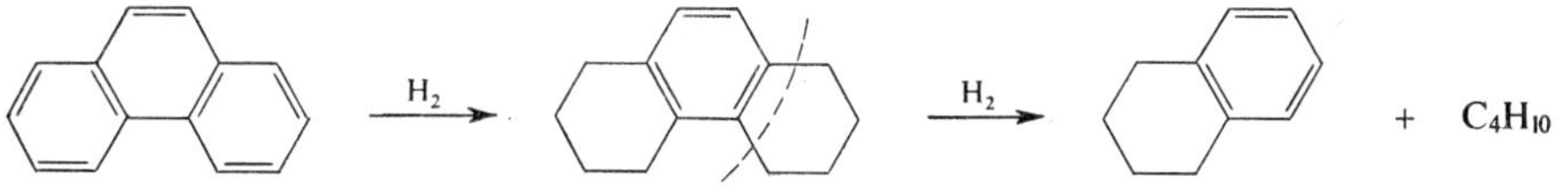

H_2 → H_2 → + C_4H_{10}

(XVIII)

Reaction (XIX), which leads to C_6-C_8 naphthenes, is very intensive.

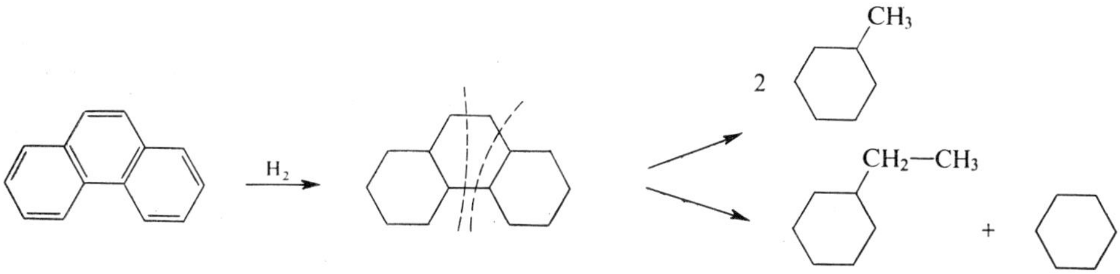

H_2 → 2 CH_3 ; CH_2—CH_3 +

(XIX)

The main process, however, is a very unusual reaction, in which bicyclic hydrocarbons, mainly tetralin are formed, without the formation of an equivalent amount of alkanes. Sullivan et al. tried to explain this reaction in their paper [1987].

Anthracene is hydrocracked on NiS/Al_2O_3–SiO_2 slightly less readily than phenanthrene: total anthracene conversion was 94% at 352 °C, 82 atm and LHSV = = 2; conversion to hydrocarbons other than those having an anthracene configuration being about 20% [1987]. Nearly three times more of light alkanes (C_1-C_7) were obtained in this process compared to the hydrocracking of phenanthrene. According to the authors [1987] the reason for this difference is rather the higher reaction temperature than a basic difference in the reaction mechanism. Tetralin again is the main product, but the occurrence of cyclisation reactions similar to those taking place in the case of phenanthrene is less probable.

Hydrocracking of anthracene, octahydroanthracene and octahydrophenanthrene on MoS_2 at 480 to 490 °C (H_2 pressure 100 atm in the cold state) resulted in a complicated mixture of reaction products containing, besides hydroderivatives of the initial substances saturated to different degrees, bicyclic and monocyclic aromatic substances and naphthenes [1685, 1686].

Sullivan et al. studied the hydrocracking of pyrene on NiS/Al_2O_3–SiO_2 [1987]. At 82 atm, 349 °C and LHSV = 2, overall conversion of the hydrocarbons was nearly quantitative, 50% being conversion to hydrocarbons other than tricyclic ones. The composition of the hydrocracking products is the analogue of the case of phenanthrene, especially in that tetralin is by far the prevailing product and the amount of alkanes formed is small. The main monocyclic naphthene formed is methylcyclopentane, as opposed to the C_7 cycloalkanes in the case of phenanthrene. The formation of bicyclic hydrocarbons from pyrene by normal splitting according to the reaction scheme (XX)

$$\text{pyrene} \xrightarrow{H_2} \text{bicyclic hydrocarbon} + \text{alkane} \qquad \text{(XX)}$$

should result in the formation of at least one alkane. Again, however, as in the case of phenanthrene, practically no alkanes are obtained, which shows that splitting reactions similar to the case of phenanthrene must be involved (cf. p. 141) [1987].

Otherwise, however, polycyclic aromatic compounds of the pyrene and perylene type are hydrocracked less readily and they are likewise efficient deactivating components in hydrocracking [416] (cf. also p. 109). The activity of sulphide catalysts is generally greater than the hydrocracking efficiency of the respective oxides [1052].

Side chains of C_3 and higher on the aromatic ring are split rather easily; lower alkylaromatic compounds on the other hand are rather more resistant.

In mild hydrocracking of n-butylbenzene in the presence of NiS/Al_2O_3–SiO_2 (288 °C, 82 atm, LHSV = 2) the main reactions were dealkylation and migration of alkyl groups. Cyclisation proceeds to a very limited extent only, the main products of this reaction being methylindanes. Tetralin is probably not formed, because its formation would necessitate the formation of an unstable primary ion (cf. the reaction scheme of decylbenzene hydrocracking) [1987]. Flinn et al. [532] also showed, that hydrodealkylation to benzene is the main reaction in n-butylbenzene hydrocracking on the same catalyst.

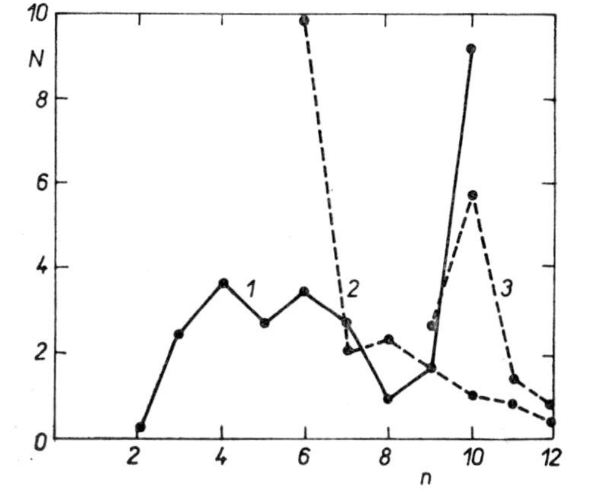

Fig. 51. n-Decylbenzene hydrocracking in the presence of NiS/Al_2O_3 [1987] (LHSV = 8, temperature 288 °C, pressure 82 atm). *n* — Number of carbon atoms. *N* — Hydrocracking products (moles of a certain hydrocarbon formed from 100 moles raw-material).

1 — Alkanes. 2 — Monocyclic hydrocarbons. 3 — Condensed bicyclic hydrocarbons.

Hydrocracking of n-decylbenzene in the presence of NiS/Al_2O_3–SiO_2 takes place in a different manner [1987]. Under the same conditions as in the case of n-butylbenzene (only with LHSV = 8) the hydrocarbon reacted to about 30%. Of the amount reacted, 35% was converted to C_{10}-alkanes and benzene and 39% cyclicised to bicyclic hydrocarbons, especially tetralin. The distribution of hydrocracking products by the number of carbon atoms and of individual hydrocarbon groups is shown in Fig. 51. The formation of tetralin may be explained by means of three basic mechanisms:

n-decylbenzene + H_2 → n-butylbenzene + C_6-alkanes
  ↳ tetralin + H_2 (XXIa)

n-decylbenzene → hexyltetralin + H_2
  ↳ $\xrightarrow{H_2}$ tetralin + C_6-alkanes (XXIb)

n-decylbenzene → tetralin + C_6-alkanes (XXIc)

The mechanism (XXIb) is thermodynamically highly improbable. The path (XXIa) is rather improbable for the same reasons; besides, the above-mentioned experiment with n-butylbenzene showed that a very small amount of tetralin is formed. Therefore, the third mechanism appears to be practical:

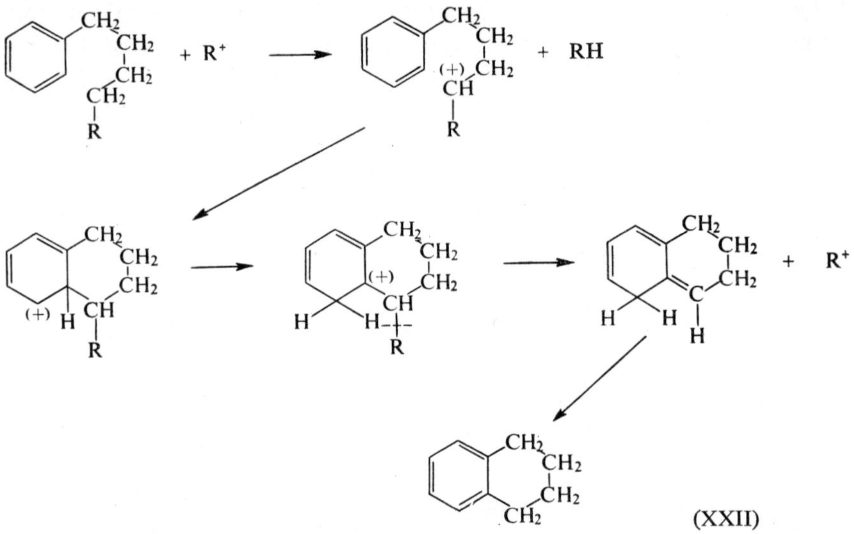

(XXII)

Hydrocracking of alkylbenzenes involves a series of subsequent reactions: e.g in the case of *p*-cymene in the presence of WS_2–V_2S_3, gradual demethylation to toluene takes place [1603] or, alkyl transfer takes place in hydrocracking of ethylbenzene on the catalyst 6434 [2247, 2249] etc.

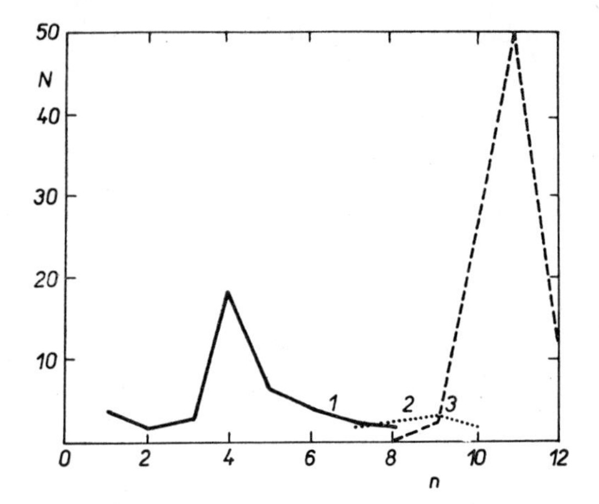

Fig. 52. Composition of hexamethylbenzene hydrocracking product on a NiS/ /Al_2O_3–SiO_2 catalyst [1986] (temperature 349 °C, pressure 16.3 atm, 1.68 moles H_2/mole hexamethylbenzene, LHSV = = 8 l/l hour).
N — Hydrocracking products (moles of the respective hydrocarbon formed from 100 moles raw material). *n* — Number of carbon atoms in the molecule. 1 — Paraffins. 2 — Naphthenes. 3 — Aromatic hydrocarbons.

Even in the hydrocracking of benzene itself, anomalous reactions of the alkylation and condensation type were observed. In hydrocracking on the catalyst 6434 (420 °C, 300 atm) naphthenes and aromatic compounds of C_7 and higher as well as diphenyl were found besides conventional hydrocracking products [955]. Similarly, in hydrocracking of benzene on the MoS_2–CoS-china clay catalyst (temperatures 450 to 550 °C, pressure 200 atm) bicyclic naphthenes $C_{12}H_{22}$ (probably a mixture of dicyclohexyl and its isomerisation products [1694]) were found besides C_7 naphthenes, although hydrocracking on the MoS_2-china clay catalyst gave normal products [1693].

An important hydrocracking (hydrodealkylation) reaction was observed with polymethylbenzenes with at least 4 methyl groups. On hydrocracking in the presence of NiS/Al_2O_3–SiO_2, powerful conversion took place at temperatures above 320 °C, accompanied by alkyl groups being split off mainly in the form of isobutane. For example, hexamethylbenzene [1986] formed products, the composition of which is indicated in Fig. 52. Sullivan et al. [1986] assume that the reaction takes place on the basis of gradual isomerisation of six-membered rings to five-membered ones and vice versa, substituents being regrouped in the process. In this way, a higher substituent is formed especially with the tert-butyl configuration, which is finally split off in the form of isobutane (scheme in Fig. 53). The reaction takes place on acid catalyst centres since it is catalysed by the carrier proper, also.

From the reaction scheme it follows, that the main products of this "paring" reaction should be isobutane and xylenes in the case of hexamethylbenzene. Besides branched pentanes and hexanes, the component first-named actually prevails, but only a minimum amount of xylenes was recovered; the amount of dimethylcyclohexanes was very small too. The reason is, that intensive transfer of methyl groups

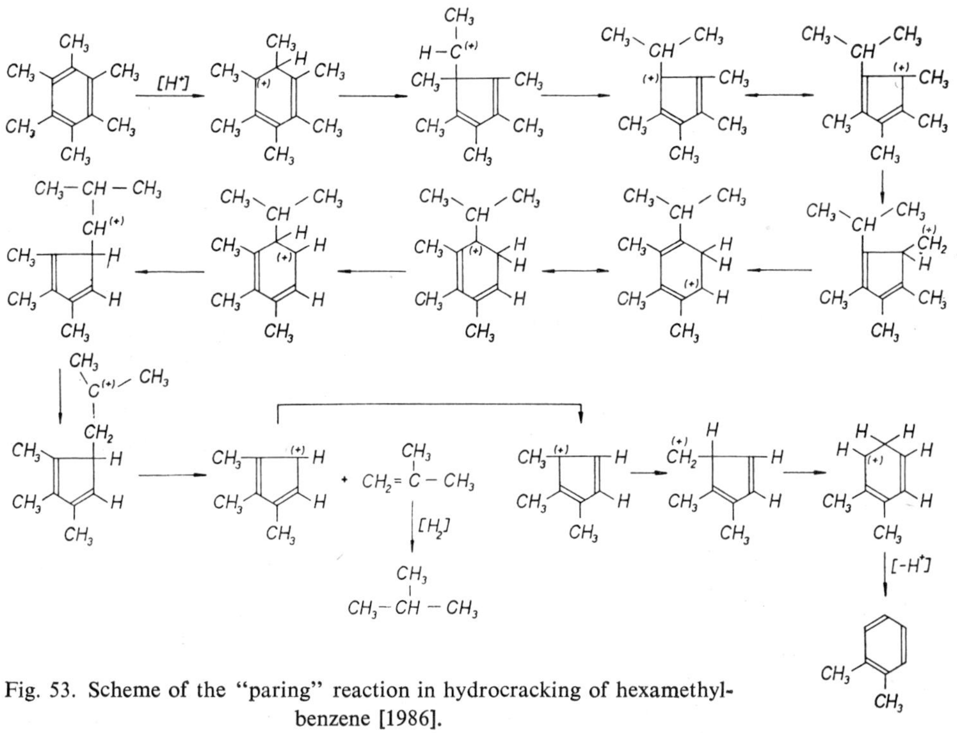

Fig. 53. Scheme of the "paring" reaction in hydrocracking of hexamethylbenzene [1986].

from the higher homologues to xylene takes place in the reaction, so that C_{11}, C_{10} and C_9-aromatic compounds are the main products (see Fig. 52).

The "paring" reaction is typical for aromatic compounds of C_{10} and higher. It is characterised by the fact, that a very small amount of ethyl- and propylbenzenes is formed. This fact indicates that cleavage and desorption from the catalyst does not set in before the side chain has grown to such a length that isobutane can be split off.

In hydrocracking of hexaethylbenzene on the NiS/Al_2O_3–SiO_2 catalyst, however, considerable differences were found [1987]. At 293 °C and 8 atm pressure, the hydrocarbon reacted practically quantitatively, but the product composition depended on the contact time to a large extent. With a short contact time (LHSV = 16) the reaction product contains only ethane, ethylene and polyethylbenzenes (mainly tetraethylbenzene). Thus, the primary reaction is nearly exclusively deethylation. With a longer contact time (LHSV = 2) other products, too, are formed to a large extent:

1. Alkanes higher than ethane are formed, especially butanes: the isobutane/n-butane ratio is about 0.8, which is very low.
2. The hydrocracking product contains a certain amount of C_{10} to C_{12} tetralines and indanes.
3. Among alkylbenzenes, hydrocarbons with an even number of carbon atoms prevail, although compounds with an odd number are formed too.
4. Hydrogenolysis of the aromatic ring does not take place during the process.

The first stage is formation of the carbonium ion and deethylation. This tendency to lose an ethyl group is characteristic of hexa- and pentaethylbenzene, but not of ethylbenzene itself, which is very stable under hydrocracking conditions. The reasons are probably of a stereochemical nature, since the C_{ar}—C_2H_5 bond is weakened when ethyl groups are closely packed around the ring. The ethyl cation is desorbed in the form of ethylene and is hydrogenated to ethane (from the large ethylene concentration in the product in the case of a short contact time follows, that the hydrodealkylation rate on a sulphide catalyst is considerably greater than the rate of hydrogenation). The ethyl cation may also be stabilised by the addition of a hydride anion (formed by transfer from another hydrocarbon molecule) or by molecular hydrogen by means of direct heterolytic cleavage (cf. [569]).

In the "paring" reaction of hexamethylbenzene, removal of branched alkanes takes place without ring splitting. It has been proved that growth of the alkyl group to C_4 and its abstraction in the form of isobutane is characteristic of this reaction. Differing from this deethylation is the most probable reaction in the course of hydrocracking of hexaethylbenzene. Here the "paring" reaction is a side reaction only, taking place only with a partly dealkylated product, as follows from the increased content of alkanes higher than C_2 with a prolonged contact time. More n-butane than iso-butane is formed by the "paring" reaction of hexaethylbenzene, while the reverse is true of hexamethylbenzene hydrocracking [1986]. According to Sullivan et al. [1987] the formation and separation of the n- and iso-butyl substituent is more probable in hydrocracking of hexaethylbenzene compared to that of hexamethylbenzene, where the methyl group only is transfered. Cyclisation and formation of tetralins and indanes may also take place during the reaction, in which e.g. the sec-butylbenzene ion is an intermediate.

Kalechits et al. [977a, 1004a, 1004b] hydrocracked durene and isodurene in the presence of the NiS/Al_2O_3–SiO_2 catalyst. They state, that the chemistry of this reaction is considerably more complicated than would appear from the scheme reported by Sullivan et al. [1986, 1987].

6.3.2.

Cracking of hydrocarbons

Few data are available on cracking of pure hydrocarbons in the presence of sulphide catalysts. Zakharenko [2252] studied the process of cetane cracking in the presence of the WS_2 catalyst (5058) and WS_2 on aluminosilicate (type 6434) at atmospheric pressure and 527 °C. From the results illustrated in Fig. 54 there is a considerable difference between the composition of cracking products obtained with the two

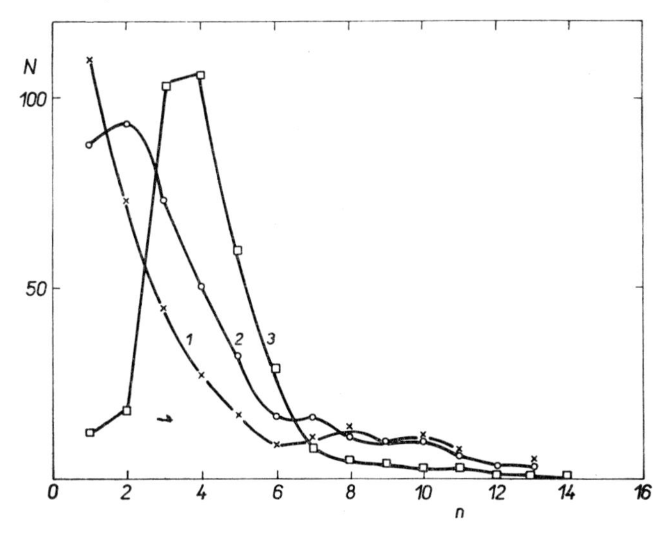

Fig. 54. Cetane cracking. n — Number of carbon atoms in the product. N — Moles product per 100 moles reacted cetane. 1 — Catalyst 5058 at 527 °C, 1 atm, feed conversion 49% [2252]. 2 — Catalyst 6434, 527 °C, 1 atm, feed conversion 32% [2252]. 3 — Catalyst UOP-B (SiO_2–Al_2O_3–ZrO_2), 500 °C, feed conversion 53.5% [646].

catalysts. While the result of cracking on a typical aluminosilicate corresponds almost exactly to the theoretical composition calculated from the ionic course of the reaction [646], sulphide catalysts are mainly characterised by a high content of the lowest hydrocarbons (C_1 and C_2). Another characteristic is the greater yield of hydrocarbons higher than C_7, which is similar to the case of cracking on non-acid catalysts (e.g. on active carbon [646]) which proceeds by a radical mechanism. At these high temperatures, radical reactions participate especially in cracking on a catalyst of relatively low acidity, i.e. WS_2, while the hydrocracking catalyst forms a transition to typically cracking aluminosilicate catalysts. Similarly, in cracking of cumene at 527 °C [2251] demethanisation was the main process observed with the catalysts WS_2, 8376 and Co–Mo/Al_2O_3, while dealkylation took place mainly in the presence of the hydrocracking catalyst 6434, benzene being formed in the reaction.

Mo and W sulphides were tested as components of catalysts for disproportionation of propylene to ethylene and butenes [745a, 745b, 745c, 745d].

Dehydrocracking of paraffinic hydrocarbons to liquid n-olefins is a very important reaction from a technologie point of view (cf. p. 353). Huibers and Waterman [818] found, that the catalysts WS_2 and WS_2–NiS can be employed to obtain higher n-olefins by means of dehydrocracking solid paraffins.

The selectivity of these catalysts and of the catalyst 8376 was studied in detail by

Table 32.

n-Hexane Dehydrocracking [1215] (LHSV = 3.76 l/l/hour).

Temperature (°C)	Catalyst	Total conversion (%)	Proportion on individual reactions (% by hexane reacted)			
			Cracking	Dehydrogenation	Dehydrocyclisation	Isomerisation
470	WS_2	5.5	9	70	9	12
	WS_2–NiS	6.5	7	67	—	26
	WS_2–NiS–Al_2O_3	8.0	28	45	—	27
510	WS_2	8.0	22	62	10	6
	WS_2–NiS	11.5	20	68	—	12
	WS_2–NiS–Al_2O_3	12.0	45	41	—	14
550	WS_2	21.0	30	54	15	1
	WS_2–NiS	17.5	37	53	—	10
	WS_2–NiS–Al_2O_3	24.0	47	41	2	10
590	WS_2	34.0	63	28	8	1
	WS_2–NiS	27.0	73	25	—	2
	WS_2–NiS–Al_2O_3	41.0	59	27	10	4

Lebedev et al. [1215] on the simple example of n-hexane dehydrocracking. The authors worked with a flow apparatus at atmospheric pressure, temperature of 470 to 590 °C and LHSV of 0.8 to 19.3. A typical example is presented for all three catalysts in Table 32. A characteristic feature of this process is the high selectivity of the carrier-free catalysts at 470 °C, which however is decreased with the rising conversion so that even at a temperature of more than 550 °C there is no difference between the various catalysts. The isomerisation activity (skeletal isomerisation) of the catalysts varies in the series WS_2–NiS/Al_2O_3 > WS_2–NiS > WS_2. Another important factor is the representation of the individual n-hexenes. The equilibrium ratio (1-hexene/2-hexene + 3-hexene) at 470 to 590 °C varies only slightly in the 0.115 to 0.135 range. From results obtained with the catalysts WS_2, WS_2–NiS and WS_2–NiS–Al_2O_3, the relative amount of α-olefins is far greater and rises with the temperature, while it depends only slightly on space velocity. This ratio, however, differs greatly with the three catalysts tested, as shown in Table 33; it is least with the

Table 33.

The Isomerisation Properties of Sulphide Catalysts in n-Hexane Dehydrocracking [1215]. (Temperature 510 °C, LHSV = 3.75 l/l/hour.)

Catalyst	Proportion of skeleton isomerisation (% by reacted hexane, cf. Table 32)	Ratio $\frac{\text{1-hexene}}{\text{2-hexene + 3-hexene}}$
WS_2	6	0.35
WS_2–NiS	12	0.21
WS_2–NiS–Al_2O_3	14	0.19

catalyst 8376, which likewise catalyses skeletal isomerisation most efficiently under the respective reaction conditions. Other studies report the results of n-decane and n-tetradecane dehydrocracking [1215b, 1215c].

The catalyst NiS/SiO_2 was used for dehydrocracking n-hexadecane at 550 °C and LHSV of 0.65 to 0.9 [2264]. When the carrier was promoted with a small amount of alkali oxide (0.6% Li_2O), a catalyst with a high degree of selectivity for α-olefins was obtained, although the cracking activity decreased slightly. The reaction product obtained with the use of the catalyst without the addition of Li_2O contained a large amount of isomeric olefins with a different position of the double bond and different skeleton. The mixed catalyst NiS–MoS_2 supported on Al_2O_3 (with 0.6% Li_2O) decreased the yield of olefins and considerably increased the extent of splitting to gases. The optimum catalyst contains 8.5% NiS and 0.6% Li_2O on silicagel:

it produces 45% olefins (related to the raw material) with 74% selectivity for α-olephines at 550 °C and a space velocity of 0.9 l/l hour.

Activation of the catalyst 8376 by means of γ-radiation was also tested in the case of cetane dehydrocracking [2263].

6.4.

Hydrogenation of alcohols and ethers

6.4.1.

Hydrogenation of alcohols

Hydrogenolysis of alcohols proceeds readily, particularly, when the hydroxyl is activated in a certain way. The hydroxyl group is very easily eliminated when bonded to an aromatic ring: e.g. benzyl alcohol readily converts to toluene in the presence of sulphide catalysts [286b, 906, 1163, 1397, 2175] and also low-temperature metal catalysts (e.g. Pd [1974]).

Hydrogenolysis of the alcoholic group is far more difficult when not activated. Drastic treatment is necessary and the process is only possible in the presence of certain catalysts. Out of the sulphide catalysts, nickel was used in several cases to hydrogenate alcohols, and also platinum, V_2O_5/Al_2O_3, Co/Al_2O_3 etc. (for a review of this topic, see [1231, 2184]). Oxide catalysts, as far as these can be employed at all, usually give low alcohol yields [286a].

Only sulphide catalysts are generally applicable. When the reaction is carried out in autoclaves, temperatures of 300 to 350 °C and hydrogen pressures of over 100 atm are generally used in the presence of MoS_2 and WS_2 catalysts. Typical results are summarised in Table 34.

During World War II, hydrogenation of aliphatic alcohols, which are formed as by-products of methanol synthesis, was a process of technical importance. When the catalyst 8376 is used at 350 °C and 300 atm, a wholy aliphatic gasoline is obtained with a high octane number and high sensitivity to tetraethyllead [1101].

Metal sulphides are typical polyfunctional catalysts, being characterised by hydrogenating properties as well as by dehydrating, splitting and isomerisation activities, so that when alcohols are hydrogenated in the presence of these catalysts, olefines and products of hydrocracking and isomerisation reactions are formed as well as saturated hydrocarbons.

To date, no kinetic data are available to clarify, unambiguously, the mechanism of hydrogenation of alcohols to saturated hydrocarbons in the presence of sulphide catalysts. On hydrogenating benzyl alcohol in the presence of MoS_3, Chang and Itabashi [286b] found that the reaction order is zero with respect to the alcohol,

Table 34.

Examples of Hydrogenation of Alcohols on Sulphide Catalysts.

Alcohol hydrogenated	Catalyst[b])	Reaction conditions			Product	Yield (% theor.)	Reference
		Tempera-ture (°C)	H_2 pressure (atm) or initial pressure (atm/t °C)	Reaction time (min)			
2-Methyl-2-propanol	MoS_2	320	120/20	30	Isobutane	59	[1163]
2-Methyl-2-butanol[a])	MoS_2	350	50	120	Isopentane	35	[1397]
1-Heptanol	MoS_3	300	105/ 0	105	n-Heptane	67.6	[906]
1-Octanol	MoS_2	330—340	122/20	45	n-Octane	85	[1163]
1-Octanol	WS_2	340—350	120/20	30	n-Octane	95.8	[1159]
1-Dodecanol	MoS_3	350	106/ 0	120	n-Dodecane	89.3	[906]
1-Hexadecanol	MoS_2	320	110/20	30	n-Hexadecane	83	[1163]
1-Hexadecanol	WS_2	340—350	120/20	30	n-Hexadecane	93.4	[1159]
Benzyl alcohol	MoS_2	220	104/ 0	90	Toluene	91.2	[906]
Triphenylcarbinol	MoS_2	320	120/20	30	Triphenylmethane	61	[1163]
Cyclohexanol	MoS_3	250	107/0	50	Cyclohexane	78.2	[906]

[a]) Experiment carried out in xylene.

[b]) 5 to 30% catalyst was used.

while a first-order reaction is involved with regard to hydrogen. Lozovoi [1260] found, when comparing the rates of hydrogenation of some alcohols and phenols (MoS_2 catalyst), that the reaction rate decreases in the order: secondary cyclic alcohols $\gg$ secondary aliphatic alcohols $>$ primary aliphatic alcohols $>$ phenol. Lozovoi judged the reactivity of the individual alcohols by the amount of water liberated by the hydrogenation process, so that the relative rates may be distorted to some extent due to the different rates of dehydration of the alcohols involved.

As shown by a number of investigations, dehydration of the initial substance plays an important role in the overall hydrogenation reaction mechanism of alcohols. A number of sulphide catalysts, e.g. NiS [1148], ZnS [1105, 1776], MoS_2 [1167] and especially WS_2 [1167, 2061] (cf. p. 282) have distinct dehydration properties, so that with those alcohols which are easily dehydrated (e.g. cyclohexanol [53, 1167]), "hydrogenolysis" proceeds by a slower hydrogenation of olefin formed in the preceding, rapid step of dehydration: the saturated hydrocarbon yield therefore depends on the conditions, especially temperature, at which the reaction is being carried out. Dehydration, however, does not proceed down to olefins only, since especially at lower temperatures (below 300 °C) ethers are also formed in the presence of metal sulphides [1167, 1384, 2180], so that some of the saturated hydrocarbons are formed by hydrogenolysis of ethers.

The significance of dehydration as an important step in the overall hydrogenolysis reaction is also indicated by the results of hydrogenation of those alcohols, which are generally dehydrated with simultaneous change of their carbon skeleton or, which cannot be dehydrated in the original structure. In hydrogenation of pinacol and pinacolyl alcohol, Amemiya [26, 29] found that in the presence of MoS_2, 2,3-dimethylbutane is mainly formed. Later, Landa et al. [1165] showed that on hydrogenation of pinacolyl alcohol, pinacolone and pinacol in the presence of MoS_2, a mixture of 2,3- and 2,2-dimethylbutanes is always formed, the isomer first-named prevailing. Pinacolone rearrangement accompanies molybdenum sulphide catalysed hydrogenation of pinacols derived from cyclopentanone and cyclohexanone [2175, 2185a].

The rearrangement likewise accompanies the hydrogenation of borneol and isoborneol in the presence of MoS_2 under mild conditions (200 to 250 °C, hydrogen pressure about 100 atm), in which isocamphane was obtained as a main hydrogenation product [1026].

CH_3, H, OH $\rightleftarrows$ CH_3, OH, H $\xrightarrow[\text{[MoS}_2]]{-H_2O}$ CH_2 $\rightleftarrows$ H_2C, CH_2, CH_3, CH_3 $\xrightarrow[\text{[MoS}_2]]{H_2}$ H_2C, CH_3, CH_3, CH_3 (XXIII)

The result of hydrogenation of neopentyl alcohol in the presence of MoS_2 (340 °C, initial H_2 pressure 130 atm), in which approximately equal amounts of 2,2-dimethylpropane and 2-methylbutane were isolated [1165], is evidence of the fact that the ionicdehydration course of the reaction is highly probable. Here we see an analogy to the rearrangement of isobutyl alcohol on dehydration [2208]:

$$CH_3{-}\underset{H_3C}{\overset{H}{C}}{-}\underset{H}{\overset{H}{C}}{-}OH \xrightarrow{[+H^+]} \left[CH_3{-}\underset{H_3C}{\overset{H}{C}}{-}\underset{H}{\overset{H}{C}}{-}\overset{(+)}{O}H_2\right] \xrightarrow{[-H_2O]} CH_3{-}\underset{H_3C}{\overset{H}{C}}{-}\underset{H}{\overset{H}{C}}{}^{(+)}$$

$$CH_3{-}\underset{H_3C}{\overset{H}{C}}{-}\underset{H}{\overset{H}{C}}{}^{(+)} \xrightarrow{[-H^+]} \begin{matrix} CH_3 \\ CH_3 \end{matrix}\!\!>C{=}CH_2$$

$$CH_3{-}\underset{H_3C}{\overset{H}{C}}{-}\underset{H}{\overset{H}{C}}{}^{(+)} \longrightarrow CH_3{-}\underset{(+)}{\overset{H}{C}}{-}\underset{H}{\overset{H}{C}}{-}CH_3 \xrightarrow{[-H^+]} CH_3{-}CH{=}CH{-}CH_3 \qquad \text{(XXIV)}$$

The same applies to the methyl-homologue of isobutyl alcohol, i.e. neopentyl alcohol, with which the primary carbonium ion is stabilised in a hydrogen atmosphere either directly to neopentane (the hydrogenation may be explained by transfer of a hydride anion from a different hydrocarbon molecule or, by the effect of molecular hydrogen with direct heterolytic cleavage similar to the process on the NiS catalyst [569, 1987]), or the primary ion is isomerised to the more stable tertiary ion and hydrogenated to isopentane after loss of a proton (XXV) (see p. 154).

The source of protons needed for the process is the acid surface of the sulphide catalyst.

Sulphide catalysts may be used with advantage to prepare large amounts of hydrocarbons by hydrogenating the respective alcohols of simple structure e.g. n-alcohols [1159, 2053], as well as of ketones and esters, which form simple alcohols on reduction. In the case of alcohols (or ketones, esters and acids) of more complicated structure, the situation is more complicated since the respective saturated hydrocarbon is obtained in different yields in addition to the products of isomerisation and hydrocracking due to the polyfunctional character of sulphide catalysts; the hydrocarbon required must be isolated from the reaction mixture. Since the thermal stability of the initial substances and, in part, of the products decreases with increasing molecular weight (or the number of groups in the molecule capable of being

$$CH_3-\underset{H_3C}{\overset{H_3C}{\underset{|}{\overset{|}{C}}}}-\underset{H}{\overset{H}{\underset{|}{\overset{|}{C}}}}-OH \xrightarrow{[+H^+]} \left[CH_3-\underset{H_3C}{\overset{H_3C}{\underset{|}{\overset{|}{C}}}}-\underset{H}{\overset{H}{\underset{|}{\overset{|}{C}}}}-\overset{(+)}{O}H_2\right] \xrightarrow{[-H_2O]} CH_3-\underset{H_3C}{\overset{H_3C}{\underset{|}{\overset{|}{C}}}}-\underset{H}{\overset{H}{\underset{|}{\overset{|}{C}}}}{}^{(+)}$$

$$CH_3-\underset{H_3C}{\overset{H_3C}{\underset{|}{\overset{|}{C}}}}-\underset{H}{\overset{H}{\underset{|}{\overset{|}{C}}}}{}^{(+)} \xrightarrow{[H_2]} CH_3-\underset{CH_3}{\overset{CH_3}{\underset{|}{\overset{|}{C}}}}-CH_3$$

$$CH_3-\underset{H_3C}{\overset{H_3C}{\underset{|}{\overset{|}{C}}}}-\underset{H}{\overset{H}{\underset{|}{\overset{|}{C}}}}{}^{(+)} \longrightarrow CH_3-\underset{(+)}{\overset{H_3C}{\overset{|}{C}}}-\underset{H}{\overset{H}{\underset{|}{\overset{|}{C}}}}-CH_3 \xrightarrow{[-H^+]} (CH_3)_2C{=}CH-CH_3 \xrightarrow{[H_2]} (CH_3)_2CH-CH_2-CH_2 \qquad (XXV)$$

hydrogenated), and also since the degree of complication of the reaction mixture formed increases as well, it follows that pure hydrocarbons will be obtained by hydrogenating alcohols or other oxygen-containing substances in the presence of sulphide catalysts only when highly efficient methods of isolation and identification are used. However, the contemporary state of isolation and identification techniques allows even high-boiling hydrocarbon mixtures to be separated. For example, a new hydrocarbon, diisopropylphenylmethane was obtained in a single step from the hydrogenation product of diisopropylphenylcarbinol in the presence of MoS_2. This was achieved by a combination of rectification, capillary gas chromatography, efficient preparative gas chromatography and spectral identification methods (infrared, mass and NMR spectroscopy). The amounts obtained are only limited by the availability of the initial oxygenous substance [2184].

This method is easily applied to the rapid preparation of some hydrocarbons, although in more complicated cases, classical multi-step syntheses will have to be employed [1145].

With some types of alcohols, there are difficulties of a more fundamental character. This mainly involves the above-mentioned cases of compounds which are subject to rearrangement during dehydration, and the hydrogenation of which results in mixtures of isomers which are frequently difficult to separate.

Principal difficulties also occur in the hydrogenation of some alcohols of low thermal stability which condense readily, e.g. the terpenic alcohols citronellol, geraniol and linalool, which form large amounts of resinous components in the presence of MoS_3 even under mild reaction conditions (160 to 200 °C [1030]). Similarly, diols undergo hydrocracking reactions far more easily, the yield of corresponding hydrocarbons being diminished in this way [1175]. Another group, the highly branched tertiary alcohols and ditertiary glycols, are intensively hydrocracked and partly isomerised during hydrogenation in the presence of sulphide catalysts. In the hydrogenation of tertiary alcohols and ditertiary glycols with two tertiary butyl groups on the carbinol carbon atom, one (in the case of ditertiary glycols, two) tertiary butyl group is split off on hydrogenation in the presence of MoS_2 (temperature around 300 °C, initial H_2 pressure 120 atm), so that a hydrocarbon containing four (or eight) less carbon atoms, compared to the initial compound, results [1170]. This result agrees with the mechanism of hydrocracking reactions on sulphide catalysts (cf. the preceding chapter), splitting of the C–C bond in these highly branched chains being particularly easy on the carbon atom attached to least hydrogen. Here, active splitting centres on the catalysts are in sites where non-stoichiometric sulphur is localised, which are characterised by increased acidity due to the acceptor character of these defects [1589] and therefore also by an enhanced ability to catalyse ionic reactions. Steric reasons (the presence of two large tertiary butyl groups on the carbinol carbon atom) likewise contribute to the higher rate of splitting of these compounds as compared to analogue alcohols and glycols with less branched alkyl groups [2184].

Tertiary alcohols and glycols with less branched alkyl groups are more stable in respect of hydrogenation in the presence of sulphide catalysts and, a satisfactory yield of the respective saturated hydrocarbon can be obtained from them by means of an efficient separation method (e.g. in hydrogenation of 3-methyl-3-isopropyl-3-octanol or the above-mentioned diisopropylphenylcarbinol, the respective hydrocarbons were isolated in the pure form by means of preparative gas chromatography) [2184].

We may conclude therefore, that due to the polyfunctional character of sulphide catalysts, the formation of other products, formed due to simultaneous hydrocracking and isomerisation, must be expected besides the required hydrocarbon having the same arrangement of the carbon chain as in the initial material, when alcohols (and oxygenous compounds which react to form alcohols) are hydrogenated in the presence of metal sulphides. The degree to which these side reactions participate depends on the degree of branching of the initial alcohol and its thermal stability.

6.4.2.

Hydrogenation of ethers

This reaction takes place in the presence of sulphide catalysts particularly easily with ethers of the benzyl type. In the presence of MoS_2, dibenzyl ether formed a nearly 90% yield of toluene at temperatures as low as 200 °C [906]. Hydrogenolysis of this compound is similar in the presence of FeS [1003].

Aliphatic ethers are hydrogenated less readily and at temperatures, at which decomposition may also occur with formation of the corresponding aldehyde or alcohol. Hydrogenolysis with splitting-off of the alcohol may also take place:

$$RCH_2CH_2OCH_2CH_2R + H_2 \rightarrow RCH_2CH_2OH + RCH_2CH_3\,, \qquad \text{(XXVI)}$$

$$RCH_2CH_2OCH_2CH_2R \rightleftarrows RCH = CH_2 + RCH_2CH_2OH\,, \qquad \text{(XXVII)}$$

$$RCH_2CH_2OCH_2CH_2R \rightarrow RCH_2CH = O + RCH_2CH_3\,. \qquad \text{(XXVIII)}$$

That these reactions do in fact occur is indicated by the composition of the hydrogenation product of diethyl ether in the presence of the catalyst 5058 at 270 °C and atmospheric pressure, which contained mainly ethylene with some ethane. The decomposition of diethylether in a nitrogen atmosphere likewise shows that splitting off of the alcohol is the main reaction under these conditions. In both cases, the liquid reaction product contained small amounts of ethanol and in the second case acetaldehyde was also found [2180].

The hydrogenation of phenolic ethers and cyclic ethers will be discussed in the following sections.

6.5.

Hydrogenation of phenols

Hydrogenation of phenols in the presence of sulphide catalysts is one of the important reactions which takes place in hydroprocessing of tar and shale materials. The reaction may be controlled in such a way, that phenols are totally dehydroxylated producing hydrocarbons, which is the basic requirement in hydrorefining of the products of coal carbonisation or hydrogenation to motor fuels [416, 793, 1091, 1101] (section 7.3.).

At the present time, phenols have become an important and widely needed chemical raw material. There is a constant shortage of lower phenols especially and, therefore, hydrogenation processes in which higher phenols are dealkylated or polyvalent phenols are selectively dealkylated or dehydroxylated to phenol or its lower homologues are nowadays of great interest [408, 951, 963, 1113].

In earlier studies, optimum reaction conditions and catalysts for reducing phenols to hydrocarbons were investigated. Among sulphide catalysts, the highest activity was found with MoS_2 (or MoS_3). This was particularly useful in hydrogenating raw phenols, containing substances which act as catalyst poisons on other catalysts [199, 519, 1051, 1398, 1399, 1718, 2070, 2072, 2074]. Tropsch [2073] compared a number of oxide and sulphide catalysts in cresol hydrogenation, finding that Mo, W and Co sulphides are the most active ones. Besides these catalysts, some thio-molybdates were tested (e.g. Co, Cr, Sn) [1336] as well as Cu_2S on active carbon [1057].

6.5.1.

The mechanism of hydrogenation of phenols

The composition of the products of hydrogenation of phenols, especially the naphthenes content (and the aliphatic hydrocarbons formed by hydrocracking) depends on the hydrogenation activity of the catalyst employed and on the reaction conditions. At low hydrogen pressures, aromatic substances are chiefly formed. For example, a cresol mixture was hydrogenated in the presence of the MoS_2 catalyst at 430 °C, 10 atm pressure and a space velocity of 0.5 l/lhour. To maintain a long lifetime of the catalyst, the feed contained 1% CS_2, and under these conditions, 89% conversion to toluene was achieved; only 6% methylcyclohexane being formed [279]. The content of naphthenic hydrocarbons in phenol hydrogenates rises with increasing pressure [13, 47, 1048, 1394], so that a totally saturated hydrocarbon product can be obtained with an efficient hydrogenating catalyst (e.g. 8376) [376]. The course of hydro-cracking reactions is speeded up by an increase in temperature, paraffinic hydro-carbons being formed especially at 400 °C and in the presence of catalysts with splitting activity [408].

Hydrogenation of phenols on platinum or nickel catalysts is a subsequent reaction with cyclohexanone as a reaction intermediate. This reaction will stop either at the respective cyclohexanol step [2124] or, it will continue to the respective naphthenic hydrocarbon (at higher temperatures, when the cyclohexanol obtained is unstable, it dehydrates to the respective cyclohexene, which is hydrogenated in the following step to the respective cyclohexane [32, 33, 1048, 1920].). Hydrogenation of phenol to cyclohexanol in the presence of Raney nickel (temperature 113 to 174 °C) repre-sents a classical consecutive reaction, the kinetics of which was studied in detail by Jungers and Coussemant [360, 947].

The course of hydrogenation of phenols on sulphide catalysts is more complicated and differs from the reduction mechanism of these substances in the presence of active low-temperature catalysts. The cause for this difference is the fact, that at high temperatures, where sulphides are catalytically active, side reactions are speeded up which, at low temperatures, would not take place at all.

As we have mentioned already (cf. e.g. low-pressure hydrogenation of cresols on MoS_2 performed by Cawley et al. [279]), the reaction products of hydrogenation of phenols in the presence of sulphide catalysts are characterised by different contents of aromatic hydrocarbons. Therefore, the following overall reaction scheme of hydrogenation may be assumed:

OH $\xrightarrow{[H_2]}$ + H_2O (XXIX)

OH $\xrightarrow{[H_2]}$ OH $\xrightarrow{[H_2O]}$ $\xrightarrow{[H_2]}$ (XXX)

Cyclohexene, formed as intermediate, can either isomerise to form a five-membered ring [52] or, it may moreover form the respective aromatic compound and naphthene by disproportionation [2208]:

3 → + 2 (XXXI)

Roberti [1750, 1753] and Polozov [1657] showed that (when phenol is hydrogenated on sulphide catalysts (CoS, MoS_2) at pressures of over 100 atm) cyclohexane is

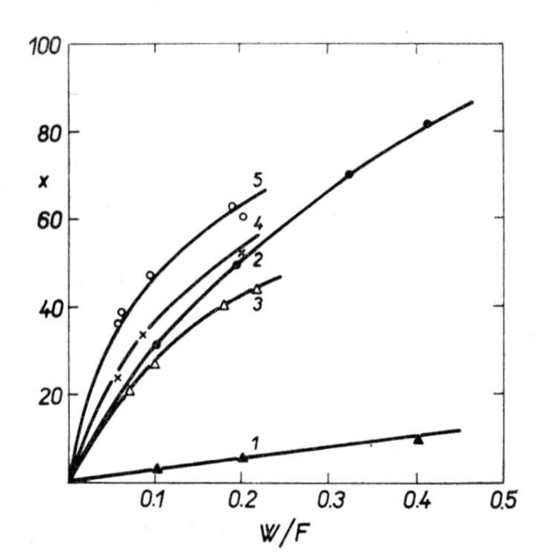

Fig. 55. Dependence of the hydrogenation of phenol and benzene (x-degree of conversion, %) on the reciprocal value of liquid hourly space velocity (W/F, hour) in the presence of the catalyst 8376 [13]. 1 — Benzene hydrogenation (300 °C, 300 atm). 2 — Phenol hydrogenation (300 °C, 300 atm, for the hydrogenate composition, see Table 35). 3 — Phenol hydrogenation (320 °C, 150 atm). 4 — Phenol hydrogenation (320 °C, 200 atm). 5 — Phenol hydrogenation (320 °C, 300 atm).

formed with cyclohexanol as an intermediate, not by consecutive benzene hydrogenation. Moldavskii and Livshits [1394] proved that hydrogenation of phenol and *o*-cresol takes place by the two reaction paths mentioned previously. These authors worked with rotating autoclaves in the presence of MoS_2 at 330 to 370 °C and different hydrogen pressures. At 25 atm, the hydrogenation rate of the benzene ring was considerably lower than the rate of the hydrogenolysis reaction, and 90% of the phenol was converted to benzene, which fully agrees with later data published by Bobyshev et al. (192). With increasing hydrogen pressure, there was an increase in the content of the naphthenic component C_6H_{12} and C_5H_{10} (25 atm – 10%, 50 atm – 17%, 100 atm – 30%) while the cyclohexene content decreased. *o*-Cresol was hydrogenated in a similar manner, but the methylcyclohexene yield was lower than the cyclohexene yield in phenol hydrogenation. In agreement with Polozov [1657], no cyclic alcohols were found in the hydrogenates. Benzene and toluene were found to be only slightly hydrogenated at the same temperatures and a pressure of 125 atm (max. 4%). This fact, together with the presence of cyclic olefins is evidence of the formation of naphthenes mainly through the respective methylcyclohexanol as an intermediate. The definite content of aromatic substances at low pressures is evidence of the direct hydrogenolysis process (XXIX) and, at the same time, this indicates that the disproportionation reaction (XXXI) does not play an important role. Ando [45] arrived at the same conclusion concerning phenol hydrogenation

Table 35.
Hydrogenation of Some Phenols on the 8376 Catalyst (temp. 300 °C, pressure 300 atm) [13].

Phenol hydrogenated	LHSV	Degree of conversion (%)	Composition of hydrogenation products (% by wt.)		
			Aromatic hydrocarbons	Unsaturated hydrocarbons	Naphthenes
Phenol	1.25	92.0	0.7	0.3	99.0
	2.4	80.1	2.9	0.9	96.2
	2.9	69.6	1.4	0.8	97.8
	4.8	52.1	3.4	0.8	95.8
	10.0	31.8	5.5	1.0	93.5
p-Cresol	2.65	85.9	3.8	0.8	95.4
	4.50	68.1	4.7	0.8	94.5
	9.4	44.1	7.4	0.9	91.7
	14.0	34.7	9.3	1.2	89.5
o-Cresol	2.5	81.1	3.2	0.9	95.9
	4.8	54.0	5.1	1.0	93.9
	10.0	30.1	7.8	1.2	91.0
	15.0	22.5	8.8	1.4	89.8
n-Butylphenol	2.5	86.0	8.2	1.2	90.6
	4.85	60.0	13.1	1.7	85.2

on sulphurised MoO_3, and Klimov and Bogdanov [1048] in hydrogenations on the WS_2 catalyst.

Later, Moldavskii together with Alekseeva [13] returned to the problem of the mechanism of hydrogenation of phenols, and compared the rate of this reaction with the rate of hydrogenation of aromatic compounds under the same con-

Table 36.

Relative Rate of Hydrogenation of Phenols to Aromatic Hydrocarbons [192].
(Autoclave experiments, MoS_2 catalyst, temperature 350 °C, starting H_2 pressure 31 atm.)

Phenol hydrogenated	Relative rate	Phenol hydrogenated	Relative rate
Phenol	100	2,4-Xylenol	70.2
o-Cresol	60.8	Thymol	65.8
m-Cresol	108	Carvacrol	44.9
p-Cresol	126	α-Naphthol	160
3,4-Xylenol	62.5	β-Naphthol	208
3,5-Xylenol	65.5	Thiophenol	2845

ditions. These authors worked with a flow-system in the presence of the refining catalyst 8376. From the results illustrated in Fig. 45 and the composition of the hydrogenate (Table 35) it follows conclusively, that (at elevated pressures of 100–150 atm) the hydrogenation rate of the phenolic benzene ring (and the following dehydration and hydrogenation to the naphthene) is considerably higher than that of hydrogenolysis of the phenolic hydroxyl group, as well as that of hydrogenation of the aromatic hydrocarbon. The hydrogenation rate of benzene in the presence of the catalyst 8376 mentioned by Moldavskii and Alekseeva [13] appears to be rather low compared to the other data published (cf. e.g. data reported by Schöngut and Vybíhal [1826, 2155]). However, the formation of naphthenes through the intermediate step of cyclic alcohols is considerably faster even when these elevated rates of hydrogenation of aromatic compounds are accepted. From Table 35, it follows, that the differences in the hydrogenation rates of phenol homologues are less than one order of magnitude. This agrees with data reported by Bobyshev et al. [192] for the conditions of the "hydrogenolysis" course of the reaction (Table 36). Naphthols are more reactive and, the low stability of thiophenol is also mentioned for the sake of comparison.

In the studies so far discussed, which dealt with the hydrogenation mechanism of phenols, only trace amounts of cyclic alcohols were found in the reaction products. The reason for this is that under the reaction conditions needed for high degrees of conversion to hydrocarbons, dehydration of cyclic alcohols is by far the most rapid reaction. Recently, Dahlke and Günther [369[hydrogenated technical cresol in the presence of the catalyst 8376 under specific conditions, and they succeeded in obtaining high yields of methylcyclohexanols. Under relatively very mild reaction conditions (236 °C, 300 atm, LHVS 0.4 l/l hour), the neutral reaction product, separated from the non-reacted raw material, contained 57% methylcyclohexanols as well as a hydrocarbon component (31.3 methylcyclohexane, 6.1% methylcyclohexenes and 5.6% toluene).

In other respects, hydrogenation of phenols in the presence of sulphide catalysts is very sensitive to temperature. For example, the apparent activation energy of cresol hydrogenation in the presence of the catalyst 8376 was found to be 56.2 kcal/mole (300 atm, 2.7 l/l hour, temperature interval 253–304 °C) [700].

6.5.2.

Hydrogenation of phenolic ethers

Phenolic ethers are a frequent and very important component of the products of the thermal processing of coal and shale. They are therefore one of the important types of non-hydrocarbons which influence the course of refining reactions. Differing from benzyl-type ethers (cf. p. 156) they are more resistant to hydrogenation, since efficient hydrogenolysis only takes place at temperatures of over 300 °C. For example, diphenyl oxide formed a mixture of benzene, cyclohexane (10%) and phenol in the presence of MoS_2 at 350 °C and 100 atm [1397]. With FeS as catalyst, a similar composition of the reaction product was obtained [1003].

Table *37.*
Hydrogenation of Phenolic Ethers on the 10927 Catalyst [2076].
(Autoclave experiments, 5% catalyst, H_2 pressure 160 atm/20 °C, temperature 485 °C, reaction time 3 hours.)

Phenolic ether hydrogenated	Energy of oxygen-alkyl bond (kcal/mole)	Hydrogenate composition[a]) (% by mol. of ether reacted)		Ratio $\frac{\text{benzene}}{\text{phenol}}$
		Benzene	Phenol	
Anisol	86.0	54.2	23.6	2.3
Phenetole	83.0	38.9	34.7	1.1
n-Propyl phenyl ether	81.0	30.4	35.4	0.86
		(25.6)	(53.6)	(0.48)
Tert-butyl phenyl ether	77.3	22.2	51.3	0.4

[a]) Data in brackets relate to experiments at 50 atm.

In the presence of WS_2 (temperatures 325 to 425 °C, initial H_2 pressure 140 to 150 atm, 5% catalyst, autoclave experiments) anisol is mainly converted to phenol at temperatures of 320–340 °C; at higher temperatures, hydrogenation to aromatic hydrocarbons takes place as well as partial reduction to naphthenes [1048]. Similarly, *p*-tolyl methyl ether formed methylcyclohexane as the main product in the presence of the highly efficient hydrogenation catalyst (WS_2–NiS, 310 to 340 °C, 170 atm [1741]).

A 73% yield of n-heptane was obtained by hydrogenating n-heptyl phenyl ether in the presence of MoS_3 [906].

The composition of the hydrogenation product of phenolic ethers depends to a large extent on the strength of bonds with oxygen. It was found in hydrogenating alkyl phenyl ethers with different alkyl groups in the presence of the catalyst 10927 at 485 °C, that the phenol yield is the greater, the higher the alkyl group of the mixed ether. It is to be seen from Table 37, that the increase of the phenol yield agrees with the decrease of the bond strength of the alkyl-oxygen bond. The partial hydrogen pressure also has a significant influence on the selectivity of the reaction. Kalechits [2076] assumes, that under these conditions radical reactions participate to a great extent, explaining this effect by the reaction of atomic hydrogen (formed by interaction of molecular hydrogen with any radical which is formed in the course of the hydrogenolysis process) with the benzene ring of a phenolic ether:

$$C_6H_5OCH_2CH_2R + H^{\bullet} \rightarrow C_6H_5(H)(OCH_2CH_2R)^{\bullet} \rightarrow C_6H_6 + {}^{\bullet}OCH_2CH_2R \qquad \text{(XXXIIa)}$$

$$^{\bullet}OCH_2CH_2R + H_2 - RCH_2CH_2OH + H^{\bullet}. \qquad \text{(XXXIIb)}$$

The higher the hydrogen pressure, the greater the probability of addition of atomic hydrogen and, the hydrogenolysis of the bond in the α-position, i. e. the phenol yield decreases with the increasing hydrogen pressure (cf. Table 37).

6.5.3.

Hydrogenation and hydrodealkylation of higher phenols

As stated at the beginning of this chapter, highly selective hydrogenation processes are being developed, the purpose being to enable conversion of higher alkylphenols and polyhydric phenols to low-boiling products containing the largest possible amount of phenols or cresols. Catalytic hydrodealkylation of higher phenols probably proceeds by a mechanism which is different from that of hydrodealkylation of alkylaromatic substances. The hydroxyl group mainly influences the strength of the bond between the alkyl group and the aromatic ring, and furthermore hydrocracking proceeds in the presence of water vapour, which is formed as the product of partial dehydroxylation of the phenol and which is able to influence the selectivity of the process to a large extent [1534, 1720]. Since destructive hydrogenation conditions are usually employed for hydrodealkylating phenols (the usual temperatures of phenol dealkylation are above 400 °C), some authors [963] believe that radical reactions take part in the process, by means of which some of the rules of the hydrogenation of higher phenols may be explained.

Hydrodealkylation of alkylphenols depends on the one hand on the type of phenol involved, and on the other on the size and structure of the alkyl group. Dealkylation of phenols with C_3 and higher alkyls is considerably easier than with methyl- and ethyl-derivatives. Splitting is particularly easy in the presence of sulphide catalysts in those cases, when the alkyl group is bound to a benzene ring by means of a quaternary carbon atom. For example, in hydrogenation of *p*-tert-butylphenol (total hydrogenation to hydrocarbons), dealkylation proceeded to more than 80% in the presence of the catalyst 8376 at 320 °C, while dealkylation was strongly suppressed in the case of *n*-butylphenol (mixture of the *o*- and *p*-derivative). Phenolic compounds of the bis-(*p*-hydroxyphenyl)-alkane or -cycloalkane type with quaternary as well as tertiary carbon atoms, which are formed by condensation of phenol with carbonyl compounds, are split with equal ease. For example, bis-(4-hydroxyphenyl)-alkanes of the type

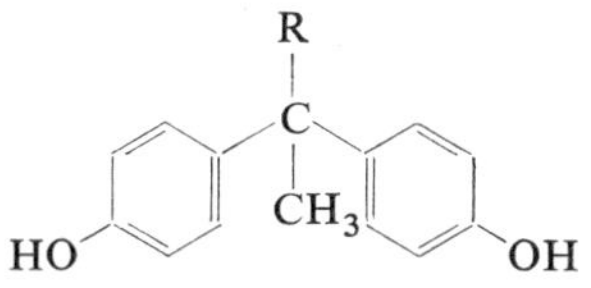

R
C
CH_3
HO OH

where R = H, CH_3 or C_2H_5 readily convert to phenol and the respective *p*-alkylphenol at temperatures of 275 to 330 °C in the presence of the catalyst 8376 [2140, 2144] or of the mixed sulphide catalyst NiS/Cr_2O_3–Al_2O_3 promoted with Cu, W or Mo sulphides [505, 1827, 1828, 1829]. The catalyst 8376 was also found suitable for the hydrogenation of residues and resins obtained in diphenylolpropane synthesis and phenol regeneration from these waste materials [2141, 2142, 2143].

Quantitative data relating to hydrocracking of pure higher phenols in the presence of sulphide catalysts are rarely found in the literature. Kalechits et al. [951, 963] determined the relative rates of hydrogenation and splitting of a number of different phenols under hydrocracking conditions (485 °C, 300 atm, autoclave experiments) in the presence of the catalyst 10927. From results stated in Table 38, the substitution of a phenol with another alkyl or hydroxyl group increases the extent of reduction to hydrocarbons as well as of splitting under the reaction conditions employed. Comparison with the results of hydrogenation of the same phenols to hydrocarbons in the presence of MoS_2 catalysts (cf. Table 36) indicates, that the reaction mechanism

Table 38.

The Relative Rates of Hydrogenation and Splitting of Some Phenols [963].
(Autoclave experiments, 5% catalyst 10927, temperature 475—485 °C, pressure 260 to 330 atm.)

Phenol hydrogenated	Relative rate of		Ratio reduction products / splitting products
	reduction	splitting	
Phenol	32.5	—	—
m-Cresol	43.0	20.4	2.9
o-Cresol	53.6	30.5	2.6
p-Cresol	65.5	44.0	2.4
3,5-Xylenol	50.1	45.0	1.2
Thymol	100.0	100.0	0.6
p-Benzylphenol	100.0	100.0	2.4
p-Phenylphenol	27.9	1.7	22.5
α-Naphthol	98.0	—	14.6
β-Naphthol	92.1	—	17.8
p-Propylphenol	57.8	67.3	0.7
5-Dihydroindanol	45.0	39.2	1.3
Tetrahydro-β-naphthol	68.0	46.6	2.0
Catechol	93.0	—	—
Resorcinol	100.0	—	—
Hydroquinone	88.0	—	—

was quite different in the two cases. This difference is obviously due to the widely different reaction conditions as well as to the different characters of the catalysts employed. The catalyst MoS_2 which is highly efficient in hydrogenation was used at 350 °C; the hydrocracking catalyst 10927 with a low hydrogenation activity was used at around 480 °C.*

On the grounds of the relative rates of reduction and splitting in the presence of the catalyst 10927, the phenols tested form the following series in the order of the decreasing splitting intensity (i.e. ratio of the amount of reduction products: amount of cleavage products – cf. Table 38): thymol < *p*-propylphenol < 3,5-xylenol < < 5-dihydroindanol < tetrahydro-β-naphthol < *p*-benzylphenol < *p*-cresol < < *o*-cresol < *m*-cresol < α-naphthol < β-naphthol < *p*-phenylphenol. From this series it follows, that the raw material best suited to selective dealkylation are alky-

*) The method of comparing relative reaction rates on the basis of autoclave experiments is objective only if the reactions are carried out in the strictly kinetic region and macrokinetic factors are eliminated. Since the majority of these studies contain no conclusive evidence of the fact that the experiments were done in the kinetic region, comparison of the results reported by different authors must be made very carefully. Diffusion processes are very frequently the rate-determining factors particularly in experiments performed in a mixed phase [2055, 2137] and kinetic conclusions based on these experiments may be quite erroneous.

phenols with a larger number of substituents on the ring, particularly with large and branched substituents. Bicyclic phenols are reduced far more quickly than they are split, since – as with hydrocracking of polycyclic aromatic compounds – the ring which bears the hydroxyl group must be hydrogenated first. Experiments show, that the main cleavage products are monoalkylphenols with an alkyl group in the *meta*-position:

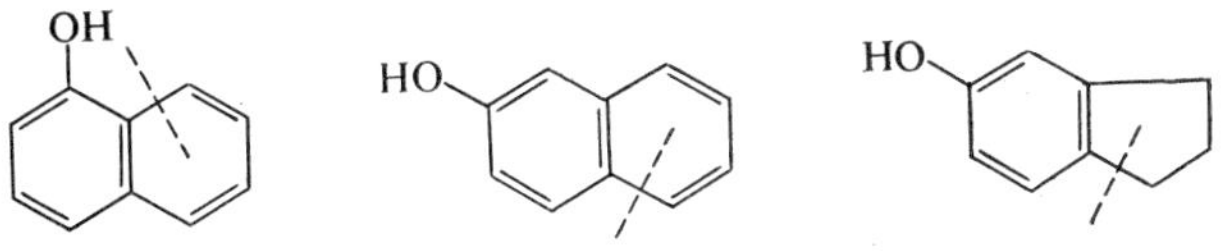

Perna and Pelčík [1604] studied the hydrodealkylation of phenols from low-temperature generator tar. Out of a number of catalysts tested, a mixture of vanadium and tungsten sulphides was found to be best. This catalyst was employed at 460 to 485 °C and an initial hydrogen pressure of 30 atm; compared to oxides, its yield of lower phenols is high while reduction to hydrocarbons is slight. Perna and Pelčík confirmed the earlier results of Rapoport and Masina [1720] (who tested MoS_3 as well as other catalysts in hydrodealkylation of phenols) concerning the favourable effect of ammonia and pyridin bases on the selectivity of the hydrodealkylation reaction. Dyakova et al. [444] and Salimgareeva et al. [1792] likewise employed a mixture of vanadium and tungsten sulphides as hydrodealkylation catalyst for higher phenols.

In other studies, especially patents, metal sulphides (NiS, MoS_2, FeS, CdS, ZnS, CaS, BaS etc.) are mentioned as components of hydrodealkylation catalysts for the conversion of alkylphenols to the lowest phenols [934, 1349, 1350, 1351, 1352, 1790, 2082].

6.5.4.

Hydrogenation of polyhydric phenols

The main reaction in hydroprocessing polyhydric phenols is their partial hydrogenolysis to monohydric phenols. The most reactive phenol of this group is hydroquinone, which however forms the greatest amount of high-molecular condensation products at the same time. Hydrogenation of dihydric phenols takes place at 400 °C very intensively even in the absence of a catalyst, the product being a mixture of neutral oils, phenols and substances similar to acid asphaltenes. Water is liberated, too [1048].

Hydrogenates of dihydric phenols, obtained by hydrogenating in the liquid phase with the 10927 catalyst [408, 963, 1048, 2014] were similar in character, with the exception of a lower content of asphaltic and resinous substances.

In the presence of hydrogenation-efficient sulphides, hydrogenation of dihydric phenols proceeds readily at temperatures as low as 250 °C. Under mild reaction conditions and with short reaction times, small amounts of condensed by-products are formed, particularly when catechol is hydrogenated. In the presence of MoS_2, the main products of catechol hydrogenation between 250 and 300 °C are phenol and cyclohexene; at 250°, cyclohexanol was also recovered as an intermediate product [1174]. It was shown by means of comparative experiments, that the strength of the C_{aryl}—O bond is considerably greater after etherification, so that for example veratrol is more difficult to hydrogenate (only 14% reacted at 250 °C under otherwise identical conditions, compared to 41% conversion achieved with catechol) [1174]. The presence of dicyclohexyl and *p*-cyclohexylphenol was proved in the reaction products from pyrocatechol, and this was also observed with other catalysts [1113], indicating that the course of the condensation and polymerisation reactions is analogous to the reactions observed in phenol hydrogenation [1394]. Klimov and Bogdanov [1048] who studied the hydrogenation of dihydric phenols in the presence of WS_2 explain the enhanced reactivity of these substances by the greater participation of their quinonoid form. This is also the reason for the formation of larger amounts of condensation and polymerisation products compared to the hydrogenation of monohydric phenols.

High percentage conversion to hydrocarbons is achieved by hydrogenating dihydric phenols under pressure in the presence of the catalysts 5058 and 8376. At lower temperatures (300 to 370 °C) partial dehydroxylation was brought about on these catalysts, the respective phenols being formed, but at the same time polymerisation and condensation products boiling at over 360 °C were formed [408].

Hydrogenation of dihydric phenols takes place readily in the presence of the catalyst 10927 under liquid-phase hydrogenation conditions [408, 963]. Selective dehydroxylation to monohydric phenols may be achieved in this process to an extent of 40–65%. Selectivity rises to about 90% in the presence of fixed-bed catalysts and after dilution with hydrocarbons [1113]. Kubička et al. tested a number of sulphides supported on carriers for this process [1113, 1115, 1126, 1127]. In Table 39, the activities of the individual catalysts employed are stated from the point of view of degradation of dihydric phenols, together with data relating to the ketones content of the hydrogenation products. A remarkable feature is the low activity of FeS and CuS catalysts supported on Al_2O_3 as well as unsupported NiS. The yield of ketones is low in those cases where alumina was used as carrier. 1,2-Cyclohexanediol and 2-hydroxycyclohexanone were identified as reaction intermediates, and in addition alkylation reactions take place, since *p*-cyclohexylphenol and 3- and 4-cyclohexylpyrocatechols were identified in the reaction products. The yield of ketones and resinous fractions may be decreased considerably by addition of ammonia to the pyrocatechol feed.

Selective hydrogenation of dihydric phenols may also be carried out in the presence of some industrial catalysts. An especially useful one is the aromatisation catalyst (DHD-catalyst) which partly dehydroxylated pyrocatechol to phenol at 350 to 390 °C

Table 39.

Hydrogenation of Technical Catechol on Different Sulphide Catalysts [1113, 1115]. (Flow reactor, LHSV 0.25 to 0.5, pressure 70 atm.)

Catalyst (metal content, % by wt.)	Content (% by wt.) of non-reacted catechol (ketone content) in the hydrogenation product at the respective temperature (°C)		
	300	400	450
NiS/Al_2O_3 (11.22)	57.6 (1.5)	40.4 (2.2)	20.2 (2.3)
NiS/SiO_2 (10.75)	38.6 (14.0)	26.7 (13.4)	7.6 (12.8)
NiS	61.5 (13.4)	47.4 (13.5)	48.4 (9.2)
CoS/SiO_2 (9.7)	76.1 (2.5)	19.8 (6.7)	2.7 (7.0)
CoS/Al_2O_3 (10.5)	43.2 (4.0)	2.6 (1.1)	— (0.7)
CuS/Al_2O_3 (11.95)	85.3 (1.4)	58.8 (1.6)	32.3 (2.6)
FeS/Al_2O_3 (8.62)	75.4 (0)	56.1 (0)	43.1 (0)

Note: In the analysis, ketones were regarded as cyclohexanone.

and 60 atm with a yield of 75% of the theoretical value [408, 933]. A spent DHD-catalyst, containing MoO_3 as well as MoS_2 on Al_2O_3 additionally activated by impregnation with ammonium molybdate or $NiSO_4$ followed by reduction, can also be used for selective hydrogenation of polyhydric phenols [1113, 1125, 1127]. The catalyst 8376 is less selective than the DHD-catalyst, especially in high-pressure operations [1116, 1119, 1122]. Catalysts containing CoS or NiS on Al_2O_3 are only active at higher temperatures (above 450 °C), which is an advantage since dealkylation of higher alkylphenols takes place at the same time and a greater amount of the required *m*-cresol is obtained [408, 1113, 1126]. Sulphide catalysts may also be used for partial hydrogenation of dihydric phenols under high hydrogen pressure (300 atm); in this case, operating temperatures are lower (300 to 325 °C) or higher throughput values are used. The catalyst 8376 may also be employed for this reaction [1122, 1816], the activity of this catalyst being satisfactory even when it is spent and partly deactivated with arsenic [1112, 1113]. Selective hydrogenation of polyhydric phenols is accompanied by dealkylation of higher alkylphenols to a limited extent only [1122].

6.6.

Hydrogenation of carbonyl compounds

The carbonyl group of aldehydes and ketones can be hydrogenated in the presence of sulphide catalysts to methyl or methylene groups and, partial hydrogenation to the hydroxyl group is also possible. In the first case, carbonyl compounds either react to form hydrocarbons, or the carbonyl group is reduced selectively, while other functional groups capable of being hydrogenated are retained (e.g. the phenolic hydroxyl group). In the second case, hydrogenation of aldehydes and ketones to the respective alcohols predominates.

6.6.1.

Total hydrogenation of the carbonyl group to methyl or methylene group

Hydrogenation of aldehydes and ketones in the presence of sulphide catalysts to hydrocarbons takes place as a stepwise process through the respective alcohols. Since the rate of reduction of the carbonyl group is considerably higher than the rate of hydrogenation of alcohols, the latter reaction will determine the overall rate of hydrogenation of carbonyl compounds. At a sufficient rate, hydrogenation of aldehydes and ketones to alcohols takes place at temperatures of 160 to 200 °C in the presence of metal sulphides [1165, 1383].

In the presence of sulphide catalysts, the carbonyl group is hydrogenated even at atmospheric pressure. Results reported by Bernardini and Brill [149], who hydrogenated some aldehydes and ketones at atmospheric pressure in the gas phase in the presence of WS_2 and MoS_2 catalysts, confirm the dehydration mechanism involved in hydrogenation of oxygen compounds to saturated hydrocarbons [1165, 1167].

To achieve total hydrogenation to the respective saturated hydrocarbons or alkyl-aromatic substances, however, temperatures above 300 °C and higher hydrogen pressures must be employed in order to hydrogenate the intermediate alcohols. In respect of the types of initial compounds, aromatic carbonyl compounds with the keto-group in the α-position relative to the benzene ring were found to be particularly reactive, their hydrogenation taking place quite readily; e.g. in the presence of MoS_2, acetophenone forms ethylbenzene in high yields at 180 °C (the initial hydrogen pressure being 105 atm) while benzophenone forms diphenylmethane in the presence of MoS_3 at 230 °C and a hydrogen pressure of 110 atm [906]. The use of MoS_2 supported on active carbon necessitates higher operating temperatures (min. 300 °C) [391].

In the presence of sulphide catalysts, simple ketones form the respective hydrocarbons in high yields. For example, acetone, methyl isopropyl ketone, mesityl oxide, cyclohexanone, cyclopentanone, chalcone, 1,3-diphenyl-1-propanone, acetophenone etc. were hydrogenated in autoclave experiments on MoS_2 or WS_2 to form 80 to 90% yields of the respective hydrocarbons at temperatures of 300 to 330 °C and initial hydrogen pressures of 100 to 130 atm/20 °C [906, 1159, 1163, 2175]. Some hydroxyketones may also be hydrogenated. For example, benzoin formed dibenzyl in the presence of MoS_2 at 330 °C and a H_2 pressure of 130 atm/15 °C in a yield of over 80%. On the other hand, very intensive hydrocracking was involved in hydrogenation of diacetonyl alcohol under roughly the same reaction conditions, since only about 12% 2-methylpentane was obtained together with a large amount of propane [1163, 2175].

Much attention has been devoted to pinacolone hydrogenation. Moldavskii and Nizovkina [1408] obtained a 75% yield of 2.3-dimethylbutane from pinacolone in the presence of MoS_2 at 340 to 350 °C and an initial hydrogen pressure of 80 atm in xylene solution. On the other hand, Amemiya [27] states that a high yield of 2,2-dimethylbutane is formed in this reaction. Landa et al. [1165] found that a mixture of 2,3- and 2,2-dimethylbutane is obtained by hydrogenating pinacolone, the first-named isomer predominating. As we have shown earlier, [p. 152] a mixture of these isomers is always obtained, irrespective of whether the initial compound is pinacolone, pinacol or pinacolyl alcohol, because hydrogenation of pinacolyl alcohol by an ionic dehydration mechanism is always involved in the final stage:

$$(CH_3)_2C(OH)\text{—}C(OH)(CH_3)_2 \xrightarrow{[+H^+]} \left[(CH_3)_2C(\overset{(+)}{OH_2})\text{—}C(OH)(CH_3)_2\right] \xrightarrow{[-H_2O]} (CH_3)_2\overset{(+)}{C}\text{—}C(OH)(CH_3)_2 \rightarrow$$

$$\rightarrow \left[CH_3\text{—}C(CH_3)_2\text{—}C(CH_3)\text{=}\overset{(+)}{O}\text{—}H\right] \xrightarrow{[-H^+]} CH_3\text{—}C(CH_3)_2\text{—}CO\text{—}CH_3 \xrightarrow{[H_2]} CH_3\text{—}C(CH_3)_2\text{—}CH(OH)\text{—}CH_3$$

$$CH_3\text{—}C(CH_3)_2\text{—}CH(OH)\text{—}CH_3 \xrightarrow{[+H^+]} \left[CH_3\text{—}C(CH_3)_2\text{—}CH(\overset{(+)}{OH_2})\text{—}CH_3\right] \xrightarrow{[-H_2O]} CH_3\text{—}C(CH_3)_2\text{—}\overset{(+)}{CH}\text{—}CH_3$$

$$CH_3\text{—}C(CH_3)_2\text{—}CH_2\text{—}CH_3 \xleftarrow{[H_2]} CH_3\text{—}C(CH_3)_2\text{—}CH\text{=}CH_2 \xleftarrow{[-H^+]} CH_3\text{—}C(CH_3)_2\text{—}\overset{(+)}{CH}\text{—}CH_3$$

$$CH_3\text{—}C(CH_3)_2\text{—}\overset{(+)}{CH}\text{—}CH_3 \rightarrow CH_3\text{—}\overset{(+)}{C}(CH_3)\text{—}CH(CH_3)\text{—}CH_3 \xrightarrow{[-H^+]} (CH_3)_2C\text{=}C(CH_3)_2 \xrightarrow{[H_2]} (CH_3)_2CH\text{—}CH(CH_3)_2 . \quad \text{(XXXIII)}$$

A considerable excess of 2,3-dimethylbutane in the hydrogenation products of the three oxygen compounds named is evidence of the much higher rate of stabilisation of the secondary carbonium ion by means of isomerisation to the tertiary ion.

Similarly, camphene was obtained along with isoborneol by a retropinacol rearrangement on hydrogenation of camphor in the presence of MoS_2 at 300 °C and 77 to 128 atm [1025] (cf. p. 152). Low-temperature hydrogenation (160 to 240 °C) of some other terpenic ketones (carvomenthone, carvone, pulegone and menthone) in the presence of MoS_2 was studied similarly [1028].

Hydrogenation of aldehydes to hydrocarbons in the presence of sulphide catalysts is accompanied by aldolisation reactions leading to the formation of a hydrocarbon with twice the number of carbon atoms as the initial aldehyde:

$$2\,RCH_2CH{=}O \longrightarrow RCH_2\underset{OH}{CH}\overset{R}{C}HCH{=}O \xrightarrow{[H_2[} RCH_2\underset{OH}{CH}\overset{R}{C}HCH_2{-}OH \longrightarrow$$

$$\xrightarrow{[H_2]} RCH_2CH_2\overset{R}{C}HCH_3\,. \qquad \text{(XXXIV)}$$

The hydrogenation products of propionaldehyde, n-butyraldehydè, oenanthal, n-heptanal and benzaldehyde in the presence of WS_2 or MoS_2 at 320 to 330 °C include 2-methylpentane, 3-methylheptane, 6-methyltridecane and dibenzyl besides propane, n-butane, n-heptane and toluene, conversion of aliphatic aldehydes to the respective monomeric and dimeric hydrocarbon being roughly equal [1161, 1162, 1163, 2176]. Dimeric products formed in benzaldehyde hydrogenation are particularly easily formed [286b]. On the other hand, when previously prepared aldols were hydrogenated in the presence of WS_2 and MoS_2 catalysts, considerable hydrocracking took place and saturated hydrocarbons were formed, corresponding to the original aldehyde. Nevertheless, hydrogenation of the aldols of ethanal, propanal, pentanal, heptanal, octanal, nonanal, decanal and undecanal at 320 to 340 °C and an initial hydrogen pressure of 100 to 130 atm resulted in the formation of the respective dimeric paraffins, i.e. n-butane, 2-methylpentane, 4-methylnonane, 6-methyltridecane, 7-methylpentadecane, 8-methylheptadecane, 9-methylnonadecane and 10-methylheneicosane, the yields of 20 to 48% being acceptable from the preparative point of view [1175].

On hydrogenation of some terpenic aldehydes, e.g. citral or citronellal in the presence of MoS_3, large amounts of resinous condensation products were obtained, particularly at temperatures over 200 °C [1029].

The easy reducibility of the carbonyl group in the α-position with respect to the benzene ring was utilised for rapid and easy preparation of alkylphenols from the respective hydroxy aldehydes and hydroxy ketones. For example, 3-hydroxyacetophenone, 3-hydroxypropiophenone, 3-hydroxybutyrophenone, 2-hydroxy-5-methylacetophenone, 4-hydroxy-3-methylacetophenone, 2-hydroxy-5-methylbenzophenone, 2-hydroxy-4-ethylacetophenone, 4-hydroxy-2-ethylacetophenone, 2-hydroxy-4-propylacetophenone, 4-hydroxy-2-butylacetophenone and m-hydroxybenzaldehyde formed the respective phenols on hydrogenation in the presence of MoS_2 and WS_2, yields being 55 to 90%. The phenols were prepared in an autoclave with initial hydrogen pressure of 100 atm and short reaction times, usually 10 to 30 min. Higher yields were achieved when working in solvents (cyclohexane or gasoline b. p. 90 to 125 °C). MoS_2 was found to be the more active catalyst, optimum yields being achieved in the temperature range of 240 to 300 °C, while temperatures 30 to 60 °C higher were needed with WS_2 [1164, 1173].

This method is also applicable to the preparation of dihydric phenols. For example, 2,4-dihydroxyacetophenone, 2,5-dihydroxyacetophenone and 3,4-dihydroxyacetophenone formed 4-ethylresorcinol (max. yield 79%), ethylhydroquinone (78%) and 4-ethylpyrocatechol (max. yield 38%), resp., by hydrogenation in cyclohexane solution at 270 to 280 °C, using a short reaction time (20 min) and in the presence of MoS_2 [1174].

Finally, vanillin-type compounds (vanillin, ethylvanillin, *o*-ethylvanillin and isoethylvanillin) formed high yields (70–80%) of the respective cresols (2-methoxy-4-methylphenol), 2-ethoxy-4-methylphenol, 2-ethoxy-5-methylphenol and 2-ethoxy-6-methylphenol) at the optimum temperature of 250 °C (hydrogen filling pressure 100 atm) in the presence of WS_2 or MoS_2 [1671, 1671a, 1673].

6.6.2.

Hydrogenation of carbonyl compounds to alcohols

One of the progressive processes of modern petrochemical industry is hydroformylation (oxo process), which provides aldehydes as intermediates for the production of primary alcohols which are required in increasing amounts for the softener and detergent industries.

Selective hydrogenation of oxo-aldehydes to alcohols is done conventionally on metallic and oxide catalysts, which are sufficiently selective and provide pure products. However the original olefinic raw material for hydroformylation frequently contains sulphur compounds (e.g. higher olefins obtained from shales [1007], gasoline obtained by thermal cracking of atmospheric residuum [1225] etc.), which are not really detrimental to the oxosynthesis proper [1172], but which remain in the aldehydes produced and poison the highly sensitive catalysts used to hydrogenate the aldehydes to alcohols. In such cases, either the raw material must be purified at great expense, or resistant sulphide catalysts must be used [352, 1723]. The use of such catalysts, however, is subject to two considerable disadvantages. Firstly, the selectivity of sulphide catalysts is lower than that of metal or oxide catalysts [1007], and secondly, the alcohols obtained always contain a slight amount of sulphur (of the order of several ppm) which gives them an unpleasant smell, so that additional purification of the hydrogenation products becomes unavoidable. In spite of this, the literature dealing with the application of sulphide catalysts in hydrogenating oxo-aldehydes is very comprehensive, although data on industrial applications are not yet available.

The main problem involved in the use of metal sulphides for the hydrogenation of carbonyl compounds to alcohols is, without doubt, securing the utmost possible selectivity of the process by limiting the extent of hydrogenolysis of the hydroxyl group formed. In consequence the process must be carried out at the lowest possible temperature but this, on the other hand, requires the use of sulphide catalysts with the maximum hydrogenation activity.

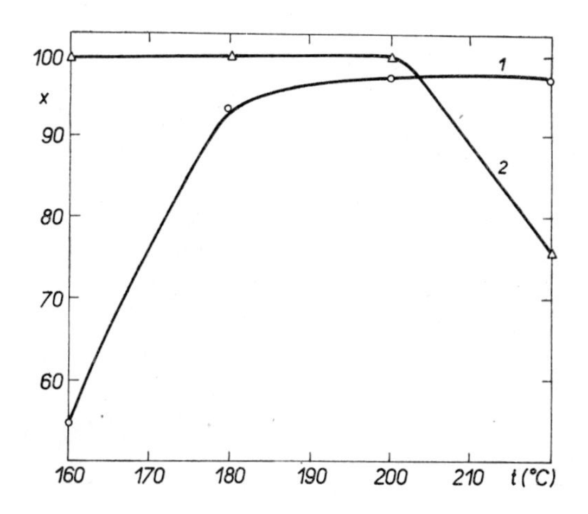

Fig. 56. Activity and selectivity of $WS_2.2\,NiS$ catalyst in hydrogenation of aldehydes (feed: C_6-C_8 aldehydes from the olefinic fraction of gasoline obtained in coking of Romashkino atmospheric residuum; pressure 150 atm, LHVS = = 1 l/l hour, hydrogen to oil ratio 1600 Nl/kg feed) [1225]. 1 — Overall conversion of aldehydes (%). 2 — Conversion of aldehydes to alcohols (%).

The mixed catalyst $WS_2.NiS$ (catalyst 3076) was found to be the most active one, its use being mainly described by Soviet authors [1007, 1224, 1225, 1226, 1228, 1228a, 1229, 1230, 1779, 1780, 1941]. The activity and selectivity of this catalyst is sufficiently high at pressures of more than 150 atm and temperatures in the range of 180 to 200 °C (Fig. 56), but selectivity decreases rapidly above 200–210 °C due to the hydrogenolysis and dehydration to hydrocarbons [1225]. With increased space velocities, the temperature may be raised a little; e.g. with a space velocity of 3 l/l hour the process may be run at 220 °C, at which temperature conversion of aldehydes reaches 94 to 95% and selectivity 97%. Similarly good results were recorded with these catalysts on hydrogenating butyraldehydes [1229] and mixture of C_{12}—C_{17} aldehydes [1226].

Lower selectivity was recorded by Ketslakh and Rudkovskii [1007] with the catalyst $WS_2.2\,NiS$ in the hydrogenation of aldehydes obtained by hydroformylation of olefins made from shale tar. Similarly with the catalyst $NiS–MoS_2$, a maximum selectivity of 79% was obtained when hydrogenating butyraldehyde (with practically total aldehyde conversion) [1383]. The catalyst 8376 similarly exhibited lower selectivity compared to the unsupported $WS_2.NiS$ catalyst [1224, 1422]. Moreover, when aldehydes are hydrogenated in the presence of this catalyst, a large amount of high-boiling condensation products is formed, which decreases the overall yield [2067].

MoS_2 was also recommended as a suitable catalyst for hydrogenating aldehydes to alcohols [492, 1325, 1356, 1359, 1953]. Under optimum reaction conditions (temperature 230 to 250 °C, pressure 70 to 210 atm) a high degree of selectivity of the reduction process was achieved with total conversion [1355, 1748]. MoS_2 supported on active carbon was also found to be an active and highly selective catalyst with a high degree of stability [495, 1094, 1326, 1358, 1950, 1951, 1952].

Molybdenum disulphide may also be used for selective reduction of the carbonyl group in hydroxyketones and hydroxyaldehydes as well as in various aldolisation products [1470], and furthermore to hydrogenate the condensation residues after hydrogenation of oxo-aldehydes [1357].

Metal sulphides (ZnS, CdS) also catalyse hydrogen transfer between aldehydes and alcohols [1278].

6.7.

Hydrogenation of acids, their esters and anhydrides

Hydrogenation of acids and esters proceeds by a number of consecutive reactions according to the scheme:

$$RC\begin{smallmatrix} \nearrow O \\ \searrow OR' \end{smallmatrix} \xrightarrow{[H_2]} \left[RCH\begin{smallmatrix} \nearrow OH \\ \searrow OR' \end{smallmatrix} \right] \xrightarrow{[-R'OH]} RCH{=}O \xrightarrow{[H_2]} RCH_2{-}OH \xrightarrow{[H_2]} RCH_3 \qquad \text{(XXXV)}$$

where R is a hydrogen atom or alkyl (aryl) group. Compared to the carbonyl group, the carboxyl group is hydrogenated in the presence of sulphide catalysts at a reasonable rate only at higher temperatures (above 240 to 250 °C) i.e. under conditions when dehydration already begins to play an important role.

Therefore it will be very difficult, when using sulphide catalysts, to conduct the reaction in such a manner as to make it proceed to alcohols with high selectivity; and in consequence, these catalysts are of neither technological nor preparative significance for hydrogenating acids and esters to alcohols. Baroni's data [111] dealing with the use of Mo, W, V, Zn, Co, Ni and some other sulphides for selective hydrogenation of acids, esters and glycerides, as well as data by Pine and Ellert [1649a] concerning the high selectivity of MoS_2 supported on active carbon in hydrogenation of neoheptanoic acid methylesther to alcohols are thus quite unique.

The low selectivity of sulphide catalysts was confirmed by a number of experiments dealing with hydrogenation of undecylenic, stearic, oleic and cinnamic acids in the presence of MoS_2 [1178]. Effective hydrogenation of the carboxyl group sets in at temperatures above 240 °C. In this process, certain differences were found with the above acids in the temperature coefficient of carboxyl hydrogenation; while with undecylenic and cinnamic acids this coefficient varied only slightly in the 200 to 300 °C range, conversion increased considerably in the case of oleic and stearic acid in the 230 to 260 °C temperature interval. Alcohol yields are very low due to their dehydration and subsequent hydrogenation, partial esterification of the non-reacted acids by the alcohols formed taking place at the same time [1178]. The yield of alcohols cannot be increased even by suitable adjustment of the reaction conditions. In the case of some experiments with oleic acid, the reaction time, temperature and the amount of catalyst were varied. A somewhat higher selectivity in hydrogenating the product of oxidation of paraffins was recently achieved by using sulphurised MoO_2 supported on active carbon [800].

Decarboxylation causes great difficulty in the process of hydrogenating acids under the relatively extreme conditions demanded by sulphide catalysts. The extent of decarboxylation is rather low when heat-stable aliphatic acids and esters are hydrogenated, so that these may easily by hydrogenated to give high yields of hydrocarbons. For example, in the presence of MoS_2 or MoS_3 at temperatures of over 300 °C (initial hydrogen pressure usually over 100 atm) acetic, palmitic [906, 1163, 2175], lauric [284] acids as well as ethyl propionate, ethyl undecylenate [1163, 2175], methyl laurate [284], cetyl palmitate and methyl palmitate [990, 991] formed the respective saturated hydrocarbons in yields of about 90%. The catalyst 5058 is equally efficient [1159], while Ni and Co sulphides are less active and more extensive decarboxylation takes place in their presence [284, 285]. According to some Japanese authors [284, 285] the decarboxylation activity of MoS_3 is particularly low. Benzoic acid esters form toluene readily in the presence of MoS_2 and WS_2 [1159, 1163, 2175] as well as phenylacetic acid forms ethylbenzene [906].

On hydrogenation in the presence of WS_2, free benzoic acid was decarboxylated to only a small extent in spite of the high reaction temperature (370 to 400 °C) [1048]. In hydrogenation of some less stable acids, however, (e.g. cinnamic acid, some hydroxyacids and hydroxyesters [1176, 1178]), decarboxylation is quite significant. Decarboxylation occurs to a considerable extent in the hydrogenation of phthalic acid and phthalic anhydride. In the presence of MoS_2 and WS_2 catalysts several authors obtained *o*-xylene yields ranging from minimum values up to 90% at 250 to 303 °C and hydrogen pressures of 70 to 120 atm/20 °C [663, 906, 1159, 1397, 1409, 2175, 2244]. At 200 °C, phthalic anhydride was partially hydrogenated to *o*-toluic acid in the presence of MoS_3 [906]. Differing from phthalic anhydride, phthalide readily formed *o*-xylene in yields of 85% in the presence of MoS_2 at 330 °C and an initial hydrogen pressure of 120 atm [1409].

With more complicated esters, containing other reducible functional groups in their molecules, side reactions take place, particularly decarboxylation and hydrocracking. Hydroxyesters obtained by the Reformatski reaction, provided the respective saturated hydrocarbons in yields of only 50% in the presence of MoS_2 and WS_2 at 320 to 340 °C, decarboxylation being the main competing reaction [1176].

Methyl pivalate formed a 64% yield of isopentane on hydrogenation in the presence of MoS_2 at 340 °C and a hydrogen filling pressure of 130 atm, only a trace of neopentane being formed. At a lower temperature (320 °C) neopentyl alcohol was found as reaction intermediate in a yield of 7.5% [1165]. The considerable decrease in the neopentane yield in this case, compared to the hydrogenation of neopentyl alcohol (cf. p. 153) may be explained by the possibility of isomerisation of pivalic aldehyde to methylisopropylketone, which increases the relative amount of isopentane in the hydrogenate [2208]:

$$(CH_3)_3C{-}COOCH_3 \xrightarrow{[H_2]} \left[(CH_3)_3C{-}CH{<}^{OH}_{OCH_3}\right] \xrightarrow{[-CH_3OH]} (CH_3)_3C{-}CH{=}O$$

$$(CH_3)_3C{-}CH{=}O \longrightarrow CH_3{-}CH(CH_3){-}COCH_3 \xrightarrow{[H_2]} (CH_3)_2CHCH_2CH_3.$$

$$(CH_3)_3C{-}CH{=}O \xrightarrow{[H_2]} (CH_3)_3C{-}CH_2OH \xrightarrow{[H_2]} (CH_3)_3C{-}CH_3. \qquad \text{(XXXVI)}$$

Although the carboxyl group is more difficult to hydrogenate compared to the carbonyl group, it may be selectively reduced to the methyl group in the molecule of aromatic hydroxyacids, the respective phenols thus being obtained from these substances. This reaction, however, is accompanied by decarboxylation, so that e.g. hydrogenation of methyl-*p*-hydroxybenzoate in the presence of MoS_2 (295 °C, hydrogen filling pressure 100 atm) results in a mixture containing 81.5% *p*-cresol and 18.5% phenol [1164]. Similarly, salicylic acid was converted to *o*-cresol in the presence of MoS_3 at 300 °C in a yield of 51.5% [906]. Reports on hydrogenation of natural esthers in the presence of sulphide catalysts include hydrogenation of sunflower oil on a special NiS catalyst [2081a] and hydrogenation of tallow in the presence of $NiS–WS_2/Al_2O_3–SiO_2$ [1834b].

6.8.

Hydrogenation of oxygen-containing heterocyclic compounds

Oxygenous heterocyclic compounds are a regular and significant component of coal, shales and the products of their thermal processing [1172]. Oxygen is similarly bound in heterocyclic compounds in petroleum and petroleum fractions, although in the latter case the presence of nitrogen and sulphur heterocycles is more important. In petroleum the presence of oxyallo-2-betulenc was proved; and in addition to oxygen bound in other types of compounds, heterocyclic oxygen atoms are found in petroleum resins and asphaltenes [2136]. These oxygenous heterocyclic compounds are hydrogenolysed in the course of refining processes and either simpler oxygen compounds are formed, or the heterocycles undergo hydrogenolysis to hydrocarbons. Here the manner of hydrogenolysis of the bonds linking the heterocyclic oxygen will be mainly influenced by the character of the hydrogenated substance (the presence of other heteroatoms, the degree of aromaticity of the oxygenous heterocycle, the molecular weight of the compound, etc.) and, furthermore, hydrogenolysis will be influenced by the reaction conditions and the refining catalyst employed.

Many features of the hydrogenolysis chemistry of natural heterocyclic substances may be predicted from a knowledge of the mechanism of hydrogenation of individual compounds. Unfortunately, unlike nitrogen and sulphur containing heterocyclic

Table 40.

Comparison of the Activity of Various Sulphide Catalysts in Hydrogenolysis of Fundamental Oxygen-, Sulphur- and Nitrogen-Containing Heterocycles [658].

(Flow experiments at atmospheric pressure, the degree of hydrogenolysis was measured by the amount of H_2O, H_2S and NH_3 formed.)

Substance hydrogenated	Degree of hydrogenolysis (%) on the catalyst												
	$MoO_2 . MoS_2$					MoS_2				Ni_3S_2			
	Initial concentration (ppm)	Temperature (°C)				Initial concentration (ppm)	Temperature (°C)			Initial concentration (ppm)	Temperature (°C)		
		225	250	275	300		150	175	200		200	350	400
Furan	1090	—	26	—	77	1200	11.0	32.6	76.0	1100	0	0	10
Tetrahydrofuran	—	—	—	—	—	895	13.0	37.0	85.8	870	0	25	55
Thiophene	460	10	18	38	69	260	—	20	60	280	0	15	39
Tetrahydrothiophene	—	—	—	—	—	260	—	40	80	280	41	100	100
Pyrrole	540	—	6	—	16	—	—	—	—	540	0	0	0

compounds, basic research into the hydrogenation of oxygen compounds, particularly in the presence of sulphide and oxide catalysts which are most important from a technical point of view is only just beginning.

A qualitative picture of the stability of oxygen heterocycles compared to other heterocyclic compounds may be obtained from the work of Griffith et al. [658]. From the results given in Table 40 it follows that the result of hydrogenation depends mainly on the nature of the sulphide catalyst employed. The MoS_2 catalyst is considerably more efficient than Ni_3S_2, the difference in hydrogenation stability being slight in the presence of the first catalyst (thiophene is a little more stable), while nickel subsulphide catalyses thiophene hydrogenolysis more effectively than furan degradation. Similarly, the resistance of heterocyclic compounds depends on the type of catalyst employed under hydrogen pressure (when subjected to hydrorefining conditions) [1181]. Pyrrole is considerably more stable.

Tetrahydroderivatives of the basic heterocyclic compounds are less stable, the differences in the degree of hydrogenolysis again being substantially greater in the presence of Ni subsulphide than with the catalyst MoS_2 (Table 40).

The greater stability of aromatic heterocyclic compounds with respect to hydrogenolysis is also shown by autoclave experiments at elevated pressures, carried out in the presence of the catalyst 5058 [309, 1177]. Totally hydrogenated aromatic oxygen heterocycles are cyclic ethers in character, so that their hydrogenolysis takes place more readily. Addition of a benzene ring to furan increases to some extent the resistance of the heterocyclic compounds to hydrogenolysis.

Much attention has been given to the hydrogenation chemistry of furan derivatives. For example, Orlov [1548] achieved 95% total conversion of the initial compounds on hydrogenating dibenzofuran at 450 to 470 °C, including 85% conversion to a hydrocarbon mixture composed of benzene, cyclohexane, methylcyclopentane and higher bicyclic naphthenic and aromatic hydrocarbons, e.g. phenyl cyclohexane. Mach and Khadshinov [1271] reported similar results.

On the basis of some experiments with hydrogenation of dibenzofuran in the presence of MoS_2, Hall and Cawley [720] point out the high stability of this substance, assuming that the reaction takes place at a lower temperature (300 to 350 °C) with the formation of 2-cyclohexylphenol or, to a lesser degree, of 2-phenylcyclohexanol. The products were phenylcyclohexane, phenol and cyclohexane. These authors believe that at high temperatures, the initial product is 2-hydroxydiphenyl which later converts to diphenyl, phenylcyclohexane, dicyclohexyl, benzene, cyclohexane and other products of hydrocracking, isomerisation, condensation and polymerisation.

Landa et al. [1187] studied this problem too and these authors conclude from a detailed analysis of dibenzofuran hydrogenation products obtained in the presence of MoS_2 at 200 to 340 °C (hydrogen filling pressure 100 atm), that the most probable reaction course is expressed by the scheme (XXXVII):

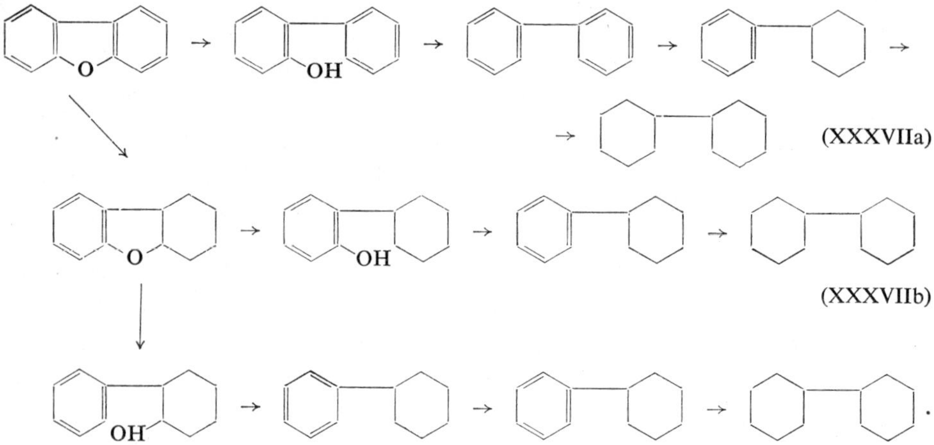

(XXXVIIa)

(XXXVIIb)

(XXXVIIc)

Landa et al. [1187] also studied the hydrogenation chemistry of coumarone by means of gas-chromatographic analysis of the hydrogenation products. On the basis of the results obtained, especially from the absence of β-phenylethyl alcohol in the reaction products, the authors conclude that hydrogenolysis proceeds mainly through *o*-ethylphenol:

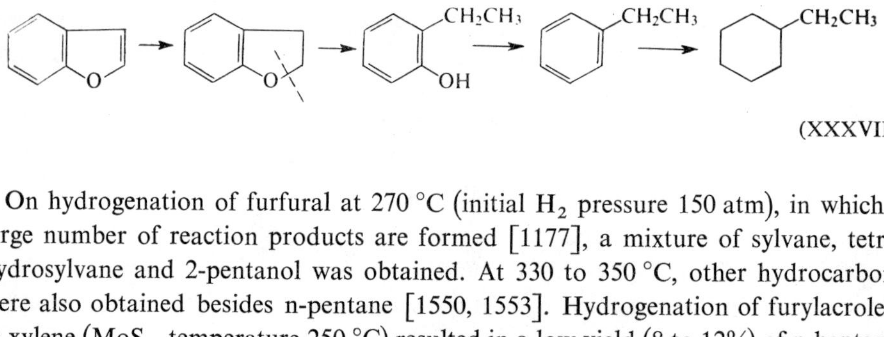

(XXXVIII)

On hydrogenation of furfural at 270 °C (initial H_2 pressure 150 atm), in which a large number of reaction products are formed [1177], a mixture of sylvane, tetrahydrosylvane and 2-pentanol was obtained. At 330 to 350 °C, other hydrocarbons were also obtained besides n-pentane [1550, 1553]. Hydrogenation of furylacrolein in xylene (MoS_3, temperature 250 °C) resulted in a low yield (8 to 12%) of n-heptane. A somewhat higher yield (30 to 35%) of n-heptane was obtained by hydrogenation of furylethylcarbinol on the mixed sulphide catalysts MoS_2–NiS or MoS_2–CoS [1550, 1552].

In connection with this problem, Petrov [1613] mentions the possibility of synthetising paraffins by means of metal sulphide–catalysed hydrogenation of easily available furylalkylcarbinols of the general formula:

$$C_4H_3O-CH(OH)-C_nH_{2n+1}\,.$$

6.9.

Hydrogenation of sulphur compounds

Hydrogenation of sulphur compounds is the most important field of application of sulphide catalysts. Research into catalytically active sulphides for this application is far wider, however, since other catalysts, e.g. oxides or metals, are also converted to sulphides under desulphurisation reaction conditions (cf. p. 43).

This fact is of prime importance particularly in technical hydrocracking, hydrorefining and hydrodesulphurisation processes (cf. Chapter 7), since natural raw materials (coal, tars, petroleum fractions, shale products etc.) nearly always contain differing concentrations of sulphur compounds as their most important non-hydrocarbon components. For example a number of mercaptans, disulphides, heterocyclic compounds (cyclic sulphides, thiophenes) were identified in petroleum fractions [1347] as well as sulphoxides, sulphones, sulphonic acids, thiol esters etc. [2136]. On the other hand, the character of sulphur compounds in coal, heavy tars, petroleum residues, petroleum resins, asphaltenes etc. is known only to a very meagre extent and frequently we do not know even the nature of the bond between sulphur and the rest of the organic molecule.

Differing from other refining reactions, catalytic hydrodesulphurisation of technically significant crude materials has been studied more deeply. This is due to the fact that this reaction was investigated with a number of typical sulphur compounds, kinetic data being obtained in many cases in addition to a deep study of chemical and thermodynamic aspects of the reaction. A number of these data have already been evaluated in McKinley's work [1347] and in monographs by Obolentsev and Mashkina [1518] and Perchenko and Sergienko [1602]. Here, therefore, we shall first briefly survey the earlier findings and then amplify them in the light of more recent results.

6.9.1.

Chemistry and thermodynamics of hydrogenation of sulphur compounds

Either total hydrogenolysis of sulphur compounds takes place in the presence of sulphide catalysts, hydrogen sulphide and a hydrocarbon residue being formed, or partial hydrogenolysis or simply reduction of functional groups capable of hydrogenation in the molecule of the sulphur compounds takes place. In the case of total hydrogenolysis, the most important types of sulphur compounds react according to the following reaction schemes:

$$RCH_2CH_2SH + H_2 \rightarrow RCH_2CH_3 + H_2S, \qquad \text{(XXXIX)}$$

$$RCH_2CH_2SCH_2CH_2R' + 2\,H_2 \rightarrow RCH_2CH_3 + R'CH_2CH_3 + H_2S, \qquad \text{(XL)}$$

$$RCH_2CH_2S.S.CH_2CH_2R + 3\,H_2 \rightarrow 2\,RCH_2CH_3 + 2\,H_2S, \qquad \text{(XLI)}$$

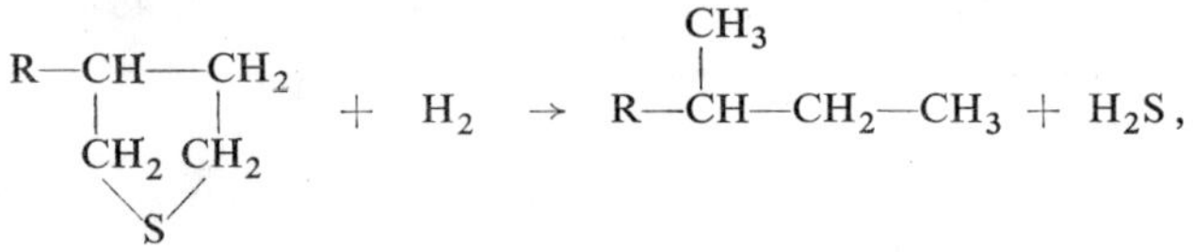

$$\text{R—CH—CH}_2\text{ (ring: CH}_2\text{, CH}_2\text{, S)} + H_2 \rightarrow R\text{—}\underset{}{\overset{CH_3}{|}}\text{CH—CH}_2\text{—CH}_3 + H_2S, \qquad \text{(XLII)}$$

$$\text{(bicyclic: CH}_2\text{—CH—CH}_2\text{, CH}_2\text{ S, CH}_2\text{—CH—CH}_2) + H_2 \rightarrow \text{(1,2-dimethylcyclopentane: CH}_3\text{, CH}_3) + H_2S, \qquad \text{(XLIII)}$$

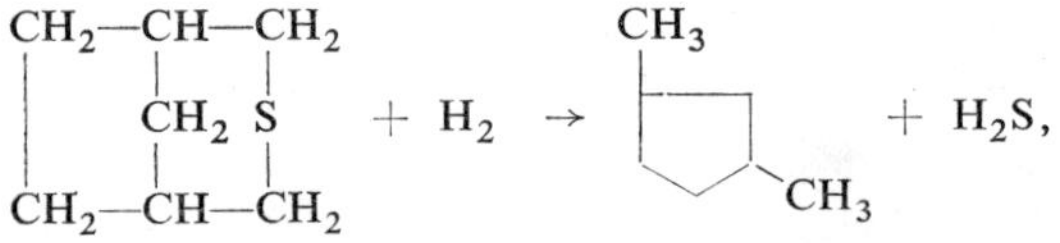

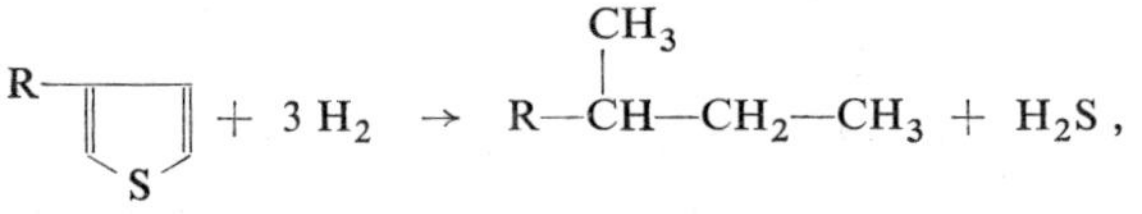

$$\text{R-(thiophene, S)} + 3\,H_2 \rightarrow R\text{—}\overset{CH_3}{|}\text{CH—CH}_2\text{—CH}_3 + H_2S, \qquad \text{(XLIV)}$$

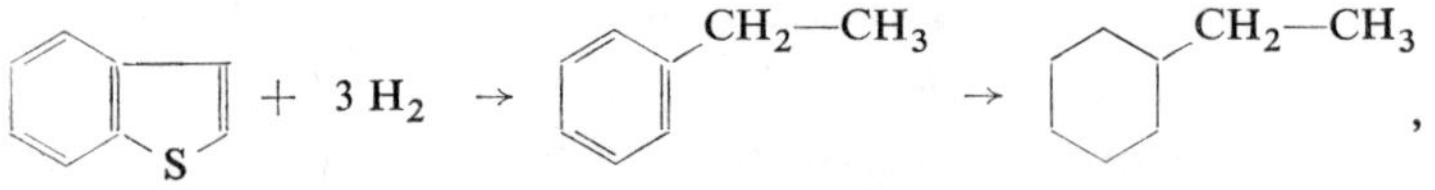

$$\text{(benzothiophene, S)} + 3\,H_2 \rightarrow \text{(benzene)—CH}_2\text{—CH}_3 \rightarrow \text{(cyclohexane)—CH}_2\text{—CH}_3, \qquad \text{(XLV)}$$

$$\text{(dibenzothiophene, S)} + XH_2 \rightarrow \text{(biphenyl)} + H_2S \;\downarrow\; \text{(cyclohexylbenzene)} \;\downarrow\; \text{(bicyclohexyl)}. \qquad \text{(XLVI)}$$

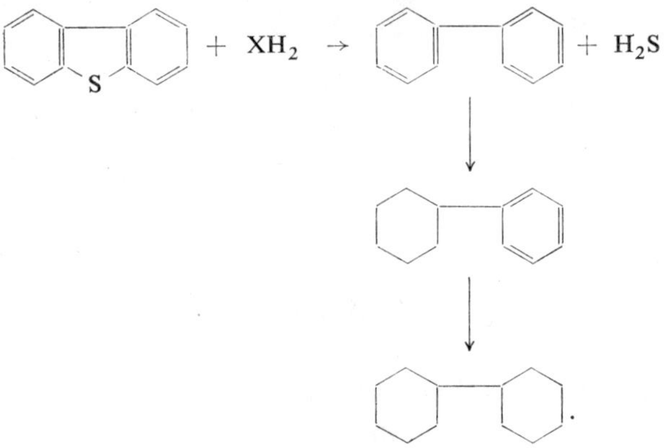

In the case of partial hydrogenolysis or reduction of sulphur compounds in the presence of sulphide catalysts, reactions of the following type are involved:

$$RCH_2CH_2.S.CH_2CH_2R' + H_2 \rightarrow RCH_2CH_2SH + R'CH_2CH_3, \qquad \text{(XLVII)}$$

$$RCH_2CH_2.S.S.CH_2CH_2R \rightarrow 2\,RCH_2CH_2SH, \qquad \text{(XLVIII)}$$

$$R\text{–}\overset{\frown}{\underset{S}{\square}} + H_2 \rightarrow R\text{–}\underset{|}{\overset{CH_3}{CH}}\text{–}CH_2CH_2SH, \qquad \text{(XLIX)}$$

$$R\text{–}\underset{S}{\square}\ (\text{thiophene}) + 2\,H_2 \rightarrow R\text{–}\underset{S}{\square}, \qquad \text{(L)}$$

$$\text{(benzothiophene)} + H_2 \rightarrow C_6H_5\text{–}CH_2CH_2SH \qquad \text{(LIa)}$$

$$\searrow\ C_6H_4(CH_2CH_3)(SH), \qquad \text{(LIb)}$$

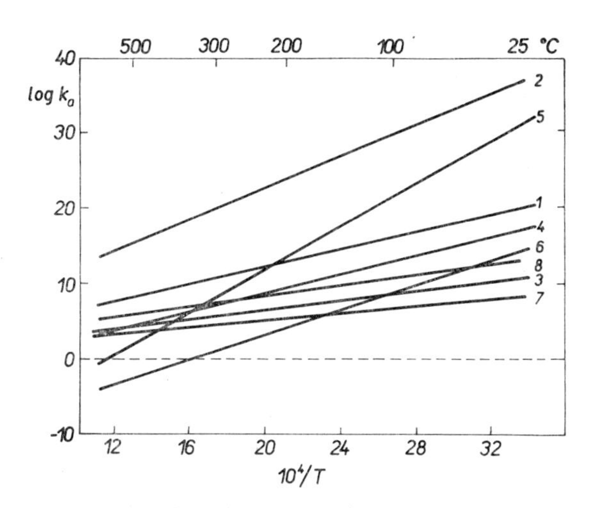

Fig. 57. Equilibrium constants (k_a) of hydrogenation reactions which are typical for sulphur-containing substances [1347, 2154].

1 – $CH_3CH_2SCH_2CH_3 + 2\,H_2 \rightarrow 2\,C_2H_6 + H_2S$.

2 – $CH_3CH_2S.SCH_2CH_3 + 3\,H_2 \rightarrow 2\,C_2H_6 + 2\,H_2S$.

3 – $CH_3CH_2CH_2CH_2SH + H_2 \rightarrow n\text{-}C_4H_{10} + H_2S$.

4 – (thiolane ring, S) $+ 2\,H_2 \rightarrow n\text{-}C_4H_{10} + H_2S$.

5 – (thiophene ring, S) $+ 4\,H_2 \rightarrow n\text{-}C_4H_{10} + H_2S$.

6 – (thiophene ring, S) $+ 2\,H_2 \rightarrow$ (thiolane ring, S).

7 – $CH_3CH_2SCH_2CH_3 + H_2 \rightarrow CH_3CH_2SH + C_2H_6$.

8 – $CH_3CH_2S.SCH_2CH_3 + H_2 \rightarrow 2\,C_2H_5SH$.

$$\text{(thiophene 1,1-dioxide ring, } SO_2\text{)} + 3\,H_2 \rightarrow \text{(thiolane ring, } S\text{)} + 2\,H_2O\,, \qquad \text{(LII)}$$

$$RCH_2CH_2SO_3H + 3\,H_2 \rightarrow RCH_2CH_2SH + 3\,H_2O\,. \qquad \text{(LIII)}$$

In total hydrogenolysis of sulphur compounds to hydrocarbons, processes of the types (XLVII) to (LII) are stages in a consecutive reaction and they serve only in exceptional cases to prepare new sulphur compounds; e. g. reactions of the type (LI) may be used to prepare arylthiols in the presence of active Group VI sulphides or Group VIII polysulphides [1209].

In thermodynamical respects, hydrogenation reactions of sulphur compounds are all favourable (the temperature dependence of the equilibrium constants of some typical desulphurisation and hydrogenation reaction is shown in Fig. 57). Particularly with high hydrogen pressures, the equilibrium of hydrogenolytic reactions is shifted far to the side of hydrocarbon formation.

At temperatures of 300 to a maximum of 500 °C, which is the temperature range most often used in technical hydrodesulphurisation reactions performed on sulphidic catalysts, conversion of the majority of sulphur compounds to hydrocarbons is almost complete. The desulphurisation reactions are highly exothermic, the heat effect of the hydrogenolytic reaction being roughly proportional to the hydrogen consumption.

Detailed thermodynamic data of desulphurisation reactions have been dealt with in the relevant monographs (for example [1347, 1518, 2154]).

6.9.2.

Review of hydrogenation reactions of sulphur compounds in the presence of sulphide catalysts

6.9.2.1.

Hydrogenation of simple sulphur compounds

As a guide to the large amount of literature and patent data, the most important information about the reaction conditions, catalysts employed and products of hydrogenation and desulphurisation reactions of individual sulphur compounds is summarised in the Tables 41 to 46. Only the most important references are stated, referring to studies in which the catalysts were metal sulphides or oxidic catalysts directly sulphurised by the initial sulphur compound or, by hydrogen sulphide formed by hydrodesulphurisation of the compound involved. The hydrogenation and desulphurisation reactions of carbon disulphide, carbonyl sulphide, mercaptans,

Table 41.

Hydrogenation of Carbon Disulphide and Carbonyl Sulphide.

Substance hydro-genated	Catalyst employed	Reaction conditions					Reaction products		Re-fer-ence
		Tempera-ture (°C)	Pressure H_2 (atm)	Reaction time in a stationary system, min (space velocity in flow reactor, l/l/hour or g/g/hour)	Molar ratio $H_2 : CS_2$	Solvent	Product identified (purpose of experiment)	Yield (% of theory)	
CS_2	Mo or U sulphides or oxy-sulphides	> 400	1	—	—	—	CH_4	quant.	[239]
CS_2	MoS_2	170—350	1—184	(Various)	1 : 2—5 : 1	—	CH_4, CH_3SH, $(CH_3)_2S$	—	[520]
CS_2	MoS_2	150—250	20	—	—	Cyclo-hexan	CH_4	—	[276]
CS_2	Ni_3S_2	100—250	1	(Various)	Very high	—	CH_4, CH_3SH	—	[366]
CS_2	Ni_3S_2	420	1	—	—	—	(Study of fuel gas desulphurisation)	—	[503]
CS_2	FeS, NiS, CoS	220—400	1	(Various)	—	—	(Measurement of CS_2 hydrogenation kinetics)	—	[922]
CS_2	Sulphurised Co–Mo/ /Al_2O_3	—	1	(Various)	—	—	(Measurement of CS_2 hydrogena-tion kinetics)	—	[601]
COS	Ni_3S_2	> 150	1	(Various)	—	—	(Measurement of COS hydrogena-tion kinetics)	—	[367]

Table 42.

Hydrogenation of Mercaptans.

Substance hydrogenated	Catalyst employed	Reaction conditions					Reaction products		Reference
		Temperature (°C)	H_2 pressure (atm)	Reaction time in a stationary system, min (space velocity in flow apparatus, l/l/hour or g/g/hour)	Molar ratio H_2/mercaptan	Solvent	Product identified (purpose of experiment)	Yield (% of theory)	
Methyl mercaptan	Ni_3S_2	200—300	1	(Various)	Very high	—	CH_4, H_2S (reaction kinetics)	—	[367]
Methyl mercaptan	MoS_2, WS_2	130—270	Subatm.	10—130	2 : 1 (initial)	—	CH_4, $(CH_3)_2S$, H_2S (reaction mechanism)	—	[2217]
Ethyl mercaptan	MoS_2, WS_2	170—300	Subatm.	10—130	5 : 1 (initial)	—	C_2H_6, $(C_2H_5)_2S$, H_2S, C_2H_4 (reaction mechanism)	—	[1022]
Ethyl mercaptan	Ni_3S_2	140—250	30	—	High	Cyclohexane	(Study of catalyst activity)	—	[1311]
Ethyl mercaptan	MoS_2	230	30	120	High	Kerosine	$(C_2H_5)_2S$, C_2H_6, H_2S	—	[1396]
Ethyl mercaptan	Sulphurised Co–Mo	225—407	1	(Various)	High	Heptane	(Hydrodesulphurisation kinetics)	—	[603]
n-Butyl mercaptan	Sulphurised Co–Mo	270—372	1	(Microreactor pulse technique)	High	—	(Hydrodesulphurisation kinetics and mechanism)	—	[399]
n-Butyl mercaptan	MoS_2	180—340	100	30	10 : 1	—	H_2S, C_4-hydrocarbons (measure-	—	[1184]

sec-Butyl mercaptan	MoS_2	180—340	100	30	10 : 1	—	ment of the resistance of different	—	[1184]
iso-Butyl mercaptan	MoS_2	180—340	100	30	10 : 1	—	mercaptan types to hydrodesulphu-	—	[1184]
tert-Butyl mercaptan	MoS_2	180—260	100	30	10 : 1		risation)	—	[1184]
3-Methyl-1-butanethiol	MoS_2	230	30	120	High	Kerosine	H_2S, C_5-hydrocarbons, sulphides	—	[1396]
1-Hexanethiol	MoS_2	180—340	100	30	10 : 1	—	H_2S, n-C_6H_{14}	—	[1184]
1-Heptanethiol	MoS_2	260	125—217	540—720	High	Hexane	H_2S, heptane	58—75	[351]
1-Octanethiol	MoS_2, CoS, Co–Mo/ /Al_2O_3	240—260	125—217	540—720	High	Hexane	H_2S, octane, octyl sulphide	—	[351]
tert-Octyl mercaptan	MoS_2, CoS, NiS_x, V_2S_5	230—240	125—217	540—720	High	Hexane	H_2S, octanes	—	[351]
1-Octanethiol	MoS_2	180—340	100	30	20 : 1	—	H_2S, octane (dependence of hydrodesulphurisation on temperature)	—	[1184]
1-Dodecanethiol	MoS_2	180—340	100	30	20 : 1	—	H_2S, dodecane (dependence of hydrodesulphurisation on temperature)	—	[1184]
1-Octadecanethiol	Co–Mo/ /Al_2O_3	375	50	(Various)	8 : 1	Tridecane	(Study of hydrodesulphurisation)	—	[786]
Cyclohexyl mercaptan	MoS_2	190—340	100	30	10 : 1	—	H_2S, cyclohexane	—	[1184]
Cyclohexyl mercaptan	MoS_2	260	125—217	540—720	High	Hexane	H_2S, cyclohexane	—	[351]

Continued Table 42.

Substance hydrogenated	Catalyst employed	Reaction conditions					Reaction products		Reference
		Temperature (°C)	H_2 pressure (atm)	Reaction time in a stationary system, min (space velocity in flow apparatus, l/l/hour or g/g/hour)	Molar ratio H_2 mercaptan	Solvent	Product identified (purpose of experiment)	Yield (% of theory)	
Benzyl mercaptan	MoS_2	80—220	100	30	10 : 1	—	H_2S, toluene, dibenzyl sulphide, dibenzyl, methylcyclohexane	—	[1184]
Thiophenol	MoS_2	100—260	100	30	10 : 1	—	H_2S, benzene, cyclohexane	—	[1184]
Thiophenol	MoS_2	230	30	120	High	Kerosine	H_2S, hydrocarbons	—	[1396]
Thiophenol	8376	220	200	0—180	High	Cetane	(Hydrodesulphurisation kinetics)	—	[1602] [1860]
Thiophenol	MoS_3	300—450	20—40	120	—	Hexane	(Study of the hydrodesulphurisation reaction mechanism)	—	[2232]
Thiophenol	Re_2S_7	300	252	540	High	Ethanol	Benzene, cyclohexane	60—65 35—40	[226]
Thiophenol	MoS_2	< 200	100	—	—	—	Benzene	—	[908]

sulphides, polysulphides and sulphur heterocycles with one or more sulphur atoms in the ring are mentioned separately in Tables 41 to 46.

6.9.2.2.

Hydrogenation of sulphur compounds containing oxygen or nitrogen

In a number of cases, sulphide catalysts were employed for partial hydrogenation of oxygen-containing sulphur compounds to compounds of lower oxidation states. Polysulphide derivatives are hydrogenated very easily. Aliphatic hydroxythiols were prepared by hydrogenation of hydroxypolysulphides in the presence particularly of active polysulphides of the iron-group metals; for example, 1,7-dihydroxy-3,4,5-trithiaheptane was hydrogenated in the presence of cobalt trisulphide at 150 °C and pressure of 98 to 175 atm, giving a good yield of mercapto-ethanol [1599]. Similarly, thioglycollic acid was obtained in good yield at 130 °C by reduction of dithioglycolic acid in the presence of MoS_2 with a hydrogen pressure of 50 atm [911].

Hydrogenation of sulphones, which are relatively very stable, was studied too. Itabashi [911a] found that hydrogenation in the presence of molybdenum sulphide takes place with an initial hydrogen pressure of 100 atm at a temperature of over 300 °C, total desulphurisation taking place only at 370 °C. Small amounts of the respective sulphides and mercaptans were found as intermediates in the hydrogenation of di-n-octyl-, dibenzyl- and diphenylsulphones. It was found possible to hydrogenate butadienesulphone and its alkylhomologues in the presence of MoS_2, WS_2 or CoS in such a selective manner that mainly the respective thiophenes were formed [211]. Similarly, sulpholane and its homologues can be selectively hydrogenated to tetrahydrothiophenes in the presence of the catalyst NiS/Al_2O_3 [1417].

In some cases, sulphide catalysts were employed in the hydrogenation of derivatives of sulphonic, sulphinic and thiosulphuric acids. For example, benzene sulphonic, *p*-toluene sulphonic and *p*-toluene sulphinic acids were hydrogenated to aromatic hydrocarbons in the presence of Mo sulphide at 375 °C with an initial hydrogen pressure of 100 atm [913, 914]. Itabashi [914] assumes the following system of consecutive reactions to take place in this hydrogenation process:

p-toluene sulphonic acid → *p*-toluene sulphinic acid → *p*-thiocresol → toluene. (LIV)

When this reaction is conducted under mild conditions in the presence of highly active catalysts, a high yield of the respective thiol is obtained. For example 82% n-dodecyl mercaptan was obtained from dodecyl thiosulphate at 150 °C and a H_2 pressure of 140 to 175 atm in the presence of cobalt sulphide [1210] or, aromatic thiols are formed by selective hydrogenation of alkali sulphinates in the presence of PtS_2 [2085a]. According to Lazier and Signaigo [1209, 1210] the respective

Table 43.
Hydrogenation of Sulphides.

Substance hydrogenated	Catalyst employed	Reaction conditions					Reaction products		Reference
		Temperature (°C)	H_2 pressure (atm)	Reaction time in a stationary system, min (space velocity in flow apparatus, l/l/hour or g/g/hour)	Molar ratio H_2: sulphur compound	Solvent	Product identified (purpose of experiment)	Yield (% of theory)	
Dimethyl sulphide	MoS_2	290	Subatm.	0—120	2 : 1	—	(Hydrogenation kinetics)	—	[2217]
Diethyl sulphide	WS_2	210—270	Subatm.	Different times	2 : 1	—	(Kinetics and mechanism of sulphide hydrogenation)	—	[1022]
Diethyl sulphide	MoS_2	230	30	120	—	Kerosine	(Determination of the relative	—	[1396]
Di-n-propyl sulphide	MoS_2	230	30	120	—	Kerosine	degree of hydrodesulphurisation in relationship to the structure of the initial substance)	—	[1396]
Di-n-butyl sulphide	MoS_2	180—340	100	30	High	—	H_2S, C_4-hydrocarbons,	—	[1184]
Di-isobutyl sulphide	MoS_2	180—340	100	30	High	—	C_4-mercaptans (measurement of	—	[1184]
Di-sec-butyl sulphide	MoS_2	180—340	100	30	High	—	the relative hydrodesulphurisat-	—	[1184]
Di-tert-butyl sulphide	MoS_2	180—340	100	30	High	—	ion rate)	—	[1184]

Diallyl sulphide	MoS_2	230	30	120	—	Kerosine	(Measurement of the relative hydrodesulphurisation rate)	—	[1396]
Diisoamyl sulphide	MoS_2	230	30	120	—	Kerosine		—	[1396]
Di-n-hexyl sulphide	8376	220	200	0—1440	High	Cetane	(Study of hydrodesulphurisation kinetics)	—	[1602] [1860]
Di-n-hexyl sulphide	MoS_2	180—340	100	30	High	—	(Influence of alkyl-length on the relative rate of hydrodesulphurisation)	—	[1184]
Di-n-octyl sulphide	MoS_2	180—340	100	30	High	—		—	[1184]
Di-n-dodecyl sulphide	MoS_2	180—340	100	30	High	—		—	[1184]
Di-n-heptyl sulphide	MoS_2	240—260	126—217	540—720	—	Hexane	H_2S, n-heptane	69	[351]
Dicyclohexyl sulphide	MoS_2	180—340	100	30	High	—	H_2S, cyclohexyl mercaptane	—	[1184]
2,8-Dimethyl-5-thianonane	Sulphurised Co–Mo	325—425	40	(Different space velocity)	High	Cetane	(Study of hydrodesulphurisation kinetics and mechanism)	—	[1515]
2,4,6,8-Tetramethyl-5-thianonane	Sulphurised Co–Mo	375	40	(Different space velocity)	High	Cetane	(Study of hydrodesulphurisation kinetics and mechanism)	—	[1517]
1,5-Diphenyl-3-thiapentane	8376	220	200	0—840	High	Cetane	(Study of hydrodesulphurisation kinetics)	—	[1602] [1860]

Continued Table 43.

Substance hydrogenated	Catalyst employed	Reaction conditions					Reaction products		Reference
		Temperature (°C)	H_2 pressure (atm)	Reaction time in a stationary system, min (space velocity in flow apparatus, l/l/hour or g/g/hour)	Molar ratio H_2: sulphur compound	Solvent	Product identified (purpose of experiment)	Yield (% of theory)	
Dibenzyl sulphide	8376	220	200	0—120	High	Cetane	(Study of hydrodesulphurisation kinetics)	—	[1602] [1860]
Dibenzyl sulphide	MoS_2	230	30	120	—	Kerosine	(Measurement of the relative hydrodesulphurisation degree)	—	[1396]
Dibenzyl sulphide	Sulphurised Co–Mo	325—425	40	(Different space velocity)	High	Cetane	(Study of hydrodesulphurisation kinetics and mechanism)	—	[1517]
Fenyl butyl sulphide	8376	220	200	0—720	High	Cetane	(Study of hydrodesulphurisation kinetics)	—	[1602] [1860]
1,3-Dimethyl-1-(phenylthio)-butane	Sulphurised Co–Mo	375	40	(Different space velocity)	High	Cetane	(Study of hydrodesulphurisation kinetics and mechanism)	—	[1517]

Benzyl octyl sulphide	Co–Mo	375	50	(Various)	—	Tridecane	(Study of hydrodesulphurisation)	—	[786]
Phenyl decyl sulphide	Co–Mo	375	50	(Various)	—	Tridecane	(Study of hydrodesulphurisation)	—	[786]
Diphenyl sulphide	8376	220	200	0—540	High	Cetane	(Study of hydrodesulphurisation kinetics)	—	[1602] [1860]
Diphenyl sulphide	Sulphurised Co–Mo	325—425	40	(Different space velocity)	High	Cetane	(Study of hydrodesulphurisation kinetics and mechanism)	—	[1517]
Diphenyl sulphide	Re_2S_7	280—300	160	120	—	—	Benzene	94	[226]
Diphenyl sulphide	MoS_3	300	250	150	—	Ethanol	Benzene	40	[226]
Allyl phenyl sulphide	CoS_X	175—200	175	102	—	Ethanol	Thiophenol	32	[226]

Table 44.

Hydrogenation of Disulphides.

Substance hydrogenated	Catalyst employed	Reaction conditions					Reaction products		Reference
		Temperature (°C)	H_2 pressure (atm)	Reaction time in a stationary system, min (space velocity in flow apparatus, l/l/hour or g/g/hour)	Molar ratio H_2: sulphur compound	Solvent	Product identified (purpose of experiment)	Yield (% of theory)	
Diethyl disulphide	MoS_2	230	30	120	High	Kerosene	(Study of hydrodesulphurisation mechanism)	—	[1396]
Dibutyl disulphide	NiS_x	150	35	—	—	Dioxane	Butyl mercaptan	73	[510]
Di-n-butyl disulphide	MoS_2	140—300	100	30	13 : 1	—	(Study of hydrodesulphurisation chemistry, influence of structure on the relative reaction rate)	—	[1186]
Di-isobutyl disulphide	MoS_2	140—300	100	30	13 : 1	—		—	[1186]
Di-sec-butyl disulphide	MoS_2	140—300	100	30	13 : 1	—		—	[1186]
Di-tert-butyl disulphide	MoS_2	100—300	100	30	13 : 1	—		—	[1186]
Di-n-hexyl disulphide	MoS_2	180—340	100	30	20 : 1	—		—	[1186]
Di-n-octyl disulphide	MoS_2	180—340	100	30	25 : 1	—		—	[1186]
Di-n-octyl disulphide	Co–Mo	375	50	—	High	Tridecane	n-Octane	Quant.	[786]

Di-n-octyl disulphide	MoS_3	130—300	100	—	—	—	(Study of hydrogenation chemistry; at lower temperatures mercaptans are chiefly formed)	—	[910]
Di-n-decyl disulphide	MoS_2	180—340	100	30	30 : 1	—	(Study of hydrodesulphurisation mechanism)	—	[1186]
Di-n-dodecyl disulphide	MoS_2	120—200	100	—	—	—	n-Dodecyl mercaptan	95	[912]
5,10-diethyl-7,8-dithiatetradecane	CoS_X	155	63	45	—	Dioxane	2-Ethyl-1-hexanethiol	Quant.	[510]
Dicyclohexyl disulphide	CoS_X	150	140	—	—	Dioxane	Cyclohexyl mercaptan	93	[510]
Dibenzyl disulphide	CoS_X	125	140	—	—	Benzene	Toluene Benzyl mercaptan	8 55	[510]
Dibenzyl disulphide	MoS_3	130—300	100	—	—	—	(Study of hydrodesulphurisation mechanism)	—	[910]
Diphenyl disulphide	Re_2S_7	165—195	154	60	—	Methylcellosolve	Thiophenol	Quant.	[226]
Diphenyl disulphide	MoS_3	130—300	100	—	—	—	(Study of hydrodesulphurisation mechanism)	—	[910]
Diphenyl disulphide	Platinum-group sulphides	—	—	—	—	—	Thiophenol	(Very high)	[430] [2088] [2089]
Di-p-tolyl disulphide	MoS_3	130—300	100	—	—	—	(Study of hydrodesulphurisation mechanism)	—	[910]
Diphenyl disulphide	Co and Ni sulphides	140—200	150	—	—	—	Thiophenol	99.8	[733a]

Table 45.

Hydrogenation of Sulphur-Containing Heterocycles (one atom in the ring).

Substance hydrogenated	Catalyst employed	Reaction conditions					Product identified (purpose of experiment)	Reference
		Temperature (°C)	H_2 pressure (atm)	Reaction time in a stationary system, min (space velocity in flow apparatus, l/l/hour or g/g/hour)	Molar ratio H_2: sulphur compound	Solvent		
Tetrahydrothiophene	Ni_3S_2	200—400	1	(260 ppm sulphur compound in the feed)	High	—	(Influence of temperature on the degree of desulphurisation)	[658]
Tetrahydrothiophene	MoS_2	150—200	1	(dtto)	High	—	(Influence of temperature on the degree of desulphurisation)	[658]
Tetrahydrothiophene	MoS_2, WS_2	300—350	Subatm.	0—180	—	—	H_2S, C_4H_8, C_4H_{10}, thiophene (hydrodesulphurisation kinetics)	[1023]
Tetrahydrothiophene	MoS_2	180—340	100	30	10 : 1	—	(Measurement of the temperature dependence of the degree of hydrodesulphurisation)	[1184] [1185]
Tetrahydrothiophene	Sulphurised Co–Mo	270—372	1	(Microreactor pulse technique)	High	—	H_2S, C_4H_{10}, C_4H_8 (measurement of hydrodesulphurisation kinetics)	[399]
2-Propylthiophane	Sulphurised Co–Mo	325—425	40	(Various)	High	Cetane	(Hydrodesulphurisation kinetics)	[1517] [1518]

2-Phenylthiophane	Sulphurised Co–Mo	325—425	40	(Various)	High	Cetane	(Hydrodesulphurisation kinetics)	[1515] [1518]
2-Phenylpropylthiophane	Sulphurised Co–Mo	375	40	(Various)	High	Cetane	(Hydrodesulphurisation kinetics)	[1517] [1518]
Thiolene	Sulphurised Co–Mo	270—372	1	(Microreactor pulse technique)	High	—	H_2S, C_4H_{10}, C_4H_8 (hydrodesulphurisation kinetics and mechanism)	[399]
Thiophene	MoS_2	300—450	80	—	—	Benzene	H_2S, thiophane (hydrogenation mechanism)	[1393]
Thiophene	Co, Ni, Cd, Mo, Cu, Mn sulphides	210—250	15—60	40—315	—	Benzene	(Hydrogenation kinetics)	[1395]
Thiophene	MoS_2	—	—	—	—	—	H_2S, C_4H_{10}	[581]
Thiophene	MoS_2, MoS_2–CoO	305—450	—	(Various)	100 : 15 to 100 : 85	Benzene	—	[2041]
Thiophene	MoS_2	300	1	—	—	—	(Removal of thiophene from gases)	[1036]
Thiophene	MoS_2/Al_2O_3	232	238	150 hr.	—	—	Thiophane	[740]
Thiophene	MoS_3	300—450	20—40	120	—	Benzene	(Study of hydrodesulphurisation)	[2232]
Thiophene	MoS_2	180—340	100	30	—	—	(Butyl mercaptan, thiophane, octane (study of hydrodesulphurisation))	[1184]
Thiophene	MoS_2	300	168	150	—	—	Thiophane (7%), H_2S	[226]
Thiophene	MoS_2	200—300	1	(0.5)	—	—	Thiophane, butyl mercaptan, di-n-butyl sulphide	[276]
Thiophene	MoS_2, WS_2	272—305	Subatm.	0—140	10 : 1	—	(Hydrogenation mechanism and kinetics)	[1023]
Thiophene	NiS/Al_2O_3—SiO_2, 8376, 3076, Co–Mo	250—490	2.5—100	(1.5—10)	—	Hydrocarbon mixture	(Study of hydrodesulphurisation)	[964]

Continued Table 45.

Substance hydrogenated	Catalyst employed	Reaction conditions					Product identified (purpose of experiment)	Reference
		Temperature (°C)	H_2 pressure (atm)	Reaction time in a stationary system, min (space velocity in flow apparatus, l/l/hour or g/g/hour)	Molar ratio H_2: sulphur compound	Solvent		
Thiophene	Sulphurised Fe_2O_3–Cr_2O_3	300	1	(Low)	High	—	H_2S, C_4H_{10} (study of the mechanism of hydrogenation)	[919]
Thiophene	Re_2S_7	230—260	140	225		—	Thiophane (70%)	[185] [226]
Thiophene	Re_2S_7	200	—	—	—	—	(Preparation of high yields of thiophane)	[85]
Thiophene	Sulphurised Co–Mo	300—375	15	(Various)	Various	Shale gasoline	(Study of hydrodesulphurisation kinetics in the presence of olefins)	[721]
Thiophene	CoS–MoO_3	200	1	(Various)	—	Benzene	(Hydrodesulphurisation mechanism and kinetics)	[1590]
Thiophene	Sulphurised Co–Mo	274—400	1	(Microreactor pulse technique)	High	—	(Study of hydrodesulphurisation mechanism and kinetics)	[1066] [1565]
Thiophene	Sulphurised Co–Mo	220—460	—	—	—	—	(Testing the hydrodesulphurisation activity of different Co–Mo catalysts)	[1283]

Thiophene	Sulphurised Co–Mo	350	10	(Various)	—	Heptane	(Hydrodesulphurisation kinetics)	[602]
Thiophene	Sulphurised Co–Mo catalysts	240—450	1	(0.125)	High	—	(Measurement of the hydrodesulphurisation activity of catalysts)	[1283]
Thiophene	Sulphurised Cr_2O_3	400	1	(Microreactor pulse technique)	High	—	(Mechanism of desulphurisation)	[1565] [1566]
2-Methylthiophene	Sulphurised Co–Mo	170—362	1	(Microreactor pulse technique)	High	—	(Hydrodesulphurisation mechanism and kinetics)	[398]
3-Methylthiophene	Sulphurised Co–Mo	170—362	1	(Microreactor pulse technique)	High	—	(Hydrodesulphurisation mechanism and kinetics)	[398]
tert-Butylthiophene	WS_2–NiS	379	49	(5)	5.1	—	Octanes	[78]
2-Octylthiophene	Sulphurised Co–Mo	325—425	40	(Various)	High	Cetane	(Hydrodesulphurisation mechanism and kinetics)	[1515] [1518]
2-Benzylthiophene	WS_2–NiS	300	180	—	—	—	n-Amylbenzene	[2075]
2,5-Diethylthiophene	Sulphurised Co–Mo	375	40	(Various)	High	Cetane	(Hydrodesulphurisation mechanism and kinetics)	[1517] [1518]
2,5-Dibutylthiophene	Sulphurised Co–Mo	325—425	40	(Various)	High	Cetane	(Hydrodesulphurisation mechanism and kinetics)	[1515] [1518]
2,3,4,5-Tetraphenylthiophene	MoS_2	180—340	100	30	33 : 1	—	(Temperature dependence of desulphurisation)	[1184] [1185]
2-Stearylthiophene	WS_2–NiS	300	180	—	—	—	n-Docosane	[2075]

Continued Table 45.

Substance hydrogenated	Catalyst employed	Reaction conditions					Product identified (purpose of experiment)	Reference
		Temperature (°C)	H_2 pressure (atm)	Reaction time in a stationary system, min (space velocity in flow apparatus, l/l/hour or g/g/hour)	Molar ratio H_2: sulphur compound	Solvent		
2-Acylthiophenes	CoS_X	225	100—150	240—780	—	Acetic acid	2-Alkylthiophenes, 2-alkylthiophanes	[252]
Benzothiophene	MoS_2	180—340	100	30	13 : 1	—	(Temperature dependence of desulphurisation)	[1184] [1185]
Benzothiophene	MoS_3	300—450	20—40	120	High	Isooctane	(Study of hydrodesulphurisation mechanism)	[2232]
Benzothiophene	Sulphurised Co–Mo	375	40	(Various)	High	Cetane	(Hydrodesulphurisation kinetics)	[1518]
Benzothiophene	Sulphurised Co–Mo	—	—	—	—	—	(Study of hydrodesulphurisation mechanism)	[281]
Benzothiophene	8376	220	200	0—1320	High	Cetane	(Hydrodesulphurisation kinetics)	[1602] [1860]
3-Methylbenzothiophene	Sulphurised Co–Mo	325—425	40	(Various)	High	Cetane	(Hydrodesulphurisation kinetics)	[1518]
Trimethylbenzothiophenes	Sulphurised Co–Mo	260—370	7—40	(1.5—10)	High	Light circulating oil	(Hydrodesulphurisation kinetics of cracked petroleum fractions)	[570]
Dibenzothiophene	MoS_2	180—340	100	30	20 : 1	—	(Temperature dependence of desulphurisation)	[1184] [1185]

Dibenzothiophene	8376	220	200	0—1320	High	Cyclohexane	(Hydrodesulphurisation kinetics)	[1602] [1860]
Dibenzothiophene	Sulphurised Co–Mo	325—425	40	(Various)	High	Cetane	(Hydrodesulphurisation kinetics)	[1518] [1517]
Dibenzothiophene	Sulphurised Co–Mo	260—370	7—40	(1.5—10)	High	Light circulating oil	(Hydrodesulphurisation kinetics of cracked petroleum fractions)	[570]
Dibenzothiophene	Sulphurised Co–Mo	345—395	11.4	(Various)	High	Gasoline	(Hydrodesulphurisation kinetics in the presence of nitrogenous bases)	[2218]
Dibenzothiophene	Sulphurised Co–Mo	375	50	—	High	Tridecane	Dicyclohexyl (35%), phenylcyclohexane (65%)	[786]
Dibenzothiophene	Sulphurised Co–Mo	—	—	—	—	—	(Study of hydrodesulphurisation mechanism)	[281] [787]
Octahydrodibenzothiophene	Sulphurised Co–Mo	325—425	40	(Various)	High	Cetane	(Hydrodesulphurisation kinetics)	[1517] [1518]
Diacenaphtylenethiophene	MoS_2	180—340	100	30	33 : 1	—	(Temperature dependence of hydrodesulphurisation)	[1184] [1185]

Table 46.

Hydrogenation of Sulphur-Containing Heterocycles (with several S atoms in the ring).

Substance hydrogenated	Catalyst employed	Reaction conditions					Product identified (Purpose of experiment)	Reference
		Temperature (°C)	Pressure H_2 (atm)	Reaction time in a stationary system, min	Molar ratio H_2: sulphur compound	Solvent		
Thianthrene	MoS_2	180—340	100	30	20 : 1	—	Diphenylsulphide, benzene, cyclohexane (hydrodesulphurisation temperature dependence)	[1184] [1185]
Thianthrene	Sulphurised Co–Mo	375	40	Various	High	Cetane	(Hydrodesulphurisation kinetics and mechanism)	[1517] [1518]
1,4-Dithiane	MoS_2	180—340	100	30	15 : 1	—	Diethylsulphide (temperature dependence of desulphurisation)	[1184] [1185]
Trithiane	MoS_2	180—340	100	30	15 : 1	—	H_2S, dimethylsulphide (temperature dependence of desulphurisation)	[1184] [1185]
trans-1,2-Cyclohexylene trithiocarbonate	MoS_3	150	140	720	High	Benzene	trans-1,2-(Methylenedithio) cyclohexane	[1298a]
Ethylene trithiocarbonate	MoS_3	150	133	720	High	Benzene	1,3-Dithiolane	[1298a]

thio-derivatives may be obtained by means of selective hydrogenation of alkyl thiocyanates, derivatives of mono-, di- and trithiocarbonic acids, sulphamides etc. in the presence of active Mo and Co sulphides and polysulphides.

A high degree of selectivity was achieved with the use of sulphide catalysts in the hydrogenation of compounds of the type of 2′-keto-3,4-imidazolido-2-ω-substituted alkylthiophenes, which are intermediate products in biotin systhesis. For example, 2′-keto-3,4-imidazolido-2-(3-phenoxypropyl) thiophene formed a good yield of 2′-keto-3,4-imidazolido-2-(cyclohexylpropyl)tetrahydrothiophene in the presence of the MoS_2/Al_2O_3 at 190 to 200 °C and a pressure of 182 atm [291]. On the other hand, some 2-thenoylderivatives may be hydrogenated selectively to the corresponding thenylderivatives in the presence of Co and Mo sulphides [308a] or Pd and Re sulphides [1787a, 1787h].

6.9.3.

Mechanism and kinetics of hydrogenation of sulphur compounds

The best possible knowledge of the course and relative reaction rates of hydrogenation of the individual types of sulphur compounds is particularly important from the practical point of view. It is known for example, that thiophenic derivatives are more resistant to hydrodesulphurisation than other types of sulphur compounds and furthermore, that the rate of desulphurisation depends on the molecular weight and structure of the initial substance. Therefore, particularly in recent years, attention has been given to systematic studies of the mechanism and kinetics of hydrogenation of the most important sulphur compounds in the presence of sulphide catalysts.

6.9.3.1.

Hydrogenation of carbon disulphide and carbonyl sulphide

The hydrogenation kinetics of carbon disulphide and carbonyl sulphide is of outstanding importance in hydrodesulphurisation of gases. Crawley and Griffith [366] studied the hydrogenation of CS_2 at atmospheric pressure on the Ni_3S_2 catalyst at temperatures of 100 to 250 °C. The CS_2 concentration in hydrogen varied from 0.28 to 1.7 g/Nm^3, i.e. of the same order of magnitude as the typical content of organic sulphur compounds in carbonisation town gas [1605]. This hydrogenation takes place as a consecutive reaction, with methylmercaptan as reaction intermediate:

$$CS_2 \xrightarrow{[H_2]} CH_3SH \xrightarrow{[H_2]} CH_4 \,. \qquad \text{(LV)}$$

The composition of the hydrogenation product, obtained at 175 °C with different amounts of CS_2 in the feed is detailed in Table 47. With lower CS_2 concentrations in the feed, hydrogenation takes place by a pseudo-first-order reaction. Fischer and Koch [520] similarly found that CH_3SH and dimethylsulphide are formed in CS_2 hydrogenation at elevated pressure (the reaction was carried out with less than the stoichiometric amount of H_2, temperature 200 to 225 °C); trithiane also proved to be present in the reaction product.

Ivanovskii et al. [922] measured the hydrogenation kinetics of CS_2 in the presence of Fe, Co and Ni sulphides in a circulation reactor at atmospheric pressure and 200 to 400 °C (the CS_2 vapour concentration in hydrogen varied from 1 to 5%). In agreement with results reported by Crawley and Griffith [366], it was found that the reaction products (H_2S and CH_4) do not retard the reaction and that the reaction rate may be expressed by the relationship:

$$r = kc^n_{CS_2} \quad \text{(LVI)}$$

where r is the reaction rate, c_{CS_2} is the carbon disulphide concentration in the original gas, k is the rate constant (related to a unit surface of the catalyst) and n is the exponent, the value of which is 0.5 with FeS and NiS and 0.6 with CoS. The activity of FeS (measured by the reaction velocity constant) is considerably lower (10 to 15 times) than the activity of the other two sulphides, of which CoS is the more active.

Table 47.

CS_2 Hydrogenation on the Ni_3S_2 Catalyst at 175 °C [366].
(H_2 flow-rate 473 ml/min.)

CS_2 feed (mg/hour)	Degree of CS_2 conversion (%)	Amount of CH_3SH formed (mg/hour)	Amount of H_2S formed (mg/hour)
9.2	70	1.50	5.57
22.6	39	2.86	6.97
36.3	22	2.75	6.19
48.5	8	1.84	3.09

The apparent activation energy determined from kinetic data is 17.4 kcal/mole with FeS, while for the other two sulphides it is roughly the same value (21.3 and 21.6 kcal/mole with NiS and CoS, resp.). The slight difference in activation energies indicates analogous reaction mechanisms on the two latter sulphides.

Ghosal et al. [601] studied the hydrogenation kinetics of CS_2 in the presence of a sulphurised Co–Mo catalyst in heptane solution; the technique employed in this measurement was the same as that used in measuring the hydrogenation kinetics of ethyl mercaptan and thiophene [602, 603] (p. 226).

It was found from the kinetic studies of atmospheric-pressure hydrogenation of carbonyl sulphide on Ni_3S_2, that the reaction proceeds above 150 °C as a first-order reaction with respect to COS. The two sulphur compounds, CS_2 and COS, are hydrogenolysed equally readily at low concentrations, while at higher concentrations COS is hydrogenated at a considerably greater rate [367].

6.9.3.2.

Hydrogenation of mercaptans, sulphides and polysulphides

In most cases, mercaptans are intermediates in the hydrogenation of higher sulphur compounds to hydrocarbons, in analogy to alcohols. The mechanism of hydrogenolysis of C—S bonds in mercaptans in the presence of sulphide catalysts is not yet fully understood. Amberg et al. [399] conclude from desulphurisation experiments with a number of compounds on a sulphurised Co–Mo catalyst, that hydrogenolysis of the C—S bond takes place on weakly acid centres with formation of H_2S and an olefinic bond, which is hydrogenated in the subsequent reaction step to form a saturated bond.

The chemistry of hydrogenolysis of mercaptans depends not only on the catalyst employed, but also on the structure of the reacting mercaptan. Cope and Farkas [351] found on hydrogenating n-octyl mercaptan in the presence of MoS_2, that dioctyl sulphide was an important reaction intermediate (240 °C, with sufficiently long reaction time) the yield of this being roughly 22%. On the other hand, tert-octyl mercaptan formed a high yield of octenes under the same conditions, while the sulphide was not identified. Some C—S bonds are considerably weakened and hydrogenolyse easily; e.g. benzyl mercaptan hydrodesulphurisation takes place in the presence of the MoS_2 catalyst very efficiently at temperatures below 200 °C [908, 1184, 1185]. The C—SH bond splits under milder conditions than the analogous C—OH bond [908].

Attention was also given to the influence of the structure of the alkyl group on the ease of hydrogenolysis of the C—SH bond. Landa and Mrnková [1184, 1185] found, on hydrogenating different mercaptans in the presence of MoS_2, that the length of the alkyl group has only a minor influence on the rate of splitting of the C—SH bond. On the other hand, highly branched tertiary mercaptans undergo hydrogenolysis more readily; typical results will be found in Table 48.

It was found by means of atmospheric-pressure hydrogenation of methyl mercaptan on the catalyst Ni_3S_2] [367], that the reaction attains a measurable rate above 200 °C and, that it is pseudo-first-order with respect to the mercaptan. Activated adsorption on nickel subsulphide also begins to play a role at this temperature [654]. Compared to CS_2, CH_3SH is more difficult to hydrogenate, therefore CH_3SH may be recovered as an intermediate of carbon disulphide hydrogenation under mild conditions [366].

Table 48.

Comparison of the Degree of Hydrogenolysis of Typical Mercaptans at 200 °C [1184]. (Autoclave experiments, MoS_2 catalyst, H_2 filling pressure 100 atm, reaction time 30 min.)

Mercaptan hydrogenated	% Desulphurisation
Isobutyl mercaptan	2.0
n-Butyl mercaptan	3.9
sec-Butyl mercaptan	9.4
tert-Butyl mercaptan	30.0
n-Hexyl mercaptan	6.2
n-Octyl mercaptan	10.9
n-Dodecyl mercaptan	5.9
Cyclohexyl mercaptan	17.9
Phenyl mercaptan	65.7[a])
Benzyl mercaptan	96.5

[a]) The experimental temperature was 205 °C.

A detailed study of atmospheric-pressure hydrogenation of the lowest mercaptans on MoS_2 and WS_2 catalysts was reported by Kemball et al. These authors hydrogenated methyl mercaptan [2217] at subatmospheric pressure in a stationary system, the basic experiments being done with partial pressures of about 55 torr CH_3SH and 113 to 115 torr H_2. Reactions were evaluated with the use of mass-spectrometric means, and the activation of individual bonds was studied by means of exchange reactions with deuterium. On both catalysts, mercaptan hydrogen is exchanged more quickly than the hydrogen atom of a C—H bond, this difference being much greater in the case of WS_2. In the case of exchange of hydrogen in H_2 and H_2S for deuterium, a considerably greater difference was again observed with WS_2 than with MoS_2 [2216].

The results of measurements indicate that WS_2 activates chemical bonds in the following order with a high degree of difference: H—H > H—S > C—S > C—H.

Under the reaction conditions employed (130 to 320 °C) two basic processes take place in methylmercaptan hydrogenation, namely elimination of hydrogen sulphide:

$$2\,CH_3SH \rightarrow (CH_3)_2S + H_2S \qquad \text{(LVII)}$$

and hydrogenolysis

$$CH_3SH + H_2 \rightarrow CH_4 + H_2S\,. \qquad \text{(LVIII)}$$

On the WS_2 catalyst, formation of the sulphide takes place readily in the temperature range of 130 to 230 °C, while hydrogenation to methane only plays a more important role above 230 °C. The progress of CH_3SH hydrogenation in the presence of WS_2 at 250 °C is illustrated in Fig. 58. On analysing kinetic data, Wilson and Kemball

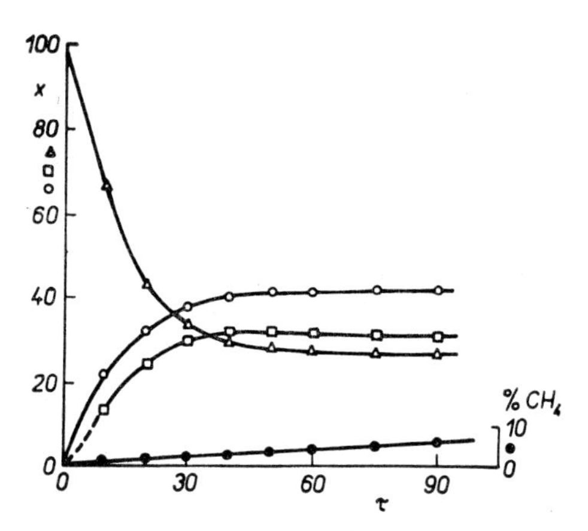

Fig. 58. Reaction of methyl mercaptan with hydrogen on WS_2 catalyst at 250 °C (initial reaction mixture in partial pressures: 55 torr CH_3SH, 115 torr H_2) [2217]. x — Reaction product composition: △ — CH_3SH, ○ — H_2S, □ — $(CH_3)_2S$ (stated in % of total sulphur content of the gas phase); ● — % CH_4, related to overall carbon content of the gas phase; τ — reaction time (min).

[2217] arrived at the conclusion that the rate-controlling process in the reaction (LVII) is formation of the C—S bond in the sulphide formed, while the hydrogenation reaction (LVIII) is controlled by methane desorption. In the case when the MoS_2 catalyst is used, the difference lies in the fact that the two reactions take place in the same temperature range and the values of their activation energies differ far less than with the WS_2 catalyst (Table 49). The order of bond activation on the MoS_2 catalyst is the same as on WS_2.

Table 49.

Comparison of Activation Energies and Frequency Factors of Reactions in Low-Presssure Methyl Mercaptan Hydrogenation [2217].

Catalyst	Reaction	Temperature range (°C)	Activation energy E (kcal/mole)	Frequency factor log A (molecules/ s.cm^2)
WS_2	$2\,CH_3SH \rightarrow (CH_3)_2S + H_2S$	130—230	15.9	17.8
	$CH_3SH + H_2 \rightarrow CH_4 + H_2S$	230—270	28.8	21.8
MoS_2	$2\,CH_3SH \rightarrow (CH_3)_2S + H_2S$	230—270	19.7	19.5
	$CH_3SH + H_2 \rightarrow CH_4 + H_2S$	230—270	22.9	20.4

Under the same reaction conditions, the authors named above also hydrogenated ethyl mercaptan [1022], finding that three basic reactions take place in the temperature range 170 to 270 °C:

$$2\,C_2H_5SH \rightarrow (C_2H_5)_2S + H_2S\,, \qquad \text{(LIX)}$$

$$C_2H_5SH \rightarrow C_2H_4 + H_2S\,, \qquad \text{(LX)}$$

$$C_2H_5SH + H_2 \rightarrow C_2H_6 + H_2S\,. \qquad \text{(LXI)}$$

In addition, other reactions are also possible:

$$(C_2H_5)_2S \rightleftarrows 2\,C_2H_4 + H_2S\,, \qquad \text{(LXII)}$$

$$(C_2H_5)_2S + H_2 \rightarrow C_2H_4 + C_2H_6 + H_2S\,, \qquad \text{(LXIII)}$$

$$(C_2H_5)_2S + 2\,H_2 \rightarrow 2\,C_2H_6 + H_2S\,. \qquad \text{(LXIV)}$$

It follows from a thermodynamic analysis [1022, 2183] that the equilibrium in the hydrogenation reactions (LXI), (LXIII), (LXIV) will be shifted far to the hydrocarbon side under the reaction conditions used. The course of the reaction (LX) is influenced to a large extent by thermodynamic factors under the operating conditions, so that equilibrium is established during the reaction between mercaptan, sulphide and H_2S. At low pressures, the conditions should be favourable for the reactions

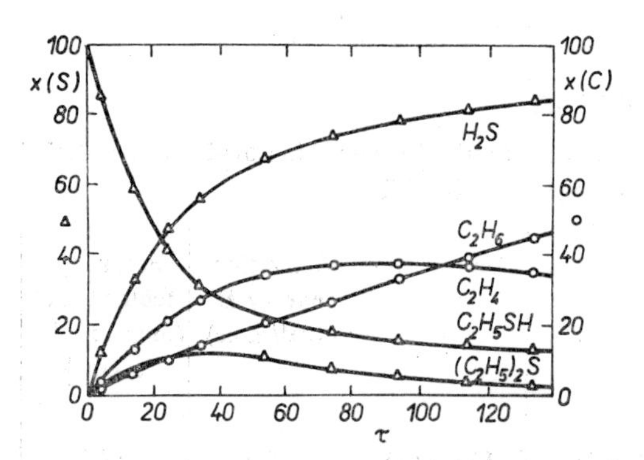

Fig. 59. Reaction of ethyl mercaptan with hydrogen on the WS_2 catalyst at 250 °C [1022]. Initial mixture composition in partial pressures: C_2H_5SH 21 torr, H_2 110 torr. Reaction product composition: $x(S)$ — amount of C_2H_5SH, $(C_2H_5)_2S$ and H_2S in % of the total sulphur content of the gas phase; $x(C)$ — amount of C_2H_6 and C_2H_4 in % of the total carbon content in the gas phase. τ — reaction time (min).

(LX) and (LXII), but from a thermodynamic point of view these reactions are more disadvantageous than the reaction (LIX). CdS was also found to be a good catalyst for the reaction (LX) [200]. The typical course of hydrogenation of C_2H_5SH on WS_2 is shown in Fig. 59.

The following differences were observed when MoS_2 was used for hydrogenation of C_2H_5SH:

a) WS_2 was more active in elimination of hydrogen sulphide to form the sulphide;
b) the ethylene: ethane ratio was considerably higher on MoS_2 compared to WS_2;
c) WS_2 was found to be the more efficient hydrogenation catalyst.

In the following Table 50, activation energies are listed for the individual reaction steps involved in ethyl mercaptan low-pressure hydrogenation.

Table 50.

Activation Energy and Frequency Factors of Most Important Reactions Taking Place in Low-Pressure C_2H_5SH Hydrogenation on WS_2 and MoS_2 Catalysts [1022].

Reaction	Activation energy E (kcal/mole)		Frequency factor log A (A is expressed in molecules/ /s.cm^2)	
	MoS_2 ($t = 200$ to $300\ °C$)^a)	WS_2 ($t = 170$ to $270\ °C$)^b)	MoS_2 ($t = 200$ to $300\ °C$)	WS_2 ($t = 170$ to $270\ °C$)
$2\ C_2H_5SH \rightarrow (C_2H_5)_2S + H_2S$	13	14	15.8 ± 0.4	15.8 ± 1.0
$C_2H_5SH \rightarrow C_2H_4 + H_2S$	20	19	19.8 ± 0.5	18.5 ± 0.8
$C_2H_5SH + H_2 \rightarrow C_2H_6 + H_2S$	20	18	18.8 ± 0.5	17.9 ± 0.8

The error in experimental determinations of E is: ^{a}) $\pm$ 1 kcal/mole, ^{b}) $\pm$ 2 kcal/mole.

Comparison of the kinetic data of low-pressure hydrogenation of the two basic mercaptans allowed Kemball et al. [1022] to draw the following conclusions:

1. Formation of sulphide (LIX) is slower on both catalysts compared to CH_3SH. The reaction rates differ more in the case of WS_2.

2. In contrast to the above, hydrogenolysis of ethyl mercaptan is quicker on both catalysts (comparison of the reactions (LVIII) and (LXI)). The difference is again more marked in the case of WS_2.

Hydrogenolysis of sulphides on sulphide catalysts is in general more difficult than for mercaptans. Compared to the analogous ether bond, the simple sulphidic bond in C—S—C is weaker and it is hydrogenolysed more readily [909]. Hydrogenolysis of sulphides is a stepwise reaction, mercaptans being formed as intermediates [909, 1184, 1185, 1518, 1602, 1860, 1862]. The amount of mercaptans formed depends both on the type of substance hydrogenated and on the reaction conditions. The maximum mercaptan concentration in the hydrogenation product of sulphides is achieved with short contact times, the mercaptan content tending to zero with prolonged contact times. A typical course for the process is illustrated in Fig. 60, showing the result of hydrogenation of 2,4,6,8-tetramethyl-5-thianonane in the presence of a sulphurised Co–Mo catalyst.

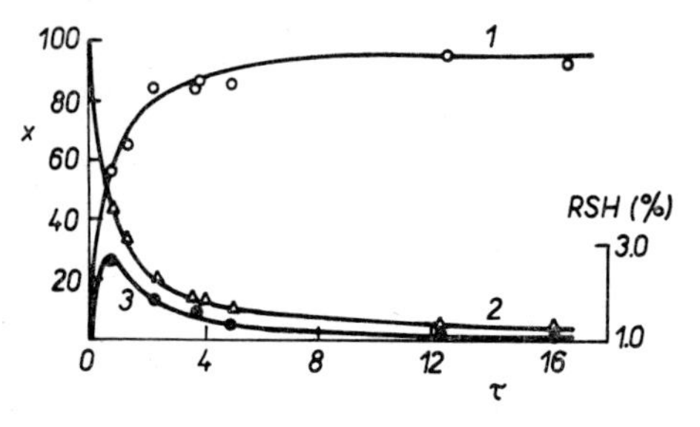

Fig. 60. Hydrogenation of 2,4,6,8-tetramethyl-5-thianonane on a sulphurised Co–Mo catalyst (375 °C, P_H = 33.3 atm, sulphur compound dissolved in cetane) [1517]. 1 — Hydrogen sulphide sulphur. 2 — Sulphide sulphur. 3 — Mercaptan sulphur (RSH). (All in % of theoretical values per unit of sulphur compound taken, x). τ — hypothetical contact time (s) calculated from the relationship

$$\tau = \frac{V_K \cdot P \cdot 273.2 \cdot 3600}{(V_H + V_S) \cdot T}$$

where: V_K — volume of reactor not filled with catalyst; V_H — hydrogen volume; V_S — volume of feed vapour (both in terms of STP); T — absolute temperature in the reactor; P — total pressure.

However, the course of the reaction cannot be predicted unambiguously from the results obtained. Besides the consecutive reaction course:

$$R.S.R' + H_2 \rightarrow RSH + R'H \qquad RSH \xrightarrow{[H_2]} RH + H_2S \qquad \text{(LXV)}$$

there is also the possibility of direct hydrogenolysis without the intermediate of the mercaptan [1518]:

$$RSR' + 2\,H_2 \rightarrow RH + R'H + H_2S. \qquad \text{(LXVI)}$$

With mixed alkylarylsulphides, the predominating reaction course involves splitting of the C_{aliph}—S bond and formation of a thiophenol [1602, 1860]:

$$C_6H_5\text{—}S\text{—}CH_2CH_2R + H_2 \rightarrow C_6H_5\text{—}SH + RCH_2CH_3 \qquad C_6H_5\text{—}SH \xrightarrow{[H_2]} C_6H_6 + H_2S. \qquad \text{(LXVII)}$$

It was found, by comparing the hydrogenation of a number of sulphides with alkyls of different size and differently branched (all dibutylsulphides were hydrogenated as well as di-n-hexyl-, di-n-octyl-, di-n-dodecyl-, dicyclohexyl- and diphenylsulphide) [1184, 1185], that the influence of the alkyl group on the strength of the C—S—C bond is roughly the same as with mercaptans, i.e. that the bond linking a tert-alkyl

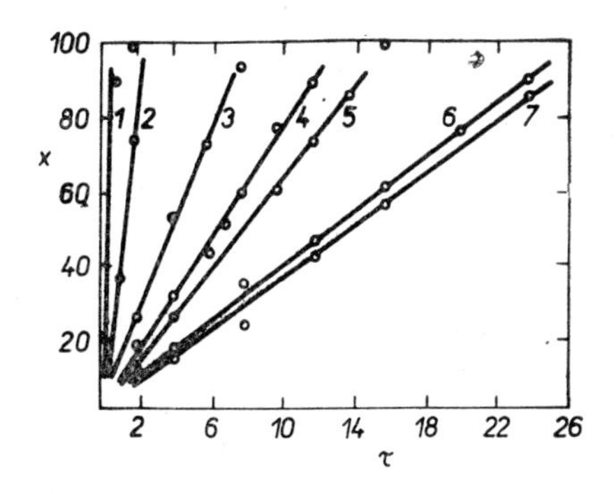

Fig. 61. Reactivity of some sulphur compounds in hydrogenolysis on the WS_2–NiS/Al_2O_3 catalyst [1602]. (Autoclave experiments, a 0.5 mol/l cetane solution of the sulphur substance was used, temperature 200 °C, working pressure 200 atm, ratio of catalyst amount: sulphur = 2 : 1). x — Degree of conversion in %. 1 — Dibenzyl sulphide. 2 — Thiophenol. 3 — Diphenyl sulphide. 4 — n-Butyl phenyl sulphide. 5 — 1,5-Diphenyl-3-thiapentane. 6 — Benzothiophene. 7 — Di-n-hexyl sulphide.

with sulphur is considerably weakened, the influence of the length of the alkyl chain being less. The C—S bond is specially reactive in dibenzyl sulphide, the hydrogenolysis of which takes place more rapidly, by an order of magnitude, than with other sulphides (cf. Fig. 61 and Table 51) [1860]. From results illustrated in Fig. 61, it follows that under the reaction conditions employed hydrogenolysis takes place over a wide range of concentrations as a zero-order reaction, i.e. the catalyst surface is constantly saturated with the sulphur compound. It was also found that the hydrogenation products do not retard the reaction [1602]. The relative rates of some other sulphides are listed in Table 51.

Table 51.

Comparison of Relative Hydrogenolysis Rates of Sulphides and of Some Other Sulphur Compounds on the WS_2–NiS/Al_2O_3.
(Results read off Fig. 61, relative rate of diphenyl sulphide = 1.0) [1860].

Substance	Structural formula	Relative rate of hydrogenolysis
Dibenzyl sulphide	$C_6H_5CH_2.S.CH_2C_6H_5$	10.0
Thiophenol	C_6H_5–SH	3.3
Diphenyl sulphide	$C_6H_5.S.C_6H_5$	1.0
n-Butyl phenyl sulphide	$C_6H_5.S.CH_2CH_2CH_2CH_3$	0.62
1,5-Diphenyl-3-thiapentane	$C_6H_5.CH_2CH_2SCH_2CH_2C_6H_5$	0.52
Benzothiophene	(structure: benzene ring fused to thiophene ring, S)	0.29
Di-n-hexyl sulphide	$(n\text{-}C_6H_{13})_2S$	0.26

Disulphides undergo hydrogenolysis very easily [1186, 1187a]. Splitting under hydrogen pressure starts even at 120 °C, and at 200 °C the reaction is practically complete. Under these conditions, a high yield of the respective mercaptans may be recovered, since hydrogenolysis to hydrocarbons takes place to a limited extent only [505a, 505b, 910, 912, 1186)].

6.9.3.3.

Hydrogenation of sulphur-containing heterocyclic compounds

Heterocyclic sulphur compounds, particularly thiophene derivatives, are amongst those organic substances which are very resistant to hydrogenolysis. From a number of comparative measurements, involving different catalysts and different reaction conditions, it appeared that the stability of sulphur compounds in hydrodesulphurisation decreases in the series thiophenes > thiophanes > sulphides > mercaptans [1518, 1602]. This means, however, that sulphur compounds of rather similar molecular weight will frequently undergo hydrogenolysis with widely different rates (the difference in rates is often up to two orders of magnitude). This observation is of wide practical significance in hydrodesulphurisation of sulphur-containing materials, since due to different types of sulphur compounds contained in two fractions of roughly the same boiling point, hydrodesulphurisation will have to be carried out under quite different reaction conditions.

The hydrogenation mechanism of sulphur-containing heterocyclic compounds, particularly thiophene, has been given much attention in the literature, in proportion to the great technical importance of this reaction. The conclusions drawn by various authors, however, differ considerably in a number of cases; this being mainly due to the diverse reaction conditions and catalysts employed in the different experiments, and not least of all to the different efficiency of analytical techniques employed in evaluating the hydrogenation products.

Moldavski et al. [1393, 1395] started a systematic study of the hydrogenation mechanism of thiophene on sulphide catalysts as long as 35 years ago. In autoclave experiments at pressures of 80 atm and temperatures of 300 to 450 °C in the presence of MoS_3, these workers found tetrahydrothiophene in the reaction mixture. From the presence of mercaptans, determined by quantitative means, they conclude that the following reaction steps are involved:

$$\text{thiophene} \xrightarrow{[H_2]} \text{tetrahydrothiophene} \xrightarrow{[H_2]} CH_3CH_2CH_2CH_2SH \,. \qquad \text{(LXVIII)}$$

On determining the basic kinetic data, Moldavskii et al. found that the apparent reaction order is zero with respect to thiophene, due to the high adsorptivity of thiophene on the sulphide catalyst [1395]. Similarly, a very low reaction order was observed in the hydrogenation of thiophene in the presence of the mixed catalyst CoS–MoO_3 [1590]. On the basis of later work, Cawley and Hall [276] assume the same course for thiophene hydrogenation as Moldavskii et al. In atmospheric-pressure hydrogenation of thiophene in the presence of MoS_2, these authors moreover identified butylmercaptan and di-n-butylsulphide in the reaction product.

The reaction scheme for thiophene hydrogenation proposed by Orlov and Broun [1549] (atmospheric-pressure hydrogenation in the presence of Ni–Co/Al_2O_3–MgO, CoO and MoO_3 oxide catalysts sulphurised in the course of the reaction) involves disulphides as well as mercaptans and sulphides; but this scheme does not seem to be reliable due to the imperfect analytical techniques employed.

Rhenium heptasulphide is a highly selective catalyst for partial thiophene hydrogenation. Broadbent et al. [226] obtained a 70% yield of thiophane at 230 to 260 °C and 140 atm, but in contrast to this, a thiophane yield of only 7% was obtained with molybdenum sulphide under atmospheric conditions. In no case was butylmercaptan isolated from the reaction mixture. The fact that a stepwise course predominates in thiophene hydrogenation was recently confirmed by Landa and Mrnková [1184, 1185], who found thiophane and mercaptan to be the main products, apart from C_4-hydrocarbons, in high-pressure hydrogenation in the presence of MoS_2 over a wide temperature range of 200 to 320 °C. Furthermore, n-octane was identified in the reaction mixture. These authors believe that the latter compound is formed by dimerisation of butadiene formed as an intermediate, followed by hydrogenation of the dimer. Earlier, Komarewsky and Knaggs [1068] and then Amberg et al. [1565] recorded the formation of butadiene as an intermediate. Octane formation may obviously also be interpreted by dimerisation of butenes.

Griffith et al. [658], who hydrogenated thiophene in the presence of MoS_2, MoS_2–MoO_2 and Ni_3S_2 catalysts in a flow reactor at atmospheric pressure, derived a different explanation for the chemistry of thiophene hydrogenation. The comparison of the efficiency of these catalysts was mentioned earlier (Table 40, p. 176). From this comparison it follows that thiophane is hydrogenolysed considerably more rapidly than thiophene. For example, at 200 °C (260 ppm of the sulphur compound in H_2) 80% tetrahydrothiophene was hydrogenolysed on MoS_2, but only 60% thiophene. An even more marked difference shows in results obtained with the Ni_3S_2 catalyst. At 200 and 350 °C, tetrahydrothiophene was hydrogenolysed to the extent of 41 and 100%, resp., while thiophene was not hydrogenolysed at 200 °C and the reaction proceeded only to 15% at 350 °C. These authors found the reaction to be of pseudo-first-order with respect to thiophene in the presence of Ni_3S_2, while it is a fractional order (0.2 to 0.6) in the presence of $MoS_2.MoO_2$, the value increasing with rising temperature or decreasing thiophene concentration in the feed. With MoS_2 at 200 °C, the reaction order was found to be roughly zero with respect to thiophene.

Griffith et al. [658] base their explanation of the reaction mechanism on the fact that they found neither thiophane nor mercaptan in the reaction products (in this connection, Obolentsev and Mashkina [1518] criticise the insufficiently exact technique used by Griffith et al. for sulphur compound analysis). The products identified were butanes and butenes as well as H_2S. Adsorption measurements [95] showed that hydrogen is adsorbed considerably more strongly than thiophene on MoS_2, while on the catalysts $MoS_2 . MoO_2$ the reverse is true. At the same time, MoS_2 was found to be the most active of the catalysts tested, and therefore the authors [658] believe that hydrogen chemisorption is one of the rate-controlling processes in thiophene hydrogenation. They support this conclusion by the fact that, although the adsorption isobars of thiophene, furan and their tetrahydro-derivatives differ widely [658], the rate of hydrogenation of these substances on MoS_2 differs only little. Chemisorption and hydrodesulphurisation of thiophene and tetrahydrothiophene start at roughly the same temperature (above 150 °C), and therefore Griffith et al. believe that chemisorption of the two reactants is essential for the reaction to take place and, therefore, that adsorption of the sulphur compound will be the rate-determining process.

Griffith et al. [658] interpret the course of thiophene hydrogenation on MoS_2 as follows: The structure of molybdenum disulphide is hexagonal. The plane of Mo ions, which are located at the vertices of equilateral triangles at a distance of 3.15 Å, becomes the uppermost layer after the surface layer of sulphur ions has been removed by hydrogenation. Taking the length of the Mo—C bond as 2.14 Å [1579] we may assume two-point adsorption of thiophene by means of vicinal carbon atoms, the bond angle Mo—C—C being roughly 112°5′. At the same time, hydrogen adsorbed on the neighbouring active centres will cause partial hydrogenation and formation of a "semihydrogenated" molecule of thiophene. This is followed by repeated two-point adsorption of the partially hydrogenated thiophene molecule, this time by means of a sulphur atom, followed by splitting of the C—S bond and desorption of the cleavage product in the form of butene, which is finally hydrogenated to butane:

S (thiophene) + Mo Mo → CH—CH (S ring) adsorbed on Mo Mo $\xrightarrow{[H]}$ CH_2—CH (S ring) adsorbed on Mo Mo → CH, CH_2, CH, CH—S adsorbed on Mo Mo → (LXIX)

$$\xrightarrow{2H_2} H_2S + CH_3CH_2CH{=}CH_2 \xrightarrow{H_2} CH_3CH_2CH_2CH_3 .$$

These authors find evidence for the suggested reaction course in the fact that cyclopentadiene added to thiophene decreases the reaction rate of thiophene considerably.

The similar structure of cyclopentadiene causes competing adsorption and decreases the number of active centres which are available for thiophene. Ethylene, COS and H_2S exert a far smaller inhibition effect.

The length of the Mo—C bond as stated by Griffith et al. [658] is considered to be too low [1385a]. A weak σ-bond is probably formed between a carbon (or sulphur) atom and two neighbouring metal atoms with valence angles to the carbon atom close to tetrahedral ($109\frac{1}{2}°$). The Mo—C (or Mo—S) distance is roughly 2.5 Å [1385a].

The hydrogenation mechanism of tetrahydrothiophene is explained by Griffith et al. [658] either by means of single-point adsorption through a sulphur atom or, by two-point adsorption of a C—S bond (the rest of the process being the same as in the case of thiophene).

The geometrical arrangement in a NiS lattice should permit a similar reaction course [660]. However, this sulphide is unstable under hydrogenation conditions, converting to the subsulphide Ni_3S_2. Hydrogenation of thiophene on Ni_3S_2 takes quite a different course according to Griffith et al. [658], due to the different surface geometry. The symmetry of the subsulphide is rhombohedric and there are no doublets of Ni ions on the planes of this crystal at a distance which would permit two-point adsorption of thiophene. Therefore, Ni_3S_2 is an efficient catalyst of the hydrogenation of simple sulphur compounds (CS_2, COS), which are adsorbed by one sulphur atom only, but is it inefficient in hydrogenation of thiophene-type substances. In spite of this however, there is some degree of hydrogenation (cf. Table 40). Griffith et al. explain this limited activity by the fact that catalytically-active centres are formed by some sulphur atoms being removed by hydrogenation, the sulphur atom of thiophene then reacting with the free Ni atom thus obtained. The above authors believe this view to be corroborated by the fact that addition of H_2S or CS_2 decreases the reaction rate considerably by blocking the active metal centres.

In agreement with results reported by Griffith et al. [658], Kalechits et al. [964] failed to find any tetrahydrothiophene in the reaction products of thiophene hydrogenation in the presence of oxide and sulphide catalysts (NiS–WS_2, 8376, NiS/ /Al_2O_3–SiO_2). Supporting the views of Yurev [2246] and Komarewsky and Knaggs [1068], Kalechits [964] suggests that thiophene hydrogenolysis takes place without preliminary hydrogenation to thiophane

$$\text{thiophene} + H_2 \rightarrow [CH_2{=}CHCH{=}CHSH] \xrightarrow{H_2} CH_3CH_2CH_2CH_3 + H_2S. \tag{LXX}$$

The temporary conjugated system formed, with greatly weakened C—S bond, easily undergoes hydrogenolysis, so that finally butane and H_2S are formed. Although the authors did not confirm this reaction course by any other specific experiments, it may explain the higher rate of hydrogenolysis of thiophene compared to the rate of hydrogenation of olefins e.g. on a Co–Mo catalyst or on some sulphurised Ni and Ni–W catalysts (cf. also p. 328).

Ivanovskii et al. [919, 920, 921] studied the hydrogenolysis of thiophene on a sulphidic iron-chromium catalyst with the use of labelled atoms. They proved by X-ray examination that the catalyst was a solid solution of sulphur in FeS. The sulphur in the catalyst is capable of isotopic exchange, i.e. the content of the ^{35}S isotope may be controlled either during the process of catalyst preparation, or by means of isotopic exchange with hydrogen sulphide:

$$Fe^{35}S + H_2{}^{32}S \rightarrow Fe^{32}S + H_2{}^{35}S . \qquad \text{(LXXI)}$$

Ivanovski et al. proved that sulphur dissolved in the catalyst is very reactive, in fact more reactive than sulphidic sulphur, and that thiophene hydrogenolysis takes place on the catalyst surface at sites where sulphur was dissolved. Thiophene is hydrogenolysed in two steps. The first step involves reduction of the dissolved sulphur to hydrogen sulphide, which is followed by bonding of thiophene sulphur to the reduced Fe-centre, splitting of the thiophene ring and finally, desorption of the hydrocarbon residue after hydrogenation in the form of butane:

$$FeS + H_2 \rightarrow (Fe) + H_2S , \qquad \text{(LXXIIa)}$$

$$(Fe) + C_4H_4S + 3\,H_2 \rightarrow FeS + C_4H_{10} . \qquad \text{(LXXIIb)}$$

Ivanovskii et al. proved that there is a close connection between the amount of dissolved sulphur and catalytic activity.

Amberg et al. recently published a thorough study of the mechanism of thiophene hydrogenation in the presence of sulphurised Co–Mo and sulphurised Cr_2O_3–catalysts. These authors worked with a flow microreactor, and in addition they used the pulse microreactor technique (the volume of the catalyst bed was 1 to 5 ml, hydrogen flow 6 to 9 l/hour, reactant microdoses were 5 to 10 µl) [1565]. Products were identified by gas chromatography.

In explaining the mechanism of thiophene hydrogenolysis on both catalysts, the authors base their views on the absence of thiophane and mercaptans in the reaction products, which they proved by experimental means [1565]. In hydrogenation on sulphurised Cr_2O_3 at 415 °C (pulse dosage 10 µl), n-butane, l-butene, *trans*-2-butene, *cis*-2-butene and butadiene were identified as reaction products [1566]. Similarly to Griffith et al. [658] these authors believe that the first step in the complex reaction in the presence of sulphurised Cr_2O_3 is splitting of the C—S bond rather than total hydrogenation of the thiophene ring [1565]. In agreement with Komarewsky and Knaggs [1068], Amberg et al. assume that butadiene is formed as a reaction intermediate (cf. the reaction scheme (LXXIII), the numbers in brackets indicate relative reaction rates):

$$\text{thiophene} \xrightarrow[\text{[2H}_2\text{] (18)}]{-H_2S} C{=}C{-}C{=}C \xrightarrow{[H_2]\ (110)} \begin{cases} C{=}C{-}C{-}C \\ \textit{trans}\ C{-}C{=}C{-}C \\ \textit{cis}\ C{-}C{=}C{-}C \end{cases}$$

$$C{=}C{-}C{-}C \underset{(>100)}{\overset{(80)}{\rightleftarrows}} C{-}C{=}C{-}C\ (\textit{trans}) \underset{(100)}{\overset{(100)}{\rightleftarrows}} C{-}C{=}C{-}C\ (\textit{cis}) \qquad \text{(LXXIII)}$$

$$\text{each butene} \xrightarrow{[H_2]\ (2.5)} C{-}C{-}C{-}C; \quad \text{thiophene} \dashrightarrow C{-}C{-}C{-}C$$

The results indicate that the possibility of several hydrogenation steps taking place on one centre (dotted lines in the scheme (LXXIII)) cannot be excluded. Analysis of the butene fraction showed a content of 1-butene greater than the equilibrium concentration [1565]. This observation agrees with data by Vernon and Richardson [2131], who found that 1-butene is the primary product of thiophene desulphurisation on MoS_2 promoted by different amounts of cobalt. From adsorption measurements on sulphurised Cr_2O_3 it followed that thiophene and butenes are adsorbed weakly but relatively rapidly. Moreover, the influence of H_2S on the rate of desulphurisation is small. It is probable therefore, that adsorption of thiophene, desorption of butenes and desorption of H_2S are not controlling factors for the rate of thiophene hydrodesulphurisation, and that the rate-determining step on sulphurised Cr_2O_3 is adsorption of hydrogen, which is adsorbed weakly [1567].

In connection with this, Mitchell [1385a] mentions that hydrogen bound strongly and weakly on a sulphide catalyst should be distinguished. Some kinetic measurements indicate, that weakly bound hydrogen only takes part in the hydrodesulphurisation reaction. It is possible, that one type is bound to metal atoms while the other is bound to sulphur atoms on the catalyst surface [1385a].

On the sulphurised Co–Mo catalyst, which is more efficient than sulphurised Cr_2O_3, 82% conversion of thiophene was achieved at 400 °C with 5 μl doses. The hydrogenation products contained neither butadiene nor products of structural isomerisation; however, butadiene was not found as a reaction intermediate even at 272 °C [1565]. This is a consequence of the fact that the hydrogenation rate of butadiene is several times greater than that of butenes. Strongly adsorbed H_2S considerably decreases the extent of hydrogenation of olefins (cf. p. 89), but it does not influence butadiene hydrogenation even at a substantially lower temperature (around 200 °C). This is another reason why butadiene was not identified among the reaction products of thiophene hydrogenation on the sulphurised Co–Mo catalyst [1565]. However, butadiene was identified as an intermediate in the course of the reaction by means of IR spectroscopy [398] and, furthermore, it is a normal intermediate of thiophene hydrodesulphurisation on a sulphurised Co–Mo catalyst, sulphurised Cr_2O_3 as well as on MoS_2 with very low thiophene conversion [1066]. The presence of butadiene in the reaction products of thiophene hydrogenation on a number of sulphide catalysts was also confirmed by Mann [1283].

On the basis of kinetic measurements, Owens and Amberg [1565] assume the following reaction scheme for the sulphurised Co–Mo catalyst:

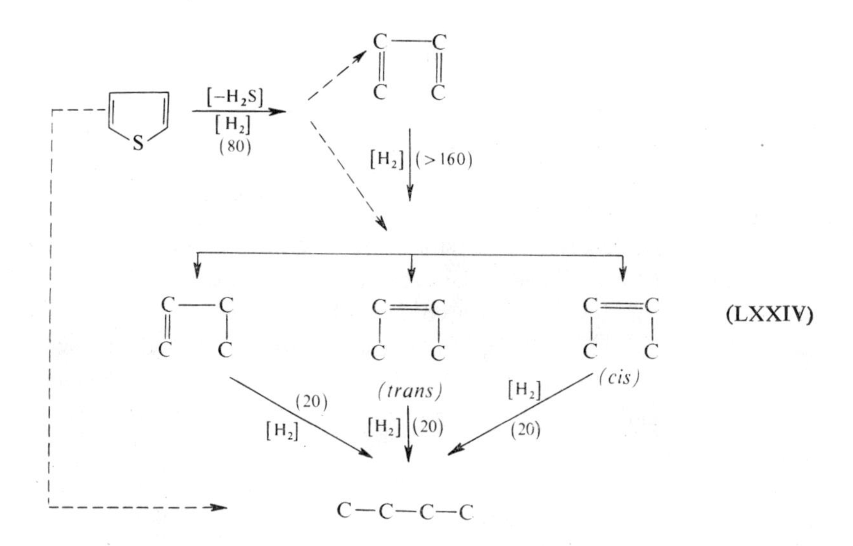

(LXXIV)

Thiophene homologues are also hydrogenated according to a similar mechanism [398]. The primary reaction is splitting of the C—S bond with diene formation, followed by hydrogenation to an olefin and a paraffin. From Table 52 it follows that there is a great difference in the activation energies of thiophene and of its methylhomologues and, at the same time, a great difference in the desulphurisation rates of 2-methyl- and 3-methylthiophene.

Other conclusions about the mechanism of hydrogenolysis of thiophenic compounds may be drawn from the kinetic data obtained. If hydrogen adsorption were the only rate-determining step, then changes in the reactant structure, e.g. substitution of a thiophenic ring by a methyl group or a change in its position in the heterocyclic ring could not cause such pronounced alterations in the reaction rate and activation energy. Similarly, Amberg et al. [398] do not consider adsorption of thiophenes to be the rate-determining step. It is thus clear, that the surface reaction, i.e. splitting of the C—S bond in the thiophene molecule, is one of the rate-controlling processes on

Table 52.

Desulphurisation Rates of Thiophenes in Atmospheric Flow Apparatus [398].
(1 g sulphurised Co–Mo, 55 ml H_2/min containing 1.67, 1.09 and 1.00% thiophene, 2-methylthiophene and 3-methylthiophene, resp.)

Temperature (°C)	Reaction rate (μmoles reacted substance/g catalyst/s)		
	Thiophene	2-Methylthiophene	3-Methylthiophene
170	0.8	1.1	2.7
180	1.5	1.7	4.2
190	2.8	2.7	6.7
200	4.9	4.1	10.0
210	8.7	6.1	15.6
225	18.8	10.7	27.8
E (kcal/mole)	24.4 ± 1.6	18.3 ± 1.8	18.8 ± 1.7

a sulphurised Co–Mo catalyst. The considerable difference in reactivity of the two methylthiophenes is probably caused by the different strengths of the inductive effect of the methyl group on the electron density of the α-carbon atom and thus also by the different capacity of this carbons atom to bind protons or to react with the acid centre of the catalytic surface [398].

Table 53.

Yield and Composition of C_4-Hydrocarbons in the Hydrogenation of Thiophene and of its Possible Hydrogenation Intermediates [399].
(1 g sulphurized Co–Mo, 50 ml H_2/min, microreactor pulse technique, dosage 2 μl.)

Temperature (°C)	Thiophene		Thiolene		Thiophane		n-Butyl mercaptan	
	% C_4 [a])	% satur.[b])	% C_4	% satur.	% C_4	% satur.	% C_4	% satur.
270	41.9	30.6	49.1	23.8	64.3	20.5	94.8	35.9
280	54.3	30.6	—	—	81.1	41.4	98.3	56.1
298	—	—	—	—	91.1	53.3	—	—
308[c])	57.5	28.9	74.9	33.9	93.9	35.4	100	46.2
313	63.9	30.8	—	—	96.4	69.6	100	72.1
365	—	—	—	—	100	98.0	—	—
372[c])	88.1	28.2	77.3	45.4	100	47.1	100	61.2

[a]) % C_4 — % total C_4-hydrocarbons/sulphur compound.
[b]) % satur. — % butane in C_4 hydrocarbons.
[c]) In these experiments, the respective sulphur compound was fed in 7 min after the feeding of 5 μl thiophene. In the other experiments, the catalyst was made free of adsorbed substances by reduction with hydrogen for 1 hour before the experiment proper.

For a more detailed assessment of the suggested reaction mechanism, Desikan and Amberg [399] carried out comparative hydrogenation experiments in the presence of a sulphurised Co–Mo catalyst with all sulphur compounds which may be considered as possible intermediate products of thiophene hydrogenation, i.e. thiolene (2,3-dihydrothiophene), thiophane and n-butyl mercaptan, and detailed analyses of the hydrocarbon fractions of the respective hydrogenation products were performed. The results are summarised in Table 53. The composition of the C_4-olefinic fraction formed on hydrogenation of these substances at 308 °C is shown in Table 54.

Table 54.

Composition of C_4-Olefins (in %) Obtained by Hydrogenating Individual Sulphur Compounds [399].

(Sulphurised Co–Mo catalyst, microreactor pulse technique, 50 ml H_2/min, sulphur compound dosage 2 µl, temperature 308 °C.)

Isomer	Substance hydrogenated				
	Thiophene	Thiolene	Thiophane	n-Butyl mercaptan	Equilibrium composition at 308 °C
1-Butene	19.5	17.8	17.7	16.0	14.0
trans-2-Butene	42.1	44.0	44.0	44.0	53.5
cis-2-Butene	38.4	38.2	38.3	40.0	32.5

The main products of thiophane hydrodesulphurisation are butenes and butane. Other products are n-butyl mercaptan and thiophene [399]. It is clear from a comparison of the actual composition of butenes with the equilibrium composition, that *trans*-2-butene cannot be the primary product of hydrogenolysis of any of the sulphur compounds hydrogenated. Compared to the equilibrium composition, an excess of 1-butene and *cis*-2-butene is obtained and it is not possible to decide from the distribution alone which of these two isomers could have formed first. However, both hydrogenation of butadiene on sulphurised Cr_2O_3 [1565] and hydrodesulphurisation of thiophene on MoS_2 promoted with Co [2131] indicate that 1-butene is the primary product. As *n*-butanethiol is one of the principal intermediates in thiophane conversion on sulphurised Co–Mo [399], 1-butene can be expected to be the initial isomer in this case as well.

Experiments have shown, that with two consecutive thiophane pulses, injected at intervals of 7 minutes, the hydrogenation activity of the catalyst did not decrease and that the butane to butenes ratio in the product was unaltered. When, however, a thiophene pulse was injected between two thiophane pulses, the hydrogenation activity of the catalyst decreased considerably in the second thiophane experiment (Table

53). When an equivalent amount of H_2S was dosed between two thiophane pulses, the butane content of the hydrogenate was similarly decreased. This indicates that H_2S formed by thiophane hydrogenolysis cannot compete with thiophane in adsorption on hydrogenation sites, while preadsorbed H_2S markedly decreases the hydrogenation rate of olefins. Hence it follows that thiophene efficiently deactivates the catalyst due to its intensive adsorption on strongly acidic catalyst centres which determine the hydrogenation rate of olefins. This is also the reason why the C_4-hydrocarbon fraction of thiophene hydrogenation contains a smaller amount of butane. In contrast to hydrogenation, predosage of thiophene has little influence on the rate of desulphurisation (cf. Table 53) [399].

A detailed analysis of the thiophane hydrogenation product showed thiophene to be present [399]. This compound, however, is also formed on hydrogenation of thiolene, so that thiolene is probably an intermediate of thiophene formation in hydrogenation of thiophane on a sulphurised Co–Mo catalyst. The main hydrogenation intermediate, however, is butyl mercaptan. From the results obtained it followed that ring-opening on hydrogenation is one of the processes which determine the overall rate of the hydrodesulphurisation reaction of this substance.

Desikan and Amberg [399] also tried to find out whether the two basic reactions, i.e. desulphurisation of the sulphur compound and hydrogenation of olefins, take place on the same or on different centres at the catalytic surface. It was found that alkali hydroxides decrease the desulphurisation, as well as the hydrogenation activity of a sulphurised Co–Mo catalyst, proving that the two reactions take place on acid catalyst centres. Strong bases, however, deactivate all acid centres, since Desikan and Amberg [399] used pyridine to differentiate the effect of weakly and strongly acid centres, dosing this substance into the reactor before thiophene hydrogenation. The results are summarised in Table 55. A part of the pyridine pulse was adsorbed weakly and desorbed rapidly (the pyridine desorption shows a sharp peak

Table 55.

Influence of Pyridine on the Activity of a Sulphurised Co–Mo Catalyst in Thiophene Hydrodesulphurisation [399].

(Microreactor pulse technique, temperature 367 °C, 50 ml H_2/min, 5 μl pyridine, 2 μl thiophene.)

Time after pyridine injection (min)	Pyridine amount on the catalyst (ml (STP))	Ratio C_4 hydrocarbons/thiophene (% mol)	Butane content in C_4-hydrocarbons (%)
2.2	1.00	84	24.8
10.0	0.62	81	21.6
22.0	0.31	83	21.9
28.4	0.18	77	21.2
Fresh catalyst	0.00	92	44.0

soon after the beginning of desorption, followed by a long tail), i.e. this portion was adsorbed on weakly acid centres, on which desulphurisation takes place. Incomplete regeneration of the original hydrodesulphurisation activity (in Table 55, shown by the experiment nearly 0.5 hour after the beginning of pyridine injection) indicates, that desulphurisation in part also proceeds on the strongly acid centres, where olefin hydrogenation mainly takes place.

Table 56.

Influence of Pyridine on Thiophane Hydrodesulphurisation in the Presence of Sulphurised Co–Mo Catalyst [399].

(Temperature 358 °C, 50 ml H_2/min. microreactor pulse technique.)

Injection procedure	Yield of C_4-hydrocarbons (% theory)	Degree of hydrogenation of olefins (%)
1 μl thiophane on fresh catalyst	100	90
1 μl thiophane + 2 μl pyridine	31	59
1 μl thiophane 6.8 min later	100	47

Table 57.

Influence of Pyridine on n-Butyl mercaptan Hydrodesulphurisation in the Presence of Sulphurised Co–Mo Catalyst [399].

(Microreactor pulse technique, temperature 247 °C, dosage 2 μl mercaptan into 50 ml H_2/min.)

Injection procedure	Mercaptan conversion (%) to	
	C_4-hydrocarbons (total)	n-Butane
On the fresh catalyst	68.9	40
7 min after injection of 5 μl pyridine	58.3	18.2
40 min after injection of 5 μl pyridine	52.5	18.2

Proof of the fact that desulphurisation takes place on weakly acid centres, is also obtained by experiments with thiophane, presented comprehensively in Table 56 [399]. Hydrogenation of n-butyl mercaptan similarly indicates that the two reactions take place on different catalyst centres (Table 57) [399].

Amberg [399] summarises the view on the hydrodesulphurisation mechanism of thiophene and of other substances tested on sulphurised Co–Mo catalysts on the basis of the results obtained. The reaction takes place on two types of acid centres. Strongly acid centres are sufficiently electrophilic for hydrogenation of olefins to take place on them. H_2S, thiophene and pyridine are intensively adsorbed on these centres, deactivating them in respect of olefin hydrogenation. Desulphurisation of the sulphur compound and dehydrogenation of thiophane to thiophene also occur on some of these centres. The weakly acid centres, which can only be deactivated by strongly basic substances, catalyse the hydrodesulphurisation of sulphur compounds. For thiophene ring-opening on hydrogenation, Amberg assumes the formation of H_2S and butadiene either in one concerted action (LXXV) or through several surface intermediates (LXXVI):

$$\underset{S}{\square} \xrightarrow{[2H_2]} \begin{array}{c}C-C\\ \| \quad \| \\ C \quad C\end{array} + H_2S\,, \qquad \text{(LXXV)}$$

$$\underset{S}{\square} \xrightarrow{[H_2]} \left[\begin{array}{c}C-C\\ \| \quad \| \\ C \quad C\\ \diagdown \\ SH\end{array} \rightleftarrows \begin{array}{c}C-C\\ | \quad \| \\ C \quad C\\ \diagdown\!\!\diagdown \\ S\end{array}\right]$$

$$\downarrow [H_2]$$

$$\left[\begin{array}{c}C-C\\ | \quad \| \\ C \quad C\\ \diagdown \\ SH\end{array}\right] \xrightarrow{[-H_2S]} \begin{array}{c}C-C\\ \| \quad \| \\ C \quad C\,.\end{array} \qquad \text{(LXXVI)}$$

Thiophane ring opening proceeds according to the scheme (LXXVII):

$$\underset{S}{\square} \xrightarrow{[H_2]} \begin{array}{c}C-C\\ | \quad | \\ C \quad C\\ \diagdown \\ SH\end{array} \xrightarrow{-H_2S} \begin{array}{c}C-C\\ \| \quad | \\ C \quad C\,.\end{array} \qquad \text{(LXXVII)}$$

Kolboe [1066a] likewise considers a decyclisation mechanism, in which the primary step would be dehydrosulphurisation of the initial heterocycle (thiophene, thiophane) and formation of an acetylenic or olefinic thiol.

Butadiene and butenes, formed on strongly acid centres, are immediately hydrogenated further to butenes and n-butane. When formed on weakly acid centres, they either immediately desorb as such, or migrate to strongly acid centres where they are hydrogenated.

Amberg [399] suggests the following probable overall scheme for hydrodesulphurisation of thiophene, and other sulphur compounds significant in this reaction, on the basis of experimental data (full-drawn arrows indicate reactions on strongly acid centres, dotted arrows correspond to reactions on weakly acid ones, ? = route not definitely identified):

$$\begin{array}{ccccc}
\begin{array}{c} CH_2\text{—}CH_2 \\ | \quad\quad | \\ CH_2 \quad CH_2 \\ \diagdown S \diagup \end{array} & \overset{?}{\rightleftarrows} & \begin{array}{c} CH\text{—}CH_2 \\ \| \quad\quad | \\ CH \quad CH_2 \\ \diagdown S \diagup \end{array} & \longrightarrow & \begin{array}{c} CH\text{—}CH \\ \| \quad\quad \| \\ CH \quad CH \\ \diagdown S \diagup \end{array} \\
\downarrow & & \downarrow ? & \swarrow & \downarrow\downarrow \\
\begin{array}{c} CH_2\text{—}CH_2 \\ | \quad\quad | \\ CH_2 \quad CH_3 \\ \diagdown SH \end{array} & \dashrightarrow & \begin{array}{c} CH\text{—}CH_2 \\ \| \quad\quad | \\ CH_2 \; CH_3 \\ \text{and isomers} \end{array} & \longleftarrow & \begin{array}{c} CH\text{—}CH \\ \| \quad\quad \| \\ CH_2 \; CH_2 \end{array} \\
 & & \downarrow & & \\
 & & \begin{array}{c} CH_2\text{—}CH_2 \\ | \quad\quad | \\ CH_3 \quad CH_3\,. \end{array} & &
\end{array} \qquad \text{(LXXVIII)}$$

(A solid arrow also leads from tetrahydrothiophene directly to thiophene.)

In this work, Amberg proved that Hammar's earlier suggestion [721] that different catalyst centres operate for desulphurisation and hydrogenation of olefins is correct in principle. Mann [1283] also confirmed this view in a later study.

Kieran and Kemball [1023] amplified the knowledge of the mechanism of thiophene hydrogenation by a kinetic study of this reaction under the catalytic influence of tungsten and molybdenum sulphides. These authors used a stationary apparatus at subatmospheric pressure, identifying the hydrogenation products by means of mass spectrometric analysis. The reaction proceeded in the presence of the WS_2 catalyst, at temperatures in excess of 270 °C, and experiments with this catalyst were carried out at 272 to 305 °C. It was found that the basic reactions taking place under these conditions are:

$$C_4H_4S + 3\,H_2 \rightarrow C_4H_8 + H_2S\,, \qquad \text{(LXXIX)}$$

$$C_4H_4S + 4\,H_2 \rightarrow \text{n-}C_4H_{10} + H_2S\,, \qquad \text{(LXXX)}$$

$$C_4H_4S + 2\,H_2 \rightarrow C_4H_8S\,. \qquad \text{(LXXXI)}$$

A typical result obtained at 305 °C is shown in Fig. 62. Kinetic analysis showed that the apparent order of the reactions (LXXIX) and (LXXX) is zero with respect to thiophene. The reaction (LXXXI) was found to be of first order with respect to thiophene.

It was found on hydrogenating tetrahydrothiophene on WS_2 [1023] that the reactions involved are analogous to those which occur in the case of thiophene:

$$C_4H_8S + H_2 \rightarrow C_4H_8 + H_2S\,, \qquad \text{(LXXXII)}$$

$$C_4H_8S + 2\,H_2 \rightarrow C_4H_{10} + H_2S\,. \qquad \text{(LXXXIII)}$$

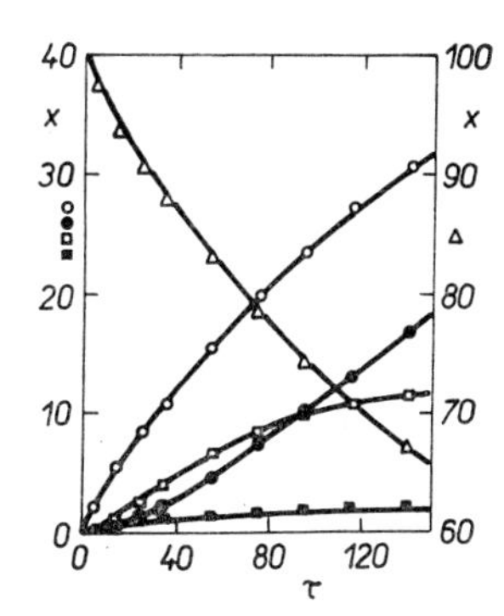

Fig. 62. Reaction of thiophene with hydrogen on the WS_2 catalyst at 305 °C (initial mixture in partial pressures: 13 torr thiophene, 130 torr H_2) [1023]. x — Product composition (%).

○ — H_2S; △ — C_4H_4S; ■ — C_4H_8S } expressed in % of total sulphur content.

□ — C_4H_8; ● — C_4H_{10} } expressed in % of total carbon content.

τ — Reaction time (min).

Kieran and Kemball likewise confirmed the observation reported by Desikan and Amberg [399] concerning partial dehydrogenation to thiophene:

$$C_4H_8S \rightarrow C_4H_4S + 2\,H_2\,. \qquad \text{(LXXXIV)}$$

Rather a significant difference compared to thiophene hydrogenation is the lower butane yield, but in other respects the two reactions are very similar to each other. The two sulphur compounds give practically the same hydrocarbon product, the composition of which is close to the equilibrium constitution. n-Hydrocarbons only are formed, no structural isomerisation being observed. Butadiene was not recovered under the reaction conditions applied.

The MoS_2 catalyst used by Kieran and Kemball was considerably less active than WS_2 [1023]. In the presence of this MoS_2 catalyst, but under otherwise similar conditions, thiophene hydrogenation started only at temperatures above 350 °C. At this temperature, the catalyst was deactivated rather quickly due to surface contamination by condensation products.

On the basis of the results achieved, the authors arrived at the conclusion that the rate-controlling process in hydrogenation of thiophene and tetrahydrothiophene is the splitting of the C—S bond in the ring and formation of a straight chain. Neither butyl mercaptan nor any unsaturated sulphur compounds with four carbon atoms was identified among the reaction products; and no butadiene was isolated, due to its high hydrogenation rate (which is roughly 30 times greater than that of thiophene decomposition [1023]).

Recently, Lipsch and Schuit [1243d] also investigated the mechanism of thiophene hydrodesulphurisation on the cobalt–molybdenum catalyst. As mentioned earlier (section 4.5, p. 72), the authors assume one-point adsorption of thiophene by a S–Mo bond on an anion vacancy on the surface of octahedral MoO_3. On the same vacancy, butene as well as H_2O and H_2S are also adsorbed. This may explain the deactivating influence of these substances on thiophene desulphurisation and hydrogenation of butenes. Since thiophene is bound more strongly than butene, H_2O and H_2S cause stronger deactivation of the process of butene hydrogenation. Thiophene hydrodesulphurisation then proceeds by the following steps:

1. Thiophene is adsorbed on an anion vacancy (formed by reduction of MoO_3 with hydrogen) and a S—Mo bond is formed. Hydrogen is adsorbed on oxygen atoms adjacent to the vacancy.
2. One hydrogen atom from a neighbouring hydroxyl group is transferred to the adsorbed thiophene, and one C—S bond is broken.
3. A second hydrogen atom from a different neighbouring OH group migrates to the adsorbed thiophene and a second C—S bond is broken.
4. 1,3-butadiene is desorbed (in agreement with data by Kolboe and Amberg [1066]), and is very rapidly hydrogenated on the same site or a different one.
5. The original vacancies are regenerated by reacting with two hydrogen atoms, hydrogen sulphide being desorbed.

The authors [1243d] assume a similar mechanism for the case of a sulphided catalyst, in which MoO_3 is converted to the sulphide. They state, that under certain circumstances this fact may increase the catalyst activity.

It is important to know the chemistry of hydrogenation of higher thiophenes from the point of view of desulphurisation of higher petroleum fractions. The behaviour of benzothiophenes is particularly important as these are regularly contained in the sulphur components of kerosines, diesel fuels, cracking circulation oils and oil fractions. Cawley [281] assumes that benzothiophene hydrogenolysis in the presence of a Co–Mo catalyst involves a stepwise reaction course, with thiophenol as reaction intermediate:

$$\text{benzothiophene} \xrightarrow{[H_2]} \text{dihydrobenzothiophene} \xrightarrow{[H_2]} \text{2-ethylthiophenol } (CH_2CH_3, SH) \xrightarrow{[H_2]} C_6H_5CH_2CH_3 + H_2S. \qquad \text{(LXXXV)}$$

Therefore three moles of hydrogen are consumed for each mole of benzothiophene. Calculated in terms of specific hydrogen consumption, this mechanism would mean the consumption in hydrodesulphurisation of a petroleum fraction of about 0.2%

by wt. hydrogen for every 1% by wt. sulphur. However, the aromatic hydrocarbon formed may undergo further partial hydrogenation, especially under more drastic refining conditions [1184].

The hydrogenolysis of dibenzothiophene is more problematic. In thit case, a simple reaction scheme involving the consumption of two moles of hydrogen

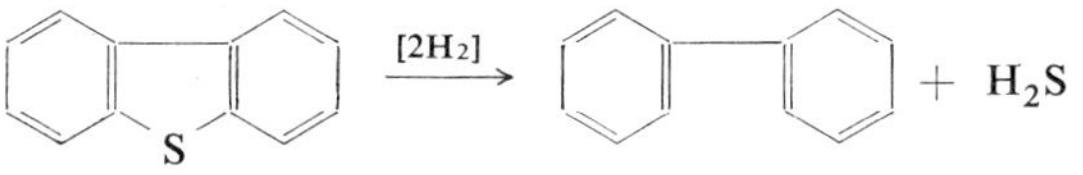

$$\xrightarrow{[2H_2]} \quad + \; H_2S \qquad \text{(LXXXVI)}$$

is believed to be rather improbable by some authors. Dibenzothiophene hydrogenolysis is more difficult than that of thiophene or benzothiophene [570], therefore Cawley [281] believes that hydrogenolysis of the thiophenic ring is preceded by hydrogenation of one of the two benzene rings, after which the thiophenic ring is split:

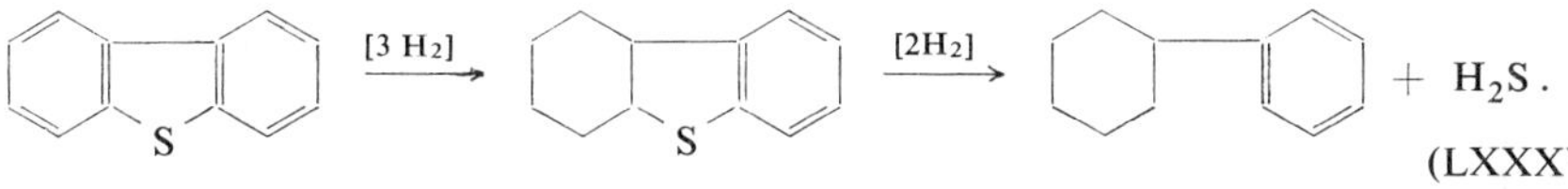

$$\xrightarrow{[3\,H_2]} \quad \xrightarrow{[2H_2]} \quad + \; H_2S. \qquad \text{(LXXXVII)}$$

This course of the hydrodesulphurisation reaction, however, would require five moles of hydrogen for every mole of dibenzothiophene, which in desulphurising petroleum fractions would mean a far grater specific hydrogen consumption.

Hoog [787] confirms that the hydrogen consumption in hydrodesulphurisation of petroleum fractions containing dibenzothiophene is greater than would correspond to the simple scheme (LXXXVI). The product of dibenzothiophene hydrogenation on a Co–Mo catalyst, as well as on MoS_2 [1184], contained phenylcyclohexane and even dicyclohexyl as well as diphenyl. On the other hand, Obolentsev and Mashkina [1513, 1514, 1518] do not agree with Cawley's [281] and Hoog's [787] conclusions, since in their experiments diphenyl was the only hydrocarbon formed in dibenzothiophene hydrogenation on a sulphurised Co–Mo catalyst under a wide range of reaction conditions (325 to 425 °C, total pressure 10 to 60 atm, partial hydrogen pressure 9.4 to 33.3 atm, with various space velocities). No sulphur-containing intermediate was isolated, and the authors favour the view that dibenzothiophene hydrogenation proceeds according to the reaction scheme (LXXXVI).

The hydrogenation mechanism of heterocyclic substances with several sulphur atoms is little understood. Diethyl sulphide was found to be an intermediate of 1,4-dithian hydrogenation on MoS_2 [1184]. For the course of thianthrene hydrogenation in the presence of a sulphurised Co–Mo catalyst, Obolentsev abd Mashkina [1517, 1518] assume the following reaction scheme:

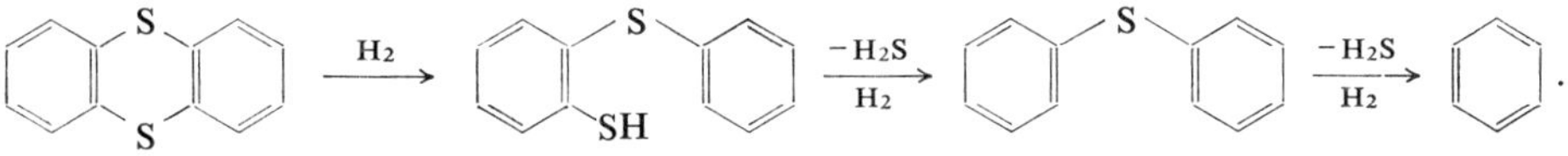

$$\xrightarrow{H_2} \quad \text{SH} \quad \xrightarrow[H_2]{-H_2S} \quad \xrightarrow[H_2]{-H_2S} \quad . \qquad \text{(LXXXVIII)}$$

Landa and Mrnková [1184] confirmed the presence of diphenyl sulphide as a reaction intermediate of thianthrene hydrogenation on MoS_2 and cyclohexane was found in addition to benzene. Similarly, dimethyl sulphide was obtained as reaction intermediate in trithiane hydrogenolysis on MoS_2 [1184].

In hydrogenation of aromatic high-molecular weight sulphur compounds, there prevail side reactions leading to formation of carbonaceous residues, which deactivate the catalyst. E. g., on hydrogenation of diacenaphthylene-thiophene [1184] in the presence of MoS_2, H_2S was eliminated at temperatures as low as 200 °C, but a large amount of condensation products were formed at the same time.

6.9.3.4.

Hydrogenation kinetics of sulphur compounds

Several important kinetic relationships have been derived for the hydrogenation of pure sulphur compounds, and some of these expressions are quite generally valid, i.e. they are applicable to different types of sulphur compounds and their different mixtures in natural materials, e.g. petroleum fractions, fractions of shale tar etc.

Ghosal et al. [603] based their work on earlier findings reported by Wilson et al. [2218] and used the following kinetic equation for hydrodesulphurisation of sulphur compounds:

$$-\left(\frac{F}{W}\right)\ln\frac{X}{X_0} = k_H\, e^{-E/RT}\, X_H^m\, P \qquad \text{(LXXXIX)}$$

where F – raw material feed (i.e. sulphur compound + solvent + H_2, gmole/hour), W – catalyst weight (g), X_0 – input concentration of the sulphur compound (mole fraction), X – output concentration of the sulphur compound (mole fraction), X_H – mole fraction of H_2, k_H – rate constant (gmoles/(hour) (g catalyst)), E – activation energy (cal/mole), m – coefficient, P – total pressure (atm.).

The authors studied the kinetics of ethyl mercaptan hydrogenation (in roughly 0.5% heptane solution) at temperatures of 225 to 430 °C, atmospheric pressure and different LHSV values. A sulphurised $Co–Mo/Al_2O_3$ catalyst activated with Ni was used. It was found by evaluation of the kinetic data, that the rate equation (LXXXIX) for this case takes the form:

$$-\left(\frac{F}{W}\right)\ln\frac{X}{X_0} = 1.35\, e^{-3974/RT}\, X_H\,. \qquad \text{(XC)}$$

It was verified, by experimental means, that the space velocity at which the same degree of mercaptan desulphurisation is achieved, is directly proportional to the partial hydrogen pressure in the pressure range of 0.19 to 0.67 atm. Ghosal et al. [603] believe that this linear relationship holds up to partial hydrogen pressures of 10 atm.

Ghosal et al. also found a similar kinetic relationship in the case of CS_2 [601] and thiophene [602] hydrogenolysis. The kinetic relationships thus obtained permit a very exact estimate of the operating conditions needed for complete desulphurisation of individual sulphur compounds.

Hammar [721] performed a detailed kinetic study of thiophene hydrogenation. This author determined the desulphurisation selectivity of a number of sulphidic catalysts in desulphurisation of highly unsaturated shale gasoline (containing 41% olefins, 27% aromatic substances and sulphur compounds up to a total sulphur concentration of 1.14%, this including traces of mercaptan sulphur, 0.14% sulphidic sulphur and 1.0% thiophenic sulphur). The kinetic measurement was performed with a sulphurised Co–Mo catalyst, activated with chromium, at temperatures of 300 to 375 °C and a pressure of 15 atm. Hammar assumes a stepwise scheme:

$$\text{thiophene} \xrightarrow{H_2} \text{dihydrothiophene} \xrightarrow{H_2} \text{tetrahydrothiophene} \xrightarrow{H_2} CH_3CH_2CH_2CH_2SH \xrightarrow{H_2} \rightarrow n\text{-}C_4H_{10} + H_2S\,. \qquad \text{(XCI)}$$

This author considers the rate-determining step to be thiophene hydrogenation to dihydrothiophene, assuming that further hydrogenation of dihydrothiophene is very rapid [721]. Hammar compared the experimental results with values obtained from a number of kinetic equations according to the Hougen-Watson system [799]. The following kinetic relations gave good agreement, assuming the surface reaction to be the rate-controlling step:

$$r = \frac{dx_S}{d\left(\frac{W}{F_S}\right)} = \frac{dx_S}{d\left(\frac{W}{F}\right)} \frac{S}{100(1 + N)} = \frac{ka_S a_H}{(1 + K_{H_2S}a_{H_2S})^2} \qquad \text{(XCII)}$$

where x_S is the conversion of the sulphur compound, S – the content of the sulphur compound in gasoline (mol%), F_S – the sulphur compound in feed (moles/ /hour), F – total feed (moles/hour), N – H_2: gasoline ratio (moles per mole gasoline), r – reaction rate, k – apparent reaction rate constant, a_s – activity (= mol. fraction) of thiophene, K_{H_2S} – equilibrium adsorption constant of H_2S, a_H – hydrogen activity, a_{H_2S} – H_2S activity.

The experimentally determined relationship between degree of conversion and space velocity is also satisfied by a kinetic equation, derived under the assumption that the reaction involves one adsorbed component and one component in the gas phase:

$$r = k \frac{a_S a_H}{1 + K_{H_2S} a_{H_2S}} . \tag{XCIII}$$

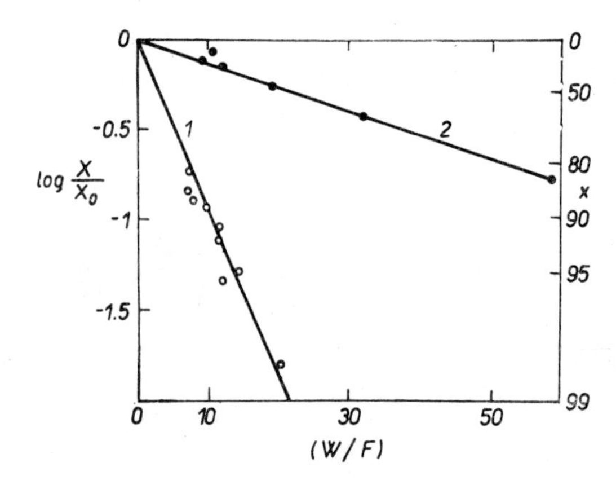

Fig. 63. Dependence of sulphur and nitrogen removal on space velocity [2218]. (Sulphurised Co–Mo catalyst, temperature 371 °C, pressure 11.4 atm, 1.2 moles H_2/mole virgin naphta; the feed contains 800 ppm dibenzothiophene and 150 ppm 3-methylisoquinoline.)
X, X_0 — Concentration and initial concentration of the sulphur or nitrogen compound. x — Degree of removal (%) of S or N. W/F — Reciprocal value of space velocity (g hour/gmole). 1 — Sulphur. 2 — Nitrogen.

The following rate equations were found suitable for the hydrogenation of olefins:

$$r = k' \frac{a_0 a_H}{(1 + K_0 a_0 + K_p a_p)^2} \tag{XCIV}$$

and

$$r = k' \frac{a_0 a_H}{1 + K_0 a_0 + K_p a_p} \tag{XCV}$$

where a_0, a_p are the activities of the olefin and paraffin, resp., and K_0, K_p are the respective adsorption equilibrium constants. The form of the kinetic equations for the two types of substances differs in the adsorption term.

According to Hammar [721], as stated earlier (p. 90), adsorbed H_2S has a distinct influence on the reaction rate of thiophene, while the presence of hydrocarbons plays no role. On the other hand, Hammar found that H_2S has none or only a slight influence in the case of hydrogenation of olefins (which does not agree with results published by other authors, cf. p. 89). On the grounds of these findings, Hammar [721] arrives at the conclusion (which seems to be justified to a large extent, as shown in the preceding chapter, p. 222), that the two reactions take place on different types of centres at the catalyst surface and, therefore, there is a theoretical possibility of highly selective catalysts being prepared.

Wilson et al. [2218] measured the hydrodesulphurisation kinetics of dibenzothiophene dissolved together with 3-methylisoquinoline in gasoline (b.p. 100 to 200 °C). Moreover, the gasoline employed contained 230 ppm of its own sulphur compounds, which were hydrogenolysed to more than 99% in all experiments and therefore need not be considered. The concentration of sulphur compounds was 800 ppm including the sulphur compounds already contained in the gasoline, the concentration of the nitrogen compound was 150 ppm. A sulphurised Co–Mo catalyst with stabilised activity was used.

The results of all the experiments performed indicated that desulphurisation, as well as denitrogenation, takes place as a first-order reaction (the experimental results are shown in Fig. 63), so that the relationship between conversion and space velocity may be expressed as:

$$\log \frac{X}{X_0} = \frac{k}{\left(\frac{F}{W}\right)} \tag{XCVI}$$

where X, X_0 are the concentration of the sulphur or nitrogen compound in the hydrogenation product and raw material, resp., k is the reaction rate constant, (F/W) is space velocity (gmole/g hour).

The temperature dependence of the rate constants of hydrodesulphurisation and hydrodenitrogenation satisfies the Arrhenius relation and enables the values of activation energies to be determined. In the case of hydrodesulphurisation, $E = 3.8$ kcal/mole and in the case of hydrodenitrogenation, $E \sim 20$ kcal/mole. The low activation energy of hydrodesulphurisation indicates the influence of diffusion, which retards the reaction.

The complex equation [2218] which includes the influence of temperature, space velocity and partial hydrogen pressure takes the following form in the case of hydrodesulphurisation:

$$\log \frac{S}{S_0} = \frac{-k_S\,\mathrm{e}^{-E/RT}}{\frac{N_H}{N_{HC}}\left(\frac{F}{W}\right)}. \tag{XCVII}$$

For hydrodenitrogenation, the following applies:

$$\log \frac{N}{N_0} = \frac{-k_N\,\mathrm{e}^{-E'/RT}\,P_H}{\frac{N_H}{N_{HC}}\left(\frac{F}{W}\right)}. \tag{XCVIII}$$

S, S_0, N, N_0 denotes the concentration and initial concentration of the sulphur and nitrogen compounds, resp., k_S, k_N – rate constants of desulphurisation and denitrogenation (functions of overall pressure, the value of k_S is 1.81 at 11.4 atm), N_H, N_{HC} – moles hydrogen and hydrocarbons, resp., E, E' – activation energy, P_H – partial hydrogen pressure. The same first–order kinetic equation holds for the pure substances as well as for refining of light petroleum fractions (cf. p. 239).

In hydrogenolysis of sulphur compounds, a typical irreversible reaction is involved, which is usually carried out with a large excess of hydrogen, so that its concentration remains practically constant. Obolentsev et al. [1515, 1516, 1518, 1519] therefore applied Frost's equation for irreversible unimolecular reactions, finding that it is applicable to the hydrogenolysis kinetics of individual sulphur compounds over rather a wide range of reaction conditions. Frost's equation takes the form:

$$\left(\frac{F}{W}\right) \ln\left[\frac{1}{1-y}\right] = \alpha + \beta\left(\frac{F}{W}\right) y \qquad \text{(XCIX)}$$

where (F/W) is space velocity (mmole/g hour), α is a coefficient, proportional to the rate constant of the unimolecular reaction, β is the constant of Frost's equation, including the adsorption coefficient of the substances involved, y is the degree of conversion.

The result of hydrogenation of dibenzothiophene to diphenyl and H_2S in the presence of a sulphurised Co–Mo catalyst, expressed in terms of Frost's equation, is illustrated in Fig. 64 for different temperatures [1518]. Evidently, the value of the coefficient β, which characterises the stability of ratios of adsorption coefficients, is equal to 1 and is the same for all temperatures. It was found furthermore that this value also does not vary with the partial hydrogen pressure and that with other

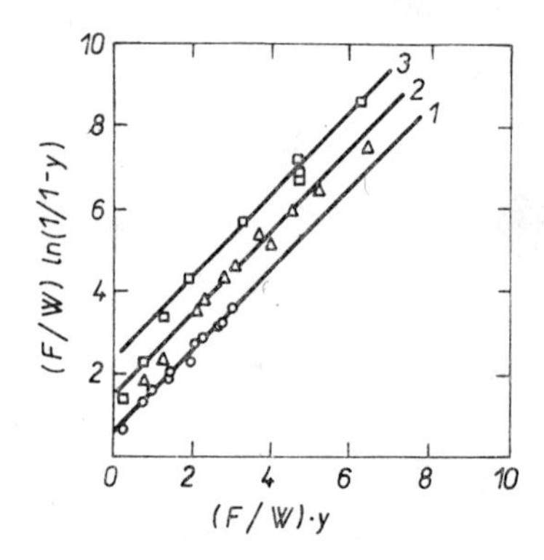

Fig. 64. Hydrogenation kinetics of dibenzothiophene, expressed on the basis of Frost's equation [1518]. (The catalyst was sulphurised Co–Mo, total pressure 40 atm, P_H = 33.3 atm.) 1 — 325 °C. 2 — 375 °C. 3 — 425 °C.

sulphur compounds as well, $\beta \doteq 1$. This indicates a considerably stronger adsorption of the reaction products compared to the initial compounds. This was confirmed by an experiment in which diphenyl was added, which distinctly retarded the reaction rate. A kinetic analysis indicates that the retarding effect of diphenyl is far greater than that of hydrogen sulphide [1518].

Obolentsev and Mashkina showed [1518], that the hydrodesulphurisation kinetics of a number of other sulphur compounds may also be interpreted by means of Frost's equation. The reaction rates of hydrodesulphurisation of individual sulphur compounds in the presence of a sulphurised Co–Mo catalyst were compared in terms of the magnitude and temperature relationship of the coefficient α (assuming that the adsorption coefficients are constant over the respective temperature range) and values of the apparent activation energy were established. The results are summarised in Table 58. The relative rate constants α_{rel} in Table 58 for individual substances were determined in comparison to the rate constant (coefficient α) of dibenzothiophene. The data presented in Table 58 confirm the high resistance of dibenzothiophene to desulphurisation, which was discussed earlier, and on the other hand the unusually weakened C—S bond in dibenzylsulphide. The apparent activation energy of hydrogenolysis of the substances involved differs only very little, its value being 8 to 11 kcal/mole.

The influence of diffusion on the hydrogenolysis of sulphur compounds in the presence of a sulphurised Co–Mo catalyst is significant, too. This catalyst has a large specific surface (usually over 200 m^2/g, cf. Table 78) so that in some cases, the retarding influence of inner diffusion on the reaction rate may play a decisive role. This effect may be especially strong in the case of large-grained industrial catalysts. Obolentsev and Mashkina [1518] determined the retarding influence of inner diffusion on dibenzothiophene hydrogenation with two Co–Mo catalyst samples (mean particle diameter 3 and 0.6 mm) at temperatures of 325, 375 and 425 °C. The constants in the above equation (the coefficient α is proportional to the rate constant) and the value of the Thiele modulus h and the effectiveness factor E were determined on the basis of the physico-chemical properties of the catalyst and after kinetic evaluation by means of Frost's equation. The results are presented in Table 59. It is seen from this table, that the reaction rate is 2 to 3 times higher on the fine-grained catalyst than on the larger particles. On the catalyst with the greater particle size, the reaction proceeded in the transition region at 325 °C and in the inner diffusion region at higher temperatures. On the fine-grained catalyst, the reaction takes place in the transition region. Hence it follows that the reaction rate and activation energy are decreased by inner diffusion, especially on the catalyst with larger particles.

Frost's equation is satisfactory in the region of high LHSV values, while at lower space velocities it does not quite elucidate the actual situation. Furthermore, the influence of the hydrogen partial pressure on the hydrodesulphurisation rate could not be expressed in terms of this equation. Obolentsev and Mashkina [1516, 1518] therefore used Kazeev's empirical equation [1005], which expressed very closely the results of desulphurisation experiments with individual pure substances dissolved in cetane over the entire range of space velocities and partial hydrogen pressures

Table 58.

The Relative Rate Constants α_{rel} and Apparent Activation Energy of Hydrodesulphurisation of Some Sulphur Compounds [1516, 1518].

(The evaluation was done on the basis of Frost's equation, the rate of hydrogenolysis of dibenzothiophene with $\alpha_{rel} = 1$ being taken as the basic value.)

Substance	α	α_{rel}	E (kcal/mole)
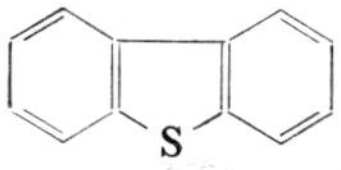 S	1.24	1.0	10.8
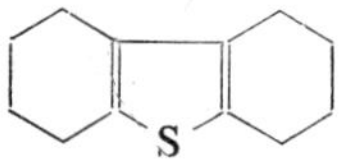 S	1.77	1.4	8.9
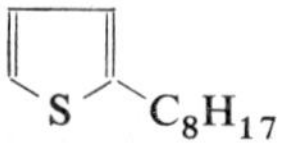 S C_8H_{17}	1.70	1.4	9.8
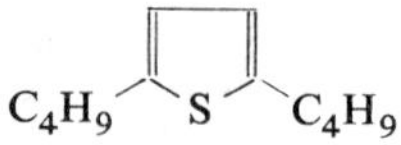 C_4H_9 S C_4H_9	1.70	1.4	7.6
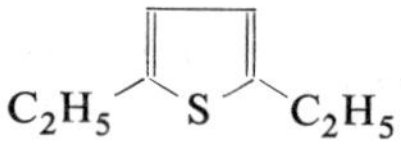 C_2H_5 S C_2H_5	1.82	1.5	—
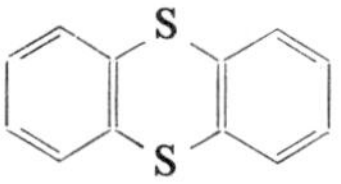 S S	2.27	1.8	8.3
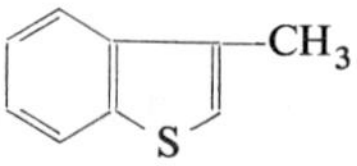 CH_3 S	2.70	2.2	8.5
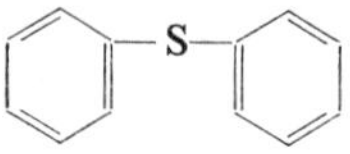 S	3.25	2.6	8.6
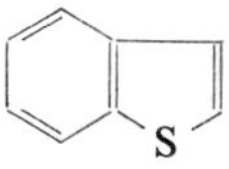 S	3.80	3.1	—
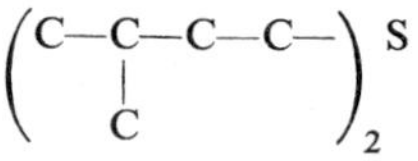 (C—C(C)—C—C—)$_2$S	4.30	3.5	11.0
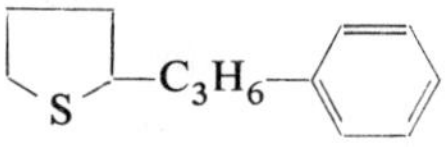 S —C_3H_6—	5.14	4.1	—
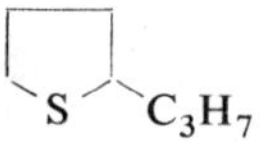 S C_3H_7	5.23	4.2	—
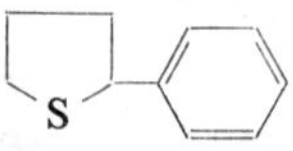 S	5.50	4.4	8.9
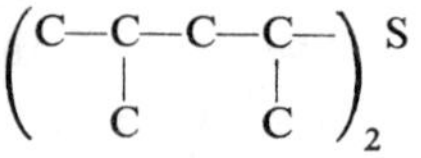 (C—C(C)—C—C(C)—)$_2$S	5.81	4.7	—

Continued Table 58.

Substance	α	α_{rel}	E (kcal/mole)
C_6H_5–S–C(C)–C–C(C)–C	5.88	4.7	—
C_6H_5–CH_2–S–CH_2–C_6H_5	9.40	7.6	11.3

used and under the influence of sulphurised Co–Mo catalysts. Kazeev's equation takes the form of:

$$\ln\left[\frac{D}{D-M}\right] = a\tau^{b} \qquad \text{(C)}$$

where τ is the hypothetical contact time (s), M the degree of conversion (%), D the degree of conversion at $\tau \to \infty$ (%); a and b are constants.

Table 59.

Influence of Catalyst Particle Size on Dibenzothiophene Hydrogenolysis Kinetics [1518]. (Sulphurised Co–Mo catalyst, $P_H = 33.3$ atm, different space velocities of the dibenzothiophene solution in cetane.)

Average mean catalyst particle size (mm)	Coefficients of Frost's equation[a])						Thiele modulus h[b])			Effectiveness factor E[c])		
	α_g			β								
	325°	375°	425°	325°	375°	425°	325°	375°	425°	325°	375°	425°
3	0.62	1.34	2.29	1	1	1	1.65	2.35	3.60	0.560	0.42	0.28
0.6	1.06	3.00	7.07	1	1	1	0.33	0.47	0.72	0.965	0.93	0.86

[a]) $\alpha_g = (F/W) \ln [1/(1-y)] - \beta (F/W) y$, where α_g is a coefficient proportional to the rate constant of a first- order reaction and to the effectiveness factor E; the other symbols are the same as conventionally used in Frost's equation [1678].

[b]) Thiele modulus is here defined by the relationship [1774]: $h = \sqrt{(a^2 k_S/(9rD)}$, where a is a measurement of length characterising the catalyst particle size (particle diameter, cm), k is the rate constant of a first-order reaction related to 1 cm^2 of the surface area ($cm\,s^{-1}$), r is the pore radius (cm), D is the diffusion coefficient (Knudsen diffusion, $cm^2 s^{-1}$).

[c]) The effectiveness factor is defined by the relationship: $E = (1/h)$ tgh (h); with $h = 0$ to 0.2, tgh $(h) \sim h$ and $E = 1$, i. e. the reactions take place in the kinetic region with no retardation due to diffusion; with $h \geqq 2$, tgh $(h) \sim 1$ and $E = (1/h) \leqq 0.5$, i. e. the inner catalyst surface is insufficiently utilised and the reaction proceeds in the inner diffusion region; with $0.2 < h < 2$, the range $0.5 < E < 1$ holds and a transition region is involved. The observed reaction rate is $r = r_0 E$ [2055, 2257].

The coefficient a characterises the kinetics of the process, and its significance is the same as that of the coefficient α in Frost's equation. The quantity b characterises the nature of the surface processes, or the type of reaction involved. The value of b remains practically unaltered with changes of the hydrogen partial pressure.

Kazeev's equation also clearly elucidates the relationship between the reaction rate and partial hydrogen pressure [1517, 1518]. The quantity D is defined by the expression $D = C + NP_H$ (the constants C and N are determined from a plot of the relationship between D and the partial hydrogen pressure P_H). Furthermore $a = kP_H^\varepsilon$, so that the relationship (C) converts to the equation:

$$\ln\left[\frac{C + NP_H}{C + NP_H - M}\right] = kP_H^\varepsilon \tau^b . \qquad \text{(CI)}$$

These relationships fit the experimental data of hydrogenolysis of sulphur compounds in the presence of a sulphurised Co–Mo catalyst over the entire range of temperatures, hypothetical contact times and partial hydrogen pressures used.

In a recent study, Frye and Mosby [570] investigated the hydrodesulphurisation kinetics of benzothiophenes on a sulphurised Co–Mo catalyst in the mixed phase*. The material desulphurised was a light circulating oil obtained by cracking, which was then treated at 260 to 370 °C under pressures of 7 to 40 atm and with space velocities of 1.5 to 10 kg/kg hour. The degree of desulphurisation was determined on the one hand from the total sulphur content of the hydrogenate, and on the other hand from the decrease of concentration of typical sulphur compounds (trimethylbenzothiophenes and dibenzothiophene) measured by means of gas chromatography. The authors [570] confirmed Hoog's [785] earlier finding, i.e. that low-boiling sulphur compounds are desulphurised more readily than high-boiling ones. The kinetics of hydrogenolysis of two types of trimethylbenzothiophenes, with different positions for methyl groups, and of dibenzothiophene were studied. With all three substances, the hydrodesulphurisation process was found to be a first order reaction. It is clear from the above measurements, that type B trimethylbenzothiophenes are hydrogenolysed about 1.6 times faster than type A trimethylbenzothiophenes (Fig. 65), although their boiling points differ only slightly. Further research into other sulphur compounds with similar boiling points indicated that hydrogenation rates may differ by a factor of up to 5, i.e. it was again confirmed that structure plays a significant role in hydrodesulphurisation. It is furthermore evident from the above measurements, that dibenzothiophene hydrogenation takes about 3.8 times as long as for type A trimethylbenzothiophenes. In addition to the structural

---

*) The conditions in the mixed phase are considerably more complicated than for a reaction conducted in the gas phase. Firstly, the conditions in both of the two phases must be considered in the kinetic evaluation, and secondly, diffusion processes are very frequently the limiting factor in the mixed phase reaction [2137] (cf. also p. 311).

effect, the decrease in the vapour pressure of the substance desulphurised also plays a significant role in this case. Frye and Mosby [570] showed by calculation, that under the operating conditions the vapour pressure of dibenzothiophene is about one third that of trimethylbenzothiophenes, so that in this case the structural effect does not predominate.

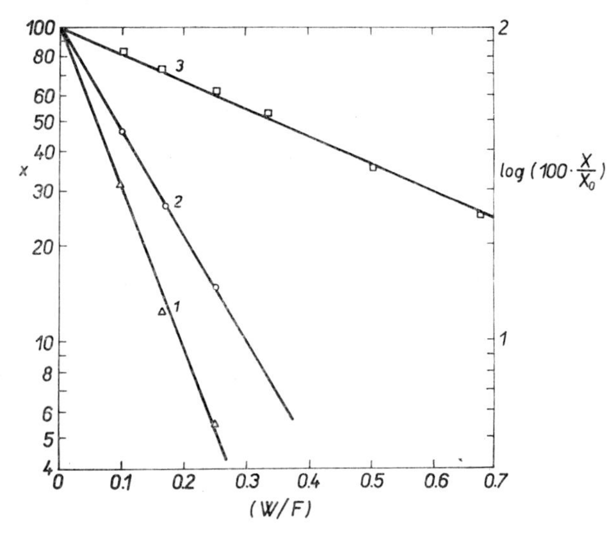

Fig. 65. Desulphurisation kinetics of typical sulphur compounds [570]. (Sulphurised Co–Mo catalyst, temperature 287 °C, pressure 15.7 atm.) X, X_0 — Molar fraction of the sulphur compound in the liquid (in a given moment and the initial value, resp.). x — % of the non-reacted compound. W/F — Reciprocal value of space velocity. 1 — Trimethylbenzothiophenes B. 2 — Trimethylbenzothiophenes A. 3 — Dibenzothiophene.

On evaluating the kinetic data, the authors of the study [570] arrived at an integrated first order reaction equation for the desulphurisation of the sulphur compound in the mixed phase, involving the influence of temperature, partial pressures of the individual components and degree of evaporation of the raw material:

$$\ln \frac{X_0}{X} = \frac{kP_0P_H}{\left(\frac{F}{W}\right)\left[1 - a + \left(a + \frac{H}{M}\right)\frac{P_0}{P}\right]\left[1 + K_{H_2S}P_{H_2S} + K_AP_A\right]^2} \qquad \text{(CII)}$$

where X_0 is the molar fraction of the sulphur compound in the liquid raw material, X — molar fraction of the sulphur compound in the liquid product, k — rate constant, P — total pressure, P_0 — vapour pressure of the sulphur compound, P_H — partial hydrogen pressure, (F/W) — space velocity (LHSV), a — degree of evaporation of the liquid (mole fraction), H — moles H_2 entering the reactor in a unit of time, M — moles of hydrocarbons plus sulphur compounds entering the reactor in a unit of time, K_{H_2S} – equilibrium adsorption constant for H_2S, K_A – mean equilibrium adsorption constant for aromatic hydrocarbons, P_{H_2S} – partial pressure of H_2S, P_S – partial pressure of aromatic hydrocarbons.

Frye and Mosby [570] found, that the temperature relationship (in the range of 260 to 370 °C) of the equilibrium adsorption constant of hydrogen sulphide is:

$$K_{H_2S} = 2.8\ e^{660/RT} \tag{CIII}$$

and for the mean adsorption constant of aromatic hydrocarbons:

$$K_A = 0.38\ e^{2800/RT}. \tag{CIV}$$

The following approximate relation is obtained from the apparent activation energy of desulphurisation, determined e.g. for type A trimethylbenzothiophenes:

$$kP_0 \sim e^{-30800/RT}. \tag{CV}$$

The apparent activation energy of desulphurisation is a combination of the following components: adsorption enthalpy of the sulphur compound, adsorption enthalphy of H_2, heat of evaporation of the sulphur compound (its value is approximately 13 kcal/mole) and the activation energy of desulphurisation proper. Since the two heats of adsorption are negative, the authors [570] estimate the activation energy of desulphurisation proper at a minimum of 18 kcal/mole. This value indicates that bulk phase diffusion is not a rate-controlling process in this case.

The equation derived above applies to every individual sulphur component or group of components behaving in a similar way. For practical purposes, therefore, relationships may be determined between desulphurisation of e. g. cracked circulating oil as a whole and desulphurisation of one of the individual sulphur compounds or group of sulphur compounds contained in the raw material. Frye and Mosby [570] believe that similar relationships apply to other processes in the mixed phase too, e. g. hydrodenitrogenation in the first hydrocracking stage etc.

6.9.3.5.

Hydrodesulphurisation kinetics of hydrocarbon mixtures

Hydrodesulphurisation of petroleum, shale-oil and tar fractions is one of the most important refining reactions, which is being developed particularly intensively and rapidly to replace classical chemical refining processes. The solution of engineering problems involved in hydrodesulphurisation, especially the determination of the optimum size of reactors, essentially requires a knowledge of the basic kinetic data concerning the reaction involved.

However, for several fundamental reasons, the hydrodesulphurisation kinetics of a mixture of sulphur compounds in natural materials is far more complicated than for hydrogenolysis of pure substances. Firstly, different types of sulphur compounds of widely differing reactivity occur in petroleum and other hydrocarbons fractions. As already stated in section 6.9.3.3., thiophenes are particularly resistant, so that during the course of the reaction the rate constant will decrease systematically with the decreasing content of easily reacting components and the increasing content of highly resistant compounds [533, 1327].

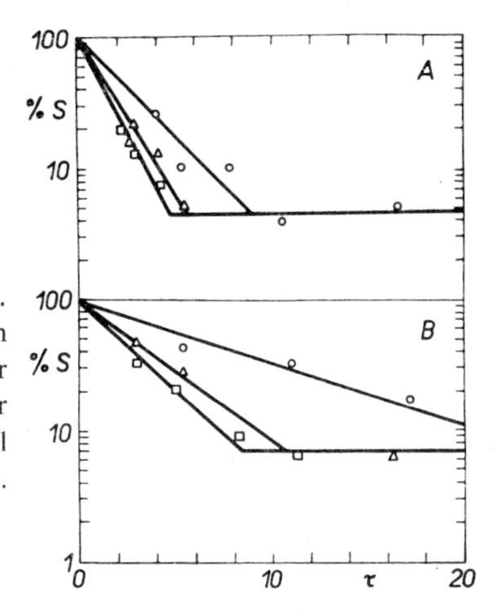

Fig. 66. Hydrodesulphurisation of gas oil fractions [785]. (Pressure 51 atm, temperature 375 °C.) *A* — Fraction b. p. 290—300 °C, sulphur content 1.20%, molecular weight 233. *B* — Fraction b. p. 330—340 °C, sulphur content 1.76%, molecular weight 268. % S — % residual sulphur in hydrogenate. τ — Hypothetical contact time. ○ — 1500 Nm^3 H_2/ton. △ — 3000 Nm^3 H_2/ton. □ — 6000 Nm^3 H_2/ton.

Frye and Mosby [570] have shown, that the hydrogenolytic properties of sulphur compounds of roughly the same boiling point may differ several-fold. Larger still becomes the possible difference in reaction rates for hydrodesulphurisation of wide range petroleum or tar fractions or, to an even greater extent, when crude materials as a whole are desulphurised. In the latter case, the sulphur compounds present differ in structure as well as in molecular weight.

Another complication which makes an exact kinetic interpretation of catalytic hydrodesulphurisation of natural materials very difficult, is the fact that the reaction is carried out in part in a mixed phase, where the retarding influence of diffusion very frequently predominates and masks the kinetics of desulphurisation reactions proper.

Finally, one of the major difficulties in elucidating the true kinetic conditions in hydrodesulphurisation of hydrocarbon mixtures is caused by the interference of some non-hydrocarbon high-molecular weight components of asphaltic and resinous character, high-molecular weight nitrogen compounds and compounds of trace metals. These compounds considerably influence the kinetics, particularly of the more rapid hydrodesulphurisation reaction, probably due to selective adsorption on the catalyst surface [533].

One of the first attempts at a kinetic expression of hydrodesulphurisation of petroleum fractions was made by Hoog [785] in a study of gas oil desulphurisation in the gas phase on a Co–Mo refining catalyst. Hoog compared the behaviour of gas oil as a whole (b. p. 260 to 350 °C, mean molecular weight 236, sulphur content 1.25%) and of two of its narrow fractions: fraction A (b. p. 290 to 300 °C, sulphur content 1.2%, mean molecular weight 233) and fraction B (b. p. 330 to 340 °C, sulphur content 1.76%, mean molecular weight 268). Hoog [785] mainly found that hydrodesulphurisation of high-boiling fractions is in principle more difficult than for low-boiling fractions, as we see from Fig. 66. Hoog [785] justified this difference in behaviour by a steric effect, as the C—S bond is the less accessible to hydrogenolysis, the higher the molecular weight. From results plotted in Fig. 66 it follows that the degree of desulphurisation is limited to 95–97%, which Hoog [785] explains by the presence of very resistant sulphur compounds, which are not hydrogenolysed under the conditions employed.

The course of hydrodesulphurisation of narrow gas oil fractions up to 95% desulphurisation may be represented in quite a satisfactory manner by the kinetic equation for a first-order reaction:

$$\ln \frac{P_s}{P_s^0} = -k_{P_H P_0} \tau \qquad \text{(CVI)}$$

where P_s – partial pressure of the sulphur compound at the end of the reactor, P_s^0 – initial partial pressure of the sulphur compound, τ – hypothetical contact time, $k_{P_H P_0}$ – rate constant, valid for a specific hydrogen partial pressure P_H, partial pressure of oil P_0 and specific temperature.

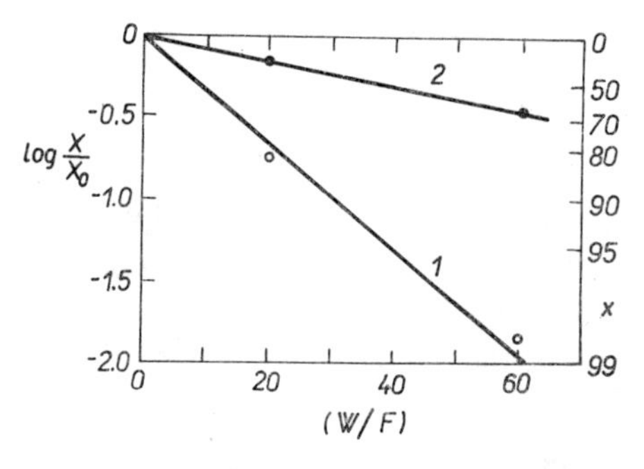

Fig. 67. Influence of space velocity on hydrorefining of cracked gasoline [2218] (partial H_2 pressure 7.3 atm.) 1 — Sulphur removal. 2 — Nitrogen removal. X, X_0 — Concentration and initial concentration of sulphur and nitrogen, resp. W/F — Reciprocal value of space velocity (g hour/gmole). x — Degree of removal of S or N (%).

This simple kinetic equation is not satisfied for gas oil as a whole, a fact which indicates the presence of sulphur compounds of differing reactivity. Hoog [785] furthermore found that the rate constant of desulphurisation decreases with the growing partial pressure of the oil. The author believes that this is caused by preferential adsorption of oil molecules on the free centres of the catalyst. The concentration of chemisorbed hydrogen on the catalyst surface will be decreased, and thus the rate of desulphurisation decreases, too. Ohtsuka et al. [1536] similarly observed that a first-order kinetic equation is satisfactory for hydrodesulphurisation of gas oil, but that this equation cannot be applied to desulphurisation of raw petroleum [1537].

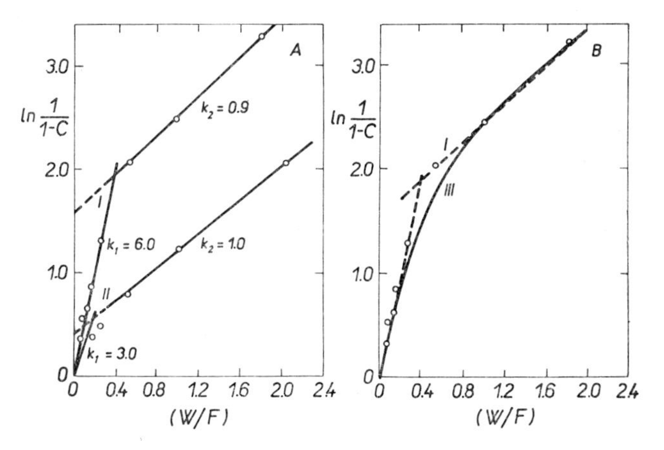

Fig. 68. Dependence of the value of ln (1/1 — C) on the reciprocal value of space velocity (*W*/*F*) in the desulphurisation of the vacuum residue from Kuwait petroleum at 70 atm [533]. *A* — Course of desulphurisation of deasphalted (I) and non-deasphalted (II) residue. *B* — Comparison of experimental data (○—○—○) with the course of desulphurisation calculated from the equation (CVIII) for the deasphalted residue (curve III).

A first-order kinetic equation, which Wilson et al. [2218] verified with pure substances (p. 229), may also be applied to the process of hydrodesulphurisation of light petroleum fractions. Fig. 67 shows evidence for the fact that this rate equation gives good agreement not only for pure sulphur compounds, but also for sulphur compounds contained in cracked gasoline.

Although hydrodesulphurisation of light petroleum distillates in the gas phase could be very well represented in a number of cases by a first-order equation, the validity of this kinetic relationship is not general. Voorhies et al. [2147] state that the rate of desulphurisation of Kuwait atmospheric residuum, as well as of the whole petroleum, may be described by a first-order kinetic equation, but in spite of this some influences play a role in the case of high-boiling fractions and particularly of distillation residues, which render the simple course of the reaction more complicated. In the case of residual fractions, complications are caused by asphaltenes, as was proved by Flinn et al. [533] in the case of hydrodesulphurisation of the vacuum residue from Kuwait petroleum. The catalyst employed in this case was a supported one, containing metals of the Groups 6A and 8 of the periodic system, and results were compared for deasphalted and non-deasphalted material at 70 atm. Fig. 68

shows that desulphurisation of the deasphalted, as well as non-deasphalted material, does not satisfy the simple first-order relationships:

$$\ln\left[\frac{1}{1-C}\right] = k\left(\frac{W}{F}\right) \qquad \text{(CVII)}$$

where $C = 1 - (S/S_0)$, S being the sulphur content of the product, S_0 – sulphur content of the feed ($100C = \%$ desulphurisation), (W/F) – reciprocal of space velocity, k – rate constant.

In this kinetic interpretation, the desulphurisation process appears to divide into two different reactions, the first of these, taking place on short contact times, being 3 to 6 times more rapid than the second one. In the case of the deasphalted vacuum residue, with the more rapid reaction which has a rate constant of $k_1 \sim 6$, approximately 80% desulphurisation is achieved, while in the second reaction, with $k_2 \sim 0.9$, desulphurisation is practically complete. It is probable, that the two reactions take place simultaneously, but the second only becomes apparent when the first one has been completed.

When the non-deasphalted residue is desulphurised, the conditions are similar, but the rate constant of the initial rapid reaction is considerably slower ($k_1 \sim 3$). It is remarkable in this case, that the second, slower reaction has practically the same rate for both the deasphalted and non-deasphalted material ($k_2 \sim 1$). With the non-deasphalted residue, considerably less sulphur is removed by the rapid reaction (roughly only 30% compared to 80% in the case of the deasphalted residue).

Desulphurisation of the two materials could not be expressed by a single equation over the entire range of space velocities, neither for a whole nor for a fractional reaction order. Flinn et al. [533] therefore assume, that desulphurisation of the deasphalted and non-deasphalted residue takes place by means of two separate first-order reactions. When the extreme reaction rate constants at the beginning and end of the reaction are employed to characterise the two separate first-order processes, satisfactory agreement with experimental data is achieved by using the equation:

$$S = S_1^0\, e^{-k_1\Theta} + S_2^0\, e^{-k_2\Theta} \qquad \text{(CVIII)}$$

where S is the sulphur content in the product, S_1^0 – initial content of type 1 sulphur (easily hydrogenolysed), S_2^0 – initial content of type 2 sulphur (difficult to hydrogenolyse), k_1 – rate constant of the desulphurisation reaction for type 1 sulphur, k_2 – rate constant of the desulphurisation reaction for type 2 sulphur, $\Theta = (W/F)$ – reciprocal of space velocity. Experimental data for the deasphalted residue are compared with desulphurisation proceeding according to equation (CVIII) in Fig. 68, showing good agreement with the exception of a transition region between the two reactions. Arey et al. [86] similarly described the kinetics of hydrodesulphurisation of vacuum oil and petroleum residue by means of two equations for first-order reactions.

The experimental results indicate the strong inhibiting effect of asphaltenes on the course of the more rapid reaction. Although the residue contains only some 20% asphaltenes, their influence on the rapid desulphurisation reaction is incomparably greater. Since the viscosity of the material decreases substantially by removal of asphlatenes, the rate of internal as well as external diffusion rises, and this change has a favourable influence on the rate of the overall desulphurisation reaction.

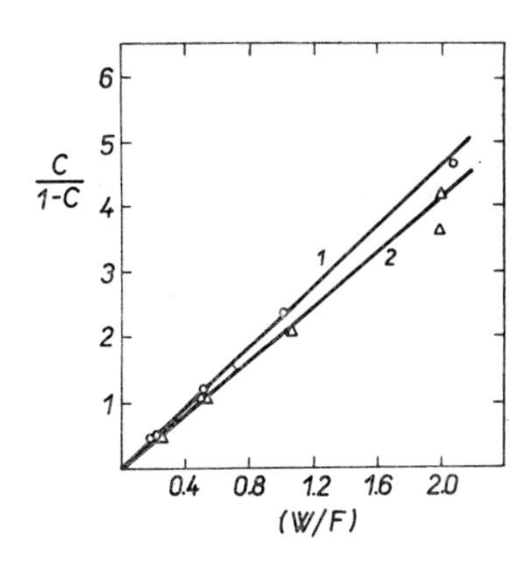

Fig. 69. Desulphurisation of the vacuum residue of Kuwait petroleum under typical working conditions, proceeding apparently as a second-order reaction [157]. 1 — Commercial desulphurisation catalyst. 2 — Improved desulphurisation catalyst at temperatures 4 to 5 °C lower (for symbols, cf. equation (CIX)).

Massagutov et al. [1327] likewise did not succeed in interpreting the hydrodesulphurisation kinetics of vacuum gas oil obtained from Arlan raw petroleum in the presence of a Co–Mo catalyst by means of a first – order reaction, although they tested this kinetic relationship in the form of Frost's [76, 567], Wilson's et al. [2218] and Hoog's [785] equations.

Kazeev's equation [1005] as well as a kinetic equation for a second-order reaction (see later, [1327]) could be successfully applied. Pestrikov et al. [1607] confirmed the applicability of Kazeev's kinetic equation to the desulphurisation of a pyridine extract of a gas oil fraction (b. p. 234 – 363 °C) on a Co–Mo catalyst.

The equation suggested by Beuther and Schmid [157] seems to be a universal relationship for hydrodesulphurisation of heavy petroleum materials taking place in a mixed phase. These authors found, after a rather unsuccessful attempt at using a rate equation for a first-order reaction (p. 240), that the desulphurisation rate may be expressed with sufficient accuracy by means of a simple equation for a second-order reaction in the form of:

$$\frac{C}{1 - C} = k \frac{1}{\left(\frac{F}{W}\right)} \tag{CIX}$$

where $C = (1 - (S/S_0)$ is relative conversion, S – sulphur content in the product (%), S_0 – sulphur content in the feed, k – rate constant, (F/W) – space velocity.

Fig. 69 shows that the kinetic relationship mentioned above is well satisfied for a wide range of space velocities (up to LHSV = 0.5) and even for as complicated a crude material as the vacuum residue of Kuwait petroleum. The explanation of the observation, that the overall desulphurisation reaction proceeds as a second-order process, is seen by the authors [157] to lie in the fact that natural materials contain sulphur compounds of differing reactivity. The hydrogenolysis of these compounds frequently takes place at widely varying rates (cf. the data by Frye and Mosby [570] and Hoog [785] mentioned earlier, pp. 235 and 237). If the hydrogenation of each type of sulphur compound were to proceed by a first-order reaction with respect to the sulphur compound, then the rate constant would decrease systematically with the gradual removal of the more reactive sulphur compounds and rising concentration of compounds which are more resistant to hydrodesulphurisation. This consecutive series of first-order reactions, taking place with a great excess of hydrogen and a gradually decreasing rate constant, appears as a whole to be a second-order reaction. Although there is no complete explanation for these kinetics, there yet remains the fact that a number of hydrodesulphurisation reactions proceed by an apparently second-order reaction, particularly in the case of the heaviest raw materials. Since the value of the rate constant can be determined experimentally as a function of temperature and overall pressure at the same time, this simple kinetic relationship enables a very exact estimate of the course of hydrodesulphurisation to be made, even under conditions for which no preliminary experiments were carried out.

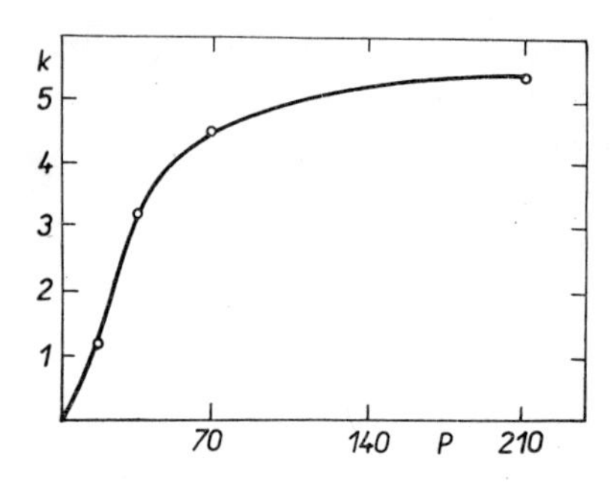

Fig. 70. The influence of pressure (P, atm) on the rate constant k of hydrodesulphurisation of petroleum residues [157]. (Atmospheric residue from Kuwait crude, 36% of the original raw material.)

The temperature relationship of the rate constant satisfies the Arrhenius relation very closely, thus permitting an estimate of the activation energy to be made and, at the same time, further justifying the use of a second-order equation to express the hydrodesulphurisation kinetics of heavy petroleum fractions.

The rate constant is also a function of the applied pressure in the hydrodesulphurisation process. Fig. 70 gives a typical relationship for the atmospheric residue of Kuwait petroleum, showing the high degree of sensitivity of the reaction to pressure variations in the range of 1 to 70 atm, while above 70 atm the effect of pressure is rather small. This characteristic course is exhibited by all heavy material at all temperatures.

The reaction rate is moreover a function of the chemical and physico-chemical properties of the raw material. Up to the present time, however, no general dependence of the rate constant on these influences has been determined. Beuther and Schmid [157] mention the relationship between the rate constant and the „size" of the petroleum residue (i.e. the proportion of the overall amount of raw material in the residue) which is important in practical respects. When the logarithm of the size of the petroleum residue is plotted against the rate constant of desulphurisation, a linear graph is obtained. This purely empirical relationship is very useful for practical purposes.

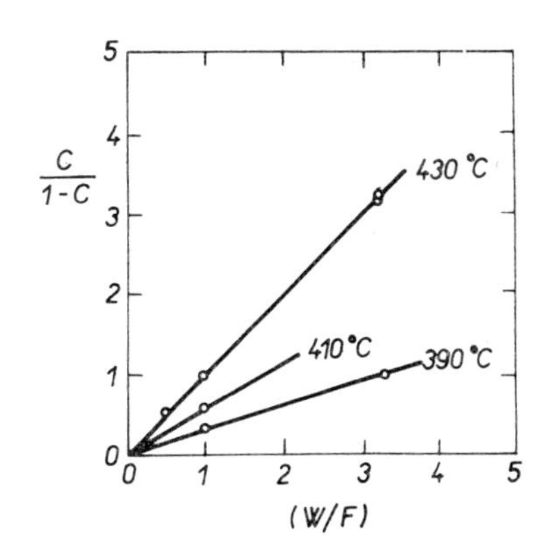

Fig. 71. Interpretation of the hydrodesulphurisation of crude petroleum (Khafji) according to a second-order kinetic equation [1537]. (Pressure 40 atm, the crude petroleum contained 2.78% S, 4.1% asphaltenes, 7.4% carbon residue, 35 ppm V, 15 ppm Ni; the initial b. p. was 32 °C, 70% distilled up to 568 °C.)

Beuther and Schmid confirmed their earlier finding [533] of the strong inhibiting effect of asphaltenes on the hydrodesulphurisation reaction of residual fractions. A rate constant of $k = 2.3$ was obtained on hydrodesulphurising a vacuum residue at a specific temperature, while this rate rose more than four times, i.e. to 9.7 when the same material was deasphaltised [157].

Ohtsuka et al. [1537] verified the high validity of the kinetic equation for a second-order reaction in the hydrodesulphurisation of a raw petroleum on a Co–Mo catalyst for a wide range of LHSV values (Fig. 71). The applicability of this equation was also tested in the case of hydrodesulphurisation of petroleum residues [23a, 1442a, 1890a, 2168a]. The kinetic equation mentioned above was also successfully applied to hydrodesulphurisation of petroleum distillates in the presence of a Co–Mo catalyst, e.g. in the case of desulphurisation of vacuum gas oil [1327] or diesel fuel and the vacuum distillate of Romashkino petroleum [2129, 2130]). A rate equation for a second-order reaction was used to evaluate the process of hydrorefining cracked diesel oil, in terms of desulphurisation as well as hydrogenation of olefins [948b, 1765a].

When petroleum residues are hydrorefined in the mixed phase, retardation by diffusion may be more important in determining the overall reaction rate than surface kinetics. In mixed-phase hydrogenation, three types of diffusion barriers are decisive: hydrogen transport across the gas-liquid interface, transport from the bulk stream to the catalyst surface and internal diffusion. In particular the diffusion of the large molecules contained in petroleum residues is far slower than in the case of refining of distillates. Adlington and Thompson [6] believe that the rate-controlling diffusion process is hydrogen transport through the gas-liquid interphase boundary. It would appear from the activation energies for desulphurisation of dibenzothiophene (slightly less than 25 kcal/mole [86, 1272]) and for vacuum residues (27 kcal/mol in the case of Kuwait residue, 39 kcal/mole in the case of a mixture of Kuwait and Venezuelan residues [533, 1808]), that retardation due to internal diffusion is no more of a problem in hydrodesulphurisation of residual fractions than it is in desulphurisation of distillates [86].

6.10.

Hydrogenation of nitrogen-containing compounds

Sulphide catalysts have been little used up to now in industrial hydrogenation of individual nitrogen compounds. Cases worth mentioning particularly are hydrogenation of aromatic nitrogen compounds to amines and of pyridine to piperidine. The application of sulphide catalysts in hydrogenating nitrogen compounds is particularly useful in those cases where the raw materials are not pure and contain catalytic poisons, particularly sulphur compounds.

Sulphide catalysts are especially important where the removal of nitrogen compounds from industrial raw materials is involved, particularly from petroleum and petroleum semiproducts and also from tars and shale oils. In these cases, however, complete hydrogenolysis of practically all nitrogen compounds present must be carried out, since nitrogen compounds act as efficient deactivating agents and even as catalyst poisons in later catalytic processing, especially cracking, hydrocracking and reforming. Problems of this type are discussed in detail in section 7.2.3. (p. 339).

Hydrogenolysis of nitrogenous heterocyclic compounds in the presence of sulphide catalysts is, in general, more difficult compared to the respective oxygen and sulphur heterocyclics. Moreover, the majority of sulphide catalysts lose their hydrogenating and hydrogenolytic efficiency in the presence of basic nitrogen compounds, and special catalysts, e.g. mixed Ni–Mo catalysts, must be employed in order to secure a high degree of denitrogenation. Besides the blocking of the acid catalyst centres, one of the most frequent causes of deactivation are polymeric products, which are formed rather easily in the course of hydrogenation of nitrogen compounds and contaminate the catalyst surface.

6.10.1.

Hydrogenation of amines

Günther [695] studied the kinetics of aniline hydrogenolysis on the industrial catalysts 5058 and 8376. A 2% aniline solution in heavy gasoline was hydrogenated at 210 to 325 °C and a hydrogen pressure of 60 to 300 atm with different space velocities of the feed. It was found that in kinetic respects, aniline hydrogenolysis is close to a first-order reaction. The value of the apparent activation energy of this reaction was 24 to 24.8 kcal/mole in the case of the catalyst 5058, which was the more active one; while with the catalyst 8376 it was 18.2 kcal/mole. The influence of the partial pressure of hydrogen on the reaction rate is surprisingly slight, and variations in the space velocity of the hydrogenation gas are similarly of little importance.

On the sulphide catalyst U 63, the reaction rate of aniline hydrogenation was likewise found to be independent of the molar excess of hydrogen [1961]. The product of aniline hydrogenolysis at 50 atm and 190 to 310 °C is cyclohexane together with a small amount of benzene [1961]. The benzene concentration, however, is considerably greater than the value which corresponds to the hydrogenation-dehydrogenation equilibrium, cyclohexane $\rightleftarrows$ benzene. Cyclohexylamine was not found in the reaction mixture. Stengler et al. [1961] found that cyclohexylamine hydrogenolysis takes place at a high rate, so that it is improbable that it could be recovered as a reaction intermediate. Since benzene is rather difficult to hydrogenate under the reaction conditions employed, the authors of the study [1961] assume that the phenyl residue, formed by splitting off the NH_2 group, is reactive to such an extent that it is readily hydrogenated to cyclohexane. The activation energy of aniline hydrogenation on this catalyst is shown in Table 63.

The main hydrogenolysis products of aniline and α-naphthyl amine in the presence of MoS_3 (H_2 pressure 40 to 100 atm, temperature 200 to 450 °C) were cyclohexane and tetralin, resp. Moreover, side products formed in the case of aniline were diphenylamine, cyclohexene, methylcyclopentane and pentane. In the hydrogenation of naphthyl amine, decalin was also formed [2234].

Other interesting side products were recorded when aniline was hydrogenolysed in the presence of WS_2. The main product of pressure hydrogenation at 450 °C in this case was cyclohexane, while N-methyl-1,2,3,4-tetrahydroquinoline was found in the reaction product at a lower temperature (250 °C) and decahydroquinoline and bicyclo-[4,3,0]-nonane were formed when the process was carried out at 450 °C [1388].

Selective hydrogenolysis of amines was also utilised to synthetise some phenols. Phenolic Mannich bases were hydrogenated at 200 to 240 °C and a hydrogen pressure of 140 atm in the presence of the MoS_2/Al_2O_3 catalyst, forming a high yield of the respective phenol. For example, mesitol was obtained in a yield of 85% by this reaction [798]:

$$(CH_3)_2N{-}CH_2{-}C_6H_2(OH)(CH_3)_2 \xrightarrow{[H_2]} (CH_3)_3C_6H_2OH \quad \text{(CX)}$$

Selective hydrogenation of some N-thenylidene- or N-furfurylidene derivatives of aromatic amines to the respective thenyl- or furfuryl amines was catalysed by rhenium heptasulphide [1787]. Some imines (Schiff's bases) of the thiophenic series were hydrogenated to the respective amines in the presence of Pd and Re sulphides [1787e, 1787g].

Some other data concerning the hydrogenation of amines will be mentioned in connection with hydrogenolysis of nitrogenous heterocyclic compounds (cf. section 6.10.4).

6.10.2.

Hydrogenation of nitrocompounds

Production of amino compounds by classical reduction methods is now being replaced to an increasing extent by catalytic hydrogenation of nitrocompounds.

Copper catalysts, which are used most often in industry [674, 824] are very sensitive to sulphur, so that e.g. nitrobenzene obtained from coke-oven benzene cannot be treated in their presence. Sulphide catalysts are particularly suitable for this purpose, but their industrial application has been very rare up to now [1600]. One example is the Allied's Natl. Aniline Division (Moundsville, USA) use of a NiS/Al_2O_3 catalyst to hydrogenate nitrobenzene to aniline [59].

An analysis of the literature shows that sulphide catalysts have been used exclusively to hydrogenate aromatic nitrocompounds. These catalysts are sufficiently selective under the reaction conditions employed to ensure that the benzene ring is not hydrogenated and other side reactions are greatly limited [341].

In the majority of cases, hydrogenation was carried out in the gas phase. Operations in the liquid phase were applied in isolated cases only. Nitroxylenes were hydrogenated in the presence of MoS_2 [621, 622] at 175 to 245 °C, and pressures of up to 42 atm in a reactor divided into four catalytic beds [621]. The reaction heat was removed by recirculating a part of the product [621, 1427].

When the suspension of a nitrocompound, which has no substituent in the *para*-position, is hydrogenated in sulphuric acid in the presence of Mo, Co, Cr, W or Pt sulphides, amino phenols as well as amines are formed. The reduction process probably involves phenylhydroxylamine, which rearranges to amino phenol in an acid medium:

$$C_6H_5NO_2 \xrightarrow{[H_2]} C_6H_5NHOH \xrightarrow{[H_2SO_4]} p\text{-}H_2NC_6H_4OH \, . \qquad \text{(CXI)}$$

The reaction is carried out at 150 °C and medium hydrogen pressures (28 to 35 atm) [674, 761].

Hydrogenation in the gas phase [18, 340, 668, 931, 1564, 1777, 1946, 2221, 2222, 2223, 2224] was carried out in a flow reactor at 250 to 450 °C and pressures of mostly 1 to 35 atm. The use of medium pressures and a large hydrogen excess is advantageous from the point of view of maintaining the catalyst activity for a long time, since tar-like polymers are formed and deposited in the course of hydrogenation of nitroderivatives [340].

The catalyst most frequently employed in hydrogenation of aromatic nitrocompounds to amines is nickel sulphide supported on alumina [132]. Winstrom used either amorphous Al_2O_3 [2223, 2224] or activated alumina [18, 2222] as carrier. These catalysts are very active and selective, and they are characterised by a long lifetime and the possibility of repeated regeneration [2221, 2224].

Jarkovský et al. [931] used the sulphurised catalyst 6524 (fresh Ni/Al_2O_3, sulphurised in an ammonium sulphide solution by gaseous H_2S after reduction) to hydrogenate nitrobenzene to aniline in the gas phase. These authors worked at atmospheric pressure, with a cooling bath temperature of 200 to 250 °C, a molar nitrobenzene-to-hydrogen ratio of 1 : 10 and a space velocity of about 300 g/l hour. Under optimum reaction conditions, no side products were formed and conversion was practically quantitative. The stability of the catalyst activity is very high (the authors estimate the lifetime between two regenerations at 1,000 to 2,000 hours from the rate of shift of the temperature maximum in the catalyst bed). The catalyst is easily regenerated by a mixture of air and steam at 300 to 350 °C.

Non-supported nickel [232, 1777] and cobalt [668] sulphides were also applied in nitrobenzene hydrogenation.

Molybdenum and tungsten sulphides have been very often applied to hydrogenation of aromatic nitroderivatives. Either unsupported sulphides are used [454, 1946] or, more frequently, sulphides are supported on Al_2O_3 [2149] and especially on activated carbon [229, 340, 341, 1323, 1944]. The high activity of sulphide catalysts is maintained by slight sulphurisation of the raw material (addition of e.g. 1% CS_2) [341, 1947]. It was found furthermore, that a small amount of water in the feed improves the activity of the catalyst and keeps the hydrogenate from darkening [1563, 1947].

Among other sulphide catalysts, the following were found to be useful in the hydrogenation of nitrocompounds: CuS/Al_2O_3 [1564], PbS [232] as well as noble metal sulphides: platinum sulphide on active carbon [430, 430a], Pd, Ir and Os sulphides [430b], rhodium sulphide on activated carbon [2088, 2089] and rhenium heptasulphide [185, 226a]. Besides nitrobenzene [18, 185, 232, 341, 668, 931, 1563, 1777, 2088, 2089, 2221, 2222, 2223, 2224] and nitroxylenes [229, 340, 621, 622, 1563, 1944, 1946, 1947, 2149], polynitrocompounds [132, 1323], alkali salts of nitrobenzene sulphonic acids [454] and halonitroderivatives [117a, 430, 430c, 651, 651a, 1561a, 1927a] were hydrogenated to the respective amines in the presence of these catalysts.

6.10.3.

Hydrogenation of nitriles

Acrylonitrile was selectively hydrogenated to propionitrile in the gas phase in the presence of an aliphatic alcohol (serving as stabiliser to avoid acrylonitrile polymerisation) on mixed sulphidic catalysts e.g. $2\,NiS.WS_2$, $NiS.MoS_2$, $CoS.MoS_2$ [1915, 1917].

Nitriles may be hydrogenated at normal pressure in the presence of WS_2. Bernardini and Brill [149] hydrogenated acetonitrile at 360 to 530 °C with a twelve-fold excess of hydrogen. The reaction involved hydrogenolysis as well as conversion to the amine. Above 360 °C, the catalyst was rather quickly poisoned by basic substances.

Hydrogenolysis of α-naphthonitrile in the presence of MoS_2 resulted in a high yield of α-methylnaphthalene at 100 atm hydrogen pressure and temperatures above 240 °C. A small amount of α-N-methylnaphthylamine was formed as by-product. Similarly, benzonitrile reacted to give toluene [286b, 907].

6.10.4.

Hydrogenation of nitrogen-containing heterocyclic compounds

The mechanism of hydrodenitrogenation of technically important crude materials (petroleum, tars, shale oils) has been little studied up to now. This is due to the fact that investigation into the chemistry, and in particular into the kinetics, of hydrogenation and hydrogenolysis of heterocyclic nitrogen compounds in the presence of sulphide catalysts is only just starting. The development of analytical methods of separation and identification, especially of gas chromatography and spectrometric methods, has been a great help in this respect.

6.10.4.1.

The chemistry of hydrogenation of nitrogen heterocycles

6.10.4.1.1.

Hydrogenation of five-membered heterocycles

Hydrogenation of simple five-membered nitrogen heterocycles was studied mainly with nickel, copper-chromium and platinum catalysts. Among sulphide catalysts, Landa et al. [948, 1188] used MoS_2 to hydrogenate 2-methylpyrrole. The reaction

Table 60.

Partial Hydrogenation of Indole on a Supported Ni–W Catalyst [534, 733].
(The solvent was a middle petroleum distillate, indole content 0.5%, hydrogenation temperature 315°C, pressure 21 atm, LHSV = 4.)

Component	ppm
Total nitrogen content in the hydrogenate	4300
Non-basic nitrogen[a])	3561
Basic nitrogen	739
Semiquantitative composition of compounds with basic nitrogen:	
Amines[b])	80
Indoline	70
Quinolines[c])	15
Non-identified higher basic compounds	554

Qualitative composition of the individual fractions:

[a]) Non-basic nitrogen compounds: indole, 3-isopropylindole, 1,3-dimethyl-2-ethylindole, 3-propylindole, 1-ethylindole, 2-tert. butylindole and carbazoles.

[b]) Amines: N-ethylcyclohexylamine, n-octylamine, β-cyclohexylethylamine, β-phenylethylamine, *o*-ethylaniline.

[c]) Quinolines: Quinoline, dimethylquinolines, 1,2,3,4.-tetrahydroquinoline.

conditions for work in an autoclave (H_2 filling pressure 100 atm, temperature 340 °C, reaction time 1 hour) were selected so as to form the greatest possible number of intermediates. At temperatures lower than 340 °C, the ring was mainly hydrogenated, while intensive hydrogenolysis takes place above this temperature. Individual compounds isolated from the lower fractions include 2-methylpyrrolidine and 2,5-dimethylpyrrole. A large amount (over 50%) of higher-boiling polymeric products was formed in the hydrogenation process. These products caused rapid deactivation of the catalyst and difficulties in isolating the hydrogenation intermediates.

The hydrogenolysis of indole was also studied under mild hydrogenation conditions, so that the course of individual stages in the overall splitting process could be judged from the intermediates isolated. Hartung et al. [733] hydrogenated a 0.5% solution of indole in middle petroleum distillate in the presence of a sulphactive supported catalyst at 315 °C, 21 atm pressure and LHSV = 4. The result of a typical analysis of nitrogen compounds in the hydrogenate is shown in Table 60 (the basic character of a compound was judged from its reactivity with a mixture of perchloric and acetic acids). The quantitative composition of basic nitrogen compounds is shown in Table 61; the major compounds formed were indoline and *o*-ethylaniline. Among the products expected from partial hydrogenation and hydrogenolysis of indole, perhydroindole and 2-ethylcyclohexylamine were surprisingly not identified. On the other hand, a number of substances were formed which are probably the products of splitting and alkylation side-reactions. The participation in alkylation side-reactions of carbonium ions, formed by partial cracking of the hydrocarbon solvent, cannot be excluded [733].

Table 61.

Quantitative Composition of Basic Nitrogen Compounds Isolated from the Indole Hydrogenate (cf. Table 60) [733].

Compound identified	Content in the basic hydrogenate fraction (% by wt.)
N-Ethylcyclohexylamine	4.8
n-Octylamine	1.0
β-Cyclohexylethylamine	4.2
β-Phenylethylamine	8.1
o-Ethylaniline	24.7
Indoline	37.9
Quinoline	1.1
Dimethylquinolines	0.7
1,2,3,4-Tetrahydroquinoline	6.9
Indole (from indole polymer)[a])	10.6

[a]) The presence of indole among the basic substances is explained by the passage of indole polymers into the basic fraction [733].

The formation of some substances identified may be explained by reactions already described. For example 2-methylindole, forms quinoline at higher temperatures [1624]. The 3-alkylindole cation, formed from alkylindoles, is the first intermediate in the process of opening of the pyrrole ring. This is followed by dehydrocyclisation and formation of 1,2-dihydroquinolines. The latter compounds are unstable, however, disproportionating to form quinolines and 1,2,3,4-tetrahydroquinolines provided they are not alkylated on the nitrogen atom [733]. Thermal conversion of non-basic nitrogen compounds in petroleum to basic compounds, especially alkylquinolines, was also observed [214].

Hartung et al. [733] reported another important finding, that hydrogenolysis of an indole derivative is more difficult, the more substituted the pyrrole ring of the indole molecule and the closer to the nitrogen atom a substituent is located.

Flinn et al. [534] assume the following scheme for the hydrodenitrogenation of carbazole on the basis of the intermediates isolated in the process of indole hydrogenation:

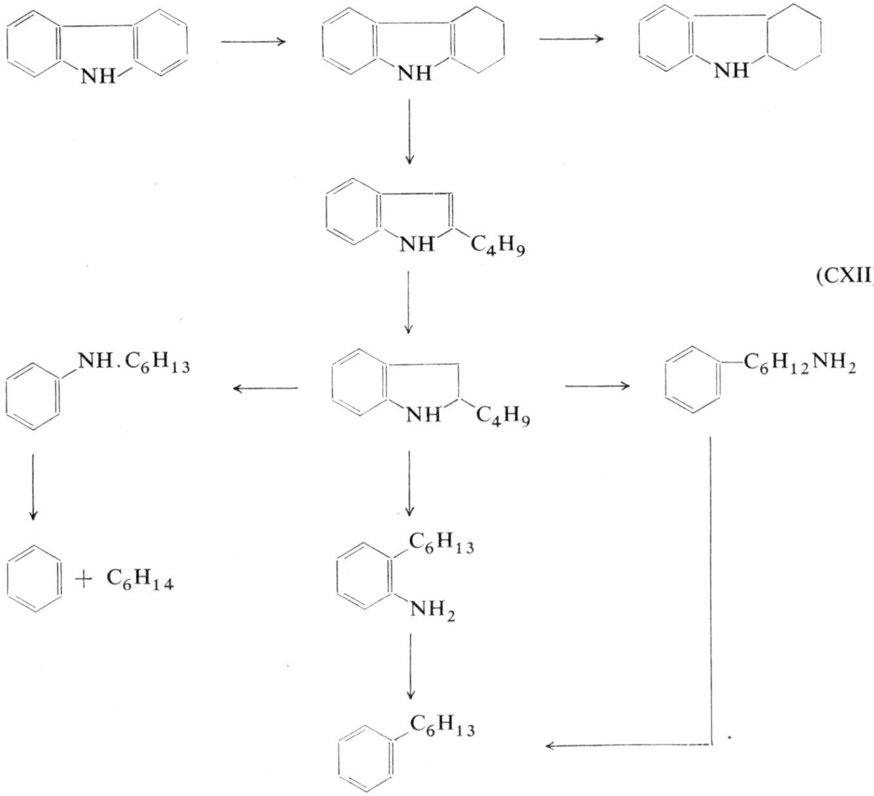

NH → NH → NH

NH C_4H_9

(CXII)

$NH.C_6H_{13}$ ← NH C_4H_9 → $C_6H_{12}NH_2$

$+ C_6H_{14}$

C_6H_{13}

NH_2

C_6H_{13}

According to experience gained in hydrogenating indole [733] the authors [534] assume that splitting of the bond between nitrogen and the aliphatic carbon atom with formation of *o*-hexylaniline, is most probable. Fission of the bond between the nitrogen and the aromatic carbon atom, with formation of a primary amine, takes place to a lesser extent.

Under more drastic reaction conditions (temperatures of 430 to 470 °C, high hydrogen pressures and reaction times of 7 to 30 hours when working in autoclaves) hydrogenation of carbazole in the presence of MoS_3 or sulphurised ammonium molybdate resulted in a number of nitrogen-containing products, e.g. alkylindoles, indole, toluidine, aniline etc., which confirm that hydrogenolysis of the pyrrole ring is preceded by hydrogenation of the aromatic ring. Simple aromatic and hydroaromatic, as well as dicyclic, hydrocarbons were obtained (dicyclohexyl, dimethyldicyclopentyl etc.) [1548, 1680, 1722]. In the presence of WS_2, hydrogenation similarly takes place readily, no catalyst deactivation by carbazole or the reaction intermediates being observed [954].

6.10.4.1.2.

Hydrogenation of six-membered heterocycles

Catalytic hydrogenation of pyridine to piperidine is a very important reaction. Technical pyridine, however, produced as a product of coal-tars processing, contains sulphur compounds so that sulphur-resistant catalysts must be used for hydrogenation. Conventional sulphide catalysts are insufficiently selective since hydrogenolytic and condensation reactions take place at the same time, decreasing the yield of piperidine. For example on the WS_2–NiS catalyst at 150 °C to 200 °C and hydrogen pressure of about 200 atm, a product was obtained by hydrogenating a pyridine–piperidine mixture, which contained, besides piperidine and hydrocarbons rather a high concentration of 1,5-dipyridylpentane [2128].

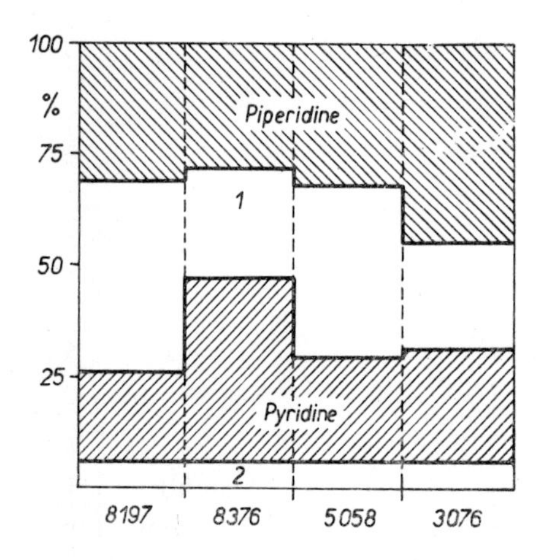

Fig. 72. Comparison of the selectivity of sulphide catalysts in hydrogenation of pyridine to piperidine. (Autoclave experiments, 300 g pyridine, 30 g catalyst, reaction time 15 hours, 280 °C, 160 to 220 atm) [1913]. % — % yield (related to amount of pyridine used). Sulphide catalysts employed: 8197 — sulphided Ni–Mo/Al_2O_3. 8376–WS_2–NiS/Al_2O_3, 5058–WS_2. 3076–WS_2 . 2 NiS. 1 — Liquid by-products. 2 — Loss.

The low selectivity of sulphide catalysts is seen in Fig. 72. The best yield of piperidine from hydrogenation of technical pyridine containing 0.3 to 0.8% S was obtained with the catalyst 3076. By-products identified were n-pentane, n-amylamine, N-n-amylpiperidine, N-cyclopentylpiperidine and 1,5-dipyridylpentane [1913]. For this reason it is more advantageous to use sulphide catalysts for hydrodesulphurisation of technical pyridine, e.g. a sulphurised 8197 catalyst [1913, 1914], and to complete the hydrogenation of desulphurised pyridine to piperidine with a metal catalyst, e.g. Ni/Al_2O_3 [1913].

Hydrogenolysis of pyridine and its homologues, with complete removal of nitrogen in the form of NH_3, is rather difficult to achieve in the presence of sulphide catalysts, particularly in comparison to oxygen- and sulphur-containing heterocyclic compounds [534, 1387, 1892, 2233]. The main hydrocarbon product is n-pentane [1755]. The formation of higher hydrocarbons on the CoS catalyst was also observed [1752]. Landa et al. [948, 1188] carried out a thorough study of pyridine hydro-

Table 62.

Hydrogenation of Pyridine on MoS_2 [948, 1188].

(2.5 l autoclave, 200 g pyridine, 7.5% catalyst, temperature 280 °C, H_2 filling pressure 100 atm/20 °C, reaction time 60 min; total pyridine conversion was 43.4%.)

Substance identified	Yield (% of pyridine reacted)
n-Pentane	1.5
Cyclopentene	3.2
Cyclopentane	2.4
N-Methylpiperidine	2.0
N-Ethylpiperidine	9.0
Piperidine	31.4
N-Propylpiperidine	4.2
N-Butylpiperidine	0.8
N-Amylpiperidine	13.9
N-Cyclopentylpiperidine	6.9
1,5-Dipiperidinopentane	10.7

genation on the MoS_2 catalyst. Pyridine was hydrogenated in an autoclave at 280 °C and a hydrogen filling pressure of 100 atm for 1 hour. These mild reaction conditions were selected in order to achieve partial pyridine hydrogenolysis in addition to hydrogenation. A survey of the products isolated as well as their proportions will be found in Table 62. Hydrogenation of pyridine to piperidine started at 220 °C, the amount of piperidine obtained increasing with rising temperature. At temperatures above 260 °C different homologues of piperidine start to appear, particularly amylpiperidine. The formation of N-alkylpiperidines may be explained by the reaction of the products of piperidine ring-opening with piperidine:

$$\text{piperidine (N–H)} \xrightarrow{H_2} CH_3CH_2CH_2CH_2CH_2NH_2 \qquad \text{(CXIIIa)}$$

$$CH_3CH_2CH_2CH_2CH_2NH_2 + HN\langle C_5H_{10}\rangle \rightarrow CH_3CH_2CH_2CH_2CH_2N\langle C_5H_{10}\rangle + NH_3. \quad \text{(CXIIIb)}$$

N-cyclopentylpiperidine and 1,5-dipiperidinopentane are formed in a similar manner:

$$C_5H_{10}NH + C_5H_{10}NH + HNC_5H_{10} \longrightarrow C_5H_{10}N{-}(CH_2)_5{-}NC_5H_{10} + NH_3 \quad \text{(CXIV)}$$

At higher temperatures, more piperidine homologues with shorter side chains are formed, e.g. N-ethylpiperidine [948, 1188].

Good yields of 1,2,3,4-tetrahydroquinoline were obtained by mild hydrogenation of quinoline in the presence of sulphide catalysts. In autoclave experiments with molybdenum sulphide at 220 to 240 °C and a hydrogen pressure of 80 to 100 atm (initial filling pressure) a 95 to 99% yield of tetrahydroquinoline was obtained [948, 1719, 1722]. Similarly, 85 to 90% conversion of quinoline to tetrahydroquinoline was achieved on the catalyst 8376 under similar conditions, the optimum temperature being 250 to 270 °C [1447]. Good results were also achieved by using the WS_2/Al_2O_3 catalyst [1448]. In a flow reactor, best results were recorded with the catalyst 8376, on which technical quinoline containing sulphur could be treated with a high catalyst lifetime [489, 986]. Optimum operating conditions were: temperature 220 to 250 °C, hydrogen pressure 200 to 300 atm, LHSV 0.15 to 0.16. The tetrahydroquinoline content of the hydrogenate was 80 to 90%. The decahydroquinoline content was very low (1 to 2%), the rest of the yield being high-boiling condensation products.

The process of quinoline hydrocracking has been studied very thoroughly, as this compound is typical of the nitrogen compounds contained in petroleum and tar materials. It was found that the course of quinoline hydrogenation depends to a certain extent on the catalyst employed. In every case however, the pyridine ring is hydrogenated, after which hydrogenolysis takes place. This follows from the fact that neither pyridine nor its homologues were isolated [485, 1075a, 1722].

Rappoport [1719, 1722] analysed the hydrocarbon and nitrogen fractions of the hydrogenate after hydrogenation of 1,2,3,4-tetrahydroquinoline in the presence of molybdenum sulphide (temperature 420 to 450 °C, hydrogen pressure 80 atm/ /20 °C). The main hydrogenolytic reaction takes place in the 1,2-position, i.e. the amino group remains on the carbon atom 9 and mainly aniline, its homologues and hydroderivatives are formed. Aromatic hydrocarbons are formed by hydrogenolysis of anilines (benzene, toluene, ethylbenzene and a small amount of propylbenzene were isolated); the respective cycloalkanes are formed in part too. Eru et al. [485, 1722] also hydrogenated tetrahydroquinoline on a molybdenum sulphide catalyst, finding that N-alkylanilines and N-alkylcyclohexylamines are formed and these authors also identified indoles in the product. The hydrocarbons formed included aromatic compounds (ethylbenzene, propylbenzene) and hydroaromatic substances (cyclohexane and cyclopentane derivatives).

On the other hand, hydrogenolysis of the pyridine ring in the 1,9-position is considered to be a typical reaction in the hydrogenation of tetrahydroquinoline in the presence of cobalt sulphide [1722].

Landa et al. [948, 1188a] performed a detailed analysis of the hydrogenation products of quinoline on MoS_2 by gas-chromatographic techniques. Quinoline was hydrogenated in an autoclave in the presence of MoS_2 at 360 °C and an initial hydrogen pressure of 100 atm for 1 hour. Under these relatively mild conditions, both partial and complete hydrogenation of quinoline took place, the cyclic structure being retained in the product. In part, quinoline was hydrogenolysed and its pyridine ring opened, and to some extend nitrogen was totally eliminated and hydrocarbons formed. Furthermore the pyridine ring was found to be degraded to a five-membered ring. The major hydrocarbons isolated were propylbenzene and propylcyclohexane, which is evidence of the fact that the primary reaction in quinoline hydrogenolysis is opening of the pyridine ring in either the 1,9- or 1,2-position, followed by hydrogenolysis of the respective amine. Lower hydrocarbons as well

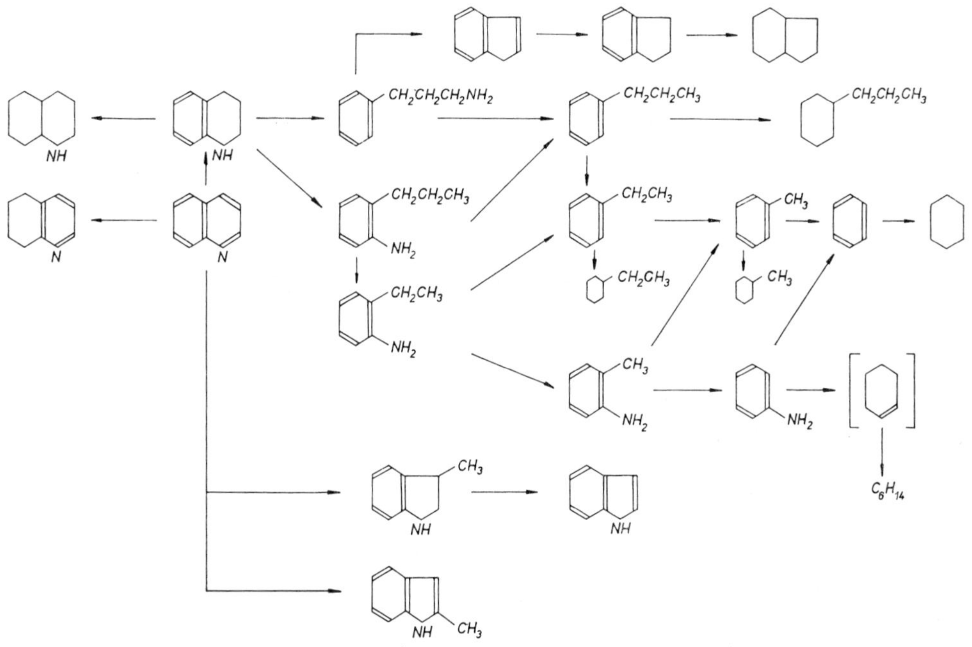

Fig. 73. Survey of compounds isolated from the product formed by hydrogenation of quinoline in the presence of MoS_2 [948, 1188a].

as lower anilines are formed by stepwise demethylation of the propyl chain. Partial cyclisation of the cleaved pyridine ring with simultaneous separation of ammonia leads to bicyclic hydrocarbons. The greater proportion of basic compounds was made up of *o*-propylaniline and its lower homologues formed by stepwise demethylation.

The primary product of quinoline hydrogenation on MoS_2 is 1,2,3,4-tetrahydroquinoline, which is formed at temperatures as low as 100 °C. It is obtained in nearly a quantitative yield under the optimum reaction conditions (240 °C); small amounts of 5,6,7,8-tetrahydroquinoline and of the perhydroderivative being also formed [948, 1188a]. As a result of isomerisation of the pyridine ring, pyrrole derivatives i.e. 2-methylindole, 3-methylindole and indole were also formed. On the other hand, the hydrogenolysis products of 3-methylindole, i.e. *o*-isopropylaniline and cumene were not proved to be present. In contrast to pyridine hydrogenation, N-alkyltetrahydroquinolines were formed by hydrogenation of quinoline. Also in comparison to pyridine, a larger yield of polymeric products was found, part of which may have been formed from indoles produced by secondary reactions [948, 1188, 1188a]. A survey of the products isolated from quinoline hydrogenate, and the reactions assumed to account for their formation, will be found in Fig. 73.

Tungsten [1387], or cobalt [1752] sulphides and the catalyst 8376 [306] were used besides MoS_2 in quinoline hydrogenation.

The hydrogenation product of isoquinoline is simpler than the quinoline hydrogenate. In the presence of MoS_2, under reaction conditions identical with those employed for quinoline, Landa et al. [948, 1188a] found that hydrogenation of the nitrogen-containing ring is rather more difficult in the case of isoquinoline. 5,6,7,8-Tetrahydroisoquinoline was present in greater amounts than the 1,2,3,4-derivative in the basic fraction, and the presence of a small amount of decahydroisoquinoline was also proved. The formation of β-phenylethylamine was not confirmed. The major hydrocarbon formed was *o*-methylethylbenzene together with toluene, ethylbenzene and the corresponding cycloalkanes.

6.10.4.2.

Hydrogenolysis kinetics of nitrogen compounds

Although removal of nitrogen compounds from the raw materials used in the fuel industry is a refining process of outstanding importance, the literature contains very little thermodynamic data relating to nitrogenous heterocyclic compounds. Hydrogenolysis of most nitrogen compounds is thermodynamically practical under the reaction conditions used in hydrorefining. However, there are considerable differences in the reaction rates of hydrodenitrogenation for individual groups of nitrogen compounds.

Denitrogenation of some low-boiling nitrogen bases takes place quite readily. On the other hand, hydrogenolysis of high-molecular weight nitrogen heterocycles (contained e.g. in heavy vacuum oils) is as a rule very difficult, and necessitates high hydrogen pressures. There are, in principle, two reasons for which hydrogenolysis of these substances is difficult [534]:

1. By-products of the hydrogenation of some nitrogen heterocycles are other heterocyclic compounds, which are considerably more resistant to denitrogenation.
2. Hydrogenolysis of the heterocyclic ring is preceded by hydrogenation of this ring which, in the case of high-molecular weight substances with bulky groups, is particularly difficult for steric reasons.

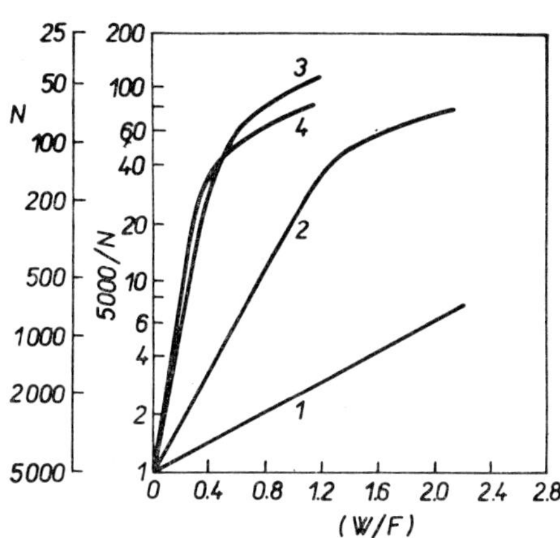

Fig. 74. Relationship between the degree of nitrogen removal and the reciprocal value of space velocity in the hydrogenolysis of nitrogen compounds on a W–Ni catalyst [534]. (0.5% solution of the nitrogen compound in middle petroleum distillate, H_2 pressure 70 atm, temperature 315 °C). 1 — Quinoline. 2 — Indole. 3 — Aniline. 4 — n-Butylamine. N — Total nitrogen content (ppm) in the hydrogenate.

The differing resistance to hydrogenolysis of individual types of nitrogen compounds is shown by the results of kinetic measurements reported by Flinn et al. [534]. The logarithm of the reciprocal of residual nitrogen concentration is plotted vs. the reciprocal of space velocity in Fig. 74 for some typical nitrogen compounds: n-butylamine, aniline, indole and quinoline. The hydrogenation experiments were carried out in the presence of the very active tungsten–nickel catalyst under the conditions stated in the figure. It is clear, that denitrogenation of all the substances tested may be considered to be a first-order reaction for a wide range of space velocities, the slopes of the individual kinetic curves representing the apparent rate constants. From the results it follows that quinoline, as a typical example of compounds with a basic nitrogen atom, is very difficult to hydrogenolyse (rate constant $k \sim 0.7$), while basic compounds of a non-heterocyclic nature (n-butylamine and aniline) are the least stable ones ($k \sim 9$ to 12). The stability of the six-membered heterocycle is due to the high resonance energy, as in the case of benzene. The non-basic five-membered nitrogenous heterocycle (in this context basic character being judged from the reactivity of the respective compound with a mixture of acetic and perchloric acids) represented here by indole, is hydrogenolysed some 4 times more rapidly ($k \sim 3.1$) than the pyridine derivative.

Another important finding is the fact that the rate of denitrogenation of compounds which readily undergo hydrogenolysis is retarded considerably at high conversion values, when the reaction rate approaches that of quinoline. The reason is, that by-products formed on hydrogenation of these compounds are substances which are difficult to hydrogenolyse, so that the reaction rate is substantially decreased. This is clearly seen from an analysis of the products from partially hydrogenated indole (Table 60) [534, 733].

On the basis of experimental results, Flinn et al. [534] conclude that the limiting step in hydrogenation of nitrogenous heterocyclic compounds is the hydrogenation of resonance-stabilised aromatic systems. These structures may be either present in the initial compounds or formed as intermediates. This is the reason for some degree of difficulty experienced in total denitrogenation of petroleum materials, since heterocyclic compounds resistant to hydrogenolysis may be formed during the reaction even though they are not contained in the raw material. In some cases (cf. e.g. the hydrogenation of 2-methylpyrrole) these substances are very easily formed and they deactivate the active centres of the catalyst. A similar effect is observed in the hydrogenation of some of the more complicated sulphur compounds, e.g. diacenaphthylene-thiophene [1184].

Stengler et al. [1961] studied the kinetics and mechanism of pyridine hydrogenolysis using a sulphurised as well as non-sulphurised refining catalyst U 63 (for its composition, see Table 3 [1429]). These authors hydrogenated a 0.135 molar solution of pyridine in gasoline at 200 to 400 °C with hydrogen pressure 50 atm and LHSV = 1 to 8. The amount of circulating hydrogen varied from 500 to 8000 Nl/1 of liquid feed. It was found that pyridine hydrogenolysis takes place in two stages. First, the ring is saturated to form piperidine which, in the second stage, hydrogenolyses to form pentane and NH_3. n-Amylamine is formed at first under the reaction conditions employed, but its hydrogenolysis is so rapid that the amine is not recovered as an intermediate. The authors [1961] found that the rate-controlling step at temperatures up to 350 °C is the opening of the piperidine ring, while at higher temperatures hydrogenation of pyridine to piperidine is the rate controlling step (this observation applies to the non-sulphidic form of the catalyst). Stengler et al. [1961] also investigated the kinetics of partial hydrogenation of pyridine to piperidine and also the hydrogenation kinetics of the reaction intermediates, i.e. piperidine and n-amylamine. Hydrogenation of pyridine to piperidine, and hydrogenolysis of n-amylamine to pentane and NH_3 are pseudo-first-order reactions, while hydrogenolysis of piperidine to n-pentane and NH_3 proceeds as a zero-order reaction. It was also found, that an excess of hydrogen has practically no influence on the reaction rate.

From the results obtained it follows that the sulphidic form of the catalyst is more active than the non-sulphidic one in terms of the conversion values achieved. Both forms of the U 63 catalyst have a very good lifetime and no deactivation of the catalyst by nitrogenous bases was observed.

Table 63.

Results of Kinetic Measurements of Pyridine Hydrogenation and Hydrogenolysis of Piperidine, n-Amylamine, Aniline and Cyclohexylamine on the U 63 Catalyst [1961].
(Flow reactor, H_2 pressure 50 atm, concentration of nitrogen compounds in gasoline 0.135 moles/l.)

Reaction	*(a)*	*(b)*	*(c)*	*(d)*	*(e)*
U 63 sulphide catalyst					
Temperature (°C)	190—280	190—280	190—340	270—310	220—280
Activation energy E (kcal/mole)	28.30 ± 2.40	40.06 ± 3.03	22.8 ± 4.2	32.1 ± 3.1	28.8 ± 3.1
Frequency factor $\ln A$ (h^{-1})	25.6 ± 2.3	37.9 ± 3.3	21.4 ± 4.2	23.9 ± 3.2	—
U 63 non-sulphide catalyst					
Temperature (°C)	250—340	220—280	190—370	280—340	220—310
Activation energy E (kcal/mole)	25.77 ± 0.02	36.90 ± 1.40	16.2 ± 1.9	28.6 ± 4.9	23.9 ± 2.4
Frequency factor $\ln A$ (h^{-1})	20.9 ± 0.2	33.7 ± 1.3	13.8 ± 4.6	20.3 ± 4.1	20.7 ± 2.3

Notation of reactions: *(a)* Aniline → benzene + NH_3
(b) Cyclohexylamine → cyclohexane + NH_3
(c) Pyridine → piperidine
(d) Piperidine → n-amylamine
(e) n-Amylamine → pentane + NH_3.

Table 63 shows a comparison of the activation energies and frequency factors for the sulphidic and non-sulphidic forms of the catalyst U 63. The table includes results for the individual reactions involved in pyridine hydrogenation (as well as in hydrogenolysis reactions of aniline, cf. p. 245). The higher frequency factors found for the sulphidic catalyst, in all the reactions considered, indicate a larger number of active centres, though the reactivity of these is inferior (in terms of higher activation energies) to that of the centres on the non-sulphidic form of the catalyst. This means, therefore, that the higher activity of the sulphidic U 63 catalyst is due to a considerably greater number of active centres.

A comprehensive kinetic study of hydrogenation of nitrogen heterocyclic compounds with a catalyst, for which no detailed data are mentioned, was reported by Cox and Berg [361]. At 390 °C and a hydrogen pressure of 17.5 atm these authors worked with different space velocities (LHSV = 0.5 to 20), finding that hydrogenolysis of six-membered heterocycles is a first-order reaction while hydrogenolysis of five-membered heterocycles may be described by a kinetic equation for a second-order reaction.

6.11.

Synthesis and reactions of organic compounds in the presence of sulphide catalysts

The most important case is the application of sulphide catalysts in the synthesis of sulphur compounds. Some of them (e.g. methylmercaptan, dimethylsulphide, mercaptanic odorants, some terpenic mercaptans) are produced on an industrial scale in the presence of these catalysts. Sulphide catalysts have also been applied in a number of cases to the synthesis of some nitrogen compounds and in condensation, dehydration and other types of reaction.

6.11.1.

Synthesis of sulphur compounds in the presence of sulphide catalysts

One of the most important applications of sulphide catalysts is the synthesis of carbon disulphide, mercaptans, sulphides, thiophene and other substances. Moreover, syntheses of organic sulphur compounds, in which one of the initial substances is H_2S or sulphur, very frequently involve the use of metals, oxides or other inorganic compounds as fresh catalysts. When the synthesis is conducted at a sufficiently high temperature, partial or complete sulphurisation of the reactive components of the catalyst takes place, so that the newly-formed sulphide becomes a significant constituent of the original catalyst. Of the large number of literature and patent data concerning this field, only those studies in which sulphide catalysts were unambiguously used will be dealt with in the following sections.

6.11.1.1.

Synthesis of carbon disulphide

The classical production of carbon disulphide is based on the reaction of charcoal at high temperatures. Catalysts employed include e.g. alkali carbonates or other salts [542]. Among sulphide catalysts, FeS was found useful in the synthesis of CS_2 from sulphur and powdered coke, charcoal and activated carbon [131].

Carbon disulphide has recently become an important petrochemical product, and has been made from natural gas or other hydrocarbons (C_2 to C_4 or directly from gas oil or fuel oil). Three basic reactions must be taken into account in the methane-based synthesis:

$$CH_4 + 2\,S_2 \rightarrow CS_2 + 2\,H_2S\,, \tag{CXV}$$

$$CH_4 + S_2 \rightarrow CS_2 + 2\,H_2\,, \tag{CXVI}$$

$$CH_4 + 2\,H_2S \rightarrow CS_2 + 4\,H_2\,. \tag{CXVII}$$

In practice the process is carried out at 700 °C, when the equilibrium of the fundamental reaction (CXV) is shifted far to the side of carbon disulphide, and the reaction (CXVI) is also favoured from a thermodynamic point of view. At 700°C the major part of the sulphur is present in the form of S_2 molecules, but reaction with the sulphur molecules S_4 and S_8 is also shifted thermodynamically to the CS_2 side. The reaction (CXVII) is rather disadvantageous and at 700 °C the equilibrium conversion value is only about 20% [538, 2051]. Active clay [2051] and oxides which are difficult to reduce [538, 556, 2056] were used as the most suitable catalysts for the synthesis from methane. Among sulphide catalysts, NiS [1522], FeS and MnS supported on Al_2O_3 [2052] were found to be suitable. V, Mn, Cu, Fe, Ni, Cr and Mo sulphides were also tested in the synthesis of CS_2 from higher paraffinic hydrocarbons [2052].

The reaction (CXVII) is thermodynamically more favourable only at temperatures above 1000 °C. In the presence of a NiS-containing catalyst, 50% conversion of methane to carbon disulphide was achieved at 1100 °C [1268].

The source of sulphur for the synthesis of CS_2 from methane may also be sulphur dioxide. At 800 °C and a molar ratio of $CH_4 : SO_2 = 3 : 4$, an 84% yield of carbon disulphide was achieved (calculated on the basis of sulphur in SO_2); PbS on pumice was found to be a suitable catalyst [1696].

6.11.1.2.

Synthesis of sulphur compounds involving addition

Catalytic addition of hydrogen sulphide or mercaptans to double bonds between carbon atoms represents one of the most important reactions in the synthesis of sulphur compounds. Addition of H_2S or mercaptans may take place in two ways:

$$RCH{=}CH_2 + R'SH \rightarrow RCH_2CH_2SR'\,, \tag{CXVIII}$$

$$RCH{=}CH_2 + R'SH \rightarrow RCH(SR')CH_3 \tag{CXIX}$$

where R and R′ are alkyl groups, aryl groups or hydrogen. The first reaction scheme corresponds to the so-called abnormal addition, i.e. a reaction which takes place contrary to the Markownikoff rule. In the second case, the reaction proceeds in agreement with this rule and the so-called normal addition is involved.

Additions of H_2S and mercaptans, however, have not been investigated in such details as the addition reactions of hydrogen halides and, therefore, the literature includes a number of discrepant data. Very often the published results cannot be verified because data relevant to the purity of the reactants, the quality of the catalyst employed etc. are missing [903]. The problem of addition of hydrogen sulphide and

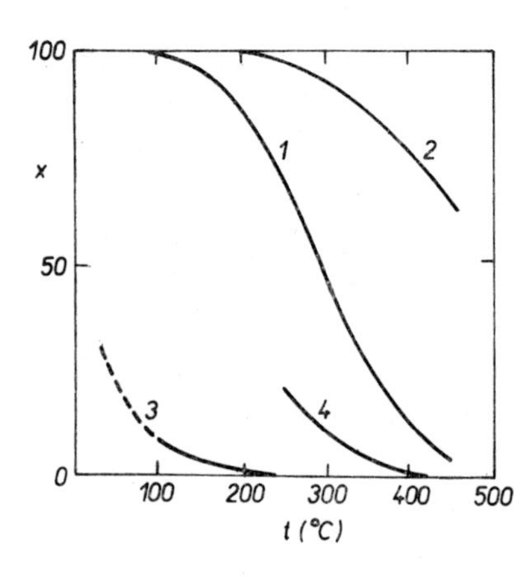

Fig. 75. The temperature dependence of the eqilibrium yield of C_2H_5SH or C_2H_5OH (x % of theory) in addition of H_2S or H_2O to C_2H_4 [2183] (the molar ratio of the initial components is 1 : 1). Ethylene thiolation [1344] : 1 — At atmospheric pressure. 2 — At a pressure of 80 atm. Ethylene hydration [2256]: 3 — At atmospheric pressure. 4 — At 80 atm pressure.

mercaptans on multiple bonds was treated comprehensively for the first time by Mayo and Walling [1338] and later by Hoffert and Wendtner [775] and Knunyants and Fokin [1060]. Schöbert and Wagner [1823], Reid [1727, 1728] and most recently, Prilezhaeva and Shostakovskii [1676] also investigated this topic. Although a number of problems have already been clarified from the structural and stereochemical aspects, as well as from the point of view of the mechanism of addition reactions in in the homogenous phase, very little is known of the mechanism of heterogenous-catalysed reactions [1676].

So-called normal addition of H_2S or mercaptans to a double bond proceeds in the absence of peroxides, and a catalysed reaction is nearly always involved. Besides acid catalysts H_2SO_4, CH_3COOH, H_3PO_4 and Friedel-Crafts catalysts [902] the reaction is also catalysed by elementary sulphur [945], metal sulphides [17] etc.

Abnormal addition, which takes place in opposition to Markownikoff's rule, is catalysed by oxygen, peroxides and light.

In the case of heterogenous-catalytic processes, especially those which are important for industrial purposes, the raw materials to be processed frequently contain oxygen and peroxides, so that the conditions under which abnormal addition products are formed exist even when catalysts, which are selective for normal addition are used. Formation of by-products should therefore always be kept in mind. These by-products will complicate the separation and isolation of the pure products.

From the thermodynamic point of view, syntheses of sulphur compounds involving addition are advantageous at lower temperatures and elevated pressures [2154, 2183]. However, addition of H_2S and mercaptans in the presence of sulphide catalysts only takes place at a sufficient rate at temperatures of more than 150–200 °C, and above these temperatures the equilibrium yield rapidly decreases (the temperature dependence of the yield of ethylmercaptan is illustrated in Fig. 75). In spite of this, the yields are considerably greater than the equilibrium yield in hydration of olefins. Increasing the molar excess of H_2S or mercaptan also has a favourable influence on the yield of the addition reaction. The yield rises rapidly with an excess of 1 to 3, while a further increase has no substantial influence [2183].

In practical operations, it is sometimes better to prepare hydrogen sulphide *in situ* from the elements instead of using ready-made H_2S:

$$R{-}CH{=}CH_2 + \tfrac{1}{2}\,S_2 + H_2 \rightarrow RCH_2CH_2SH\,. \qquad \text{(CXX)}$$

Sulphur hydrogenation and the addition reaction usually take place at a considerably lower temperature than double bond hydrogenation, so that unpleasant handling of gaseous H_2S is avoided, particularly in laboratory work.

Addition reactions are always accompanied by a number of simultaneous reactions, which either cause the formation of consecutive addition products or, lead to quite unwanted compounds. In syntheses of mercaptans by addition, the by-product is always a sulphide, formed from the mercaptan which originated in the main reaction either by addition to the olefin (only one of the reactions is considered here):

$$RCH{=}CH_2 + RCH_2CH_2SH \rightarrow RCH_2CH_2SCH_2CH_2R \qquad \text{(CXXI)}$$

or, by dehydrosulphurisation:

$$2\,RCH_2CH_2SH \rightarrow RCH_2CH_2SCH_2CH_2R + H_2S\,. \qquad \text{(CXXII)}$$

Fig. 76 shows the simultaneous equilibrium in the addition of H_2S to ethylene under different reaction conditions. It is clear from the results, that particularly with lower molar excesses of H_2S the possibility of a large amount of sulphide being formed must be considered. At the same time, the selectivity of the catalyst employed will decide to a large extent how much sulphide is formed [2183].

Nickel sulphide was used as an active and selective catalyst in the addition of hydrogen sulphide and mercaptans to olefins. A series of patents [15, 17, 1497, 1498, 2213, 2214, 2215] describes the use of nickel sulphide to prepare mercaptans and sulphides. For example, at elevated pressure and temperatures up to 200 °C a good yield of di-tert-butylsulphide was obtained by the reaction of isobutylene with tert-butyl mercaptan [17]. An olefin excess and prolonged time of contact with the catalyst leads to increased sulphide yields [1502].

Nickel sulphide supported on an aluminosilicate carrier was tested for use in the synthesis of higher mercaptans, e.g. dodecylmercaptan [1832] by addition while this catalyst supported on kieselguhr was used in the addition of H_2S to propylene [112].

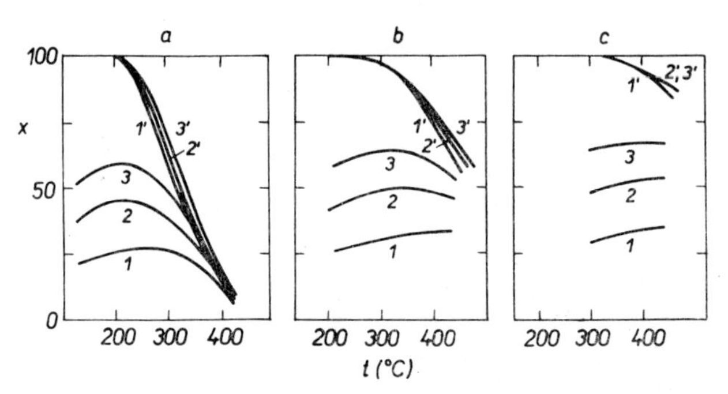

Fig. 76. Simultaneous equilibrium of H_2S addition to ethylene (reactions (CXXIII) and (CXXIV)) [2183].

$$C_2H_4 + H_2S \rightleftarrows C_2H_5SH\,. \qquad \text{(CXXIII)}$$

$$2\,C_2H_5SH \rightleftarrows (C_2H_5)_2S + H_2S\,. \qquad \text{(CXXIV)}$$

1, 2, 3 — Yield (x % theory) of C_2H_5SH at a molar H_2S excess of 1, 2, 4.
1′, 2′, 3′ — Total conversion (x % theory) at a molar H_2S excess of 1, 2, 4.
a — $P = 1$ atm, b — $P = 20$ atm, c — $P = 80$ atm (sulphide yield = total conversion — mercaptan yield).

Constantinescu et al. [347] studied the use of a number of sulphurised nickel-based catalysts for use in the addition of H_2S to ethylene. Involved were Ni/SiO_2, Ni/SiO_2–Al_2O_3, Ni/Al_2O_3, 50% Ni_2O_3 + 50% Al_2O_3, Ni/activated carbon as well as a sulphurised nickel catalyst activated with platinum or palladium [348]. The experiments were carried out at atmospheric pressure and under the optimum reaction conditions (220 °C, 22% excess of H_2S, and with an optimum space velocity for the synthesis mixture of 20 l/l hour) when a 67% yield of ethylmercaptan was obtained, only small amounts of the diethylsulphide being formed (selectivity for mercaptan formation was better than 90%). The authors [347] do not mention any data about the composition, activation, lifetime and regeneration of the most effective catalyst.

A detailed study has been made of the addition of hydrogen sulphide to ethylene, in which a number of sulphide catalysts used particularly in industry was studied [2181, 2182]. The good activity of nickel sulphide supported on Al_2O_3 was confirmed, but its selectivity was found to be lower than reported by Constantinescu et al. [347]. The influence of the reaction conditions during atmospheric-pressure addition of H_2S to ethylene was studied in detail in the case of the NiS/Al_2O_3 catalyst

[2182]. The selectivity of NiS supported on Cr_2O_3 was found to be even lower (Table 64).

Among the sulphides of the Group VIII elements, cobalt sulphide [17] and cobalt trisulphide [24, 25] were used as addition catalysts in mercaptan synthesis. In the presence of this latter catalyst, hydrogen sulphide was added to ethylene, propylene, cyclohexene, pinene etc. at temperatures of 150 to 200 °C and pressures of 70 atm in good yields. Hydrogen sulphide was prepared *in situ* from suphur and hydrogen. Ferrous sulphide has been described as a catalyst for the addition of H_2S to dienes [195, 1415, 2102]. Pyrite also exhibited a catalytic effect on the addition of H_2S to double bonds [945].

Sulphides of metals in Group VI of the periodic system also serve as active catalysts of the addition of hydrogen sulphide and mercaptans to olefins. Table 64 [2182] shows that, among the catalysts studied, pure Mo and W sulphides give the best overall conversion in the addition of H_2S to olefins, though their selectivity is low, particularly in the case of MoS_2. When MoS_2 is supported on activated carbon, a

Table 64.

Comparison of the Efficiency of Catalysts Employed in Addition of H_2S to Ethylene [2182]. (Temperature 230 °C, atmospheric pressure, space velocity C_2H_4 = 11 l/l/hour, molar H_2S excess 1.3; basis for calculating relative values: with the MoS_2/Al_2O_3 catalyst, total conversion and selectivity = 100.)

Catalyst	Catalyst efficiency[a])			Relative value	
	Total conversion to sulphur compounds (%)	Selectivity to C_2H_5SH (%)	Productivity (kg C_2H_5SH/ m^3 catalyst/ /hour)	Total conversion	Selectivity
NiS/Al_2O_3	49.6	75.0	11.34	79.4	98.4
	52.6	76.5	12.23	84.2	100.4
	53.0	72.4	11.67	84.9	95.0
NiS/Cr_2O_3	63.9	45.8	8.87	102.4	60.1
8376	30.6	76.1	7.08	49.0	99.8
MoS_2	88.5	39.3	10.58	141.8	51.5
5058	90.6	60.5	16.66	145.1	79.4
MoS_2/Al_2O_3	62.4	76.2	14.44	100.0	100.0
Sulphurised 7362	61.2	67.3	12.52	98.0	88.2
K_2WO_4/Al_2O_3[b])	8.9	39.3	1.34	14.2	61.5

[a]) Conversion values for comparison were taken from the first 12 to 14 hours of catalyst operation (this time does not include the starting period; the value for the NiS/Al_2O_3 catalyst (first line) was taken after 203 hours).

[b]) The space velocity of C_2H_4 was 13.8 l/l hour, the molar excess of H_2S was 1.5.

useful catalyst is obtained, although its inferior regeneration is a disadvantage. The catalyst 8376 is sufficiently selective, but its activity is low. In view of this, the catalyst 7362 (an industrial sulphurised Co–Mo/Al_2O_3 catalyst used in desulphurisation) appears to be the best choice, since it has both a satisfactory activity and good selectivity for mercaptan [2181, 2182]. The greatest advantage of this catalyst is its easy availability and ease of regeneration.

All the theoretically possible addition products (mercaptans as well as sulphides) were obtained in atmospheric-pressure addition of H_2S to propylene in the presence of the catalyst 5058. In this case also, technical-grade reactants were used with no special purification, so that the formation of a complicated mixture of suphur compounds must be expected when this addition process is used in industry [2182].

Sulphide catalysts, especially W and Mo sulphides, are also efficient catalysts of the addition of mercaptans to olefins and of dehydrosulphurisation reactions. In atmospheric-pressure dehydrosulphurisation of ethylmercaptan on the catalyst 5058, diethylsulphide was obtained at 230 °C in a 45% yield. The addition of ethylmercaptan to ethylene proceeded on the same catalyst under identical conditions to give nearly 60% yield [2182]. Efficient dehydrosulphurisation properties were also observed in the case of cadmium sulphide [1789].

It is advantageous to carry out the addition of H_2S to higher olefins at elevated pressure. In the presence of sulphide catalysts, the reaction with cyclohexene [1171, 1191], octenes [2182] and α-pinene [2182] results in good yields. This process was also used to prepare a number of thiols from terpenes (α- and β-pinene, dipentene, camphene etc.) under the catalytic effect of sulphurised Ni [2203].

Low-pressure addition (at 7 atm) of hydrogen sulphide to olefins (ethylene, 1-butene, propylene tetramer etc.) was also catalysed by means of sulphurised heteropolyacids supported on an aluminosilicate carrier. The optimum temperature range for this reaction is 240 to 250 °C [638].

Up to the present time, sulphide catalysts have been used on a limited scale only for industrial mercaptan synthesis. No detailed data have as yet been published for the composition of the addition catalyst used in a large plant (roughly 600 t/year) producing ethylmercaptan for odorisation purposes [69]. Higher mercaptans, particularly tert-dodecylmercaptan for use in the plastics industry, are being produced mainly in the USA and France at lower temperatures under the influence of Friedel-Crafts catalysts [110].

The advantages of the use of sulphide catalysts for these purposes are undoubted, however, and their increasing application in technical hydrogen sulphide addition processes may be expected in the future.

Mo and W sulphides may also be used to prepare selenomercaptans and selenides by means of the addition of H_2Se to olefins. The operating conditions are similar to those used in the addition of hydrogen sulphide, so that this process becomes a very simple preparative method for the basic organic selenium compounds [1171].

Among a number of different types of catalysts, metal sulphides were tested in the preparation of terpenic mercaptans which are suitable for use as flotation agents [203], and phosphorus sulphides were found useful for the addition of H_2S to α-pinene or α-fenchene [2229].

The reaction of olefins with hydrogen sulphide under rather drastic reaction conditions results in the formation of thiophenes. For example, a ferric-chromium sulphurised catalyst formed thiophene by the reaction of 1-butene with hydrogen sulphide at 600 °C, and thiophene was similarly formed on an FeS/Al_2O_3 catalyst. CS_2 was obtained as by-product in both cases. The formation of CS_2 predominated when the mixed catalyst WS_2–NiS was used at 400 to 425 °C [79]. Similarly, styrene reacted to give benzothiophene at temperatures of about 600 °C with short contact times on the FeS/Al_2O_3 catalyst, the yield being 60% [1415]. Benzothiophene was also obtained by the analogous reaction of ethylene with thiophenol on the same catalyst [1220].

Thiophenes are likewise formed by addition of H_2S to dienes. Efficient catalysts of this reaction are FeS and sulphurised Fe_2O_3 supported on Al_2O_3. For example, 1,3-butadiene gave a 60% yield of thiophene at atmospheric pressure and 600 °C. Isoprene was similarly converted to methylthiophene [647]. In the case of pyrite, a catalytic effect has been recorded in the synthesis of thiophene from butadiene and H_2S [1818].

Sulphide catalysts have also been applied to some more complicated addition reactions. For example, mixed Co–Mo sulphides catalysed the addition of H_2S to the esters of unsaturated acids (acrylic, crotonic, metacrylic), esters of 3,3′-thioether-di-carboxylic acids being obtained [806].

Cobalt sulphide was used for the reaction of ethylene with dibutyldisulphide, 5,8-dithiadodecane being formed as well as ethylbutylsulphide [1900].

A great variety of sulphur compounds is formed on addition of hydrogen sulphide to acetylene in the presence of a number of sulphide catalysts, including alkali metal sulphides and polysulphides, Cu, Fe, Ni sulphides etc. [843, 1733, 1734, 1735, 2062]. When the reaction is carried out at mild temperatures with medium pressures (10 to 20 atm) and in a solvent (water, dioxane, mono-, di- and triethylene glycol, glycerine etc.), trithioacetaldehyde is the main product. When a mixture of H_2S and C_2H_2 is passed through a solvent in which the catalyst is suspended, a mixture of sulphur compounds is obtained (ethyl mercaptan, vinylethyl sulphide, 3,6-dithiooctane etc.) [843]. At high temperatures (650 °C) thiophene is the main product of the reaction of acetylene with hydrogen sulphide under the catalytic effect of a number of sulphides (Pb, Mn, Cu, Ag, Mo, W, U, Fe, Co, Ni) [90, 91].

In a number of cases, the addition of H_2S to double bonds was speeded up by catalysts which were not sulphides in the original form, but which were capable however, of converting to sulphides, i.e. sulphurisable oxides, metals and some salts. The relevant studies are listed and analysed elsewhere [2182], and here we shall mention only a few of them for the sake of completeness. Among the oxide catalysts,

Greensfelder and Moore [647] used alumina activated with Fe_2O_3 to catalyse the addition of H_2S to double bonds. In two German patents by the I.G.F., oxides were similarly used to catalyse addition reactions. In the first case [858] basic oxides were tested for the addition of H_2S to acetylene, while in the second patent [859] oxides of alkali metals and alkali earths gave good service as catalysts of the addition of H_2S to α-olefins, as did other catalysts, e.g. carbonates, oxalates, sulphides and mercaptides.

Reactions of terpenic hydrocarbons (α-pinene, β-pinene, dipentene, camphene) with sulphur and hydrogen in the presence of active nickel as catalyst gave good yields of thiols [2203]. Ford [555] employed metallic Fe, Co or Ni in mixtures with organic peroxides for the addition of H_2S to olefins. Salts, usually chlorides of Fe, Cr, Mo, V, Mg, Mn etc., have also been used as catalysts of the addition of H_2S or mercaptans to α-olefins and acetylene [774, 1487].

6.11.1.3.

Thiolation of organic compounds catalysed by metal sulphides

In this monograph, the term "thiolation reaction" will be applied to the replacement of any functional group of an organic substance by the thiol group —SH. In practice, substitution mainly of hydroxyl, phenolic hydroxyl and etheric groups will be involved. Moreover reductive thiolations, in which thiols are mainly converted to carbonyl compounds, acids, esters and nitriles, are also important.

6.11.1.3.1.

Thiolation of alcohols, phenols and ethers

Due to the rising consumption of methylmercaptan and dimethylsulphide, catalytic thiolation of alcohols and ethers is one of the most important industrial syntheses of organic sulphur compounds. The major part of CH_3SH production is performed by methanol thiolation, the purpose being the preparation of methionine [66, 110, 540, 1364]. The rising consumption of dimethylsulphoxide, a compound with a very promising future in industry, will likewise be met mainly by thiolation of dimethylether [1697[.

From the thermodynamic point of view, thiolation reactions as a whole are more advantageous than addition reactions, particularly at the elevated temperatures (above 200 °C) needed for catalytic thiolation (Fig. 77). In addition to the basic reactions:

$$RCH_2CH_2OH + H_2S \rightarrow RCH_2CH_2SH + H_2O\ , \qquad \text{(CXXV)}$$

$$RCH_2CH_2OCH_2CH_2R + H_2S \rightarrow RCH_2CH_2SCH_2CH_2R + H_2O\ , \qquad \text{(CXXVI)}$$

$$RCH_2CH_2OCH_2CH_2R + 2\,H_2S \rightarrow 2\,RCH_2CH_2SH + H_2O\ , \qquad \text{(CXXVII)}$$

a number of other side-reactions can also take place, complicating the reaction mechanism and leading in general to more complicated products than would be obtained by simple addition of H_2S to unsaturated hydrocarbons. These side reactions include mainly dehydration of alcohols to olefins and ethers, dehydrogenation to carbonyl compounds and, also splitting reactions. When hydrogen sulphide is prepared *in situ* from sulphur and hydrogen, alcohols and ethers may be hydrogenolysed. The mercaptans and sulphides formed in the main reaction may similarly undergo hydrogenolysis to saturated hydrocarbons, which reduces irreversibly the yield of sulphur compounds. The formation of olefins by dehydration reactions is

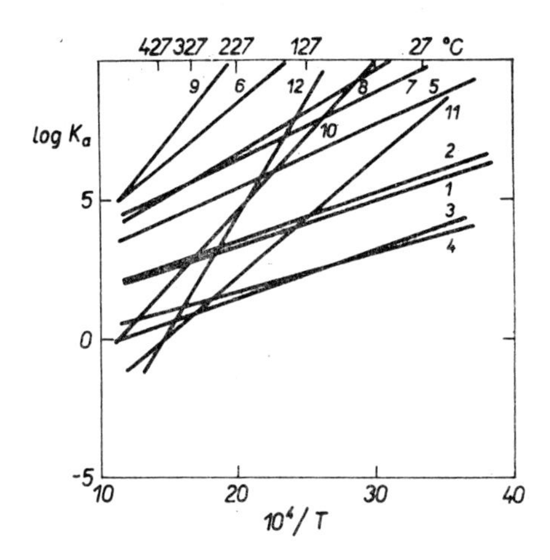

Fig. 77. Equilibrium constants (defined on the basis of activities) of the thiolation of alcohols and ethers and of reductive thiolation of carbonyl compounds and nitriles [2183].

1 — $CH_3OH + H_2S \rightleftarrows CH_3SH + H_2O$.
2 — $C_2H_5OH + H_2S \rightleftarrows C_2H_5SH + H_2O$.
3 — $n\text{-}C_3H_7OH + H_2S \rightleftarrows n\text{-}C_3H_7SH + H_2O$.
4 — $n\text{-}C_4H_9OH + H_2S \rightleftarrows n\text{-}C_4H_9SH + H_2O$.
5 — $(CH_3)_2O + 2\,H_2S \rightleftarrows 2\,CH_3SH + H_2O$.
6 — $(CH_3)_2O + H_2O \rightleftarrows (CH_3)_2S + H_2O$.
7 — $(C_2H_5)_2O + 2\,H_2S \rightleftarrows 2\,C_2H_5SH + H_2O$.
8 — $(C_2H_5)_2O + H_2S \rightleftarrows (C_2H_5)_2S + H_2O$.
9 — $C_2H_5OH + \frac{1}{2}\,S_2(g) + H_2 \rightarrow C_2H_5SH + H_2O$.
10 — $CH_3CH{=}O + H_2 + H_2S \rightleftarrows C_2H_5SH + H_2O$.
11 — $CH_3COCH_3 + H_2 + H_2S \rightleftarrows CH_3CH(SH)CH_3 + H_2O$.
12 — $CH_3C{\equiv}N + 2\,H_2 + H_2S \rightleftarrows C_2H_5SH + NH_3$.

not detrimental, provided the thiolation catalyst has some addition properties at the same time. Because of this, the demand of selectivity is even more important in the case of thiolation than it is in the preparation of sulphur compounds by means of addition, since there is a greater possibility of side reactions taking place.

Among the catalysts used in thiolation reactions, the commonest are those oxides for which the conversion to sulphides is difficult, or even improbable, under the reaction conditions employed, e.g. ThO_2 [126, 127, 128, 130, 1273, 1788, 1789], ZrO_2 [120, 844], TiO_2 [346, 423, 425, 1202]. Among other oxides, activated Al_2O_3 has been used most often. Either pure active alumina was used [129, 316, 763, 764, 1301] or Al_2O_3 promoted in various ways, e.g. by a small amount of SiO_2 [423, 424], or phosphotungstic and silicotungstic acids and their salts [385, 386] etc.

Table 65.

Influence of the Catalyst and Reaction Conditions on CH_3OH Thiolation [540].
(Atmospheric pressure, space velocity of alcohol 0.39 l/l/hour, $H_2S : CH_3OH = 2 : 1$.)

Catalyst	Temperature (°C)	Conversion (%)	Yield (% mol.)	Selectivity (%)		
				CH_3SH	$(CH_3)_2S$	$(CH_3)_2O$
Al_2O_3/K_2WO_4	373	84.7	90.5	90.5	2.9	2.1
	400	92.7	90.9	98.1	1.0	0.9
	413	89.2	87.7	98.3	1.7	0.0
	427	87.7	87.0	99.1	0.9	0.0
Al_2O_3	400	68.7	47.1	68.6	31.4	0.0

The most detailed treatment up to now of the thiolation of alcohols to mercaptans was reported by Folkins et al., who carried out a detailed study of the effect of different compounds on activated Al_2O_3, in order to find highly active and selective thiolation catalysts. Potassium carbonate and tartrate and, particularly, potassium tungstate were found to be suitable promoters for thiolation of CH_3OH and other alcohols; the last-named compound being the most efficient [546, 547, 548, 549, 550].

The optimum catalysts contain 5 to 10% of the promoter supported on γ-Al_2O_3 (surface area 100 to 200 m^2/g). They can be used in a stationary as well as fluidised catalytic bed. The lifetime of this type of catalyst is very good (decrease in the conversion of CH_3OH to the mercaptan was about 5% after 140 hours), they are easily regenerated by oxidation and resist the influence of sulphur and nitrogen compounds [540, 541]. The selective promotion effect and the results achieved with this catalyst in the thiolation of methanol, ethanol and n-octanol are illustrated in Tables 65 and 66. The yield and selectivity are practically uninfluenced by a rise in pressure.

Other promoters were tested besides those mentioned above, in particular easily sulphurised oxides of alkali metals and alkali earths [543] and also salts containing Cu, Fe, Ni or Co as cation and W, Cr, Mo, U or Mn as anion [545]. Good results were achieved in methanol thiolation by the use of the Al_2O_3 catalyst containing 0.5 to 15% SnO_2 [553].

Table 66.

Thiolation of C_2H_5OH and n-Octanol on the K_2WO_4/Al_2O_3 Catalyst [540]. (Atmospheric pressure, LHSV of alcohol 0.35 to 0.39, molar ratio H_2S : alcohol = 2.86 with ethanol and 7.73 with octanol.)

Alcohol	Catalyst	Temperature (°C)	Conversion (%)	Yield (% mol.)	Selectivity %			
					RSH	RSR	ROR	Olefins
Ethanol	Al_2O_3/K_2WO_4	289	60.1	55.9	93.0	2.5	4.3	0.2
		318	83.4	68.7	82.4	6.0	9.3	2.3
		360	83.1	66.9	80.5	4.1	7.5	7.6
		382	82.3	64.1	78.0	3.6	6.6	11.5
	Al_2O_3	317	93.0	42.6	45.8	1.7	13.5	38.8
n-Octanol	Al_2O_3/K_2WO_4	235	48.0	47.5	99.0	—	—	1.0
		260	81.6	75.3	92.3	—	—	7.7
		287	100	91.3	91.3	—	—	8.4
		318	100	77.4	77.4	—	—	18.1
	Al_2O_3	317	100	34.3	34.3	—	—	6.7

Folkins and Miller [539] also recorded the significant influence of acidity on the selectivity of thiolation catalysts: the higher the acidity of a catalyst, the easier will it form sulphides. The selectivity for mercaptans is enhanced by the introduction of promoters of alkaline character, while additions of acid components increase the yield of sulphides. For example, Folkins and Miller [551, 552, 554] used SiO_2 promoted with Cu, Ce, Ag, or Zn oxides and also acid SiO_2 or SiO_2–Al_2O_3 containing Mg, Ti or Zr to prepare sulphides by thiolating alcohols or ethers.

In the presence of catalysts employed by Folkins et al., thiolation reactions are carried out under conditions (H_2S medium, temperature of roughly 400 °C), which enable efficient sulphurisation of a number of the promoters employed, so that their sulphidic form will play an important role in the catalyst activity.

For comparison of the thiolation catalyst described by Folkins et al., a K_2WO_4/Al_2O_3 catalyst was prepared in our laboratory [2181, 2182], which was activated and sulphurised in the reactor directly before an experiment was carried out. This catalyst was compared with a number of other sulphide catalysts in the

Table 67.

CH_3OH Thiolation in a Flow Reactor [2182].

Catalyst (sulphurised)	Reaction conditions			Total CH_3OH conversion (% of theory)	Yield CH_3SH (% of theory)	Selectivity (%)		
	Temperature (°C)	Space velocity of CH_3OH (l/l/hour)	Molar ratio (H_2S : CH_3OH)			CH_3SH	$(CH_3)_2S$	$(CH_3)_2O$
5058	270	0.64	2.15	26.2	12.3	46.3	53.7	0
5058	300	0.26	2.08	83.4	33.4	40.1	59.9	0
5058	300	0.69	2.04	78.1	31.7	40.6	59.4	0
5058	300	1.41	1.54	62.1	25.1	40.4	59.6	0
8376	300	0.59	2.33	74.1	45.1	60.8	32.8	6.4
7362	300	0.54	2.53	91.2	53.8	59.0	32.1	8.9
K_2WO_4/Al_2O_3	400	0.52	2.63	73.1	68.7	94.0	6.0	0
K_2WO_4/Al_2O_3	400	0.57	2.42	74.2	71.0	95.8	4.2	0
No catalyst	300	0.57	2.18	—	0	0	0	0

Table 68.

Thiolation of C_2H_5OH in a Flow Apparatus [2182].

Catalyst (sulphurised)	Reaction conditions			Yield					
	Temperature (°C)	Space velocity of C_2H_5OH (l/l/hour)	Molar ratio H_2S : C_2H_5OH	Total conversion (%)	Yield C_2H_5SH (% of theory)	Selectivity (%)			
						C_2H_5SH	$(C_2H_5)_2S$	$(C_2H_5)_2O$	C_2-hydrocarbons
5058	220	0.53	3.72	28.5	4.1	14.3	32.6	1.75	51.2
5058	250	0.50	3.90	71.7	20.3	28.3	33.7	1.4	36.5
5058	280	0.60	3.30	98.9	21.4	21.8	26.8	1.4	49.7
K_2WO_4/Al_2O_3	360	0.71	2.80	45.0	27.5	61.1	2.2	28.8	7.7
TiO_2	320	0.78	2.53	97.1	54.8	56.4	9.2	14.8	19.4

thiolation of methanol, ethanol and their respective ethers. The experiments were done in an atmospheric-pressure flow reactor with the catalysts 5058, 8376, sulphurised Co–Mo catalyst (catalyst 7362), sulphurised K_2WO_4/Al_2O_3 and sulphurised TiO_2. Typical results are shown in Tables 67 and 68.

In the case of methanol thiolation, the optimum operating temperature for the K_2WO_4/Al_2O_3 catalyst is about 100 °C higher than with the other sulphide catalysts. Although it appears that the sulphurised K_2WO_4/Al_2O_3 catalyst prepared in our laboratory [2182] has a lower thiolation activity compared to the catalyst reported by Folkins et al. (cf. Table 64) the high selectivity of this catalyst with respect to the formation of mercaptans has been confirmed. With the other sulphide catalysts tested, selectivity varies from 40 to 60%, being least with the catalyst 5058, when it is practically independent of space velocity. Both sulphide catalysts supported on Al_2O_3 differ from the sulphurised K_2WO_4/Al_2O_3 in the relatively high ether content of the thiolation product.

It was found in respect of ethanol thiolation, that sulphurised K_2WO_4/Al_2O_3 loses in selectivity to a large extent (Table 68). This fact is evident from the results achieved in ethanol thiolation by Folkins and Miller [540] (Table 66) where, under optimum conditions (space velocity 0.35 to 0.4, temperature 320 to 380 °C, molar excess of H_2S 2.8) lower conversion and selectivity were recorded in comparison to CH_3OH thiolation and, furthermore, selectivity was found to decrease with the increasing conversion. Also in contrast to CH_3OH thiolation, a markedly high yield of diethyl ether was obtained, again in agreement with results described by Folkins and Miller [540] as well as with the results of dehydration experiments described in section 6.11.3. (p. 282). In this respect, sulphurised TiO_2, i.e. a catalyst analogous to the one used by Constantinescu [346] appears to be an equivalent thiolation catalyst [2182]. The catalyst 5058 is characterised by low selectivity for mercaptan, due mainly to the high dehydrating activity of this catalyst [2182] (cf. p. 282).

The results of dimethyl ether thiolation with the use of a sulphurised $K_2WO_4/$ $/Al_2O_3$ catalyst, prepared in our laboratory were less satisfactory than reported by Folkins and Miller [540]. Better selectivity (nearly 75%) was achieved in thiolating diethyl ether.

Metal sulphides are recommended as suitable promoters for the preparation of thiolation catalysts in one of the patents by Folkins et al. [544]. Sulphided blue Mo and W oxides were used as catalysts in thiolation of alcohols by Martin et al. [1300].

Beach and Barnett [124] achieved high methanol conversion on thiolating in the presence of the CdS/Al_2O_3 catalyst (temperature 390 to 400 °C). Dimethyl sulphide formed as a by-product was recycled and converted to mercaptan by hydrosulphurisation. Sulphides of Group VIII metals (Ni, Co and Fe) [16] are also efficient catalysts for the hydrosulphurisation of sulphides and polysulphides to mercaptans.

Conventional industrial sulphide catalysts (5058, 8376, sulphurised Co–Mo/ $/Al_2O_3$ and MoS_2) secure satisfactory conversion in the thiolation of alcohols and ethers, the optimum reaction conditions for autoclave experiments being 250 to

290 °C, reaction time 30 to 50 min, with twofold excess of H_2S; however, the selectivity of these catalysts for mercaptans is low [1442b, 2182]. This is related to the polyfunctional character of these catalysts, which besides being active in thiolation also have addition, dehydration, hydrosulphurisation and hydrogenation, or hydrogenolytic properties. Advantage may be taken of the hydrogenation properties of sulphide catalysts both in reductive thiolations (cf. the following section) and in the preparation of H_2S *in situ* from the elements, so that ready-made H_2S need not be used when thiolating alcohols or ethers [1193, 2182, 2188, 2189]. This process was utilised for the cheap and easy production of mixtures of mercaptan-sulphide odorisation agents [2178, 2179, 2189].

Some more complicated hydroxyderivatives were also thiolated successfully in the presence of sulphide catalysts. For example, Farlow et al. [507] successfully thiolated glucose to 1-thiosorbitol in the presence of cobalt trisulphide, working at 125 °C and a total pressure of 70 atm.

Antimony sulphide supported on Al_2O_3 very efficiently catalysed the thiolation of CH_3OH to dimethyl sulphide. At the same time, this catalyst was found to be useful in hydrosulphurising dimethyl sulphide to methyl mercaptan [1413]. Thiolation of furane compounds to the respective benzothiophene derivatives was likewise tested in the presence of MoS_3 [556a].

The rising importance of dimethyl sulphide as a raw material for dimethyl sulphoxide production lead to attempts at using dimethyl ether for the production of this sulphide, since large amounts of dimethyl ether are formed as a by-product of methanol production. Sulphide catalysts, which allow high yields of dimethyl sulphide to be obtained by selectively thiolating dimethyl ether, were actually prepared. These are mainly W and Mo sulphides, either alone or supported on Al_2O_3 or SiO_2 [811, 812], or catalysts containing sulphides of metals of Group VI of the Periodic Table (the latter have not been defined in any detail) which enable industrial implementation of this process [1697].

The industrial implementation of methanol thiolation to methyl mercaptan is most advanced in France, where a production of 1 t/day was started in the year 1959. In 1965, one plant was producing 2,000 t/year and another, with a capacity of 10,000 t CH_3SH/year, is under construction. The main part of the product is intended for use in methionine production [66, 110, 1364]. The process was developed by the Société Nationale des Pétroles d'Aquitaine (S. N. P. A.), and the production plant is operating in Lacq. The operating conditions are: temperature 340 to 350 °C, atmospheric pressure and molar ratio $H_2S : CH_3OH = 2$. The space velocity and composition of the catalyst have not been reported, although it is known that the catalyst base is active Al_2O_3. Data concerning the selectivity of the process have not been published.

The Standard Oil Co. (Indiana) developed a process for synthetising 98% pure CH_3SH by thiolating methanol. The reaction is carried out on an Al_2O_3 catalyst at 330 °C and 20 to 25 atm. The yield per CH_3OH applied is 64 to 71% [63].

The technological process suggested by Folkins and Miller [540] utilises a highly selective catalyst developed by these authors, i.e. activated Al_2O_3 promoted with K_2WO_4; the production volume is assumed to reach 2,260 t CH_3SH/year, with 45 t $(CH_3)_2S$ to be produced as by-product. The reaction is to be carried out under the optimum reaction conditions, i.e. 400 °C, atmospheric pressure and a twofold molar excess of H_2S. Dimethyl ether may also be thiolated by this process. The catalyst is regenerated by careful oxidation at 450 to 500 °C.

6.11.1.3.2.

Reductive thiolation

The respective thiols may be obtained by reductive thiolation of carbonyl compounds as well as of acids, their esters, anhydrides and nitriles in the presence of those metal sulphides which are efficient hydrogenation catalysts:

$$RCH_2CH{=}O + H_2S + H_2 \rightarrow RCH_2CH_2SH + H_2O\,, \qquad \text{(CXXVIII)}$$

$$RCH_2CH{=}O + 2\,H_2 + \tfrac{1}{2}\,S_2 \rightarrow RCH_2CH_2SH + H_2O\,, \qquad \text{(CXXIX)}$$

$$RCOR + H_2S + H_2 \rightarrow RCH(SH)R + H_2O\,, \qquad \text{(CXXX)}$$

$$RCH_3COOH + H_2S + 2\,H_2 \rightarrow RCH_2CH_2SH + 2\,H_2O\,, \qquad \text{(CXXXI)}$$

$$RCH_2COOR' + H_2S + 2\,H_2 \rightarrow RCH_2CH_2SH + H_2O + R'OH\,, \qquad \text{(CXXXII)}$$

$$RCH_2COOR' + 2\,H_2S + 2\,H_2 \rightarrow RCH_2CH_2SH + R'SH + 2\,H_2O\,, \qquad \text{(CXXXIII)}$$

$$RCH_2C{\equiv}N + 2\,H_2S + 2\,H_2 \rightarrow RCH_2CH_2SH + NH_4HS\,. \qquad \text{(CXXXIV)}$$

Fig. 77 shows that reductive thiolation reactions are thermodynamically advantageous over a wide temperature range (up to about 350 °C). Good yields of mercaptans are obviously only possible when active and sufficiently selective catalysts are used.

During World War II, reductive thiolation was developed by Signaigo et al. at the E. I. du Pont de Nemours Co. into virtually a general catalytic process for mercaptan preparation, especially for the preparation of technically important mercaptans. The first attempts at implementing this reaction date back to an earlier time, since in 1936—1937 the Deutsche Gold- und Silber-Scheideanstalt carried out reductive thiolation of carbonic acids in the presence of sulphide catalysts [401, 405, 406].

The catalysts employed in reductive thiolation of aldehydes and ketones were precipitated Ni and Co polysulphides, alloy-skeleton Co sulphide, a sulphurised

Mo catalyst [508, 509] and Re sulphide [372a]. Cobalt trisulphide was found to be the most selective catalyst in this group. Reductive thiolation was carried out in autoclaves at temperatures of 150 to 200 °C and pressures of 105 to 140 atm in a solvent, usually acetic acid, benzene, xylene or dioxane. Either ready-made H_2S was used, or it was produced *in situ* from the elements. For example, enanthal, 2-ethylhexanal and benzaldehyde were converted to the respective primary thiols in good yields. Similarly, acetone, 2-hexanone, 2-octanone, 8-pentadecanone, 7-heptadecanone, cyclohexanone and acetophenone formed the respective secondary mercaptans [455, 508, 509, 511]. Molybdenum sulphide also proved to be an efficient catalyst [917].

With more complicated carbonyl compounds, e.g. unsaturated ketones or diketones, both functional groups are thiolated. For example, crotonaldehyde forms 1,3-butanedithiol. Reductive thiolation of 2,5-hexandione is accompanied by cyclisation leading to 2,5-dimethyltetrahydrothiophene [508, 509].

It was found possible to thiolate the keto-group selectively, retaining the carboxyl group in the molecule, on reductive thiolation in the presence of sulphide catalysts (particularly cobalt polysulphide). For example, ethyl acetoacetate formed ethyl 3-mercaptobutyrate, levulinic acid converted to 4-valerothiolactone and 2-carbethoxycyclopentanone gave 2-carbethoxycyclopentanethiol [508, 1207]. In a similar manner, unsaturated ketoacids may be reductively thiolated in the presence of cobalt polysulphide, forming the respective dimercaptoacids [237]. The advantageous properties of Co, Fe and Ni polysulphides also make possible selective reductive thiolation of some hydroxycarbonyl compounds to hydroxythiols. For example, in dioxane solution in the presence of the CoS_x catalyst at 70 to 140 atm, 125 °C and with a reaction time of 6 hours, aldol formed a product containing 3-hydroxy-1-butanethiol besides other substances. Some sugars may be selectively thiolated in a similar manner [511, 1208].

When aldehydes or ketones are reduced in the presence of mercaptans, the reaction product is the corresponding sulphide. For example, enanthal hydrogenation in the presence of butyl mercaptan resulted in a good yield of butyl heptyl sulphide in the presence of CoS_3 at 175 °C and a hydrogen pressure of 175 atm [1899].

Carboxylic acids, their esters and salts may also be reductively thiolated. From esters, mixtures of the two respective mercaptans are obtained. For example, ethyl propionate formed a mixture of ethyl- and propylmercaptan at 100 to 300 atm hydrogen pressure and 300 °C in the presence of sulphur and a molybdenum sulphide catalyst. Ammonium thiomolybdate may also serve as catalyst in this reaction [401 405, 406]. Molybdenum sulphide likewise catalyses the reductive thiolation of lauric acid to lauryl mercaptan (temperature 250 °C, threefold theoretical excess of H_2S; hydrogen sulphide was prepared *in situ* by hydrogenating sulphur under pressure), when yields may be up to 90% [916]. Re_2S_7 was also found to bea good catalyst for reductive thiolation of carboxylic acids. The reaction was carried out in the presence of sulphur at 250 °C and hydrogen pressure 40 atm. Good yields of mercaptans were obtained from carboxylic acids having an even number of carbon atoms C_8—C_{18} [1156].

Similarly, diesters of 1,1-alkanediols (e.g. cyclohexanemethanediol diacetate) form the respective mercaptans in the presence of cobalt subsulphide (temperature 125 °C, hydrogen pressure 140 atm, H_2S excess). In this case, a 50% yield of cyclohexanemethanethiol was obtained [1585].

Anhydrides of acids may also be reductively thiolated [253, 916]. For example, a mixture of thiophene, thiophane and γ-butyrothiolactone was obtained by re-

Table 69.

Examples of Reductive Thiolation of Some Organic Compounds [508].
(Temperature 150 to 200 °C, H_2 pressure 100 to 140 atm, reaction time 1.8 hours, autoclave volume 0.4 l.)

Reacting substance[a]) (solvent)	Weight of substance (g)	Product	Yield (% of theory)
Heptaldehyde S (CH_3COOH)	100 60 40	Heptyl mercaptan	46
Benzaldehyde S	118 60	Benzyl mercaptan	70
Acetone S (CH_3COOH)	75 45 50	Isopropyl mercaptan	58
2-Octanone H_2S	114 34	2-Octanethiol	75
Crotonaldehyde S (CH_3COOH) (Dioxane)	55 75 25 40	1,3-Butanedithiol	41
Ethyl acetoacetate S (Dioxane)	100 35 54	Ethyl-3-mercaptobutyrate	26
Lauronitrile S (Xylene)	280 90 40	Dodecyl mercaptan	88

[a]) Cobalt polysulphide was used as catalyst in all experiments, excepting the one with benzaldehyde, in which a sulphurised ferrous catalyst was used.

ductive thiolation of succinic anhydride in the presence of the CoS catalyst (temperature 230 to 260 °C, hydrogen pressure 67 atm, fourfold theoretical excess of H_2S) [253].

Nitriles of acids may likewise be reductively thiolated. Cobalt trisulphide was again found to be the best catalyst for this purpose [456, 508, 1895]; but iron polysulphide [2037] and molybdenum sulphide [915] can also be used. Lauronitrile, palmitonitrile, acrylonitrile, benzonitrile etc. were reductively thiolated in good yields. The reaction was carried out in the presence of polysulphide catalysts at temperatures of 150 to 170 °C and pressures of 70 to 140 atm. Molybdenum sulphide requires a slightly higher temperature (250 °C) [915]. Hydrogen sulphide was in most cases prepared *in situ* from sulphur. Reductive thiolation of acetonitrile (catalysts MoS_2, WS_2 8376, CoS_3) which is contained in benzole fore-run may be used with advantage to prepare odorisation agents from this material [2178]. Examples of typical reductive thiolations are listed in Table 69 together with details of reaction conditions and yields.

A quite specific thiolation reaction is reductive thiolation of carbon oxides:

$$CO + 2\,H_2 + H_2S \rightarrow CH_3SH + H_2O\,, \qquad \text{(CXXXV)}$$

$$CO_2 + 3\,H_2 + H_2S \rightarrow CH_3SH + 2\,H_2O\,. \qquad \text{(CXXXVI)}$$

The reaction is catalysed by Cr, Co, Cu, Pb, Fe and V sulphides and is carried out in the presence of organic bases. For example, a mixture of CO, H_2S and H_2 (molar ratio 1 : 2 : 4) was hydrogenated in piperidine medium for 3 hours at 300 °C at a total maximum pressure of 120 atm in the presence of NiS/Al_2O_3 catalyst. Analysis of the reaction mixture showed nearly 18% conversion to methylmercaptan [1539].

6.11.1.4.

Preparation of mercaptans by hydrogenolysis of sulphides and polysulphides

In a number of cases, mercaptans and mercaptanic compounds have been prepared by hydrogenating sulphides, and particularly polysulphides, formed by the reaction of sulphur with hydrocarbons of different types. Reaction of gaseous C_2-C_4 olefins with sulphur at 170 to 250 °C and high pressures results in a sulphur product, which undergoes hydrogenation in the presence of cobalt trisulphide to give mercaptans, dithiols and alkylmercaptoalkanethiols. For example, the sulphurisation product of ethylene was hydrogenated at 150 °C for 3 to 6 hours at a total pressure of 168 atm in the presence of CoS_3, resulting in a mixture of ethyl mercaptan, 1,2-ethanedithiol, ethylmercaptoethylmercaptan $C_2H_5.S.CH_2CH_2SH$ and 3,6-dithiaoctane [1896]. Sulphurisation products from higher olefins and some dienes react similarly [1211, 1584].

This method was also used, with the aid of sulphide catalysts, to prepare technically important terpenic mercaptans derived from α-pinene, camphene [1211], β-pinene, dipentene etc. [557, 558] as well as thiophenols [1897], thionaphtols [1212] and mercaptophenols [797] (in the latter case, the hydrogenation catalyst was MoS_2/Al_2O_3 and polysulphides were prepared by means of sulphurising phenols with sulphur chloride).

6.11.1.5.

Some other reactions of organic sulphur compounds catalysed by metal sulphides

Some active metal sulphides catalyse the process of dehydrosulphurisation of mercaptans, sulphides being the reaction products. The optimum temperature for this reaction is 300 to 350 °C and the most suitable catalysts are Zn, Cd, Sn, Bi and Fe sulphides [565, 566, 1846, 1847]. Hexyl mercaptan is dehydrosulphurised in the presence of Ni sulphide in the temperature range of 150 to 450 °C, dihexylsulphide being formed, while at higher temperatures thermal splitting and condensation reactions occur (formation of saturated and unsaturated hydrocarbons, disulphides etc.) [579].

Attention was also given to dehydrogenation and cyclisation of sulphides, particularly of diethyl sulphide, which lead to thiophene formation:

$$CH_3CH_2-S-CH_2CH_3 \longrightarrow C_4H_4S\ (\text{thiophene}) + 3\,H_2\,. \qquad \text{(CXXXVII)}$$

Mashkina et al. [1302, 1304, 1306, 1984a] studied this reaction in the presence of a number of sulphide catalysts (those tested were V, Cr, Mn, Fe, Co, Ni, Cu, Zn, Mo, W, Re sulphides). The authors used a pulse microreactor technique, with a stream of He as carrier gas at 200 to 600 °C. The thiophene yields varied from 5 to 80%. The most active sulphides were found to be NiS, MoS_2, CoS, Re_2S_7 and ReS_2. At the same time, the above authors, together with Chernov [1303, 1305], studied the splitting activity of the dehydrocyclisation catalysts used, finding that it is lower the more active the catalyst is in dehydrogenation. Pumice is the most suitable carrier from the point of view of a high dehydrocyclisation activity [2103a]. An attempt was also made to apply a sulphurised Cu–Cr catalyst to this reaction [2069a].

Sulphides have been used in other cases as active catalysts in the synthesis of sulphur compounds. For example, organic dialkylpolysulphides (suitable additives for elastomers, antioxidants for lubricating oils and diesel fuels etc.) are prepared by the reaction of alkyl disulphides with sulphur under the catalytic effect of P_2S_5 [1270]. Zbirovský and Šilhánek [2255] found, that As, Sn and W sulphides and thiosalts are the best catalysts for the preparation of bis-trichloromethyl trisulphide by the reaction of trichloromethanesulphenylchloride with hydrogen sulphide. Arsenic trisulphide also catalyses the reaction of *o*-nitrophenyl- and phenylsulphenylchloride with hydrogen sulphide, though its efficiency is lower [1900b].

Sulphide catalysts have also been employed in the synthesis of thiophenes from hydrocarbons and SO_2 (or SO_3). For the reaction of butane with sulphur dioxide (temperature about 560 °C, atmospheric pressure) MoS_2/Al_2O_3 was used as catalyst; the thiophene yield varied from 17 to 30% [1096, 1097, 1098, 2049].

6.11.2.

Synthesis of nitrogen compounds

Nitriles are formed by the reaction of ammonia with olefins in the presence of metal sulphides. MoS_2 and WS_2 supported on promoted carriers (diatomaceous earth, alumina or silica gel promoted with TiO_2 or MgO) were tested as catalysts. For example, propylene reacted to give nitriles in 10–20% yield at 350—400 °C, 86% of the nitrile being surprisingly acetonitrile. Its content in the product rises with the increasing temperature [2237]. Alcohols react similarly, ZnS on pumice being the most suitable catalyst. 1-Heptanol reacted with NH_3 at 440 °C to give the respective nitrile in a 60% yield [1332, 1333].

Methyl amine was prepared by aminating methanol or dimethyl ether in the presence of aluminosilicate with 0.05 to 0.95% Re, Mo or Co sulphides [1219b]. A sulphide catalyst was also found useful in aminating aromatic compounds. In the presence of MoS_3–CuO, for example, benzene reacted with an excess of NH_3 at 400 °C to give a high yield of aniline [1807].

An important application is the use of sulphide catalysts for reductive amination of carbonyl compounds and acids. The reaction involved is analogous to reductive thiolation

$$RNH_2 + O{=}C{<} + H_2 \rightarrow RNHCH{<} + H_2O\,. \qquad \text{(CXXXVIII)}$$

In this way, lower alkylamines may be obtained from aldehydes or ketones and ammonia in the gas phase, at 320 °C and 200 atm, using the $WS_2.2\,NiS$ catalyst. This process was used, e.g. in Germany, to make technical isobutyl amine by a reaction of isobutyraldehyde, ammonia and hydrogen (molar ratio 1 : 10 : 30) at 300 °C and 220 atm pressure on a WS_2–NiS catalyst [500]. Similarly, reductive amination may be used to convert fatty acids to amines in the presence of a catalyst containing Mo and Ni sulphides. The reaction requires a high excess of hydrogen and ammonia (1 : 10) [674]. Reductive amination of aliphatic and naphthenic acids to primary amines in the presence of NiS–WS_2, NiS–MoS_2 or CoS–MoS_2 catalysts was also tested [2254b].

Table 70.

Reductive Alkylation of Amines [429].

Catalyst	Catalyst weight (g)	Amine or nitro-compound	Moles amine	Ketone	Moles ketone	Tempe-rature (°C)	Pressure (atm)	Reaction time (hours)	Products (yield, mole %)
Mo-sulphide	10	Aniline	0.75	Acetone	2.25	185	98—112	3.5	N-Isopropylaniline (91)
Ni–W-sulphide	10	Aniline	0.75	Acetone	2.25	185	98—112	2.8	N-Isopropylaniline (90) Aniline (8)
Mo-sulphide	20	N-Phenyl-*p*-phenylenediamine	0.80	Acetone	4.5	180—190	35—49	5.5	N-Isopropyl-N′-phenyl-*p*-phenylenediamine (100)
W-sulphide	16	N-Phenyl-*p*-phenylenediamine	0.75	Methyl ethyl ketone	3.0	245	133—179	2.5	N-Phenyl-N′-sec-butyl-*p*-phenylenediamine (86) Initial substance (7)
Fe-sulphide	10	Nitrobenzene	0.20	Acetone	2.72	180	112—126	3.6	N-Isopropylaniline (12) Aniline (48) Initial substance (30)
Co-sulphide	11.5	*p*-Nitroaniline	0.30	Methyl ethyl ketone	1.80	110—150 180	84—98	0.1 6	N,N′-Di-sec-butyl-*p*-phenylenediamine (94)
Ni-sulphide	15	Cyclohexylamine	1.0	Cyclohexanone	1.90	180	98—119	14	Dicyclohexylamine (95.5)
Re-sulphide	2.5	Aniline	0.75	Acetone	2.25	140	84—98	4.4	N-Isopropylaniline (90)

Dovell and Greenfield investigated the reductive alkylation of primary aromatic amines with ketones in great detail [429]. They used Mo, W, Fe, Co, Ni and Re sulphides as catalysts, carrying out the process in autoclaves at hydrogen filling pressures of 35 to 178 atm and temperatures of 100 to 245 °C with reaction times of 0.1 to 14 hours. The respective amines were obtained in high yields (Table 70). Instead of amines, the starting compounds may also be nitrocompounds which form the amine *in situ* in the presence of hydrogen [226a, 429]. It was found that besides conventional sulphides, noble metal sulphides and particularly platinum [430, 1010, 2088, 2089] and rhodium [475, 650], as well as rhenium [1787c, 1787f] sulphides are excellent catalysts of reductive alkylation of amines.

A sulphurised Co–Mo catalyst was tested for use in synthesis of unsaturated aliphatic nitriles from olefins, NH_3 and air [930b].

6.11.3.

Metal sulphides used as catalysts in dehydration, hydration, halogenation, dehalogenation and some other reactions

The dehydration properties of a number of metal sulphides are a significant manifestation of the polyfunctional character of these catalysts. As was stressed earlier (section 6.4.), the degree of dehydration plays an important role in the mechanism of hydrogenation of alcohols and of other oxygen-containing compounds with higher oxidation states, which results in the formation of hydrocarbons. These properties of sulphide catalysts may also play a significant role in industrial hydrorefining processes, particularly where removal of phenols from a number of fuel-industry raw materials is involved. The course of dehydration reactions is important in technical thiolations, where olefins and ethers may be formed from alcohols. The thiolation mechanism and kinetics of these two classes of compounds differ, and therefore, the overall rate and yield of the thiolation process is influenced. Finally, when metal sulphides are applied as hydrogenation catalysts for conversion of carbonyl compounds to alcohols, the dehydrating properties of these catalysts decrease their selectivity and cause the formation of unwanted by-products.

The important technical sulphide catalysts, particularly WS_2, are efficient dehydration catalysts. Ethanol is 70% dehydrated in the presence of the 5058 catalyst in nitrogen atmosphere at 300 °C and atmospheric pressure. Similarly, 2-methyl-2-propanol formed more than 80% yield of isobutylene at 280 °C on MoS_2 [1167]. The high dehydrating efficiency of tungsten sulphide was also confirmed by Balandin et al. [2061] and Spitsyn et al. [1939].

Other sulphide catalysts, of which the dehydration properties have been studied, are NiS [1148], MoS_2 [1649], ZnS [1107, 1776] and PbS [1105]. In all these cases, isopropyl alcohol was used as model reactant.

Besides olefins, ethers are also formed on dehydration of alcohols on sulphide catalyst at temperatures below 300 °C [1384, 1385, 2180]. In this respect, WS_2 again is the most efficient catalyst. Typical results achieved in ethanol dehydration in a flow reactor are shown in Table 71.

Table 71.

Ethanol Dehydration in a Flow Apparatus [2180, 2182].

(Experiments 1 to 11 were carried out in the presence of the catalyst 5058 in a hydrogen atmosphere; the catalyst was TiO_2 in the case of experiment 11A, K_2WO_4/Al_2O_3 in experiment 11B, and a nitrogen atmosphere was used in both cases.)

Experiment	Reaction conditions[a])			Products[b]) (% of theory)	
	Temperature (°C)	Feed (l/l/hour)	Feed (W/F)	Diethylether	Hydrocarbons
1	200	0.8	1.25	8.0	8.8
2	230	1.60	0.62	12.2	12.8
3	230	0.72	1.39	33.0	25.3
4	230	0.70	1.42	31.2	19.4
5	230	0.40	2.50	30.0	53.6
6[a])	230	0.85	1.17	26.1	16.1
7	230	1.20	0.83	22.0	—
8	270	1.95	0.51	27.0	16.0
9	270	0.84	1.19	39.0	51.0
10	270	1.64	0.60	29.7	24.2
11	290	0.78	1.27	11.5	77.2
11A	320	0.59	1.69	27.7	19.6
11B	360	0.56	1.78	36.2	5.3

[a]) In atmospheric-pressure experiments, hydrogen was passed through the apparatus at a rate of 3 l/hour. The pressure experiment (experiment 6) was carried out at 30 atm and 90 l H_2/hour was passed through the apparatus.

[b]) The balance of 100% is non-reacted alcohol and processing loss; products include a small amount of aldehydes and acids (carbonyl number a maximum of 10, maximum content of acids 0.9%).

The results obtained allow some conclusions to be drawn in respect of thiolation reactions in flow systems (cf. section 6.11.1.3.). First of all, it must be considered that the mercaptans and sulphides formed on thiolation of alcohols, result from direct thiolation of alcohols as well as from thiolation of intermediate products, i.e. olefins and ethers which are also formed in almost comparable amounts under optimum thiolation condition (about 230 °C). The same applies to the two other important thiolating catalysts, i.e. active titanium dioxide and sulphurised K_2WO_4/Al_2O_3, in the case of which dehydration to ethers takes place to a considerable extent at temperatures which are optimum for thiolation. It is characteristic that the K_2WO_4/Al_2O_3 catalyst in particular causes little dehydration to hydrocarbons. This is obviously one of the reasons for the high selectivity of this catalyst. Furthermore, addition reactions may take place between mercaptans and olefins, leading to the formation of sulphides. The ratio of the mercaptan and sulphide yields will, therefore, be mainly influenced by the selectivity of the catalyst employed.

In the case of the reverse reaction, i.e. hydration of unsaturated hydrocarbons, sulphide catalysts have also been tested. In addition to other tungsten catalysts, Boreskov et al. [202] and Odioso et al. [1523] investigated the activity of tungsten sulphide (when pure as well as when supported on Al_2O_3) in hydration of ethylene and propylene. Combinations of Group VII and VIII metal sulphides were successfully used [168, 684]. For example, a 40% yield of isopropylalcohol was achieved on the WS_2–NiS/Al_2O_3 catalyst at 255 atm and 270 °C (molar excess of steam 15 : 1). A similarly good yield of 2-octanol was obtained from 1-octene [168].

Finch and Furman [518] prepared ketones by hydrating olefins under pressure, the catalyst employed being mixtures of Groups VI and VIII metal sulphides. For example, a 38% yield of the ketone was obtained from 1-octene at 300 °C, 112 atm and sevenfold excess of steam in the presence of a NiS–WS_2 catalyst. 1- and 2-pentenes and C_6 and C_7 olefins reacted similarly.

In several cases, metal sulphides were tested as components of catalysts for hydration of acetylene to acetaldehyde [837, 1935, 2123].

In two cases, metal sulphides were also used as catalysts for halogenating and dehalogenating hydrocarbons. Polystyrene was chlorinated in the presence of Sb_2S_3 in carbon tetrachloride solution [31]. Chlorobenzene was successfully selectively dehalogenated to benzene in the course of hydrogenation on Mo and Cr sulphides [89].

A sulphurised low-pressure Fischer-Tropsch catalyst (36% Co, 2% ThO_2, 2% MgO supported on kieselguhr) was tested for use in hydroformylation. Ethylene was converted to propionaldehyde, diethyl ketone and some higher aldehydes [714].

Pine et al. used molybdenum disulphide supported on activated charcoal [496a, 1648a, 1648b, 1648c, 1649a] as catalyst of esterification. Organic esters were also prepared by reaction of halogen derivatives with mixed acid anhydrides in the presence of Sb_2S_5 [1089a].

7.

USE OF SULPHIDE CATALYSTS IN PROCESSING RAW MATERIALS IN THE FUEL INDUSTRY

In the fuel industry, sulphide catalysts have found a wide field of application, becoming – more than forty years after their first industrial application in hydrogenating coal and tars – one of the most important types of heterogeneous catalysts in terms of technical significance and consumption.

This chapter will mainly present a survey of the most important applications of sulphide catalysts in the chemical processing of coal, tars and petroleum. On the basis of the literature cited, the reader will be able to find detailed information in the original studies, patents or in one of the following specialised technological studies and monographs [119, 273, 318, 320, 416, 635, 636, 784, 793, 1033, 1034, 1035, 1101, 1347, 1518, 1602, 1722, 1736a, 1737, 1782, 1826a, 1832a, 1970, 2040, 2161, 2212, 2229a].

7.1.

Hydrogenation of coal and related materials

Destructive hydrogenation of coal is a complicated process, involving thermal splitting of high molecular weight substances in the coal mass with simultaneous hydrogenation of the degradation products. According to Storch et al. [461, 523, 524, 1969], coal hydrogenation involves the following partial processes:

1. Dissolution and extraction of substances from coal, taking place mainly below 370 °C; the rate of these processes being determined by the rate of diffusion of the dissolved fraction from the coal surface.

2. Thermal decomposition of coal substances involving the hydrogenative elimination of oxygen and other heteroatoms. In this case, the factor which determines the rate of catalytic hydrogenation of the coal mass is the diffusion of hydrogen through a liquid film to the catalyst surface. Removal of roughly the first 60% of the oxygen content is very rapid. The rest is removed at a considerably slower rate, and this reaction in particular requires a highly efficient catalyst.

3. Destructive catalytic hydrogenation at temperatures of about 400 °C and thermal decomposition of dissolved coal at temperatures above 415 °C, taking place with considerable formation of hydrocarbon gases.

4. Dehydrogenation accompanied by condensation and polymerisation of some very reactive fractions of the dissolved coal substances, leading to the formation of coke-like materials at temperatures above 440 °C.

These processes are efficiently catalysed by stannous sulphide, which compared with e.g. MoO_3 (the catalyst originally used) [1101], is more efficient both in eliminating oxygen and sulphur from the coal substances and for destructive hydrogenation of asphaltic substances [461, 1903], and also in securing efficient hydrogenation of free radicals and unsaturated compounds formed on thermal decomposition of the coal substances [1972]. On the other hand, the splitting properties of molybdenum catalysts are better than those of tin-based catalysts [1140, 1141], sulphides (and halides) being in general more efficient than oxides in these reactions [1142, 1143]. The catalytic effect of ash substances, mainly FeS_2, should also be kept in mind [773, 1722].

In the course of coal hydrogenation to form liquid products a major role is played by asphaltenes, which are the most likely intermediates in the conversion of the solid coal substances to oil-like products [1417a, 2200]. Hydrogenation of asphaltenes involves their partial splitting to gaseous products, while the majority are catalytically degraded to reactive fragments. These may either recombine to give asphaltic products or polymerise and condense to form high molecular weight substances insoluble in benzene, which are the cause of formation of coke-like substances. Finally, fragments from the degradation may also be hydrogenated to form oils [280, 643, 2202]. The yield of these products depends on the relative rates of the individual processes and, therefore, also on the activity and selectivity of the catalysts employed and the hydrogen pressure applied. Kinetic measurements of coal hydrogenation in the presence of SnS (together with NH_4Cl) have shown [1597, 2200] that asphaltenes are the reaction intermediates in the hydrogenation of coal to give oil fractions and that this is a first-order reaction with respect to the non-reacted coal mass. Similarly, the hydrogenation of asphaltenes to oil fractions is a first-order process with respect to the non-reacted initial material as well as to hydrogen over the temperature range of 400 to 430 °C [2199]. Orochko [1722] also considers the majority of reactions taking place in destructive coal hydrogenation to be first-order reactions. Here the rate of hydrogenolysis of coal asphaltenes is considerably less than that of petroleum asphaltenes [1465].

In coal hydrogenation, sulphide catalysts are of prime importance. Though other catalysts which differed in their initial form have been used, it may be shown from equilibrium data that, under the operating conditions employed in coal hydrogenation, the partial H_2S pressure is sufficient to convert a number of oxides (e.g. MoO_3) or halides (e.g. tin halides [2201]) to stable sulphides [773a, 773b]. Under the hydrogen pressure employed, sulphates e. g. $FeSO_4$ may also be reduced to sulphides [2202].

The activity of some sulphide catalysts in bituminous coal hydrogenation was investigated by Rapoport [1721, 1722] and by Orechkin [1542]. Molybdenum sulphides were found to be highly active (cf. Table 72) especially after addition of certain promoters.

Table 72.

Influence of the Catalyst on Hydrogenation of Bituminous Coal [1722].
(Temperature 400 °C, initial pressure 80 atm H_2, reaction time 1 hour, 1% of catalyst; it was hydrogenated in the form of a paste obtained by mixing coal with the heavy tar fraction in a ratio of 1 : 1.)

Catalyst	Yield (% combustible matter in the paste)			Hydrogen consumption (in % of combustible matter in the paste)	% conversion of coal matter
	Liquid products	Combustible matter in the solid residue	Gases (except H_2)		
SnS	74.23	—	—	—	79.10
MoS_3	86.53	—	—	—	—
SnS + MoS_3	92.63	—	—	—	93.00
MoS_3 + Fe_2O_3	82.94	—	—	—	—
MoS_3 + $Sn(OH)_2$	81.98	—	—	—	—
MoS_2 + NiO	86.44	2.60	4.44	2.62	44.12
MoS_3 + CaO	86.81	2.84	7.86	2.84	93.67
No catalyst	69.00	16.82	6.86	0.79	62.78

The selection of a suitable vehicle or homogenising oil is an important factor. One of the most important functions of the vehicle is the transfer of hydrogen into the coal mass [523]. Suitable solvents, or hydrogen sources, are in particular fully or partly hydrogenated aromatic compounds. When these were used, hydrogen transfer was catalysed by some sulphides (e.g. Mo and W) [410].

For industrial use, however, very cheap and easily available catalysts and vehicles must be employed. Originally molybdenum catalysts were used [1101, 1172] to hydrogenate brown coal in Germany, and tin-based catalysts were used in Billingham, England, for bituminous coal [634]. Both were soon replaced by ferrous catalysts [1101, 2040], in which ferrous sulphide is (in the operating state of the catalyst) the main component [793, 1172, 1722]. In practice, Bayermasse or Luxmasse (a commercial product containing Fe_2O_3) have been used most often and, in some cases, a part of this catalyst was replaced by impregnating coal with ferrous sulphate or its mixture with Na_2S (the latter increases splitting and neutralises HCl formed from chlorine bound in the coal) [1101].

The procedure in brown coal hydrogenation [1101] involved mixing crude coal with Bayermasse (the amount of catalyst employed depending on the operating pressure), drying, mixing with a vehicle solvent to a content of roughly 45% coal in the resulting paste and grinding. The paste thus obtained was fed into the liquid-phase reactors, and the remaining procedure was identical with that used in hydrogenations of tars (cf. p. 296).

In recent years, coal hydrogenation has been developed intensively, particularly in the U.S.A., in view of limited supplies of petroleum (Gasoline, H-Coal and Seacoke processes) [19, 70, 240, 751]. The economic success of liquid fuel production from coal (cf. also [68]) primarily demands a simple and efficient hydrogenating process [1367a] as well as a source of cheap hydrogen and the existence of a sufficient difference in the prices of coal and petroleum.

The procedure, according to the Gasoline Project (Consolidation Coal Dev.), is in the pilot-plant testing stage. Pulverised coal is fed to the extracting plant (hydrocarbon solvent from the process, pressure of 35 atm, temperatures up to 400 °C), where roughly 2/3 of the coal mass is brought into solution and the insoluble residue is subjected to low-temperature carbonisation. After filtering, evaporating the solvent and washing with hot water to remove ash residues, the extract is hydrogenated on a Co–Mo catalyst at 280 atm and temperatures of about 425 °C. It is supposed that in the future, a $ZnCl_2$-based catalyst will be employed. The main product of this process is a hydrocarbon mixture, ranging from C_4 up to fractions boiling in the heavy kerosene region; $C_3 + C_4$ hydrocarbons and phenols are also obtained [70].

The H-Coal hydrogenation process (Hydrocarbon Research Corp.) [19, 72, 73] is based on the requirement of a considerably reduced pressure. This lowers investment costs to a great extent and allows conventional, commercially manufactured equipment to be used. Another important principle of the H-Coal process is the possibility of regulating the extent of conversion of the original coal mass. It is unnecessary to liquefy coal almost completely, as it has been found more advantageous to remove coal residues (char) with a low hydrogen content, which are difficult to hydrogenate, and to use this material either to make hydrogen or as a fuel. Lastly, the newly suggested process is highly flexible, since either a primary hydrogenate only may be made or, depending on the market situation, this may be further processed in the subsequent stages of the operation.

Similarly to the H-Oil process (p. 352), the H-Coal process involves introducing finely ground coal together with a vehicle solvent into a reactor with a catalyst in an ebullating bed, in which dissolution, depolymerisation and hydrogenation take place. After passage through the catalyst layer the solid residue is removed either to be processed to give hydrogen or to be used as an industrial fuel. It may also be sold in this form.

The liquid fraction is similar to heavy petroleum in character, its density, however, is greater because of the considerably higher content of aromatic compounds. After fractionation in which heavy gasoline is removed, there remains a material with properties similar to those of fuel oil (No. 6 by the American nomenclature).

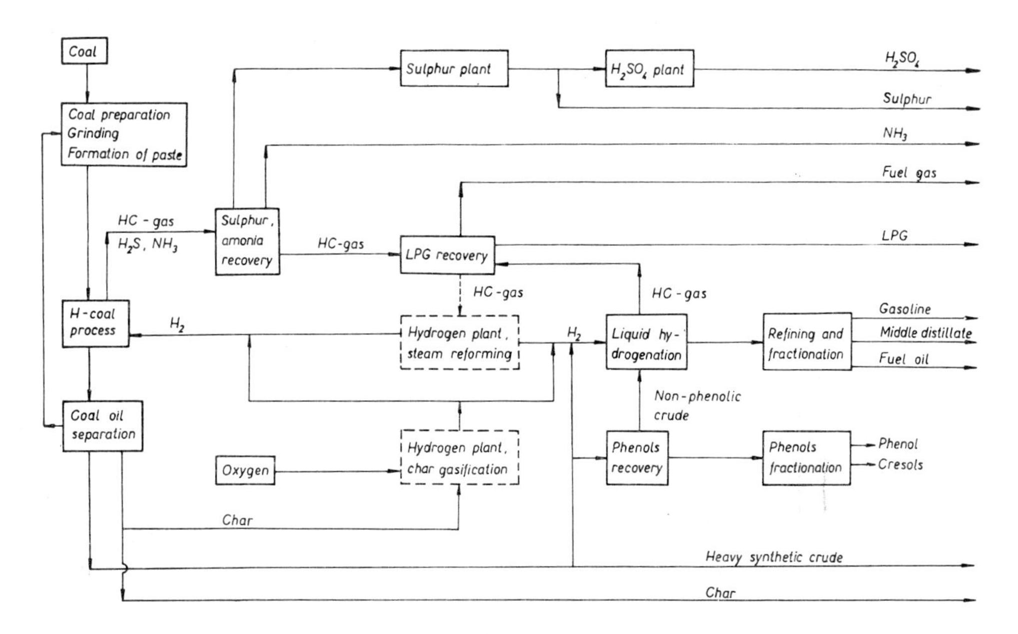

Fig. 78. Diagram of coal hydrogenation according to the H-Coal process [19].

Adequate coal conversion is achieved at 400 to 500 °C and pressures of 105 to 210 atm. An essential condition is the use of highly efficient catalysts, the composition of which has not yet been published. The catalyst is neither supported on coal, nor is coal impregnated with it [19]. For further processing of the products of the H-Coal process, it is suggested that the H-Oil or Hy–C hydrocracking and refining processes could be utilised.

Fig. 78 shows a scheme for the H-Coal process together with the subsequent applications of the individual intermediates in making fuels and petrochemical products.

Besides the catalysts employed in coal hydrogenation on the industrial scale, metal sulphides were also used in a number of cases in other research projects. The sulphides most often used are stannous sulphide [116, 236, 607, 904b, 983, 1541, 1831a, 2201], molybdenum and tungsten sulphides [211a, 374, 406a, 440, 441, 773b, 857, 866, 1144, 1494, 1746, 1990, 2103, 2122], Co–Mo catalyst [1834], iron sulphides [290, 846, 853, 886, 887, 894, 1609], mixtures of Ni sulphide with W [188] or Mo sulphides [87] as well as sodium sulphide [2033].

Sulphide catalysts were employed moreover in hydrogenating lignite [262] and peat [261, 826, 827, 828, 829, 830, 1138].

Besides destructive hydrogenation, bituminous coal was also "hydrorefined" in the liquid phase under the catalytic influence of Bayermasse and Na_2S. In this process, which is analogous to the Pott-Brosch process, so-called primary bitumen was obtained [1101].

Sulphide catalysts have also been used to study the hydrogenation of important components of the coal mass or of substances which play a significant part in coal formation, especially lignin, humic acids and montan wax.

When lignin was hydrogenated on MoS_2, a liquid product was obtained, in which the fraction boiling up to 300 °C contained aromatic and hydroaromatic hydrocarbons [1400]. Under operating conditions close to industrial liquid-phase conditions (350 to 420 °C, 150 to 400 atm H_2), lignin was hydrogenated in the presence of mixed sulphide catalysts (NiS–WS_2, FeS–CuS) to diesel fuel and a phenolic fraction in which the yield of individual phenols varied from 1 to 38% [1741]. Good yields of phenols were obtained by using MoS_3 [226b], or CoS or FeS [226c] as catalysts.

Lignins originating in cellulose production wastes were hydrogenated in a number of studies. The catalysts employed were W and Mo sulphides [1264], stannous sulphide (together with iodoform) [731] or ferrous sulphide [989, 1555, 1558]. Mixed Fe–Zn [1557] and Fe–Ni [1556] sulphide catalysts were also used. The process was carried out at about 400 °C and hydrogen pressures of 200 to 450 atm. The hydrogenates were composed of a complicated mixture of hydrocarbons and oxygen-containing compounds (alcohols, ketones, mono- and dihydric phenols).

A special process (Noguchi Process) using a highly active, sulphur resistant catalyst was developed in Japan for lignin hydrogenation [620a]. Moreover, several other sulphide catalysts were tested [1554c, 1554d].

In order better to understand the hydrogenation of coal and particularly the processes which take place at lower temperatures up to 350—400 °C, e.g. in bertinisation, in brown coal hydrogenation during the preheating period or during the hydrogenation [572, 810, 1101], as well as for the study of the structure of humates, a knowledge of hydrogenation of humic acids and humins is very important. Hydrogenation of humic acids and humins involves reduction and hydrogenolysis of oxygenous groups, which is the first stage of hydrogenation of the coal mass. Orlov et al. [1551] hydrogenated humic acids in the presence of MoS_3; in the liquid hydrogenate they found phenols and particularly naphthenic hydrocarbons. Vaiselberg [2103] also used this catalyst. Similarly, propylcyclohexane [584] was obtained as the main product of hydrogenation of humic acids (oxidised with HNO_3) in dioxane in the presence of MoS_2 (300 to 455 °C, 250 to 300 atm H_2). Landa and Eyem [1182] studied the types of compounds contained in the hydrocarbon fraction obtained by mild hydrogenation of methylated and esterified humic acids in the presence of the catalyst 5058. After 42 hours of hydrogenation in an autoclave, heteroatoms (O, N, S) had been eliminated by more than 85% and the hydrocarbon fraction consisted of polycyclic aromatic and naphthenic hydrocarbons with side chains.

With respect to the bituminous part of the coal mass, montan wax, i.e. the brown coal extract which is composed mainly of waxes and resins together with a certain amount of asphaltic and humic substances, has been studied in some detail. On hydrogenation of montan wax in the presence of Fe_2O_3 and MoS_3, a mixture of paraffinic and cyclic hydrocarbons was obtained besides a small yield of phenols in the 175 to 325 °C fraction [2103]. Hydrogenation under elevated pressures in the presence of the catalysts 8376 and 5058 led to total reduction of acids and esters at 350 °C. Oil products were formed from the dark, resinous parts of montan wax at 340 to 350 °C. In a similar way, montanic acid was converted to paraffinic C_{28} hydrocarbons when hydrogenated at 315 °C and 188 atm [409]. Landa and Rábl [1179, 1180] used MoS_2 and WS_2 to hydrogenate montan wax and its components. The hydrogenation product of montan wax acids contained cyclic hydrocarbons, though the major products were paraffins. The paraffins were mainly C_{27} to C_{29} hydrocarbons. The hydrocarbons formed by hydrogenation of the montan wax alcohols are also paraffins, corresponding to the elementary composition of C_{27} hydrocarbons.

7.2.

Hydroprocessing of petroleum and tar materials

Catalytic hydrogenation used in processing the raw materials of the fuel industry, and taking place in the presence of sulphide catalysts, is conveniently classified from the practical point of view into destructive hydrogenation and hydrorefining.

The group of destructive hydrogenation (hydrocracking) processes may be understood to include all predominantly catalytic processes taking place in the presence of hydrogen, in which substantial splitting of carbon chains in the initial molecules takes place resulting in a considerable reduction in the mean molecular weight of raw material. The reaction involved is complex, including hydrocracking proper, i.e. hydrogenative splitting of C—C bonds, taking place mainly catalytically though it can also have a thermal character. Furthermore the reaction includes splitting of asphaltic and resinous substances, hydrorefining and isomerisation together with a number of condensation, polymerisation and oligomerisation side reactions, which on the one hand lead to the formation of unwanted components in the products and, on the other hand, cause carbonaceous deposits on the catalysts, thus decreasing the catalyst activity.

At present these processes are summarily denoted hydrocracking processes, although originally this name was used only for hydrogenative splitting processes of higher distillation fractions. Within this chapter, in which we are mainly interested in analysing and recording the most important studies in which sulphide catalysts have been used, we shall use the term destructive (splitting) hydrogenation for all processes of hydrogenative splitting of tar materials and heavy petroleum residues (or, possibly, of crude petroleum), while hydrocracking will be understood to mean hydrogenative splitting proper of C—C bonds in petroleum or tar distillates (or in deasphaltisates), i.e. processes in which this is either the only or the most prevalent reaction. Hydroprocessing of heavy petroleum residues will be discussed according to its main purpose, either in the section dealing with destructive hydrogenation processes (e.g. processes according to Varga or according to Katsobashvili, section 7.2.1.) or among the hydrotreating processes if refining is the main interest, e.g. Gulf-HDS process (cf. section 7.2.3.).

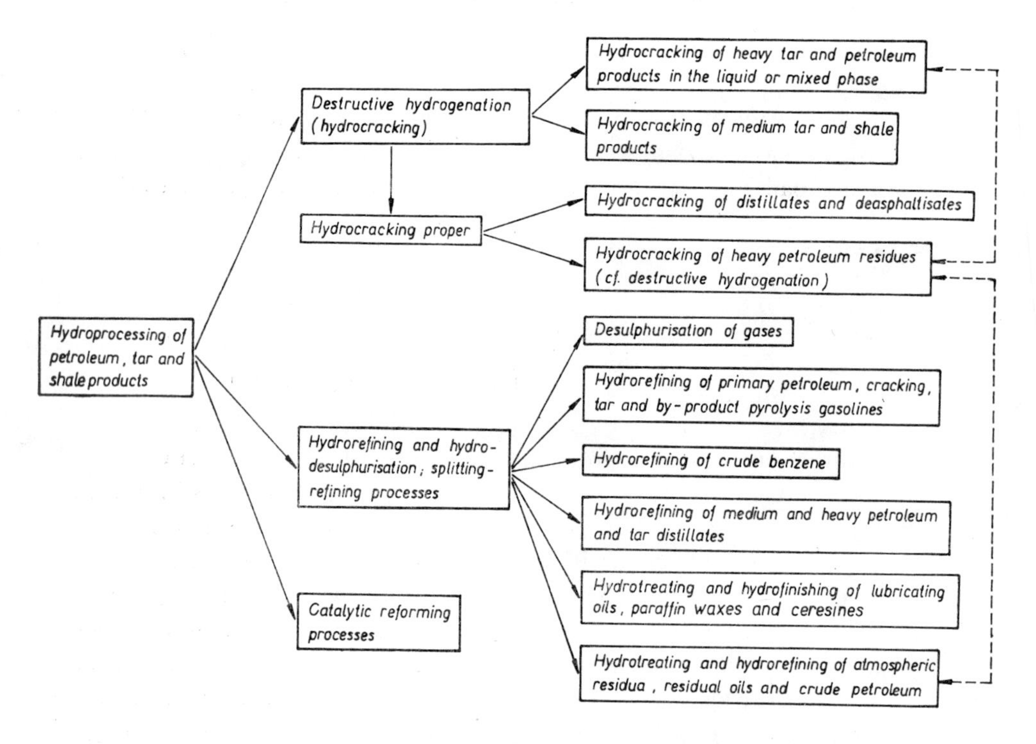

Fig. 79. Survey of technically important operations for hydroprocessing fuel-industry raw materials.

The term hydrorefining is applied to processes in which the main reactions are elimination of heteroatoms (oxygen, sulphur, nitrogen, metal atoms), reduction of highly reactive compounds (e.g. diolefins) or improvement of the colour, storage stability etc. of the material. The great majority of hydrorefining processes (provided these are not carried out under particularly mild reaction conditions) is accompanied by hydrocracking, the extent of this being determined by the severity of the reaction conditions needed to achieve a given degree of refining. Therefore, some processes which are called hydrorefining ones (e.g. DHR process, Gulf-HDS etc.) are accompanied by partial hydrocracking.

A special position is held by reforming processes which, being carried out in hydrogen medium, involve mainly dehydrogenation and isomerisation besides splitting processes and serve to considerably improve the combustion properties of the initial materials. These processes will be discussed in the section on dehydrogenation (cf. section 9.4.).

From the above analysis, it follows that the classification scheme indicated for hydrogenation processes is less perfect and less illustrative than e.g. Krönig's original classification [1101], which relates to hydrogenation of coal and tar products. It must be kept in mind, however, that after World War II the hydrogenation technique has become an important part of the technology of petroleum processing, and a large number of new processes have been developed which are frequently multi-purpose in character; these have greatly complicated the classification scheme. Fig. 79 shows schematically a survey of the most important hydrogenation processes which have been put into practice.

In the following sections, technical processes which have been implemented will be analysed and results of laboratory research into the processes mentioned will be discussed for all cases in which sulphide catalysts were employed.

7.2.1.

Destructive hydrogenation of heavy tar, shale and petroleum materials

This process is characterised by the fact that it is performed in the liquid or mixed phase with a suspended catalyst, and that its main purpose is the conversion of high-molecular weight asphaltic and resinous components of the feed to medium oils (boiling roughly up to 325 °C).

These processes may be viewed as thermal splitting of high-molecular compounds, which is influenced by the catalyst present in such a way that hydrogenation at a high hydrogen pressure hinders condensation and polymerisation of the reactive intermediates, which would lead to the formation of coke-like substances [793]. In liquid-phase processes, two basic types of reactions take place [984, 985, 992, 1722]: thermal splitting as a homogenous non-catalytic reaction, and catalytic hydrogenation of the reactive unsaturated intermediates. The activation energy for hydrogenation in the liquid phase achieves high values (e.g. roughly 70 kcal/mole in the case of atmospheric residue), proving that the rate-controlling step is evidently homogeneous thermal splitting of the high-molecular weight components of the raw material. As the temperature coefficient of this splitting is high, the process can be significantly speeded up by only a slight temperature rise.

With the majority of feeds, however, the temperature cannot be raised too much, since this would facilitate the course of dehydrogenation-condensation reactions, leading to rapid formation of deposits on the catalyst. The temperature can be raised with a simultaneous increase in the hydrogen pressure or after preliminary pre-hydrogenation of the feed, which stabilises the most easily condensable compounds. This pretreatment is suitable in the case of low-asphaltic raw materials, which may then be split on a fixed-bed catalyst in the liquid phase. In both processes, highly active Mo and W sulphides can be used and, as shown by Nemtsov [1722], the activation energy of destructive hydrogenation decreases to about one half the value determined for liquid-phase conditions with a suspended catalyst.

In principle, destructive hydrogenation of heavy petroleum and tar materials in the liquid phase could be carried out on fixed-bed catalysts [879, 891, 984]. This however is impossible with high-asphaltic materials, especially crude high-temperature tar, which contains free carbon as well as degradable components: so that in this case the catalyst would soon be completely contaminated [1034]. Even with a low content of asphaltic substances and of other impurities in the crude material the catalyst would soon be deactivated and, therefore, the reaction must be performed with a suspended catalyst, filling roughly 25% of the reaction space. To compensate for the decreasing activity of the catalyst, a part of it is continuously removed from the reactor and a corresponding amount of the fresh catalyst is added [793, 1101]. The activity of the catalyst is judged by the degree of degradation of asphaltenes, i.e. by their content in the sludge after leaving the reactor. In the steady state this content should not be more than 3% [1172]. A rise to about 5% indicates full deactivation of the catalyst [1722].

The decrease in activity of the catalyst is also shown by a decrease in the yield of the fraction boiling up to 325 °C. The rate of catalyst deactivation increases with rising temperature and molecular weight of the raw material, due to more rapid contamination with high molecular weight asphaltic substances (asphaltogenic resins, carboids and carbenes).

The selection of catalysts for destructive hydrogenation of heavy liquid materials in the liquid phase was carried out in a similar way to the case of coal hydrogenation. Finely ground MoO_3, added in amounts of 0.02% to the fresh raw material, was tested at first. The catalyst 3510 (MoO_3–ZnO–MgO) was also found to be highly active. However, the asphaltic components of the raw material and the high-molecular substances formed by dehydrogenation and polymerisation precipitate in part from the solution and are deposited on the finely ground catalyst. The amount of catalyst added is however very small for reasons of economy, so that its catalytic influence is very soon lost [1101].

This deactivating effect of asphaltic and other polymeric substances was eliminated to a large extent by supporting the active hydrogenating component of a carrier.

Kieselguhr, diatomaceous earth, pumice, fuller's earth, Bayermasse etc. were tested, while brown-coal semi-coke, especially the dust from Winkler generators was found to be satisfactory [1101]. This adsorbent however acts not only as a catalyst carrier, but also as an active component. Compared to other carriers which agglomerate in the course of the reaction to form clusters covered on the outer surface with asphaltenes, the texture of semi-coke dust from Winkler generators is such that asphaltenes may be adsorbed on the inner active surface, where they are hydrothermally split.

The transfer of hydrogen to the reaction centre is an important step, influencing the overall reaction rate of the process. For efficient distribution of hydrogen throughout the bulk of the liquid, perfect agitation as well as solubility of hydrogen in the raw material are important factors. Aromatic crudes and similarly asphaltenes dissolve hydrogen to a lesser degree, and therefore, it is important either to dilute high-asphaltic feeds, or to increase the proportion of the fraction boiling up to 325 °C in the input material [1101].

The supported catalyst (roughly 2% MoO_3 on semi-coke dust) which was used in amounts of 0.3 to 1.5% relative to the feed, permitted successful hydrogenation. In a later technical development of the catalyst, molybdenum was replaced by iron (catalyst 10927), but this catalyst had to be used in considerably greater amounts than the molybdenum catalyst. The catalyst 10927 was prepared by impregnating the brown-coal semi-coke dust (waste from Winkler generators) with a $FeSO_4$ solution and the equivalent amount of sodium hydroxide or soda to obtain a concentration of 5% Fe in the dry catalyst (a higher alkali content diminishes the catalyst activity) [317, 1172].

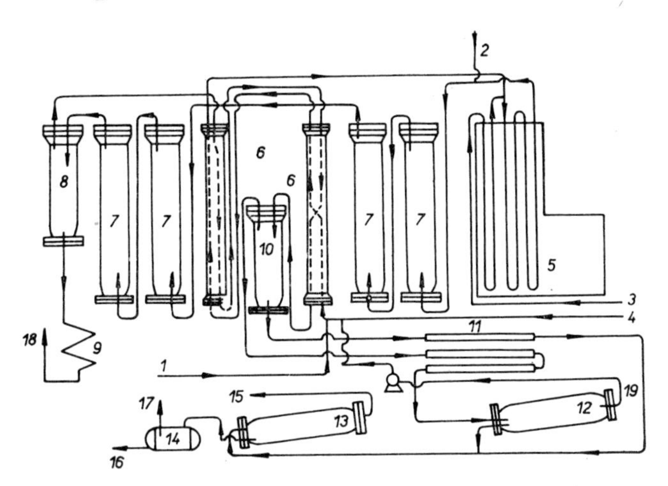

Fig. 80. Scheme of hydrogenation of tars in liquid phase [1160, 1172]. 1 — Fresh tar. 2 — Catalysts paste. 3 — Tar mixture. 4 — Fresh H_2. 5 — Preheater. 6 — Heat exchangers. 7 — Reactors. 8 — Separator. 9 — Regeneration coil. 10 — Hot separator. 11 — Water cooler. 12 — Cold separator. 13 — Partial pressure reduction. 14 — Final pressure reduction. 15 — Poor gas. 16 — Product removal to the reservoir. 17 — Rich gas. 18 — Sludge. 19 — H_2 recycle.

The liquid-phase hydrogenation unit (Sumfphase-Hydrierung) reactor involved preheating a mixture of the feed (tar, atmospheric residuum and similar materials with b. p. above 325 °C), catalyst paste and hydrogen in heat exchangers and feeding under pressures of 200 to 700 atm, depending on the nature of the material to be processed, into the reactor where destructive hydrogenation took place at 450 to 500 °C. The reaction mixture was fed to a separator, in which vapours and gaseous products were separated from the catalyst sludge. The major part of this sludge was returned (after addition of fresh catalyst paste) to the reactor, the rest was removed, the oil separated by centrifuging and the partly deactivated catalyst discarded. After passing through heat exchangers and a hot separator, the gaseous products were condensed and distilled and the fraction boiling above 325 °C was returned to the liquid phase reactor. A simplified flow diagram of the liquid-phase hydrogenation unit is illustrated in Fig. 80.

The classical method of processing residual tar and petroleum materials in the liquid phase, which was linked up with refining or hydrocracking processes, was replaced by newer methods because of a number of disadvantages. Among these modified procedures, the so-called combined hydrogenation (Kombi-Verfahren) [2099] deserves to be mentioned. This process combines the liquid phase hydrogenation and the subsequent hydrorefining of products boiling below 325 °C into a single technological procedure. The catalyst used in this case was specially developed by BASF: this was a splitting-hydrogenating aluminosilicate catalyst with addition of molybdenum and tungsten, e.g. the 9062 type [1429, 1628], the active components again being sulphides formed in the course of the reaction.

Another modification of the classical liquid-phase hydrogenation is the Varga process, the principle of which is medium-pressure destructive hydrogenation of high-molecular asphaltic substances in the liquid phase [180, 2107, 2108, 2112, 2113, 2254]. Asphaltenes are colloid-dispersed in petroleum and tar residues, and in this form they are thermally stabilised. Coagulated or adsorbed asphaltenes are decomposed at a considerably lower temperature [2083]. When the stability of this colloid system is disturbed by the addition of a suitable solvent, and when the reaction is carried out in the presence of a suitable adsorbent containing a hydrogenation component, asphaltenes are precipitated onto the adsorbent and can then be thermo-catalytically converted to polycyclic hydrocarbon oils [2136] at such low temperatures (435 to 450 °C) that coking and gasification cannot take place. When the added solvent at the same time acts as a hydrogen-transfer agent (e.g. tetralin [2107] or other hydroaromatic compounds [1197, 2083]), the process proceeds easily at medium pressures (about 70 atm) and with a considerably lower hydrogen consumption. The catalyst used is the ferrous catalyst 10927, although the molybdenum catalyst is more efficient in this case, too. The liquid-phase hydrogenation is immediately followed by treatment in a fixed-bed reactor with Ni–Mo or Co–Mo on Al_2O_3 catalyst, in which the gaseous products from the hot separator are hydrorefined. A simplified scheme of this process is seen in Fig. 81.

Newer hydrocracking methods for destructive hydrogenation of residual petroleum oils are designed as multi-purpose processes, the operation of which can be readily varied according to demands set on the products. For example, the same H-Oil equipment can be adapted to hydrocracking with formation of a large amount of low-boiling fractions by a suitable variation in the severity of reaction conditions, or the process may be operated under milder conditions mainly as a refining process (for details, see section 7.2.3.2.6). The Isomax process is likewise highly adaptable to change in the feed material. This procedure is used not only to process distillates but also for deasphaltisates and, recently, it has also been adapted to destructive hydrogenation of residual petroleum oils [1838].

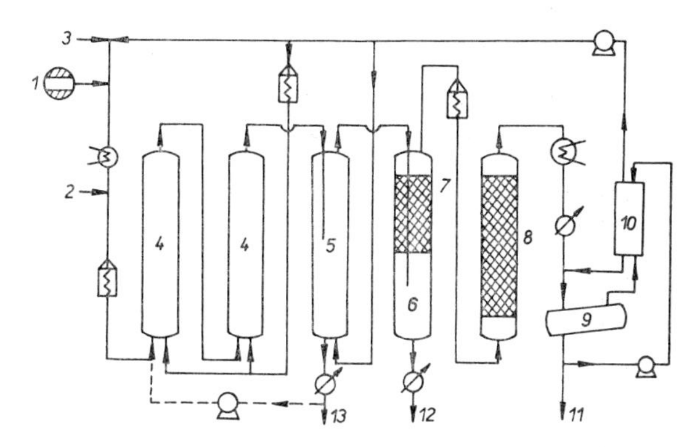

Fig. 81. Scheme of an industrial application of the Varga process as used in Böhlen, GDR [2083]. 1 — Feed. 2 — Catalyst. 3 — Make-up H_2. 4 — Liquid phase reactors. 5 — Hot separator. 6 — Second hot separator. 7, 8 — Vapour phase reactors. 9 — Separator. 10 — Gas separation. 11 — Product. 12 — Heavy oil. 13 — Residue.

Table 73 is a guide to the significant destructive hydrogenation processes for heavy residual oils, arranged according to the various literature data [2137]. Classical liquid-phase hydrogenation process and the Kombi hydrogenation are not longer used, while implementation of the Varga process continues to be hindered mainly by difficulties of an economic character.

Katsobashvili [994, 995] developed the low-pressure process for destructive hydrogenation of petroleum residues up to the pilot-plant stage. The reaction is carried out in a moving bed of Co–Mo catalyst which is regenerated continuously (Table 73).

The Houdry-Gulf and H-Oil processes and, recently, the Isomax process also, are being applied to hydrogenation of residual oils (pp. 349 and 352).

Besides these industrial processes, sulphide catalysts were used in many cases for research into the destructive hydrogenation of heavy hydrocarbon fractions. In destructive hydrogenation of low-temperature pitch and heavy tars in the liquid or mixed phase, the main objective was to obtain a maximum yield of middle oils (MoS_3 [1611], FeS [895], SnS [54] catalysts) or of lubricating oils (MoS_2 catalyst [2262]). Storch et al. [1971] hydrogenated high-temperature tar in the liquid phase at 450 °C 300 nad atm (catalyst SnS + NH_4Cl), obtaining middle oils for vapour-phase hydro-

Table 73.

Survey of the Most Important Processes of Hydrocracking (Destructive Hydrogenation) of Heavy Residual Oils [74a, 1838, 2099, 2137].

Process	Originated by	Catalyst (in fresh state)	Reaction conditions				Typical consumptions data	
			Temperature (°C)	Pressure (atm)	LHSV	Ratio H_2 : oil (Nm^3/m^3 feed)	Hydrogen[a]) (Nm^3/m^3 feed)	Catalyst (kg/m^3 feed)
Liquid phase hydrogenation	I.G.F.-A.G.	10927	to 500	300—700	0.4	4000—6000	250—300	10
Kombi hydrogenation	Scholven-Chemie, A. G.	Mo–W/Al_2O_3–SiO_2	460—480	300	—	—	350[b])	—
Varga process	Deutsch-ungarische Varga Studiengesellschaft	10927 (in the first stage)	420—450	70	0.4—1.6	400—1600	250	10
Houdry-Gulf	Houdry Process and Chemical Co., Gulf Research Dev. Co.	CoO + MoO_3 on aluminosilicate, NiO + WO_3 on Al_2O_3?	400—450	35—210	0.5—2.0	900—1800	250	6
H-Oil	Hydrocarbon Research Inc.	CoO + MoO_3 on Al_2O_3	425—455	35—140	—	—	155—300	—
Katsobashvili process	Ya. R. Katsobashvili (Academy of Sciences, USSR)	CoO + MoO_3 on Al_2O_3, NiO + MoO_3 on aluminosilicate	410—470	30	2—5	500—1500	250	—
BOC Isomax	Universal Oil Prod. Co.	Not published	Details of reaction conditions were not published					

[a]) The true hydrogen consumption differs widely according to the degree of hydrocracking.
[b]) The total hydrogen consumption in the Kombi hydrogenation [2099].

cracking. A sulphurised Co–Mo catalyst was tested for use in destructive hydrogenation of bituminous coal tars [536a].

Oils and tars from shales were hydrocracked to form gasoline and middle oil (catalysts FeS, WS_2, MoS_2 [304, 616, 1446, 1596]), or to obtain a maximum yield of monohydric phenols [617].

In the liquid or mixed phase (temperatures mainly in the 400 to 500 °C range, suspended sulphide catalyst), petroleum residues were hydrocracked to lower-boiling products, mainly gasoline and diesel fuel. Catalysts most often used for this purpose were MoS_2 [255, 256, 260, 630, 1703] and WS_2 [992, 993, 1509] as well as CoS [831] and FeS [336].

High-boiling raw materials from which salts have been completely removed, and the majority of asphaltenes and resins, may be processed in the liquid phase on fixed-bed catalysts (WS_2, WS_2–NiS) by means of which intensive destructive hydrogenation proceeds [954, 984, 985, 1204].

Attempts were also made to perform destructive hydrogenation directly with crude petroleum on fixed bed sulphide catalysts (MoS_2, WS_2, types 8376 and 6434 catalysts [891, 1157, 1647]). However, this process is difficult to apply because of rapid contamination of the catalyst surface by asphaltic deposits and difficulties in catalyst regeneration. Here the difficulties caused by severe reaction conditions are much greater than in the case of desulphurisation of petroleum and petroleum residues on fixed-bed catalysts (cf. p. 344).

Sulphide catalysts were also used to prepare naphthalene by means of destructive hydrogenation of petroleum extracts [2020].

Destructive hydrogenation in the liquid phase in the presence of sulphide catalysts was also used to obtain lubricating oils. Deasphalted materials [1133], heavy oils [1642] and also crude petroleum [1129, 1725] were hydrogenated, mainly on WS_2 catalyst.

7.2.2.

Hydrocracking of petroleum and tar fractions

The process was developed in the period between the two world wars. In Europe, mostly hydrocarbon mixtures originating from tar were treated in the German vapour-phase hydrocracking process (Benzinierung) where pretreated middle oils were processed [793, 1101]. A similar method was employed in England [318]. In the U.S.A., a high pressure hydrocracking technique was developed for petroleum fractions in Baton Rouge, Louisiana [242, 2148]. Materials from tar have been employed less frequently in recent years (the plant in England stopped operating in 1958 [320, 2161] and German hydrogenation plants closed down or replaced the Benzinierung process by the DHC process [794, 1525] and the classical vapour-phase hydrocracking was used until recently in Czechoslovakia). On the other hand, the increasing demand for gasoline mainly in the U.S.A., led to a very rapid development of modern hydrocracking processes for petroleum-derived raw materials [314, 640].

In hydrocracking, splitting (or dealkylation) of individual hydrocarbon groups takes place according to the mechanism which was described in detail in section 6.3. However, when technical hydrocarbon mixtures are being processed, the course of the catalytic cleavage process may be influenced to a large extent by the presence of even small amounts of non-hydrocarbon components, especially nitrogen-, oxygen- and sulphur-containing compounds, so that the composition of the hydrocracking product differs from the product obtained with perfectly refined feeds.

The influence of temperature on the course of hydrocracking has already been explained in section 6.3. The increased temperature causes a conversion to radical cleavage, the consequence of which is increased gas and coke formation and an increased consumption of hydrogen [1425]. Higher radical splitting is typical of catalysts with a less effective hydrogenation component, especially of the suphide type. Particularly with raw materials which contain non-hydrocarbon components (notably nitrogen containing basic compounds which act as catalyst poisons), higher temperatures are essential, i.e. conditions become more favourable to the radical course of hydrocracking. The lower hydrogenation efficiency (compared to metals) of the sulphidic component as such or formed under the infiuence of sulphur compounds from the raw material, results in a higher degree of coke formation on the active catalyst surface. To maintain the high yields of the hydrocracking process, the lower concentration of active acid centres on the catalyst surface must be compensated for by a higher reaction temperature.

Apart from temperature, the partial pressure of hydrogen also exerts a significant influence. It was found experimentally, that the dependence of the yield of the hydrocracking process on pressure passes through a maximum [710]. It may be assumed that initially, the catalyst is protonised as the hydrogen pressure rises, the concentration of Brønsted acid centres increasing at the same time. This causes the carbonium ions concentration, and also the rate of the hydrocracking reaction, to increase. On the other hand, with increasing hydrogen pressure, the hydrogenation-dehydrogenation equilibrium shifts towards saturated hydrocarbons, the rate of formation of olefinic intermediates decreasing and therefore the carbonium ions concentration decreasing also. These two opposite processes lead to the maximum observed in the pressure dependence of the hydrocracking yield [2194, 2197].

Hydrocracking catalysts

As already stated, the hydrocracking catalyst is a bifunctional catalyst containing both a splitting and a hydrogenating component. According to Beuther and Larson [159] the activity and lifetime of a hydrocracking catalyst depends on the protective effect of the hydrogenating component, the catalytic effect of this protecting the surrounding acid cracking centres from coking. The overall rate of the hydrocracking reaction depends on the number of non-coked acid centres (with catalysts having a strong hydrogenating component, the reaction on acid centre is the rate-controlling process) [2190]. On a fresh catalyst, all acid centres are "pure". In the course of time however, the centres more distant from the hydrogenation component are gradually

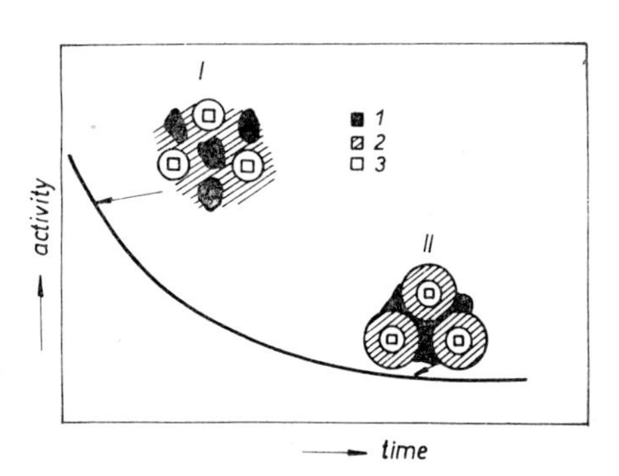

Fig. 82. Models of hydrocracking catalyst surface at early and steady-state activities [159]. I — Activity and state of the fresh catalyst. II — Hydrocracking catalyst in the "steady state". 1 — Fully coked surface. 2 — Partially coked surface. 3 — Clean active surface.

contaminated with carbonaceous deposits and become inactive. Only the acid centres in the immediate vicinity of hydrogenation centres continue to be active, and the catalyst activity attains a steady state. Thus the lifetime of a hydrocracking catalyst also depends on the degree of the "protective" effect of the hydrogenation component (Fig. 82) [159].

The hydrogenation activity of the sulphide component is far less than that of the corresponding metal. As seen from Table 74, the original activity of a metal decreases roughly 40 times when 40% of it is converted to the sulphide, assuming that the active sulphidic component is Ni_3S_2 [159].

The influence of non-hydrocarbon components and the products of their hydrogenolysis on the activity of hydrocracking catalysts is very important from the

Table 74.

Decrease of the Hydrogenation Activity of Ni/Al_2O_3 with the Sulphurisation Degree of the Catalyst [159].

S content in the catalyst (%)	Atomic ratio S/Ni	Rate constant of hydrogenation of aromatic compounds in fuel oil at $t = 288$ °C (hour^{-1})
0	0	2.5
0.014	0.0022	1.3
0.025	0.0045	1.0
1.44	0.26	0.06

technological point of view [349a]. The influence of H_2S and of sulphur-containing compounds consists of the conversion of metallic hydrogenation components of the catalyst to sulphides, thus decreasing their hydrogenating efficiency. Another consequence is an alteration of the reaction mechanism, i.e. the possibility of longer retention of ions on acid centres and slower hydrogenation of aromatic substances, leading to different hydrocracking products (formation of isoparaffinic splitting products, gasoline with a higher proportion of aromatic substances, etc.).

The influence of oxygenous compounds and water on the activity of hydrocracking catalysts is not yet clear. It is assumed [2007] that with respect to the cracking centres of the hydrocracking catalyst, water will behave as a weak base, neutralising their acid character.

Table 75.

Influence of Basic Nitrogen Compounds on the Hydrocracking Activity of the WS_2 Catalyst in Hydrocracking of Slack Wax (temperature 390 °C, H_2 pressure 100 atm, space velocity 2 l/l/hour; the feed was slack wax obtained in Fischer-Tropsch synthesis) [2174a, 2195, 2196, 2197].

Quinoline content in the raw material (%)	0.0	0.094	0.572	0.950	1.820	4.64
% Fraction up to 180 °C	43.1	18.8	5.9	3.9	3.0	3.3
% Fraction 180 to 230 °C	14.6	9.6	3.6	2.6	1.2	1.2
Content of n-paraffins (%)	40	43	49	46	55	60
% Fraction 230 to 320 °C	24.2	22.7	23.1	16.8	16.1	15.1
Content of n-paraffins (%)	37	39	45	46	60	62
Relative yield of fraction 230 to 320 °C	29.5	44	71	72	79.5	77

Nitrogen-containing bases have an extraordinarily strong deactivation effect on hydrocracking catalysts (see also Chapter 5., p. 106). Neutral nitrogen-containing substances, capable of forming basic intermediate products in the course of the reaction, have a similar effect [1201a]. It may be stated that these components act as catalyst poisons, not only substantially decreasing the catalyst activity, but also suppressing the ionic reaction mechanism, so that radical cleavage participates to a greater degree. This again influences the composition of the splitting products (giving increased content of low-molecular gaseous hydrocarbons, lower content of isoparaffins and more polymerization and coke formation). The characteristic activity decrease in the case of the catalyst 6434 was mentioned earlier (see Fig.

43). The same applies to pure WS_2, e.g. in hydrocracking of slack wax (Table 75). From the results it follows, that:

a) a very small concentration of nitrogenous bases strongly decreases the yield of hydrocracking;

b) the decreased acidity permits radical reactions to take place, leading to the formation of non-branched cleavage products (the content of n-paraffins in the product rises).

One of the most important properties of modern hydrocracking catalysts, designed for processing low-quality materials with a high nitrogen content, is therefore their low sensitivity to nitrogenous bases.

The splitting component of hydrocracking catalysts is an acid "carrier" with a sufficiently large surface and a high concentration of acid (Brønsted and Lewis) centres. Originally, fuller's earth (Terrana) activated with HF was used [416, 1626] or the American modification of this acid carrier, Superfiltrol [1598]; but various types of bentonites were also tested [318, 929]. The main disadvantage of these cracking components was their great sensitivity to nitrogenous bases, and the need of working with high pressures.

Progress was achieved in the development of hydrocracking carriers by the use of synthetic aluminosilicates, the Brønsted or Lewis acidity of which [1512, 2064] shows less sensitivity to nitrogenous bases, permitting lower pressures to be used (100 to 150 atm). The amount of strongly acid centres is further increased by the addition of halides.

Most recently, molecular sieves, with a substantially lower sensitivity to nitrogen-containing bases, are coming into use [498, 1390, 2085, 2242]. Magnesium silicate has also been used with success [1321, 1322, 2043].

Metals, especially Pt and Ni, are mainly used as active hydrogenation components, bifunctional catalysts with a strong hydrogenation component being formed [2146]. Thus, the mechanism and kinetics of the overall reaction on these catalysts will mainly be decided by processes taking place on the acid centres. The second group of hydrogenation components includes compounds (sulphides, oxides) of metals of Groups VI and VIII of the periodic table, mainly molybdenum, tungsten, nickel and cobalt, the hydrogenation effects of which are substantially less. Thus, the hydrocracking mechanism will be strongly influenced by the hydrogenation component when these catalysts are used [159]. Sulphides are especially important, since small amounts of sulphur compounds in the raw material are sufficient to form sulphides from other metal compounds [1198].

Sulphides which have been used as efficient components of hydrocracking catalysts in industrial practice, as well as in a number of research studies, include mainly WS_2 which was originally used without a carrier [80, 193, 299a, 304, 318, 338, 416, 458, 689a, 760, 794, 888, 889, 892, 1049, 1053, 1088, 1091, 1130, 1133a, 1286, 1419a, 1434, 1438, 1527, 1626, 1627, 1643, 1722, 1783, 1989, 2008, 2012, 2174, 2195], MoS_2

(also used originally without a carrier) [257, 258, 259, 583, 771, 1419, 1437, 1451, 1641, 1699, 2035, 2072, 2094] and nickel sulphide [86b, 159, 171, 248a, 249, 250, 496b, 532, 707, 779, 1084, 1198, 1203, 1370, 1561, 1577a, 1592, 1717, 1743, 1842, 1843, 1844, 1934, 1959a, 1964, 2151]. The use of other sulphides on aluminosilicate carriers was also described, e.g. CoS [1354, 1389, 1591a, 1840, 2050] or rhodium [642], chromium [727], palladium [171a, 240a, 722a, 1426b, 2240] and cadmium sulphide [1745].

Mixtures of sulphides were used in hydrocracking catalysts with aluminosilicate carriers besides single-component catalysts. In this case, the hydrogenation effect is enhanced (see Chapter 5.), and therefore the acid centres of the cracking component are better protected. Moreover these catalysts have a higher refining activity and they are more resistant when low-quality feeds with higher concentrations of nitrogen-containing compounds are being processed [1049]. Combinations employed most frequently on aluminosilicate carriers are mixtures of nickel and tungsten [96a, 162, 301, 459, 587a, 641, 687, 727b, 758, 928b, 1083, 1241a, 1346b, 1390, 1423, 1529a, 1533, 1593, 1713g, 1738, 1833, 1888f, 2006, 2050a, 2050b, 2081, 2103d, 2103g, 2156], nickel and molybdenum [10, 411a, 592, 641, 1000, 1003b, 1617, 1934a, 2003, 2091], cobalt and molybdenum [1437a] and nickel, tungsten and molybdenum [1050, 1055, 2198]. Other combinations were also tested, e.g. sulphurised Ni–Sn [1426a], MoS_2–PdS [641, 2207] and Re–MoS_2 [186, 1265].

Hydrocracking may also be carried out on sulphide catalysts supported on weakly acidic or neutral carriers. In this case the process is not of strictly ionic character. For example, hydrocracking was carried out on mixed sulphides supported on neutral carriers (or with modified acidity by addition of halogen [426, 1079, 1888a, 1888d, 2235]), e.g. on catalysts of the 8376 type [988, 1004, 1640], Co and Mo sulphides [458a, 586, 718, 817, 1073, 1391, 1534a, 1554a, 1620], and further on Ni and Mo sulphides [996], Ni and W sulphides [171c, 2057b, 2210a] and MoS_2 supported on Al_2O_3 + MgO [254].

Like other catalytic processes, hydrocracking is carried out under conditions where metals or non-sulphide compounds of metals convert to sulphides under the influence of hydrogenolytic H_2S, the sulphides then becoming important active components of hydrocracking catalysts [159].

Raw materials for hydrocracking

Among petroleum products, straight-run heavy gasolines, kerosenes, vacuum gas oils (boiling at 360 to 520 °C), deasphalted vacuum residues, recycled oils from catalytic and thermal cracking etc. are suitable for hydrocracking [2148]. Products of the hydrocracking process are fuel gases, light gasoline with a high content of branched hydrocarbons, heavy gasoline suitable for further processing by reforming, jet fuels with a low pour point, motor oils and high-quality fuel oils.

In the case of tar materials, the main hydrocracking feeds were prerefined heavy gasolines and middle oils boiling up to 325 °C [1101, 1172, 1722].

Hydrocracking of tar fractions has been most often carried out for the purpose of obtaining high-quality diesel fuels. Low-temperature tars and tar distillates have been hydrogenated on WS_2 and WS_2–NiS–Al_2O_3 catalysts [187, 1099, 1262, 1263]. Other materials hydrogenated were middle oils from coal hydrogenation (catalysts FeS–WS_2 [862, 1510] or WS_2–NiS [1001]), lignite and peat tars (catalyst MoS_2 [23, 639]) or shale oils (catalyst WS_2, WS_2/Al_2O_3, MoS_2 [272, 277] or the catalyst K-536 (Mo–Zn–Cr–S) [742b]). In a series of studies [35, 36, 37, 38, 39, 40, 41, 42, 43, 44, 46, 48, 49, 50] Ando hydrogenated low-temperature tar, oil from low-temperature tar, creosote oil and oil from shales in the presence of $(NH_4)_2MoS_4$, MoS_2 or MoO_3 + S at 460 to 490 °C and pressures of 200 atm, obtaining gasoline and diesel fuel. Carlile et al. [267] used the same molybdenum sulphide catalyst supported on Al_2O_3.

Sulphide catalysts have found a wide field of application in hydrocracking of creosote oil (the 200 to 350 °C fraction from high-temperature tar) to give high-quality gasoline [206]. According to a British process [318, 320] single-stage hydrocracking was originally carried out on a WS_2 catalyst, but the gasoline thus obtained was too saturated. Use of a type 6434 catalyst incurred the difficulty of great sensitivity of the catalyst to nitrogenous bases. The process was therefore later replaced by a two-stage one: hydrorefining was carried out on a cobalt–molybdenum catalyst and hydrocracking on a nickel catalyst. Sulphurised MoO_3 was also tested for use as catalyst in this process [2079].

In the course of hydrogenation, polynuclear naphthenes formed by hydrogenation of the respective aromatic substances are partly split. This is accompanied by hydrogenolysis of oxygen-, nitrogen- and suphur-containing compounds. Thus, anthracene oil was hydrogenated to form lower-boiling hydrocarbons [1893] while creosote oil was hydrogenated to obtain a maximum yield of ethylbenzene [1078] or of naphthalene [1744]. The process was carried out in the presence of MoS_2 at temperatures of 450 to 500 °C and pressures of 35 to 175 atm.

Sulphide catalysts have found an important application in the preparation of high quality jet fuels, which have to satisfy a number of special requirements [1172]. Here, the principle holds that the energy content of a fuel is proportional to its hydrogen content. The majority of these requirements are closely satisfied by polycyclic naphthenes, especially di-, tri- and tetracyclic naphthenes (mainly alkylated ones, since alkylation increases the hydrogen content and decreases the pour point). Creosote oil was found to be a suitable raw material which can be converted to very good jet fuels by means of destructive-refining hydrogenation. The process suggested by Letort et al. [60, 297, 1222, 1223, 1526] operates in the vapour or mixed phase at 420 to 430 °C and pressures of 200 to 250 atm in such a way as to achieve perfect refining of the raw material together with intensive hydrogenation of aromatic compounds and controlled hydrocracking. The best catalysts for this purpose have been found to be Mo or W sulphides on Al_2O_3; type 8376 catalysts [288a, 1089] and the catalysts K-536 developed by the Ruhröl A. G. [742c] have also been used. Sulphide catalysts are likewise recommended for manufacture of supersonic aircraft fuel by means of hydrogenation of the tar fraction, which is a by-product of high temperature pyrolysis of mineral oil [1338a].

In an extensive study of hydrocracking of bituminous coal tar and oil from coal hydrogenation, Qader et al. [1713a, 1713c, 1713d, 1713e, 1713f] recently investigated the mechanism and technical implementation of this reaction. It was found by autoclave experiments in the presence of WS_2 and WS_2–NiS/Al_2O_3–SiO_2 catalysts, that all partial reactions of destructive hydrogenation of the tarry material (hydrocracking of hydrocarbons and formation of gasoline, desulphurisation, denitrogenation and oxygen removal) take place as pseudomonomolecular reactions under the reaction conditions employed [1713a, 1713b, 1713d]. The authors assume, that hydrocracking and isomerisation of hydrocarbons proceed in the presence of WS_2–NiS/Al_2O_3–SiO_2 through an ionic mechanism, while removal of heteroatoms takes place through a radical mechanism. The rate-controlling step is splitting of bonds between carbon atoms or between carbon and hetero atoms [1713d].

While the significance of tars as sources of diesel fuels is decreasing, attempts are being made to use these materials for producing important chemicals, especially phenols [1114]. This mainly applies to the phenolic fraction boiling above 225 °C (from brown coal carbonisation tar, tars from brown coal gasification, etc.), which contains a large proportion of polyhydric phenols. These can be converted to monohydric phenols boiling below 225 °C by means of selective hydrogenation, in which dehydroxylation is the main reaction (cf. p. 166). In a number of recent studies, Indian authors also discuss the problem of utilisation of low-temperature tars [928d, 928e, 928f, 1940a].

Reaction conditions and processes

A fixed-bed catalyst is usually employed for work in the vapour or mixed phase. Exceptionally, e.g. in the H-Oil and Hy-C processes, suspended catalysts are used. Temperatures range, according to the activity of the catalyst and character of the material to be processed, from 250 °C in the case of highly active catalysts based on molecular sieves, e.g. in the Unicracking process [2148] up to around 500 °C when feeds with a high content of nitrogen are being processed. Pressures range widely, mostly in the medium pressure range from 35 to 105 atm [2148]. The lifetime of the catalyst depends on the impurity content of the feed, being rather high with most catalysts of technical importance (e.g. in the Unicracking process the regeneration period is roughly 4,000 hours) [2148].

Quite a number of technically important hydrocracking processes have been applied practically over the years. At first, non-refined middle oils from liquid-phase hydrogenation of coal and tars were processed in refining-hydrocracking units with fixed-bed catalysts (catalysts 3510 and 5058). The difficulties and drawbacks of these processes, e.g. the great sensitivity of the 3510 catalyst to nitrogen-containing bases and the production of over-hydrogenated gasoline with a low octane number in the case of the 5058 catalyst [416, 1101, 1429] were overcome by the use of prehydrogenation prior to hydrocracking, which was then carried out in the vapour phase on the 6434 catalyst [416, 794, 1101].

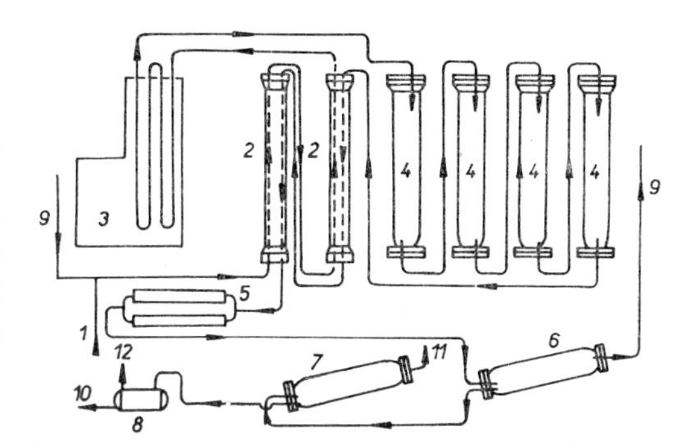

Fig. 83. Scheme of hydrocracking of middle tar oils in the vapour phase [1160]. 1 — Feed. 2 — Heat exchangers. 3 — Preheater. 4 — Reactors. 5 — Water coolers. 6 — Separator. 7 — Partial pressure reduction. 8 — Final pressure reduction. 9 — Recycle hydrogen. 10 — Product. 11 — Poor gas. 12 — Rich gas.

The classical vapour phase hydrocracking (Benzinierung) process was used to pre-refine middle oils (b.p. 200 to 325 °C) at a pressure of 300 atm and temperatures over 340 °C on the 6434 catalyst [1172, 1722]. The feed was mixed with hydrogen, from which NH_3 was first removed by washing with water, passed through heat exchangers and entered a series of reactors after having been preheated. Gasoline and kerosene were thus obtained, the higher fractions being processed to diesel fuel or returned to the hydrocracking process (see Fig. 83). The degree of hydrocracking depended on the catalyst activity, up to 60% conversion to gasoline per cycle could be obtained with a fresh catalyst; the mean conversion value was 30%.

A post-war modification of vapour-phase hydrocracking is the DHC process (Druck-Hydrogenium-Cracken). This processed gas oils and vacuum distillates as well as propane raffinates of petroleum and cracking residues to high-quality gasoline and diesel fuel at pressures between 50 and 250 atm and temperatures between 275 and 425 °C on fixed-bed catalysts with a sulphide or oxide hydrogenating component supported on synthetic aluminosilicate [794, 1525]. For single-stage hydrocracking of middle oils to gasoline, the Ruhröl A.G. developed the K-536 catalyst (Zn–Cr–Mo–S–F/terrana) [559a, 742b].

Recently, a number of important hydrocracking processe have been developed, processing various feeds ranging from heavy gasoline up to heavy residual oil, the main products being high-quality gasoline and jet fuel. The availibility of technical information has been limited up to now, especially in respect of the catalysts employed.

The most important processes developed or applied practically [2148] include Isocracking (bifunctional catalysts with a hydrogenating component containing no noble metal: according to Scott's patents [1840, 1841] this is NiS or CoS on aluminosilicate) [1747, 1836], Lomax [5], Isomax (which is a combination of the former two processes [61, 645, 1837] (for a scheme see Fig. 84), Unicracking – JHC (noble metal on Linde type Y sieve) [1601, 2138a], and the H–G process (a bifunctional catalyst with an activity maintenance of at least 18 months and balanced cracking and hydrogenating activity) [362, 537]. The Isomax process is the most important hydrocracking process in use at present (in the middle of the year 1970, Isomax processes accounted for more than 50% of the total installed capacity of hydrocracking processes for distillates [110A]); the plant capacity of Isomax units all over the world is rapidly rising. These units are designed in single- or

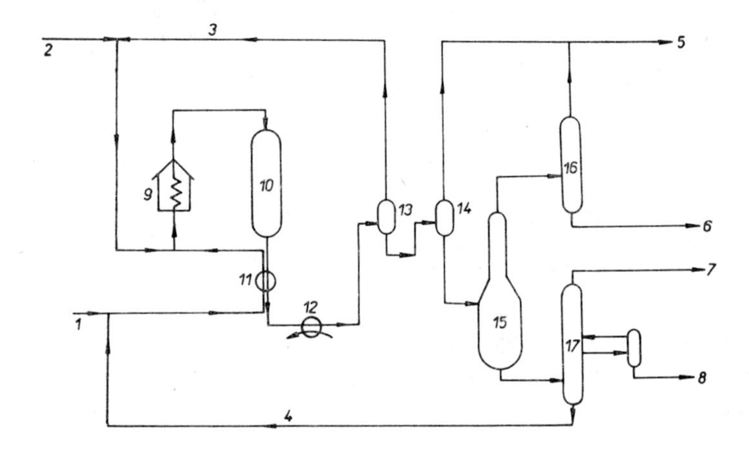

Fig. 84. Scheme of Isomax (Isocracking) process for hydrocracking medium petroleum distillates [1836]. 1 — Feed. 2 — Make-up H_2. 3 — Circulating H_2. 4 — Recycled material. 5 — Fuel gas. 6 — Butanes. 7 — Light gasoline. 8 — Gasoline (above 82 °C). 9 — Preheater. 10 — Reactor. 11 — Heat exchanger. 12 — Cooler. 13 — Partial pressure reduction. 14 — Final pressure reduction. 15 — Stabilisation column. 16 — Depropanizer. 17 — Splitter.

double-stage variants, the process being particularly important for processing residual petroleum fractions [1838]. The H-Oil process (Co–Mo/Al_2O_3 catalyst) and Hy-C process (Ni–W/Al_2O_3 catalyst) operate with a suspended, continuously regenerated catalyst which forms an ebullating or expanded bed [300, 301, 672, 752, 941, 1618, 1621, 2148]. This mode of operation enables refining and hydrocracking to be carried out in a single stage, and low-quality petroleum residues to be processed (cf. p. 352). The same purpose is served by the Gulf-HDS (Gulf Oil Co.), Texfining (Texaco) and Ultracracking (Amoco) processes [1450]. Among the European processes, the BASF-IFP hydrocracking process [74a] was succesfully developed in a single- and in a double-stage variant (cf. the DHC process as such or with the DHR process used as pretreatment, p. 307) [175], and the Katsobashvili process [994] designed for destructive hydrogenation of petroleum residues (p. 297).

Hydrodealkylation

Catalytic hydrodealkylation is a special type of hydrocracking process. It is used in modern processes, e.g. Bextol, Hydeal, Unidak, etc. [104, 1759] processes in which oxide catalysts are used and the raw materials are refined. The literature includes very few data relating to the direct use of metal sulphides for hydrodealkylation. Sweeney's patent [2020] which recommends the use of sulphides of Group VI metals for destructive hydrogenation of petroleum extracts to naphthalene was mainly concerned with hydrodealkylation of alkylnaphthalenes. In hydrodealkylation of solvent naphtha, which contains mainly *m*-xylene, 40% conversion to toluene was achieved in the presence of MoS_2 on pumice [358]. An older study by Yukhnovskii [2243] describes high-temperature (770 °C) hydrodealkylation of solvent naphtha, by means of which up to 61% of a toluene fraction was obtained. Good yields from hydrodealkylation of alkylaromatic and aromatic concentrates (also containing olefins, sulphur, nitrogen and oxygen compounds) were achieved with the WS_2/ /Al_2O_3–SiO_2 catalyst [2158]. A sulphurised commercial reforming catalyst is likewise an active hydrodealkylation catalyst [624]. Presulphurised Co–Mo catalyst was tested in hydrodealkylation and hydrocracking of the aromatic extract of karosene fraction [457a].

7.2.3.

Hydrorefining and hydrodesulphurisation

In this field, the easily available and cheap sulphide catalysts have found their widest field of application. The refining reaction involves mainly hydrogenolysis of oxygen-, nitrogen- and sulphur-containing compounds, with formation of hydrocarbons and water, ammonia and hydrogen sulphide. Moreover, organo-metallic compounds are hydrogenolysed if these are contained in the raw material [1743]. Reactive olefinic bonds, as well as some of the aromatic compounds, are hydrogenated at the same time, and partial hydrocracking takes place depending on the operating conditions and catalyst selectivity [1101].

Gum-forming and sulphur-containing substances are removed from gasolines and crude benzol while kerosenes and gas oils are hydrogenated in order to improve the combustion properties and remove sulphur compounds. In the case of catalytic recycled oils, aromatic compounds are hydrogenated and nitrogenous bases removed, fuel oils are desulphurised, middle tar oils are made free of nitrogen substances before being hydrocracked, etc.

The chemistry of the hydrorefining reactions follows from the hydrogenation and hydrocracking reactions of individual oxygenous, nitrogenous and sulphurous substances (cf. Chapter 6.). The situation is more complicated, however, in the case of hydrorefining and desulphurisation of high-molecular raw materials, e.g. crude petroleum, petroleum residues etc. With these materials, the heteroatoms are bound both in oily fractions and in resins and asphaltenes. For example, in the atmospheric residuum of Romashkino petroleum, 50% of the total sulphur is bound in oils, about 45% in resins and only 5 to 6% in asphaltenes [1121]. Although achievement of the required degree of desulphurisation of atmospheric residue does not depend on desulphurisation of asphaltenes, it is these compounds in the feed which have a decisive influence on the catalyst activity, since they are the main cause of contamination of the catalyst surface with coke. Another significant cause of deactivation of refining catalysts is the presence of ash in heavy feeds. All these influences cause the refining mechanism of raw materials to differ in many respects from the course of refining reaction of pure substances, especially at elevated temperatures when thermal splitting reactions will participate, leading to different reaction products and at the same time causing the activity of refining catalysts to decrease.

The most variegated raw materials are subjected to hydrorefining, ranging from gases up to high-boiling petroleum residues and heavy tar materials. Hydrorefining is gradually replacing classical chemical refining methods, and becomes an essential amplification of the range of industrial refining processes.

The reaction conditions of hydrorefining processes depend on the quality of the raw material to be treated as well as on the degree of refining required. Some refining processes take place with no substantial reduction in molecular weight and little change in physico-chemical properties. They either represent a final treatment of products (hydrofinishing) or, hydrorefining is followed by an additional refining process (e.g. adsorption), by means of which the final properties of the product are adjusted. In the case of processes performed under more severe refining conditions (hydrotreating), refining is accompanied by substantial changes in the molecular structure of the original material [1821].

The course of hydrorefining reactions is influenced in a decisive manner by temperature. In technical practice, hydrorefining and hydrodesulphurisation are usually carried out at temperatures of 300 °C and more, desulphurisation reactions taking place in general more readily than hydrogenolysis of oxygen substances or denitrogenation. Hydrocracking reactions start to play an important role at temperatures around 400 °C.

The rate of refining reactions is increased by a rise of the hydrogen partial pressure. This at the same time decreases the extent of coke formation and deactivation of the catalyst surface. High pressures are conventionally employed in refining tar products (200 to 700 atm) [1101], while medium- and low-pressure processes are typical in refining and desulphurisation treatment of petroleum products (20 to 70 atm) [1347].

The overall effect of the partial pressure of hydrogen is also determined by the hydrogen-to-feed ratio. The ratio of hydrogen (recycle gas) to the feed varies over a wide range from 45 to 4,000 Nm^3/ton according to the character of the refining reaction [1101, 1347], and it is also influenced by the specific hydrogen consumption of the refining process. For example, the hydrogen consumption varies from 3 to 100 Nm^3/t when light petroleum distillates are being desulphurised, while it is several times greater in the case of tar products refining (e.g. tar refining by the MTH process with hydrogen consumption values of about 700 Nm^3/t [1101]).

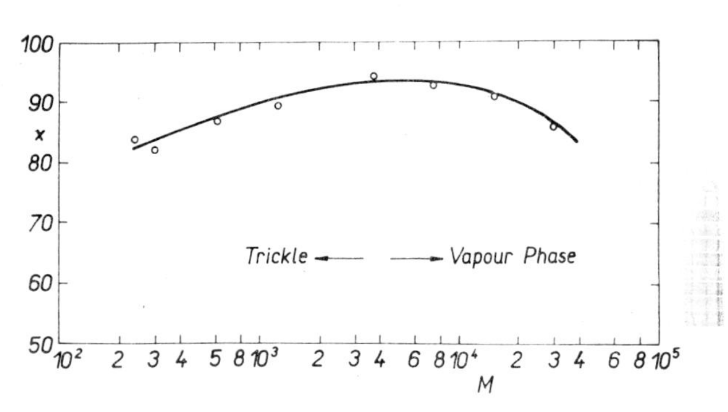

Fig. 85. Dependence of the hydrodesulphurisation degree of gas oil on the H_2: oil ratio [788]. (The feed was gas oil with 1.25% S, catalyst Co–Mo/Al_2O_3, pressure 52 atm, temperature 375 °C, LHSV 2.4). M—H_2 : oil ratio (scf/bbl). x — Desulphurisation degree (%).

At constant temperature, LHSV of the raw material and overall pressure, a variation of the H_2: oil ratio will influence the degree of evaporation of the raw material, the partial pressure and also the time of contact with the catalyst particularly in those reactions taking place in the mixed phase. All these parameters influence the reaction rate, as shown e.g. by Hoog et al. [691, 788] in the case of gas oil hydrodesulphurisation on a Co–Mo/Al_2O_3 catalyst (Fig. 85). The decreasing degree of desulphurisation observed with a decreasing H_2: oil ratio in the mixed phase may be explained by a decrease in the hydrogen partial pressure as well as by the resistance of the liquid film on the catalyst surface to mass transfer. At the moment in which all the liquid has evaporated, the reaction rate achieves its maximum and a further increase of the hydrogen excess causes a slight conversion decrease, due probably to a lowered degree of utilisation of the inner catalyst surface under the influence of the lowered partial pressure of the oil (cf. hydrogenation of aromatic substances, p. 128).

The hydrogen partial pressure finally also depends on the H_2 content in the hydrogenation gas. Because of the high price of hydrogen, all available sources are utilised for hydrorefining and particularly hydrodesulphurisation, provided they contain at least 20% hydrogen [791]. For example, benzol may be hydrorefined with the use of coke-oven gas [2101]. Similarly gases from reforming processes are widely used [1347] and, in the Autofining process, the hydrogen formed by partial dehydrogenation of the feed is actually sufficient for the needs of the process [412, 1247, 1347].

Besides hydrogen, recycled gases mainly contain saturated hydrocarbons, nitrogen and hydrogen sulphide. As shown already (p. 89), hydrogen sulphide may influence the selectivity of a catalyst and thus also the quality of the hydrogenation product. In hydrodesulphurisation, the negative influence of hydrogen sulphide is rather slight [789, 795], particularly in the case of desulphurisation of light fractions, e.g. gasoline from coking processes or in processes operating in the mixed phase (the Trickle Hydrodesulphurisation process of the Shell Oil Co.) [788]. In the case of gas oils, the conversion decrease is more marked [135, 1347].

In hydrorefining, space velocities vary over a wide range, depending on the ease with which the respective bond is hydrogenated or hydrogenolysed, e.g. LHSV is 0.6 to 1 in prehydrogenation of tars [1101], while it is considerably greater in the case of gasoline desulphurisation with a minimum extent of olefin hydrogenation, in exceptional cases being up to 30 l/l hour [330, 1347].

Hydrorefining is carried out in either the vapour or liquid (mixed) phase [790, 1101, 1347]. In practical operations, fixed-bed catalysts are mostly employed, while moving catalysts are only seldom used [1345, 1845].

7.2.3.1.

Hydrorefining catalysts

The catalyst influences the course and kinetics of the hydrorefining reaction in a decisive manner. From the thermodynamic point of view, all refining reactions are favourable. The reaction rate of catalytic refining reactions depends not only on the type of molecule involved (e. g. oxygen and sulphur substances are hydrogenolysed more readily than nitrogenous ones), but also on the molecular weight and, therefore, widely varying reaction rates are observed. With catalytic reactions in the vapour phase, the rate-controlling process is most often the surface reaction of adsorbed molecules. However, diffusion is also involved; this being particularly significant when the process is carried out in the liquid or mixed phase [795].

Raw materials processed by hydrorefining methods are multi-component mixtures of substances with different molecular weights, so that different reaction rates may be expected on hydrogenation in the vapour phase when using catalysts of different particle size. The reaction takes place mainly on the inner surface of the catalyst, which is accessible to the reacting molecules by diffusion only. Reitz [1730] studied the degree of hydrogenation in middle tar oil hydrorefining on the WS_2 catalyst (using the aniline point as a measure) as well as the degree of splitting (by the gasoline

Table 76.

Dependence of Hydrogenation of Middle Tar Oil on the Grain Size of the WS_2 Catalyst [1730]. (Reaction conditions: the middle oil had b. p. 180—325 °C, aniline point —20 °C, phenols content 15%, nitrogen content 0.6%; hydrogenation was carried out at 400 °C, pressure 250 atm, space velocity 0.35 kg/kg/hour, H_2 : oil ratio 4 Nm^3/kg.)

Reaction products		Grain size of the catalyst (mm)		
		Cylinders 10 × 10	Cylinders 3 × 4	Grains 2 — 4
Gasoline production (% in the product)		38	38	29
Amount of gaseous hydrocarbons (% of the feed)		3	4	4
Gasoline up to 180 °C	Aniline point (°C)	37	34	29.5
	Fraction up to 100 °C (%)	18.5	14.0	12.5
	Aromatic compounds content (%)	15	16	19
Residue over 180 °C	Aniline point (°C)	48.5	54.5	55.0
	End point (°C)	306	301	305

yield) in dependence on the catalyst particle size (Table 76). The course of hydrogenation of the individual raw material fractions (20 °C-fractions) is illustrated in Fig. 86. The results show that formation of gaseous hydrocarbons remains practically constant. The aniline point of the refined middle oil rises with the decreasing

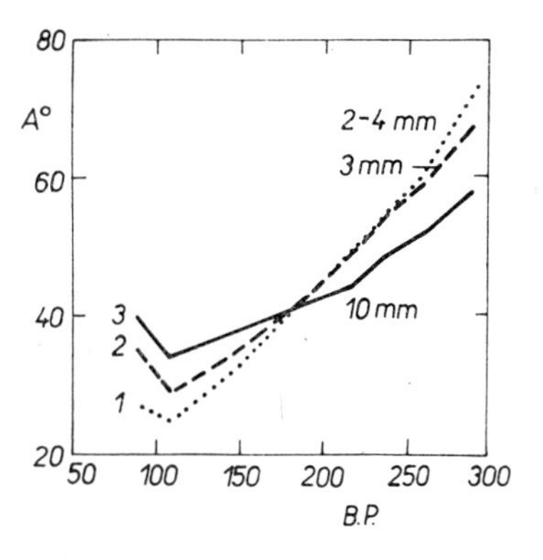

Fig. 86. Aniline points of 20-degree fractions in the hydrorefining product of middle tar oil on WS_2 catalyst (feed material cf. Table 76) [1730]. *A°* — Aniline point of the fraction (°C). *B. P.* — Average boiling point (°C) of the twenty-degree fraction. 1 — Grains 2—4 mm. 2 — Cylinders 3 mm. 3 — Cylinders 10 mm.

particle size, while the reverse applies to the gasoline fraction. The light gasoline content (up to 100 °C) in the total gasoline fraction decreases with the decreasing particle size (the overall gasoline yield, related to the raw material, varies in an insignificant manner). Therefore it is better from the point of view of middle tar oil hydrorefining to work with smaller particle sizes, since a more aromatic gasoline with higher octane number is obtained. On the other hand, the refined middle oil is hydrogenated to a greater extent and, therefore, its properties are better from the motor fuel aspect; contrarily, this property is unfavourable when the material is hydrocracked, as a gasoline of lower octane number is obtained.

Table 77.

Characteristics of the Porous Structure of Industrial WS_2 and WS_2–NiS/Al_2O_3 Catalysts [1866].

Data		Catalyst	
		WS_2	WS_2–NiS/Al_2O_3
Porosity (%)		49	61
Surface area	(m^2/g)	51.5	151.9
	(m^2/cm^3)	196	243
Pore distribution (%)	up to 80 Å	50.7	78.2
	80 to 180 Å	3.8	3.8
	180 to 380 Å	6.8	3.4
	> 380 Å	38.7	14.6

A strong dependence of the rate of hydrogenation on the particle size was observed with the type 8376 catalyst, but differing from WS_2 (cf. Fig. 86) no decreased reactivity of larger molecules was observed with larger catalyst particles [1730]. The above relationships indicate that in the course of hydrogenation diffusion processes have a significant influence, which has to be taken into consideration when selecting the proper catalyst bed to be used.

The fact, that diffusion has a retarding effect on conversion of high-boiling fractions with the rising particle size, is an important finding. Internal diffusion of large molecules into the pores of larger particles is slowed down, utilization of the internal catalyst surface decreases and thus the reaction rate diminishes, too. At the same time, catalysts 5058 and 8376 differ in their texture. Soviet authors found on comparing the industrial catalysts WS_2 and WS_2–NiS/Al_2O_3 [1866], that type 8376 catalysts contain a larger proportion of narrow pores up to 80 Å in radius (78% of the total volume), while the WS_2 catalyst contains more pores of radius above 360 Å (Table 77). These larger pores act as transport channels for the inflow of reacting substances to the operating pores and as outlets for the reaction products. The above authors calculated that the inner surface of the WS_2 catalyst with particle sizes up to 10 mm is 90% utilised. With the WS_2–NiS/Al_2O_3 of the same particle size, the degree of inner surface utilisation is only 39% and does not rise much even with particles of about 5 mm (52%). From these findings, which confirm the significance of macropores in the catalyst for the refining reactions (as already mentioned, cf. p. 55) [644, 2084] Shavolina et al. [1866] deduce that the reaction proceeds on type 8376 catalysts in only a thin layer of the catalyst surface.

A catalyst of suitable selectivity must be chosen for the hydrorefining reaction in order to satisfy the main objectives of the process. A number of pure (single-component, unsupported) sulphide catalysts are polyfunctional in character and exhibit at the same time a hydrogenating or refining, splitting and isomerising activity. This is one of the main reasons for the replacement in industrial practice of unsupported single-component sulphides by selective catalysts [793, 1101, 1429]. Unsupported catalysts continue to be used under mild conditions only for special refining purposes e.g. the 5058 catalyst is used for hydrorefining .ubricating oils under pressure [1479, 1480]. However, the literature contains a large amount of data concerning the use of pure sulphide catalysts in hydrorefining and hydrodesulphurisation of tar and petroleum products. Catalysts most frequently used in this way are WS_2 [137, 681, 693, 762, 877, 1008, 1128, 1131, 1132, 1158, 1166, 1189, 1192, 1287, 1289, 1290, 1291, 1292, 1293, 1439, 1476, 1477, 1480, 1665, 1811, 1812, 1813, 1814, 1853, 1949, 2027] and MoS_2 [34, 268, 296, 305, 307, 308, 333, 378, 512, 515, 580, 767, 1183, 1194, 1200, 1393, 1401, 1405, 1443, 1444, 1580, 1702, 1756, 1942, 2066, 2261]. Applications of NiS [84, 330, 353, 355, 665, 1466, 1932, 2039, 2109] and CoS [354, 665, 1749, 1754] are less frequent.

Other sulphides were also tested for use as catalysts, e.g. FeS [1237, 1269, 1455], SnS [1634], ZnS [215, 1633], CdS [1645], CuS [1634, 1635] and Re_2S_7 [480]. The use of ammonium thiomolybdate was also described [814, 1606].

In a number of cases, pure sulphides were supported on suitable low-acidic carriers. Besides improving the thermal stability of the hydrorefining component, the carrier also enhances its selectivity in numerous cases. Al_2O_3 is used most often. It served as a carrier of refining catalysts with MoS_2 [417, 490, 513, 1200, 1979] (under hydrorefining conditions, the reforming DHD catalyst may also be employed [1101, 1468]), NiS [369a, 408, 1454, 1773, 2025, 2026], CoS [597a], PdS [134] or Cr_2S_3 [117] as active components. Other high-porosity carriers employed for hydrorefining catalysts with a single sulphidic component include activated carbon as carrier for MoS_2 [307, 308, 514, 820, 1285, 1793], NiS [704] or Cr_2S_3 [1017] and silica gel for MoS_2 [1700, 1701], NiS [1150] and (together with MgO) for the NiS + MoS_2 mixture [1217].

The use of low-porosity carriers, e.g. pumice (for MoS_2 [93] or CoS [1751]), or of kieselguhr or diatomaceous earth (for NiS [499, 1803]) is rather unusual.

Sulphide catalysts attain highly selective properties by the influence of suitable promoters, in some cases by simultaneous supporting on suitable carriers [154, 243, 416, 1347, 1356a, 1429, 1626, 1760a]. With respect to refining and desulphurisation catalysts these problems were discussed in detail in sections 5.2 and 5.3. This analysis showed that the hydrogenation and refining activity of Mo and W sulphides as components of refining catalysts (possibly after supporting on a carrier) is considerably improved by the promoting effect of compounds of the iron-group metals [825, 1101]. The most efficient is a combination of metals of Groups VI and VIII of the periodic table [163, 165, 395, 465a, 491, 705, 732, 928a, 1011, 1012, 1055a, 1074, 1080a, 1218, 1320a, 1473, 1488, 1499, 1504, 1537a, 1594, 1817, 1874, 1876, 1888b, 1966, 2239a] and therefore it is used practically exclusively. Other combinations, e.g. of Group II sulphides with sulphides of Groups V to VIII [1471], mixtures of sulphides of Group VI metals [204], mixture of CoS + ThO_2 [315], sulphurised Pt–MoO_3 catalyst [1080b] or a combination of Cr_2O_3 with Ni [704], Cu [477, 1016], Ca [1445] and Mo sulphides [1401, 1405, 1704] (or Cr thiomolybdate [174]), have been used in exceptional cases for hydrorefining purposes.

Mixed refining catalysts containing sulphides of Groups VI and VIII metals are used in the unsupported form (cf. e.g. [932]), or supported on a carrier which is mostly activated alumina [416, 1347, 1626] or alumina stabilised by an addition of SiO_2 [161, 164, 465c, 727a, 780, 1656, 1874, 1882, 2078] (problems of catalyst carriers have been discussed in section 3.2.). These catalysts most often contain two sulphides, though modern catalysts with a high degree of activity and selectivity are frequently combinations of three sulphidic components [1429].

A combination of tungsten and nickel sulphides is the most common. Compared to pure WS_2, this mixture has the higher hydrogenation and refining selectivity. Moreover, the unsupported WS_2.NiS catalyst, and more particularly this combination when supported on γ-Al_2O_3, is characterised by a considerably lower splitting activity (cf. section 5.2.) and thus it acts as a typical refining catalyst. The catalyst 8376, containing 3 to 4.5% NiS, 25.5 to 27% WS_2 on γ-Al_2O_3 [1101, 1429] is mostly used for high-pressure hydrorefining of tar materials, and it is characterised by a high refining activity, resistance to nitrogenous bases and long lifetime. It has been applied industrially in the prehydrogenation of tars, in the TTH and MTH processes, in hydroprocessing of tar and shale oils and in a number of research studies [133a, 189, 375, 376, 416, 442, 443, 445, 694, 1100, 1101, 1227, 1379, 1380, 1529, 1626, 1714, 1731, 1801, 1907, 1916, 2013, 2017, 2211]. Švajgl used a modification of this catalyst with higher nickel content (called 8376 ANTIAS) with success for hydrorefining arsenic-containing brown-coal tars [1993, 1994, 1999, 2001, 2002, 2005].

In a number of cases, the type 8376 catalyst is also used for hydrorefining petroleum products. Important applications of this catalyst include the selective hydrogenation of pyrolysis gasoline [435, 631, 1677], refining of various petroleum fractions [431, 1762, 1887], lubricating oils [222, 517, 759, 1476, 1477, 1478, 1479, 1480, 1812, 1813, 1814, 1815, 1907], liquid and solid n-paraffins [99, 100, 2133, 2134], atmospheric residuum [1543] or resins and high-molecular weight aromatic sulphur compounds [1858, 1859, 1861]. The catalyst 8376 has been employed less frequently for hydrodesulphurisation only [1544, 2028, 2030, 2110, 2111].

The unsupported catalyst WS_2.NiS is the most efficient sulphidic hydrogenation catalyst. The commercial catalyst 3076 was used for hydrorefining the kogasin fraction, for hydrogenation of residual aromatic substances in liquid n-paraffins, reduction of higher alcohols and total hydrogenation of raw naphthalene and benzene (with thiophene) [486, 487, 626a, 1101, 1429]. In the U.S.A., the commercial catalyst $WS_2.NiS$ of the Shell Oil Co. [1347, 1743] has mainly been used for hydrorefining cracked gasolines and for hydrodesulphurisation. The composition of this commercial catalyst is close to the molar ratio $WS_2.2\,NiS$ [321], the optimum is considered to be an atomic ratio of Ni : W = 1.5 – 2 : 1 [83, 327]. This catalyst is characterised by a high hydrogenation activity, and a very low splitting activity above 425 °C. Compared to the standard cobalt–molybdenum catalyst, it also has a somewhat higher desulphurising activity [788]. The dehydrogenation properties of this catalyst have also been described [1347] (cf. p. 373).

Besides the above-mentioned technological applications, the $WS_2.NiS$ catalyst has been used in a number of other cases for hydrorefining middle tar and shale oils [1213, 1616, 1922, 1923], primary and cracked gasolines [121, 152, 274, 772, 776, 777, 803, 1002, 1507, 1949, 2160], kerosenes, diesel fuels and other middle distillates [105, 432, 561, 790, 802] and lubricating oils [133, 169, 331, 332, 400, 1281, 1614, 2150, 2153]. This catalyst has been used frequently in hydrodesulphurising straight-run and cracked gasolines [1, 2, 121, 122, 274, 321, 330, 446, 473, 626, 1890, 2021].

Of equal importance in hydrorefining and hydrodesulphurisation are catalysts based on molybdenum sulphide, promoted with a sulphide of a Group VIII metal. Among these catalysts, the cobalt–molybdenum catalyst has achieved widest use, particularly in medium-pressure catalytic desulphurisation of various petroleum products. The significance and application of this catalyst type is discussed in detail in McKinley's monograph [1347]; a few amplifying remarks shall be added here only.

The cobalt–molybdenum catalyst is used almost exclusively when supported on γ-Al_2O_3 [468, 471, 1347] or on alumina stabilised with a small amount (about 3%) of SiO_2, by means of which the carrier achieves an optimum porosity and the catalytic system achieves a high thermal stability [161, 428, 561, 1347, 1656, 1874]. The use of bauxite as a carrier for the Co–Mo catalyst has also been described [1334]. The unsupported mixture of Co and Mo sulphides or, Co thiomolybdate have been seldom used for hydrodesulphurisation [493, 675, 835, 1239, 1240, 1335, 1337, 1665, 2132].

Besides an outstanding desulphurisation activity, the cobalt–molybdenum catalyst also possesses a good refining activity with respect to oxygen containing substances. It is, however, more sensitive to nitrogenous substances, compared to tungsten and tungsten–nickel catalysts (cf. p. 96) and, therefore, it is not used for refining raw materials containing large concentrations of nitrogen-containing compounds (e.g. mixture of petroleum and brown-coal feeds) [1429].

Small amounts of nitrogen compounds in petroleum feeds do not have a significant influence on hydrorefining with cobalt–molybdenum catalysts. For hydrodenitrogenation purposes, however, these catalysts are used infrequently [725] and only then if sulphur compounds are also present in the feed. Gerald [596] actually re-

Table 78.

Composition of Some Commercial Cobalt–Molybdenum Catalysts [1743, 2137].

Property	Catalyst notation									
	Nalco		Aero HDS-2	Houdry 200-A	Oronite	Girdler G-35 B	Comox P. Spence	Cyanamide HDS 1	USSR	ČSSR 7362
	471	810								
Chemical composition (% by wt.):										
CoO	3.5	3.5	3	2	3.5	3.0[a])	min. 2.5	3	5.6	3.9
MoO_3	12.5	12.5	15	8	10.0	5.6[a])	min. 14	15	11.9	14.2
Na_2O	0.05	0.05	0.02	max. 0.5	0.01	max. 0.04	max. 0.15	max. 0.02		
Fe	0.03	0.2	0.1	—	0.07	traces	—	—	—	—
Surface area (m^2/g)	270	153	260	195	143	270	—	—	218	231
Pore volume (cm^3/g)	0.48	0.32	0.55	0.53	—	0.22	—	—	—	0.34
Apparent density (g/cm^3]	0.68	0.68	0.69	0.92	1.02	—	—	—	—	—

[a]) Content of Mo and Co [1537].

commends the deliberate addition of sulphur when carrying out denitrogenation on a Co–Mo catalyst (the amount of sulphur in the feed should correspond to the relation: $S = 25.8\ N + 0.071$, S and N being the sulphur and nitrogen contents in percentages). For efficient denitrogenation of raw materials, e.g. before hydrocracking, W–Ni or Ni–Mo catalysts are used preferably.

The activity of cobalt–molybdenum catalysts is sufficient to hydrogenate olefinic double bonds in the feed. This activity can be decreased by adding to the catalyst an alkali metal, particularly potassium, which increases the hydrodesulphurisation selectivity of the resulting catalyst [467]. Compared to tungsten, tungsten–nickel and nickel–molybdenum catalysts, the Co–Mo catalysts are substantially less active in hydrogenation of aromatic compounds [318].

The activity in splitting C–C bonds is also low with this catalyst under hydrorefining conditions [795], so that the Co–Mo catalyst is exceptionally selective for hydrodesulphurisation.

The promoting effect of Co on the desulphurisation activity of MoS_2/Al_2O_3 was discussed in detail in section 5.2. These data indicated an optimum atomic ratio of Co : Mo = 0.3 to 0.4. The composition of some conventional commercial catalysts is stated in Table 78. It can be seen from this table that the atomic Co : Mo ratio used in practice is a little higher than the optimum which Beuther et al. [154] determined by experimental means.

The problem of sulphurising cobalt–molybdenum catalysts to the active sulphide form was discussed in section 5.1. (p. 85). In the industrial application, Co–Mo catalysts are usually sulphurised, in the course of the initial working period, by the hydrogen sulphide formed from hydrogenolysis of sulphur compounds in the feed [1347, 2029]. This method sometimes leads to a more active catalyst [2111]. Hendricks et al. [754] however recommend presulphurisation in the case of oxide catalysts. A presulphurised Co–Mo catalyst, Co thiomolybdate or a preliminarily prepared mixture of the two sulphides have also been used in a number of research studies [14, 137a, 212, 244, 246, 324, 578, 597, 623, 643a, 675, 676, 791, 873, 1013, 1074, 1127a, 1239, 1240, 1299, 1334, 1335, 1385a, 1536, 1577, 1665, 1667, 1697a, 2022, 2093a, 2229b].

A promotion effect on MoS_2, similar to that of cobalt, is also exhibited by nickel. Ni–Mo sulphidic catalysts (cf. p. 96) have practically the same desulphurisation activity with an optimum promoter concentration as a cobalt–molybdenum catalyst (cf. p. 99) [1347], but they are more active in respect of refining materials with a high content of nitrogenous substances [1429, 2130] and they are characterised by a high degree of selectivity for degrading nitrogen compounds in aromatic raw materials [2004]. Therefore, they are used for catalytic hydrodenitrogenation of petroleum-derived materials, particularly those which serve as feed for catalytic cracking or hydrocracking [65, 74, 101, 302, 521, 780, 926, 928, 1576, 2078, 2096].

The catalyst $NiS–MoS_2/Al_2O_3$ was formerly used in Germany (catalyst 7846) for refining before the catalyst 8376 came into use. At present it is in use for hydrodenitro-

Table 79.

Survey of Technically Most Important Processes of Hydrorefining and Hydrodesulphurisation.

Process	Implemented by	Main application	Catalyst	Typical reaction conditions	Reference
Vapour phase hydrogenation	BASF and I.G.F. - A.G.	Hydrorefining and hydrocracking of tar products b. p. up to 325 °C, before hydrocracking to gasoline	8376 (WS_2–NiS–Al_2O_3)	Gas phase, $t \sim 400$ °C, $P \sim 300$ atm, LHSV 0.6 to 1.0, circulating gas 3500 to 4000 Nm^3/ton	[1101]
DHR-process	BASF	Modified vapour phase hydrogenation. Served for medium-pressure refining of light fuel oils, diesel fuels and jet fuels	Not published	—	[794]
TTH-process, MTH-process	I.G.F.-A.G.	Hydrorefining of tar with low content of asphaltenes and resins to make high-quality diesel fuels, paraffin wax and lubrication oils; increased splitting in the MTH process	8376 or 8197 (WS_2–NiS–Al_2O_3 or NiS–MoS_2–Al_2O_3)	Liquid (or mixed) phase, $P \sim$ 320 atm (newly, 80 atm), $t =$ = 405 to 425 °C, LHVS = 1 (with the MTH process, the temperature is about 40 °C higher, LHVS = 0.6)	[1101]
Benzol hydrorefining	BASF-Scholven, Lurgi, Koppers, Newton Chambers etc.	Refining and desulphurisation of crude benzol and in some cases, of higher aromatic hydrocarbons	Co–Mo catalysts	Gas phase, $P = 20$ to 60 atm, $t = 350$ °C	[2101]
Shell Vapour-Phase Hydrodesulphurisation	Shell Oil Co.	Hydrodesulphurisation of cracked gasolines	WS_2–NiS	Gas phase, $t = 240$ to 370 °C, $P = 35$ to 52 atm	[58] [1890]

Hydrofining	Esso Research and Eng. Co.	Hydrorefining of straight-run gasolines, cracked gasolines, diesel fuels, fuel oils etc.	Co–Mo catalyst	$t = 260$ to 425 °C, $P = 3.5$ to 56 atm	[64] [58]
Autofining	British Petroleum Co.	Desulphurisation of straight-run distillates up to b. p. 370 °C, hydrorefining of cracked distillates	Co–Mo catalyst	The process works without the addition of separately made H_2; $t = 400$ to 430 °C, $P = 7$ to 14 atm (liberation of excess H_2) or 42 atm (operation in equilibrium)	[64] [58]
Arofining	Howe-Baker Engineers, Inc.	Removing and decreasing the aromates content in distillates	Highly efficient catalyst resistent to S and N	Not published	[74a]
Hydrogen Treating	Sinclair Ref. Co.	Hydrorefining and hydrodesulphurisation of straight-run and cracked gasolines, light tar oils etc.	Co–Mo catalyst	$t = 300$ to 415 °C, $P = 14$ to 56 atm	[58]
Diesulforming	Husky Oil Co.	Hydrorefining of fractions from gasoline to vacuum oils; straight-run and cracked materials may be processed	Mo-catalyst	$t = 315$ to 427 °C, $P = 31$ to 36 atm, LHSV $= 1-2$	[58]
Ultrafining	Standard Oil Co. (Indiana)	Hydrorefining of materials from heavy gasoline to heavy circulating oils from cracking processes	Co–Mo catalyst	$t = 315$ to 425 °C, $P = 14$ to 105 atm, LHSV $= 1$ to 20	[64]
Selective Hydrogenation	British Petroleum Co.	Refining of pyrolysis gasoline	Ni on a non-active carrier	$t > 180$ °C, $P = 21$ to 42 atm, LHSV $= 0.5$ to 5	[64, 1221]

Continued Table 79.

Process	Implemented by	Main application	Catalyst	Typical reaction conditions	Reference
Pyrotol	Houdry Process and Chem. Co.	Production of high-purity benzene from dripolene	Not published	Not published	[74a]
Hydrorefining L-24-6	Lengiprogaz (USSR)	Hydrorefining of diesel fuels and gas oils	Co–Mo catalyst	$t = 375$ to $425\ °C$, $P = 50$ atm	[1981]
Hydrodesulphurisation	Lurgi	Desulphurisation of fractions from gasoline up to lubricating oils	Co–Mo catalyst	$t = 350$ to $400\ °C$, $P = 15$ to 20 atm	[1267]
Shell-Trickle-Hydrodesulphurisation	Shell Oil Co.	Hydrodesulphurisation and hydrorefining of petroleum distillates for catalytic cracking	Co–Mo catalyst	$t = 320$ to $430\ °C$, $P =$ up to 105 atm	[64]
Gulfining	Gulf Res. and Dev. Co. and Houdry Process and Chem. Co.	Hydrodesulphurisation of heavy gas oils	Not published	Not published	[74a]
Unifining	Universal Oil Product Co. + Union Oil Co. (Calif.)	Hydrorefining of primary and cracked petroleum products	Co–Mo catalyst	Not published	[64] [58]
Hydrodesulphurisation	Different companies: Standard Oil Co. (Indiana), M.W. Kellogg Co., Sun Oil Co., Phillips Petrol. Co., Sinclair-Barker Co.	Hydrodesulphurisation of petroleum fractions from straight-ran and cracked gasolines up to fuel oils	Co–Mo catalysts	Operating conditions selected according to the character of the raw material and required degree of desulphurisation	[64] [1347]

Hydrobon	UOP Co.	Hydrorefining feeds for reforming, desulphurisation of fuel oils; hydrofining of catalytic cracking feeds	Not published	Not published	[74a]
Unionfining	Union Oil Co. of Calif.	Hydrodesulphurisation and hydrodenitrogenation of petroleum fractions	High-activity catalyst containing no noble metal	Not published	[74a]
Unisar	Union Oil Co. of Calif.	Hydrogenation of aromatic hydrocarbons in petroleum stocks	Not published	Not published	[74a]
Gulfinishing	Gulf Oil Corp.	Hydrorefining of lubricating oils	Not published	Not published	[64]
Ferrofining	British Petroleum Co.	Hydrorefining of lubricating oils	Co–Mo - Fe catalyst	t = 250 to 290 °C, P = 20 atm	[64]
Comofining	Lurgi + Wintershall A. G.	Hydrorefining of lubricating oils	Co–Mo catalyst	t = 250 to 300 °C, P = 50 to 60 atm	[343]
Lube Hydrofinishing, Lube Hydrotreating	Institut Francais du Pétrole	Hydrofinishing of lubricating oils. Production of high-quality lubricating oils and waxes	Not published	Not published	[74a]
GO-fining RESID-fining	Union Oil Co. of Calif. and Esso Res. and Eng. Co.	Hydrodesulphurisation of vacuum gas oils and deasphaltisates (GO-fining) and of atmospheric residues (RESID-fining) with greatly limited hydrocracking	Not published	Not published	[74a]
Gulf-HDS	Gulf Research Dev. Co.	Hydrorefining and hydrocracking of petroleum residues to fuel oils and feeds for catalytic cracking	The composition of the operating catalyst was not published	t = 400 to 455 °C, P = 35 to 110 atm, LHSV = 0.5–2	[64]

Continued Table 79.

Process	Implemented by	Main application	Catalyst	Typical reaction conditions	Reference
H-Oil	Hydrocarbon Research, Inc.	Hydrorefining and hydrocracking of fuel oils and circulating oils from cracking, atmospheric and vacuum residues	Co–Mo catalyst in suspension	The operating conditions vary according to the raw material	[64]
Rcd Isomax	UOP Co.	Desulphurisation of high-sulphur residual oils	Not published	Not published	[74a]
RDS Isomax	Chevron Res. Co.	Production of low-sulphur components for fuel oils by desulphurising atmospheric residues	Not published	Not published	[74a]
Residue Hydrodesulphurisation	Institut Français du Pétrole	Desulphurisation and hydrorefining of atmospheric and vacuum residues or, of crude oil. Production of heavy fuel oils with limited sulphur content	Not published	Not published	[74a]

genation as well as refining of tar-derived materials and mixtures of tar and petroleum products (catalyst 8197 as well as other improved types) [191a, 191b, 1429, 1433, 1975a]. It has been used moreover for different refining and desulphurisation purposes [189b, 482, 1009, 1080c, 1087, 1536, 1914, 1965, 2102a, 2193, 2228]. Sepiolite [496] and aluminosilicate [109, 393, 1085] were also employed as carriers for this catalyst.

The use of the catalyst $FeS–MoS_2$ (supported on $Al_2O_3–TiO_2–SiO_2$) has also been described [942].

Besides refining catalysts with two active sulphidic components, three-component combinations of sulphides and oxides of metals of Groups VI and VIII have been developed recently, particularly for use in medium-pressure refining. This particularly applies to the Co–Mo catalyst, in which cobalt is partly replaced by nickel. With this secondary promotion effect there also exists an optimum composition, which gives a catalyst of maximum activity similar to the case of two-component catalysts [154, 535, 603a]. The optimum commercial desulphurisation catalyst of the Davison Chemical Co. contains 1.0% Co, 0.5% Ni and 8.3% Mo supported on Al_2O_3 [1743]. Beuther et al. [153] state the optimum atomic ratio for an industrial catalyst to be Ni : Co : Mo = 1 : 2 : 10 with a total metal content of 9 to 12% [795]. The composition of the German refining catalyst BR 86 is rather similar, only the Co content is a little higher (cf. Table 3) [1429]. Furthermore, the use of this catalyst has been described for the hydrorefining of lubricating oils (catalyst G-76 by the Girdler Co.) [67], petroleum feed before hydrocracking [536] and benzene obtained by dealkylating alkylaromatic compounds [689].

Another technically important combination is that of Mo, W and Ni sulphides as used e.g. in the German refining catalyst U 63 (cf. Table 3) [1429] or in the catalyst for medium-pressure refining of gasoline [138]. The three-component catalyst Co–Mo–Fe is used in the Ferrofining refining process (cf. p. 342).

7.2.3.2.

Raw materials for hydrorefining, and hydrorefining processes

The great variety of materials treated in hydrorefining processes led to the development of a number of technological procedures designed to achieve a certain refining effect, and often differing considerably in terms of reaction conditions and industrial arrangement as well as in the required extent of refining. A survey of the most important hydrorefining and hydrodesulphurisation processes will be found in Table 79.

We shall now briefly discuss the individual types of hydrorefining processes in which sulphide catalysts are used to speed up the reaction and to achieve the required degree of selectivity.

7.2.3.2.1.

Desulphurisation of gases

Industrial gases often contain, in addition to H_2S, other sulphur compounds (particularly CS_2, thiophene and its homologues, carbonyl sulphide, mercaptans, sulphides and other substances), which cannot be removed by means of conventional purification procedures. In some cases, particularly in the production of synthesis gases, the gas has to be purified so as to leave only very low concentrations (e.g. below 2g/1,000 Nm^3). One of the most efficient methods is catalytic hydrogenation of organic sulphur compounds to H_2S and absorption of H_2S either on the desulphurisation catalyst or in a sponge. Catalytic hydrogenation of carbon disulphide in town gas was tested for the first time by Carpenter et al., using nickel sulphide at about 400 °C [270, 271, 502]. This method was later also used in England on an industrial scale [781, 1605, 1652]. Thiomolybdates are highly efficient catalysts for removing organic sulphur compounds from town gas, particularly Co [1334, 1335, 1337] and Cu [1675] thiomolybdates; and this also applies to Mo and Ni sulphides [1018] and copper and chromium sulphides [235, 447, 1016, 1017]. Desulphurisation was carried out at atmospheric pressure, at temperatures usually of 250 to 350 °C and with a sp ace velocity up to 5000 l/l hour [1335]. Organic sulphur has also been removed on Ni sulphides from gases other than town gas, e.g. water gas [666], coke oven gas [499] and other fuel gases [665,1454, 1781] or from C_4 hydrocarbons [904a, 1921a]. FeS has also been used for this purpose [1455]. For desulphurising synthesis gases to remove carbonyl sulphide, mercaptans, sulphides and thiophene, Kono et al. [1069, 1070, 1071, 1072] tested sulphides and oxides of Ni and Co as well as of other metals.

With suitably selected reaction conditions and catalysts (catalysts containing particularly Ni, Fe and Cr sulphides) catalytic desulphurisation of olefinic gases is also possible [527, 529, 1237, 1773].

7.2.3.2.2.

Hydrorefining and hydrodesulphurisation of gasolines

Straight-run gasolines

Hydrorefining and desulphurisation of straight-run gasolines is carried out mainly in order to improve the octane number, to remove mercaptans and other sulphur compounds and to improve lead susceptibility and sensitivity. Conventionally employed reaction conditions are rather mild (temperatures of 300 to 400 °C, pressures up to 50 atm, LHSV 1 to 7 l/l/hour). The degree of sulphur removal is 98% and better, the product contains no mercaptans, its sensitivity and susceptibility to tetraethyllead increases considerably and the volume yield is around 100%. The refining processes are either based on external hydrogen sources (e.g. Hydrofining, Diesulforming, cf. Table 79) or, the process is carried out under partly dehydrogenating conditions (temperature 400 to 430 °C, pressures 7 to 14 atm with liberation of the excess circulating gas, LHSV ~5) by the Autofining process (for a scheme of the process, see Fig. 87) [61]. Most frequently, especially on the industrial scale, primary gasolines from petroleum have been refined on Co–Mo catalysts [394, 626, 1058, 1240, 1347, 1743, 2126]. Other sulphides of metals of Groups VI and

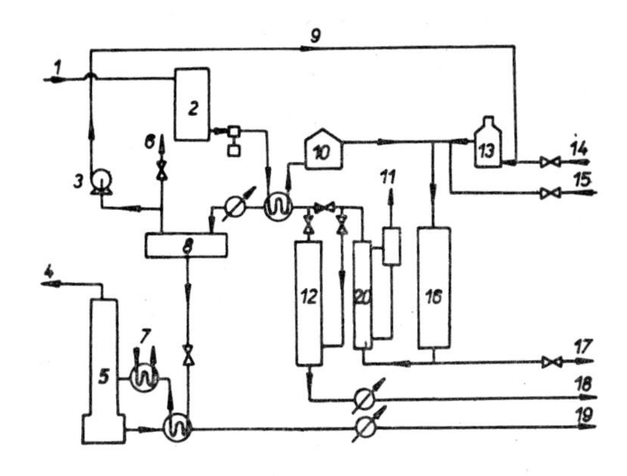

Fig. 87. Scheme of the Autofining process [61]. 1 — Feed. 2 — Feed coalescer. 3 — Recycle gas compressor. 4 — Fuel gas or gas to sulphur recovery. 5 — Stabilisation column. 6 — Fuel gas. 7 — Steam heater. 8 — Product accumulator. 9 — Recycle gas (80% H_2). 10 — Oil heater. 11 — Steam. 12 — Separation column. 13 — Gas heater. 14 — Regeneration steam. 15 — Air. 16 — Reactor. 17 — Gases from catalyst regeneration. 18 — Residue (fuel oil). 19 — Product to storage. 20 — Waste heat boiler.

VIII were also used [106, 705, 1238], particularly mixed WS_2–NiS catalysts [2, 626, 1058] or NiS–WS_2–MoS_2 [138] and furthermore Ni [309a, 923, 1059, 1310, 1949] and Mo sulphides [1405, 1700, 1701].

Heavy straight-run gasolines with higher content of sulphur, nitrogen and other impurities (e.g. arsenic compounds in the case of some raw materials [1743]) must be hydrorefined before reforming on Pt catalyst. Co–Mo refining catalysts [597, 1299, 1665, 1743] are again most often used, as well as WS_2–NiS catalyst [2021], MoS_2 [1772] and occasionally sulphides of other metals (e.g. Fe, Sn) [883]. Hydrorefining of different technical gasoline types mainly involves sulphur removal, improvement of colour and deodorisation. Besides the Co–Mo catalyst [1347] the catalyst 8376 has also been used [2271].

Refining of cracked gasolines

Cracked gasolines contain, besides sulphur compounds, a high concentration of unsaturated compounds, among which particularly diolefins, and some olefins, are frequently the cause of formation of resinous and gum-like substances and darkening of the product. Furthermore they cause deposits to form in the combustion spaces of engines, low stability on storage and difficulties in later catalytic processes, especially reforming.

In hydrorefining under non-selective conditions, sulphur compounds are removed and unsaturated hydrocarbons are hydrogenated. In this way, stable components of aviation gasolines were obtained from cracked gasolines by means of procedures developed by the Shell Oil Co. (catalyst WS_2–NiS) [1, 321] and the Standard Oil Dev. Co. (catalyst not specified) [2145]. When the WS_2–NiS catalyst was used, light cracked gasolines were processed at 340 °C, at a pressure of 50 atm and with a space velocity of about 10 l/l hour (the molar hydrogen excess was 7) [1347]. Krönig [1101] also mentions the possibility of using other refining catalysts, e.g. a DHD-type catalyst. Hydrorefining with substantial desulphurisation and hydrogenation of olefins is also very important in cases where the cracked gasoline thus modified is to be used as feed for reforming. In all technically important processes, cobalt–molybdenum refining catalysts are used. Materials especially suitable to this purpose are gasolines from thermal cracking, visbreaking and particularly from coking, as they contain a large amount of cyclic olefins which form aromatics when the pre-refined raw material is reformed [1347].

The hydrorefining processes mentioned have the great disadvantage that olefins are hydrogenated by non-selective refining of cracked gasolines due to which the octane number is decreased to an unwanted extent [1347].

Cracked gasolines may be refined, however, in a selective manner in the presence of sulphide catalysts so as to remove diolefins completely and desulphurise adequately while hydrogenating the valuable olefins to a minimum extent only. The selectivity of the process is improved initially by the presence of sulphur compounds in the feed (cf. pp. 89 and 120), since the rate of hydrogenation of olefins is distinctly lower in the presence of sulphur compounds [321, 777, 971, 1348], while conversion of diolefins to monoolefins is not influenced substantially under the reaction conditions employed. Selectivity may be further increased, especially by a suitable selection of the reaction conditions [2205]. The process is usually carried out at low pressures (starting from 5–10 atm) [121, 274, 1559], with high space velocities of the feed (up to 25–30 l/l hour) [330, 1949] and with various hydrogen to feed ratios [330]. The temperature varies over rather a wide range from 200 to 425 °C [330, 865, 1949] (exceptionally, in the 100 to 200 °C range also [743]) according to the selectivity of the catalyst employed [473].

The course of controlled refining is influenced to a great extent by the catalyst, which has to be sufficiently selective. Therefore, single-component sulphide catalysts with low selectivity are used in exceptional cases only [883, 1366, 1645], or their selectivity is enhanced by a special method of preparation or by supporting them on a suitable carrier (particularly with NiS and MoS_2) [223, 224, 353, 354, 585, 743, 782, 1402, 1803, 1870]. The secondary promotion effect of some substances on standard refining catalysts has been utilized in order to increase their selectivity. Remarkable is the increase in the desulphurisation selectivity of the Co–Mo catalyst under the influence of potassium salts (cf. Table 80) [467, 473, 1494].

Mixed sulphidic catalysts, particularly sulphides of Groups VI and VIII metals, have been used for selective hydrogenation of cracked gasolines [324, 328, 625, 1248, 1559, 1966, 2090]. The most conventional catalyst employed in medium-pressure selective refining of cracked gasoline is the unsupported WS_2–NiS catalyst [121, 122, 274, 321, 327, 330, 772, 776, 777, 1507, 1949] which was also used in the refining process developed by the Shell Dev. Co. [1].

Table 80.

Influence of Potassium on Increasing the Hydrodesulphurisation Selectivity of the Co–Mo Catalyst in Hydrorefining of Cracked Gasoline (the two catalysts were compared under optimum conditions for each catalyst) [1347].

% Desulphurisation	Non-hydrogenated olefins (%)	
	Co–Mo/bauxite	Co–Mo/bauxite promoted with potassium[a])
40	87	97
50	80	91
60	72	85
70	62	78
80	51	72

[a]) Weight ratio bauxite : Mo : Co : K = 100 : 6.5 : 0.8 : 0.4; the composition of the non-promoted catalyst was the same but no potassium was present.

Selective hydrogenation of pyrolysis gasoline

Pyrolysis gasoline is a by-product in the production of gaseous olefins by thermal splitting of straight-run gasoline or gaseous hydrocarbons. It is a very complicated mixture, mainly composed of aromatic and unsaturated hydrocarbons together with a small amount of sulphur, nitrogen and oxygen compounds. Very good high-octane gasoline can be prepared from this material, or pure aromatic compounds and diolefins may be isolated from it.

The octane number of pyrolysis gasoline is high (motor and research octane number 80 and 93, resp.), but it must be selectively refined before being added to motor fuels i.e. the main objective is to hydrogenate reactive diolefins. In addition to commercial processes for selective hydrogenation in the liquid phase, the British petroleum Co. process (nickel catalyst), the Kalthydrierung process (palladium catalyst [1102]) and the newly developed processes, e.g. HPG by the Houdry Process Corp. [1973] and

Kellogg-Hydrotreating [667, 2170], the application of sulphide catalysts to this purpose has also been tested. According to Krönig [1102] a part of the noble metal supported on the carrier could be kept in the sulphide form in the Kalthydrierung process. Soviet authors prefer to presulphide palladium catalyst [134b, 1712a]. In medium-pressure hydrogenation (30 to 100 atm) the catalyst 8376 was used [435, 631, 632, 632a, 633, 1235b]. Diolefins were found to be easily hydrogenated (the reaction taking place at 120 °C and being complete at 200 °C), but the catalyst selectivity is low. With suitable reaction conditions, however, diolefins can be removed completely (temperature 180 °C, pressure 70 atm, LHSV 1), while the octane number is maintained at the original value (81 by the motor method) [631]. A sulphidic Ni–W catalyst supported on Al_2O_3–SiO_2 [435a] and, furthermore, nickel sulphide [1219a] and sulphurised Co–Mo catalyst [412a] were also tested. The sulphidic nickel–tungsten catalyst was also found useful in hydrorefining of pyrolysis gasoline with the purpose of producing a maximum yield of benzene [999, 2170]. In a French process, the same purpose is served by a catalyst based on Groups VI and VIII sulphides [356a].

Hydrorefining of tar and shale gasolines

Highly unsaturated gasolines obtained in low-temperature carbonisation of coal and bituminous shales, which have a high content of sulphur, nitrogen and oxygen substances, are not hydrorefined as a rule. The hydrorefining is usually carried out when the tar fraction is hydrogenated as a whole (e.g. the whole fraction of b.p. up to 325 °C is treated by prehydrogenation before hydrocracking, processing of tars by the TTH process, etc.) and the refined gasoline is isolated from the hydrogenation product by fractionation [1101].

Sulphide catalysts have been used, however, for refining tar and shale gasolines in a number of research studies. For example, the catalyst 8376 was used for selective hydrorefining of low-temperature carbonisation tar [1714] or, for its complete hydrogenation for the purpose of identifying the carbon skeleton of the substances of which this gasoline is composed [694]. This catalyst was also used for hydrorefining gasoline obtained by carbonisation of shales [443]. It was found in this case that the temperature dependence of desulphurisation and hydrogenation of olefins are parallel. Substances very difficult to remove from shale gasoline are thiophene derivatives and, therefore, the process must be carried out in the presence of sulphide catalysts at elevated temperatures; therefore olefins are hydrogenated to a substantial extent and the octane number decreases [721]. When low hydrogen pressures and suitable temperatures are employed, the MoS_2/Al_2O_3 catalyst may be applied, in which case desulphurisation and hydrogenation of diolefins is accompanied by partial dehydrogenation, by means of which aromatic compounds are formed [417]. The MoS_2 catalyst was also used by other authors for hydrorefining shale gasolines

[767, 1443, 1444]. Nonnenmacher et al. [1469] state that hydrorefining of gasolines obtained by carbonisation and gasification of coal may be based on a technology developed for refining crude benzol.

7.2.3.2.3.

Hydrorefining of crude benzol

The disadvantages and uneconomical character of refining benzol with acids led to a rapid development of hydrorefining processes after World War II, the main advantages, i.e., considerably decreased loss of benzene during refining and elimination of the acid sludge which is difficult to dispose of, being particularly important when large amounts of this raw material are being treated.

Benzol from coking plants contains, besides benzene and its homologues, a number of other components, mainly unsaturated hydrocarbons (e.g., cyclopentadiene, styrene), oxygenous (phenols, coumarone, ketones), nitrogenous (pyridine and its homologues) and sulphur compounds (particularly thiophene). These impurities have to be removed, while the aromatic ring is hydrogenated at the same time to a minimum extent only [1393], irrespective of whether the benzene is to be used as a component of motor fuels or as a raw material for further chemical processing.

The problem of a suitable catalyst with a sufficient refining selectivity was solved by the introduction of sulphide catalysts [852, 1101, 1469]. Sufficiently active catalysts with high selectivity are mostly combinations of Groups VI and VIII metal sulphides [689, 1473, 1608, 1965], particularly Co–Mo catalysts [94, 486, 487, 682, 835, 1239, 1267, 1295, 1469, 1713]. Nickel sulphide was also used [704, 1466], particularly in mixture with WS_2 [325, 1965] (the latter catalyst is also suitable for hydrorefining xylenes [1916]), although the selectivity of this combination is lower due to the increased hydrogenating activity [486, 487]. In the technical arrangement of this refining process, the catalyst 3510 was also found useful (MoO_3–ZnO–MgO in the fresh state) [1101]. Other sulphides too were tested during the development of this process, e.g., Mo sulphide [268, 1393], W sulphide [681, 877], Co sulphide [315, 1751] and most recently, sulphides of Groups V, VI and VIII [1825] as well as noble metal sulphides [1100a].

In the practical application of this refining method (e.g., the processes BASF-Scholven, Lurgi, Koppers, Newton Chambers, Hydrofining developed by the Esso Res. Eng., etc.) the optimum reaction conditions are: hydrogen pressure 20 to 60 atm and temperature around 350 °C [269, 1101, 1267, 1469, 2098, 2101]. The optimum operating pressure, however, depends on the boiling point of the benzol which is to be refined. Higher pressures are obviously needed with heavy benzol [1101]. Either pure hydrogen or coke oven gas is used (Table 81). Recent development in the technology of benzol hydrorefining is the Houdry Litol process [1246], which enables either substantial refining to produce pure aromatic hydrocarbons or, a maximum benzene yield may be achieved by a combination of refining, hydrocracking and dealkylation reactions (the catalyst composition has not been published).

Table 81.

Properties of Hydrorefined Benzene [1469].

Property	Crude benzol	Refined benzol	
		Refined with coke oven gas at 40 atm pressure (partial H_2 pressure 18 atm)	Refined with H_2 at 20 atm pressure (partial H_2 pressure 16 atm)
Density (at 15 °C)	0.879	0.876	0.876
Bromine number (g/100 ml)	13	0.06	0.06
Content of non-aromatic substances (vol. %)	1.5	2.0	2.0
Total sulphur (% by wt.)[a]	0.38	0.003	0.002

[a]) The refined benzol contained no active or mercaptan sulphur, the thiophene content of the benzene fraction was 0.001% by wt.

Besides selective refining of aromatic hydrocarbons, sulphide catalysts were also used for hydrorefining aromatic compounds to the respective naphthenes, especially to prepare pure cyclohexane by hydrogenating benzene containing sulphur compounds, and by hydrogenation of naphthalene (containing benzothiophene) to decalin (cf. p. 132).

7.2.3.2.4.

Hydrorefining of middle petroleum and tar distillates

Hydrorefining and hydrodesulphurisation of this category of raw materials in the presence of sulphide catalysts serves for the final treatment of high-quality motor fuels as well as for pretreatment of low-quality materials before their further processing (e.g., by means of cracking). Since large amounts of materials are involved, the demands on the quality, lifetime and economy of production of the refining catalysts are very great.

Refining of middle tar and shale oils

Prehydrogenation of tars

The main purpose of prehydrogenation (Vorhydrierung) in hydroprocessing of tars is to pretreat the feed to be used in the hydrocracking stage. Besides hydrorefining reactions, however, splitting of middle oils to gasoline also takes place in prehydrogenation [1722]. The catalyst 5058 which was originally used for this purpose was highly efficient either in hydrogenation and refining, or splitting (the gasoline formed was saturated and had a low octane number). Therefore, this catalyst was replaced by the more selective type 8376, which causes considerably less splitting while its refining activity is high and its hydrogenating efficiency is somewhat less.

The prehydrogenation stage [416], for which the technological scheme is similar to the vapour phase hydrocracking (cf. Fig. 83) and in which hydrocarbon fractions up to 325 °C were treated [1160, 1172], was usually operated at temperatures of about 400 °C and pressures close to 300 atm. The space velocity of the feed was 0.6 to 1.0 m^3/m^3 hour and the ratio of circulating gas to feed 2500 to 4000 Nm^3/t [1101, 1172]. The selectivity of the catalyst 8376 rises with decreasing pressure, i.e., the reaction rate of hydrogenation of aromatic substances decreases while hydrogenation of olefins is practically complete [1124d]. The life-time of the catalyst 8376 is very long, but its activity decreases with time. This decrease must be compensated for by a gradual rise in the temperature (e.g., after 10,000 hours the catalyst required a refining temperature of 425 °C [1172].)The deactivating influence of arsenic on this catalyst was discussed earlier (p. 114).

The DHR process (Druck-Hydrogenium-Raffination) was developed in Germany from the prehydrogenation technology upon conversion to petroleum as basic raw material. This process was applied to the medium-pressure refining of light oil, diesel fuels and jet fuels. It moreover served as a hydrorefining step before the DHC hydrocracking process (p. 307) [794].

The TTH and MTH processes

The use of fixed-bed catalysts in hydrotreating raw materials which contain asphaltenes and resins is rendered very difficult by the fact that the splitting products of these substances polymerise rapidly to form coke-like deposits which deactivate the catalyst surface.

When, however, the labile and easily condensed substances are first converted to stable hydrocarbons under very mild reducing and refining conditions in the liquid phase, the temperature can be gradually raised up to the value needed to achieve the full activity (refining or hydrocracking) of the catalyst. This principle was applied practically in the TTH process (Tieftemperaturhydrierung) [793, 1101, 1731, 1811, 2212]. This process was used to treat tar, including the lighter fractions (middle oil and part of the light oil) at 300 to 320 atm pressure (later at medium pressures also [1172]) and space velocities up to 1 l/l hour [204, 1101]. The hydrogen consumption varies from 500 to 550 Nm^3/t. The reaction temperature of the TTH process depends on the catalyst employed. The catalyst 5058 may be used [693] and in the presence of this the maximum temperature is 375 – 380 °C or, the refining catalysts 8376 (WS_2–NiS–Al_2O_3) or 8197 (MoS_2–NiS–Al_2O_3), which necessitate a temperature some 40 °C higher [793], may be employed. This process is accompanied by a minimum degree of splitting [1101]. The fixed catalyst in the TTH process must be protected from the danger of contamination by solid substances and ash from the feed [1101, 1811] (cf. p. 111).

The TTH process was used to pretreat low-temperature tar with a slight asphalt content (below 3%) [1101, 2212], refined gasoline with a low octane number being obtained (58 Motor Octane Number only) together with high-quality diesel fuel, paraffin wax and spindle and machine oil [1101, 1172]. Other suitable raw materials are oils from shales [1101, 1731], from which very good gas oil and spindle and machine oils with excellent properties can be obtained.

The TTH process can be applied with advantage to processing petroleum materials, particularly for refining lubricating oils produced from low-quality petroleum. The catalysts applied may be 5058, 8376 or Co–Mo [1172] (cf. Chapter 7.2.3.2.5). A particular disadvantage of this process is the considerably more rapid deactivation of the catalyst than occurs in the prehydrogenation phase [2135].

When TTH-hydrogenation is carried out with a lower space velocity (about 0.6 l/l hour) and the final temperature is raised by some 40 °C, the reaction takes place in the mixed phase in the final stages of the process and the splitting activity of the catalyst becomes more effective [1101, 2212]. In this MTH process (Mittel-Temperatur-Hydrierung) the same catalysts were used and similar raw materials were treated as in the TTH process, and due to the higher degree of hydrocracking, the content of gasoline and diesel fuel was higher (and more gas was formed) at the expense of the formation of higher-boiling products.

A number of research studies in the field of hydrorefining of tar and petroleum products in the presence of sulphide catalysts (particularly when done in autoclaves) were also performed in principle under the conditions of the TTH and MTH processes [34, 204, 515, 1189, 1285, 1616, 1793, 2066].

Refining of other middle tar and shale products

Sulphide catalysts found an important field of application in hydrorefining medium tar and shale oils to high-quality diesel fuel [204, 1101, 1379, 1381], jet fuel [1101] and illuminating kerosene [1101, 1382, 1382b]. Refining of the kogasin fraction (obtained by Fischer-Tropsch synthesis) to mepasin used to produce alkyl aryl sulphonates was formerly significant [500, 1101]. The catalysts 5058 or 8376 were conventionally employed in these hydrorefining processes, though a mixture of Group VI sulphides was also used [204] as well as a sulphurised DHD-type catalyst (10 to 12% MoO_3 on γ-Al_2O_3) [179, 1468]. The use of Pt as promoter for sulphidic catalysts has also been described [1218].

The most important problem involved in refining shale products in the presence of sulphide catalysts is the efficient removal of nitrogenous substances, the concentration of which is rather high in shale-derived materials. Together with denitrogenation, considerably easier desulphurisation takes place. Kalechits et al. [307, 308] studied the activity of a number of catalysts, finding that the refining activity decreases in the series: MoS_2 > WS_2 > 8376 > Co–Mo/Al_2O_3 > MoS_2/activated carbon. It was found that nitrogen removal takes place as a first-order reaction in the presence of the catalyst MoS_2 [305, 308], the reaction rate of hydrodenitrogenation being directly proportional to the pressure (in the 10 to 260 atm range). The relationship between the velocity constant (k'), the reciprocal of the space velocity (W/F), pressure (P) and degree of nitrogen removal from the initial value N_0 to the final N(%) is expressed by the equation:

$$k'\left(\frac{W}{F}\right)P = \log\frac{N_0}{N}\,. \qquad \text{(CXXXIX)}$$

From this relationship it follows that the increase in space velocity may be compensated by a proportional increase in the hydrogen pressure at a given temperature and with a required degree of denitrogenation [307]. Deactivation of catalysts is caused by gradual surface contamination as well as by preferential chemisorption of nitrogen compounds on the active centres [305, 308], and therefore it is advantageous to hydrorefine shale oils at higher pressures, or to work with recirculation of the refined product [307]. The catalyst MoS_2 (in some cases supported on activated carbon), was used in a number of other studies dealing with refining shale oils [515, 820, 1285, 1793]. Other catalysts which have been used for this purpose are WS_2 [1853, 2066], WS_2–NiS [1922, 1923], catalyst 8376 [1101, 1227], Co–Mo catalyst [357, 1328, 1329] and chromium sulphomolybdate [174, 212]. Montgomery [1413a] employed catalysts presulphided in a special manner for hydrostabilisation and hydrodenitrogenation of shale oils.

Refining of middle and heavy petroleum distillates

In contrast to tar and shale materials, middle petroleum distillates are usually hydrorefined by means of low- or medium-pressure processes. For efficient hydrodesulphurisation of the feed, as well as for hydrogenolysis of nitrogen and oxygen compounds and for removing metals (e.g., Ni, V, Cu, As), relatively mild reaction conditions are sufficient.

Hydrorefining of kerosenes

Refined kerosene should be light in colour, have an agreeable smell, a low tendency to smoke and the lowest possible char value. Mild reaction conditions, as in the case of gasolines, are usually employed in hydrorefining, for example in refining by the Unifining process [680]. Sulphide catalysts have been used in low-pressure refining of straight-run kerosene [359, 1644, 2103E], including MoS_2 [582, 1405]. The possibility of refining kerosene with the aid of the Autofining process on the

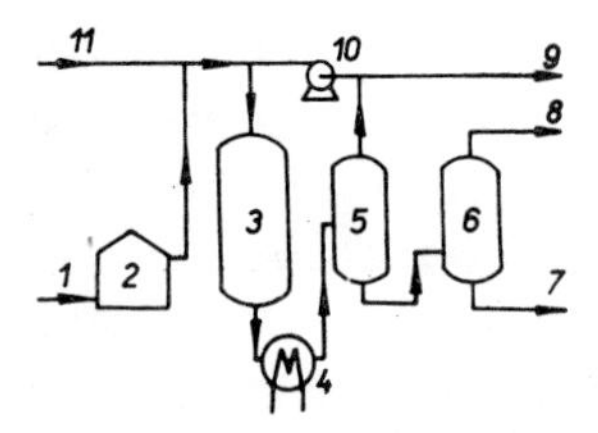

Fig. 88. Simplified scheme of the Hydrofining process (Esso Research and Eng. Co.) for hydrorefining jet fuels [64]. 1 — Feed. 2 — Furnace. 3 — Reactor. 4 — Cooler. 5 — Separator. 6 — Stripper. 7 — Product. 8 — Residual H_2S. 9 — Waste gas. 10 — Hydrogen recycle. 11 — Make-up hydrogen.

Co–Mo catalyst [1659, 1667] or WS_2–NiS catalyst [802] has also been described.

Kerosenes with a higher content of aromatic compounds have to be hydrorefined at higher pressures (50 to 300 atm) [1988] with more hydrogenation-efficient catalysts, particularly WS_2–NiS and the catalyst 8376 [2c, 431, 432, 1293a, 1887]. High-quality products are also required for the production of jet fuels (p. 305). From the point of view of hydrorefining it is important to achieve the required degree of desulphurisation (particularly to reduce the content of mercaptan sulphur to a very low value) and of demetallisation, removal of all components which might leave solid residues on combustion and removal of all gum-forming substances. The Hydrofining process is used for example in the production of jet fuels [269] (for a scheme, see Fig. 88), working with a cobalt–molybdenum catalyst. The catalyst NiS–MoS_2–Al_2O_3 [266, 1087] and WS_2 or 8376 [1293] were also tried out. Le Page et al. [1219c] criticised the application of sulphide catalysts from the point of view of hydrogenation of aromatic substances.

Refining of gas oils

The main purpose of refining these products is to obtain high-quality diesel fuels. The primary objective is desulphurisation of the raw material to the required degree (0.2 to 0.4%, down to 0.02% total sulphur in the case of special-purpose fuels), in order to remove the corrosive character of the combustion products and to improve stability on storage. Another objective is to improve the blending properties, remove coke-forming substances and acid components (naphthenic acids), enhance stability and improve the cetane number of the fuel. Straight-run gas oils are usually refined at pressures of less than 70 atm. Higher pressures (200 to 300 atm) are required for materials which contain catalytic recycle oils. In this case aromatic substances must be saturated since naphthenes are substantially better diesel fuels [1289, 1291, 1292, 1347, 1762, 2135].

The operating temperatures usually vary in the 300 to 400 °C range, with space velocities from 0.5 to 3 l/l hour. Some problems related to the desulphurisation of these materials have been discussed earlier (pp. 238 and further).

Table 82.

Refining of Diesel Fuel on Co–Mo/Al_2O_3 Catalyst by the Ultrafining Process. (The feed was crude diesel fuel, a mixture of 84% circulating oil from catalytic cracking and 16% gas oil from coking.) [64].

Property		Feed	Product
Specific gravity		0.871	0.859
ASTM distillation (°C)	Initial boiling point	187°	190°
	10%	218°	218°
	50%	257°	254°
	90%	294°	295°
	End point	330°	331°
S content (% by wt.)		1.1	0.09
Yield (% by vol.)		—	95.50
Gosoline to 205° (% by vol.)		—	6.10
Floc gum (mg/100 ml)		83	20
Cetane number		36.7	40.8

Cobalt–molybdenum catalysts are mainly used to refine diesel fuels and gas oils. For example in the Ultrafining process (Standard Oil Co., Ind.) the results given in Table 82 are achieved. A scheme of the process is shown in Fig. 89 [64].

Other processes employed to refine engine fuel and gas oils are Hydrofining [269], Unifining [2169] and the hydrodesulphurisation process by the M. W. Kellogg Co. [61]. Similar processes were developed in a number of other countries, e.g., in Germany (Lurgi) [1267], the USSR (e.g., the process L-24-6 [1963, 1981], and in Czechoslovakia [1117, 1963, 2129].

Co–Mo catalysts have also been used in a number of research and development studies, e.g., [137, 497, 791, 1011, 1012, 1128, 1132, 1439, 1440, 1504, 1595, 1656, 1762, 1885, 2022, 2023, 2024, 2027, 2029, 2030, 2109, 2129]. Catalysts which are more hydrogenation-efficient than Co–Mo catalysts have also been used, particularly for treating raw materials with a higher content of aromatic compounds and olefins. A greater degree of refining was achieved in their presence [1128, 1439], though this was sometimes at the expense of the yield of the refined product, since increased splitting of the raw material took place. WS_2 was used most often [762, 1128, 1132, 1287, 1289,

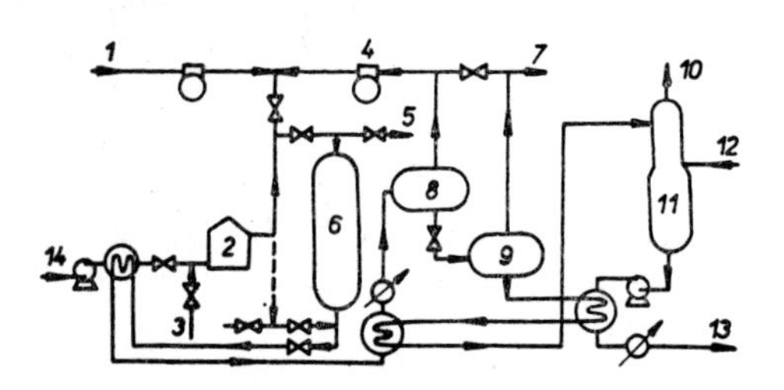

Fig. 89. Scheme of the Ultrafining process [64]. 1 — Hydrogen-rich gas. 2 — Preheat furnace. 3 — Steam. 4 — Recycle gas compressor. 5 — Regen. vent gas. 6 — Reactor. 7 — Fuel gas. 8 — High pressure separator. 9 — Low-pressure separator. 10 — Vent gas. 11 — Stripping column. 12 — Steam. 13 — Product. 14 — Raw material.

1291, 1292, 1439, 2027], but also employed were MoS_2 [137, 1194, 1484, 2068, 2109] and the catalyst WS_2–NiS (or 8376) [110b, 465b, 1214, 1762, 2030]. Recently, a sulphurised Ni–Mo catalyst was successfully applied [722b, 2012b, 2012d].

Table 83.

Comparison of Reaction Conditions Employed in Hydrodesulphurisation of Fuel Oils (Catalyst Co–Mo/Al_2O_3) [1347].

Process	Shell Trickle-Desulphurisation	Autofining	Union Oil Co–Mo Processing
Raw material	Light circulating oil from catalytic cracking	Straight-run gas oil	Heavy gas oil from coking
Raw material b. p. (°C)	230—342	208—356	400 (70%)
Sulphur content in the raw material (% by wt.)	2.90	0.72	4.15
Desulphurisation temperature (°C)	375	415	418
Operating pressure (atm)	52.5	7	31.5
LHSV (l/l/hour)	1.6	2.0	1.0
H_2 : oil ratio (Nm^3/m^3)	240	356	754
Length of catalyst hydrogenation cycle (hours)	2500	200	24 and more
Degree of desulphurisation (%)	89	50	89

Refining of light fuel oils

The main purpose of refining light fuel oils is desulphurisation to the required degree. Since these oils are formed from various raw materials, widely varying conditions are applied in their refining. Mc Kinley [1347] presents an illustrative comparison of three desulphurisation processes in which Co–Mo catalysts are employed to desulphurise different raw materials and to obtain high-quality fuel oil (Table 83). Besides the processes mentioned in Table 83, other refining processes based on Co–Mo catalysts are employed to improve the combustion properties of fuel oil and for desulphurisation [61, 62, 110a].

Light fuel oils contain various straight-run or cracked middle and heavy distillates. These components are stabilised by means of hydrorefining in order to avoid discoloration and the formation of sludges, and furthermore in order to improve the combustion properties and blending characteristics of the product and decrease its corrosivity as much as possible. With heavy distillates, adequate desulphurisation is accompanied by satisfactory denitrogenation [680].

Refining of raw materials for catalytic cracking and hydrocracking

The main purpose in refining these raw materials is the removal of catalyst poisons, e.g., nitrogen compounds and metals. Additional requirements are hydrogenolysis of sulphur- and oxygen-containing compounds and partial hydrogenation of polycyclic, aromatic compounds. Polycyclic aromatic compounds, which accumulate in the recycled gas oils in the course of catalytic cracking, are difficult to crack and cause coke-like deposits to form and have an inhibiting effect on the splitting reactions of paraffins and naphthenes [1521]. Total hydrogenation of all aromatic rings is not necessary, however, as partial hydrogenation with the retention of one aromatic ring is sufficient. The advantage of this selective hydrorefining method lies in a considerably lower hydrogen consumption and in the fact that cracked gasoline with a high octane number is obtained [691]. In general, more severe reaction conditions, particularly a higher pressure (70 to 210 atm), must be selected for hydrorefining raw materials which are to be used in catalytic cracking and hydrocracking. The temperature varies from 370 to 425 °C, while space velocities vary over a wide range up to 8 l/l hour [1347].

The most important problem is the removal of nitrogen compounds. The deactivating influence of basic nitrogen compounds on cracking catalysts increases with the rising aromatic character of the nitrogen compound, i.e., it decreases, e.g., in the series [2139]: acridine > quinoline > indole > pyridine > amines. At the same time, resistivity to hydrodenitrogenation also decreases in this series. Raw materials for hydrocracking, containing more than 100 ppm N, are usually prerefined before treatment [2227]. Either nitrogen is removed partially and a raw material of inferior quality is processed (e.g., a catalytic recycled oil with 60 ppm N is hydrogenated on the $NiS/SiO_2-Al_2O_3$ catalyst [1843]), or very efficient nitrogen removal is achieved by efficient refining. Hydrorefining must sometimes be carried out in two or three stages in order to achieve the required degree of denitrogenation [927].

Denitrogenation of lighter fractions (e.g., heavy gasoline) is quite an easy task [534]. The denitrogenation process rapidly becomes more difficult as the boiling point of the material rises. Fig. 90 shows the result of refining light fuel oil, containing 360 ppm nitrogen, obtained in catalytic cracking. To remove nitrogen down to 1 ppm, the process must be carried out at LHSV = 1.5, a hydrogen pressure of 70 atm and temperature of 370 °C. Increasing the temperature further would be inefficient, since the dehydrogenation equilibrium then starts to become important. This retards

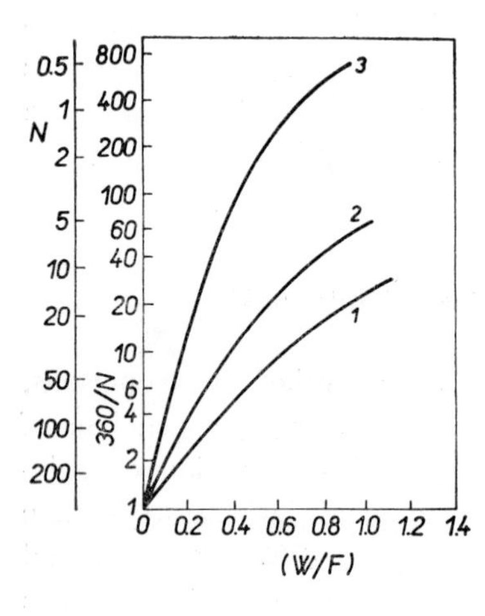

Fig. 90. Influence of pressure and temperature on hydrogenation of light fuel oil from catalytic cracking [534]. (Highly active Ni–W catalyst.) N — Nitrogen content (ppm) in hydrogenation product. (W/F)— Reciprocal value of space velocity (hour^{-1}). 1 — 35 atm, 343 °C. 2 — 70 atm, 343 °C. 3 — 70 atm, 371 °C.

the hydrogenation step needed to start the hydrogenolytic reaction of nitrogen compounds. A considerable deviation from a first-order reaction is observed in the case of denitrogenation of this raw material. The reason is that heavy fractions contain a variety of subtances, which are hydrogenolysed at widely different rates. As we see from the respective curves, a very small percentage of substances difficult to hydrogenolyse is responsible for the difficulty of complete hydrodenitrogenation [534]. Moreover, very high hydrogen pressures (up to 420 atm) are required for complete removal of nitrogen substances, e.g., in heavy vacuum gas oils or in residues.

The reason why it is so difficult to remove the last remnant of nitrogen compounds lies in the fact that the secondary products of hydrodenitrogenation are nitrogen compounds which are considerably more resistant than the original ones. Moreover the difficulty of nitrogen removal rises with the rising boiling point of the fraction refined, since the increasing size of the non-nitrogen part of the molecule lowers the rate of denitrogenation, probably because the catalytic surface is less readily accessible to a large molecule. In addition, the higher content of very stable high-molecular weight aromatic heterocycles contributes to the increasing resistance of nitrogen compounds in high-boiling fractions. Quantitative data relating to the decrease in the rate of hydrodenitrogenation with the rising boiling point of the material treated will be found in Fig. 91 [534].

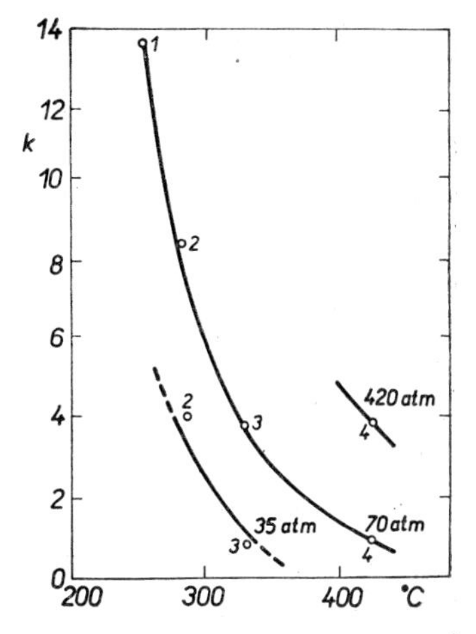

Fig. 91. Dependence of hydrodenitrogenation rate constant on the mean average boiling point of the raw material refined (reaction temperature 370 °C, highly active Ni–W catalyst) [534]. °C — Average b. p. of raw material. k — Pseudo first-order rate constant. 1 — Straight-run fuel oil. 2 — Fuel oil from catalytic cracking. 3 — Gas oil. 4 — Vacuum gas oil.

Processes used on the industrial scale to hydrorefine feeds for catalytic cracking mostly operate with cobalt–molybdenum catalysts. To remove nitrogen from middle petroleum distillates (particularly feeds for hydrocracking), cobalt–molybdenum [596, 725] and nickel–tungsten [415, 1213] catalysts are used. The type most often used, however, is nickel–molybdenum sulphidic catalysts which are characterised by a very low sensitivity to nitrogen compounds, lower than, e.g., Co–Mo catalysts [11, 780, 926, 927, 2078, 2193]. The use of Re_2S_7 as a catalyst in hydrodenitrogenation has also been described [480].

Sulphide catalysts have also been employed to remove metals from raw materials before catalytic cracking [491, 623].

7.2.3.2.5.

Hydrorefining of lubricating oils, petroleum waxes and ceresins

Hydrorefining of lubricating oils has a number of advantages compared to classical chemical or solvent refining. Firstly, a better quality product is achieved, the loss in refining is less, the process is easily adapted to changes in raw material quality, it is more economic and it is particularly suitable for treating materials obtained from sulphur-containing petroleum.

Hydrorefining in the presence of sulphide catalysts is mainly applied to straight-run vacuum oil distillates, and to deasphaltised residues, heavy cylinder oils and bright-stocks [160, 165, 608a], as well as to other materials which can be used to make good lubricating oil, e.g., synthetic oils obtained by polymerizing olefins [378], olefinic oil fractions from Fischer-Tropsch synthesis [580], petroleum extracts [2261], alkylates of recycle gas oils [332] and among the group of non-petroleum materials, e.g., shale oils [1801].

Mild hydrorefining of oils (hydrofinishing) frequently replaces final treatment with acid or clays. This process improves the colour and oxidation stability of the product, decreases the neutralisation number and particularly the content of hetero-atoms (O, N, S). The process is carried out under mild conditions, so that hydro-cracking and extensive hydrogenation of aromatic substances does not take place. The viscosity index is increased only slightly. This process is characterised by a long catalyst regeneration cycle (6 months and more) [158, 1347]. Co–Mo catalysts are used most often in industrial practice [578, 1009, 1352a], e.g., in the Comofining [343], Hydrofining (Esso) [64, 1743] and other processes. The application of WS_2 was also described [1544a].

The Co–Mo–Fe/Al_2O_3 catalyst with a long lifetime of 2.5 years is used in the Ferrofining process (British Petroleum Co. Ltd.) for mild hydrotreating [64, 370, 1821]. Keil et al. [1009] found, on comparing the refining effect of the catalysts Co–Mo/Al_2O_3 (catalyst 7362) and Ni–Mo/Al_2O_3 (catalyst 8197), that the 8197 type is more efficient in hydrofinishing pre-hydrogenated transformer oils under medium pressures (75 atm). For hydrofinishing lubricating oil, the MoS_2 catalyst was also tested, either pure [333, 512, 1200] or supported on Al_2O_3 [1979] or on activated carbon [514]. The use of nickel sulphide supported on SiO_2 has also been described [1150].

More severe refining of lubricating oil (hydrotreating) is a hydrogenation process, in which efficient refining (hydrogenation of olefins and aromatic compounds, hydrogenolysis of sulphur, oxygen and nitrogen compounds) is accompanied to a certain extent by hydrocracking and isomerisation, so that the viscosity of the re-fined oil decreases somewhat at the same time [158, 1101]. This process is utilised for treating deasphalted raw materials or distillates with no preceding refining by means of solvents or acid. The reaction conditions are more severe than in the case of hydrofinishing (temperature ~340 to 410 °C, pressures up to 320 atm). Hydro-treating is characterised by a substantial increase in the viscosity index (VI) due to efficient hydrogenation of aromatic compounds with partial splitting of polycyclic naphthenes [158, 605].

R_1, R_2 (V.I. ~ −60) $\xrightarrow{[H_2]}$ R_1, R_2 (V.I. ~ 20) $\xrightarrow{[H_2]}$

→ R_3, R_4, R_5, R_6 or R_7, R_8, R_9 etc.

V.I. ~125−140

Since a number of simultaneous reactions take place in hydrotreating, the properties of the refined product will depend to a large extent on the selectivity of the sulphide catalyst employed. Novák and Kubička [1124, 1480] compared the properties of the hydrogenation product from straight run distillate of Romashkino petroleum for which four different sulphide catalysts have been used (Fig. 92). The lowest decrease in viscosity with the achievement of the same viscosity index was recorded in the case of the catalyst 5058. The other properties of the product (colour, Conradson carbon residue, oxidation stability) were also most satisfactory for the oil refined with this catalyst. This same catalyst was found to be the most efficient one, as it permits the reaction temperature to be decreased by 10 to 30 °C compared to other catalysts, while the same

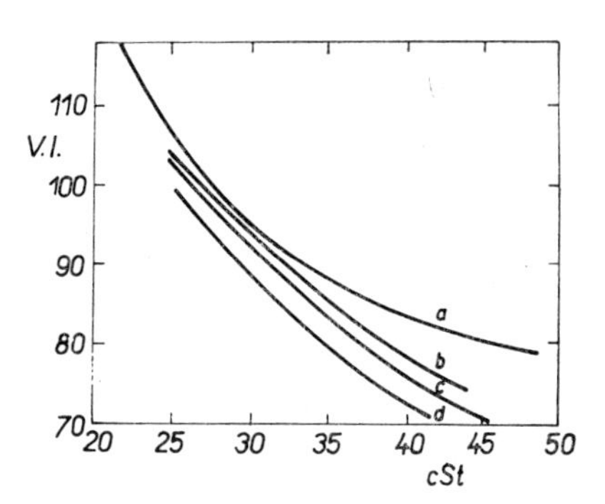

Fig. 92. Dependence of viscosity index and viscosity of oil hydrogenated on the following catalysts: *a* — 5058. *b* — 8376 + 5058 (3 : 1). *c* — 8376. *d* — 7362 (Co–Mo/Al_2O_3). V. I. — Viscosity index. cSt — Viscosity at 50 °C. (The initial oil was a vacuum distillate of Romashkino petroleum, with a viscosity of 7.5 cSt/100 °C [1480].)

viscosity index is achieved. The catalyst 5058 was used for hydrotreating of oils in a number of other cases, too [1008, 1131, 1290, 1477, 1480, 1814, 1815]. An obvious disadvantage is the high price and difficult regeneration of this catalyst. Therefore, a combination of the catalysts 5058 and 8376 (ratio of 1 : 3) was successfully used [1476, 1812, 1813, 1814].

Ni–W sulphide catalyst are frequently used for hydrotreating of lubricating oils, either non-supported [133, 158, 169, 331, 332, 400, 400a, 1281, 1614, 2150, 2153] or type 8376 catalysts [2b, 54a, 166, 222, 490, 517, 759, 760a, 760b, 1477, 1478, 1479, 1480, 1483, 1801, 1815, 1817]. Beuther et al. [161, 163, 164, 170, 606, 606A] used WS_2–NiS or CoS–MoS_2 catalyst supported on Al_2O_3–SiO_2. Activation of an aluminosilicate carrier with zirconium dioxide has also been described [451]. The successful aplication of cobalt–molybdenum catalysts in hydrorefining lubricating oils was reported by Agafonov et al. [8] and by other authors [131a, 1352a, 1436b, 2228b]. Type 8197 catalysts (MoS–NiS/Al_2O_3) [94b, 580a, 1009, 1725a, 1989a], sometimes with an SiO_2- and Fe_2O_3-activated carrier [289], and further Ni sulphides [84, 1875] or a mixed catalyst containing Pd and Mo sulphides supported on Al_2O_3–SiO_2 [301A] have also been employed.

Hydrogenation of n-paraffins, petroleum waxes and ceresins is similar to that of lubricating oils. The catalyst most often used to this purpose is a mixed Ni–W sulphide catalyst (non-supported or the type 8376). This was used in hydrotreating petroleum fractions with a high content of n-paraffins, in treating a product obtained by dewaxing diesel fuel [2133, 2134], and also in hydrogenating high-boiling fractions obtained by Fischer-Tropsch synthesis [723] and for hydrorefining slack waxes [99, 100, 167, 1877, 1879] and ceresins [167]. The catalyst 8376 is used on an industrial scale, e. g., in hydrofinishing raw petroleum wax (in Shell refinery at Stanlow, Great Britain); in this case a trickle-phase reactor is used at 120 atm, 300 to 310 °C with a LHSV of 0.7 [1821]. Petroleum waxes furthermore undergo a hydrorefining process (Esso Research and Eng. Co.) [64] with a Co–Mo refining catalyst [1743]. This catalyst was also employed in a number of other studies [97, 98, 747, 748, 1003a, 1933, 2172] as well as other catalyst, e.g., NiS–MoS_2/Al_2O_3 [57, 392, 1822, 2048], WS_2 and MoS_2 [1166, 1192].

7.2.3.2.6.

Hydrorefining of residual oils and crude petroleum

Hydrotreating of petroleum residues containing asphaltenes is one of the most difficult problems in technical catalysis.

Hydrogenation of crude petroleum or petroleum residues on fixed-bed catalysts is made very difficult by the presence of resins and asphaltenes, which rapidly contaminate the surface of catalysts so that their operating cycle and lifetime are short

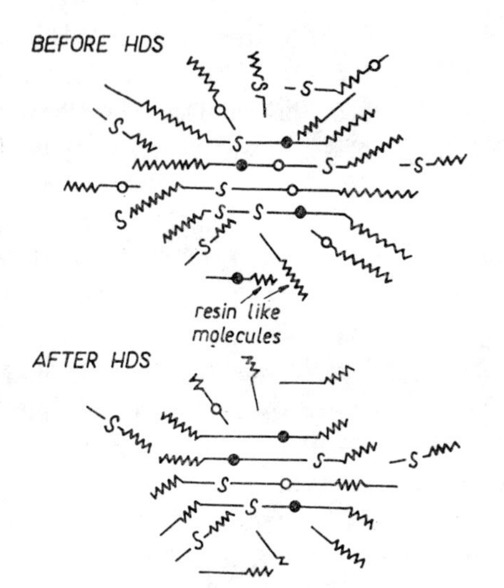

Fig. 93. Schematic illustration of qualitative changes in asphaltenes and surrounding resins during hydrodesulphurisation [157]. S — Sulphur. ○ — Vanadium. ● — Nickel. ——— — Aromatic rings. ~~~~ — Naphtenic rings.

[1646]. Metalloorganic compounds contained in the treated raw materials are decomposed on the catalyst, enhance contamination of the catalyst by deposits and cause difficulties in regeneration [2138a]. Deactivation is, as a rule, more rapid, the more severe the reaction conditions.

Asphaltenes are very difficult to desulphurise. On the basis of a detailed study by Yen et al. [2238], asphaltenes must be considered to be micellar agglomerates of true molecules with different characteristics (Fig. 93). Under normal conditions, these asphaltic micelles are stabilised by the other petroleum components. Under hydrodesulphurisation conditions, however, the asphaltene micelles are decomposed, the following being the most important changes taking place [157]:

1. the overall proportion of asphaltenic fractions decreases;
2. these fractions are partly desuphurised;
3. the aromatic nature of the asphaltenic fractions rises;
4. their metal content decreases;
5. the atomic ratio Ni : V in the asphaltenic fractions rises.

These changes are illustrated schematically in Fig. 93.

Desulphurisation of asphaltenes is characterised by the fact, that, under mild reaction conditions, sulphur is mainly removed from the outer parts of the micelles, while sulphur located inside disappears only after deep destruction under very severe desulphurisation conditions.

Two basic phenomena are the causes of catalyst deactivation during hydrodesulphurisation of petroleum residues:

1. coke deposits on the active catalyst surface;
2. accumulation on the catalyst surface of metallic deposits from organic nickel and vanadium compounds.

In order to maintain a high activity of the refining catalyst, the two types of deposits must be decreased to a minimum. For maintaining high activity, the catalyst carrier is as important [171b] as the active refining components with which it is impregnated. The overall catalytic activity is determined by the high efficiency of the active refining components, as well as by the surface area, pore radius and pore volume of the carrier. In the case of a hydrodesulphurisation catalyst, these factors are connected by the following empirical equation [157]:

$$\%\ \text{desulphurisation} = K + 0.0589A + 13.2V + 0.012R \qquad \text{(CXL)}$$

where K is a constant which depends on the reaction conditions, A is the surface area (m^2/g), R the pore radius (Å), V the pore volume (cm^3/g).

From this empirical equation, it follows that a better catalyst is obtained by increasing the surface area, pore radius and porosity of the carrier. Obviously there exists an upper limit of porosity determined by the necessary mechanical strength of the catalyst particles. At the same time, the pore radius and surface area are not independent for a specified porosity value. Therefore the most efficient catalyst carrier will represent a certain compromise between these factors.

It was recommended in older studies, that catalysts used in hydrocracking and hydrotreating of residual fractions should not have an unduly large pore diameter (over 100 Å), since metallo-organic compounds would be sorbed in larger pores, deactivating these pores quickly [2138a]. On the other hand, recent results show that the presence of pores of diameters between 100 and 1000 Å and even larger ones is essential to secure access of the raw material into the bulk of the catalyst particle where pores are smaller. When small-diameter pores only are present, catalyst deactivation is rapid, since asphaltenes which penetrate into the outer pores block the access of oil fractions into the inner pores [2138a]. A review of the published physicochemical properties of some catalysts in hydrogenation of residues is given in the study [2138a].

Coke deposition is particularly rapid at the beginning of operation of a catalyst, while the coke content rises only slowly when the steady state activity has been reached. The absolute amount of coke deposited may be decreased by increasing the partial hydrogen pressure, but this will not hinder the initial rapid coking of the most active catalyst surface.

Interesting data were obtained by comparing the physical properties of fresh and coked catalysts. A catalyst with an "equilibrium" carbon content (about 16%) had a surface and pore volume 2 to 3 times lower. The pore distribution curve, however, varied only little and the mean pore radius only decreased by some 10 to 15%. Another interesting fact was that the rate of coke deposition on the catalyst differed only a little when a vacuum residue or its deasphaltisate were treated; this being especially true at the beginning of the process. This apparently indicates that coke formation at the beginning of the process is not due to asphaltenes only, but that its cause may rather be uncontrolled hydrocracking reactions of large molecules which are firmly adsorbed on the highly active centres of the fresh catalyst [157]. Moreover, there are considerable differences in the stability of asphaltenes from different raw materials. For example, asphaltenes from the Romashkino crude oil are far more resistant to hydrogenation compared to asphaltenes of Kuwait petroleum [1124a].

The behaviour of organic metal compounds, particularly of V and Ni compounds, in the hydrorefining of residual fractions is very important. A part of the metal is doubtless contained in the form of inorganic components in suspension, and these may be removed by filtering or centrifugation. A second part of the metal content is bound in porphyrinic complexes and is clearly shown in the spectrum of the material. A third, "non-porphyrinic" part is probably quite similar in character to the porphyrinic one, but for some, as yet unknown, reasons it does not form a characteristic spectrum in the visible region [1201].

The problem of the different ease of hydrogenolysis of metal compounds in petroleum is a very significant one. Nickel compounds are always more resistant, and are more difficult than vanadium compounds to remove by means of hydrorefining. It was found in refining raw materials by the Gulf-HDS process, that practically complete removal of V and 98% removal of Ni was achieved on a fresh refining catalyst; while on an older catalyst, V was eliminated to 98–99%, and Ni only to 93% [155]. This observation applies to metal compounds in the petroleum residue as well as in de-asphaltisates (maltenic fraction). Typical results for the two types of raw materials are presented in Table 84.

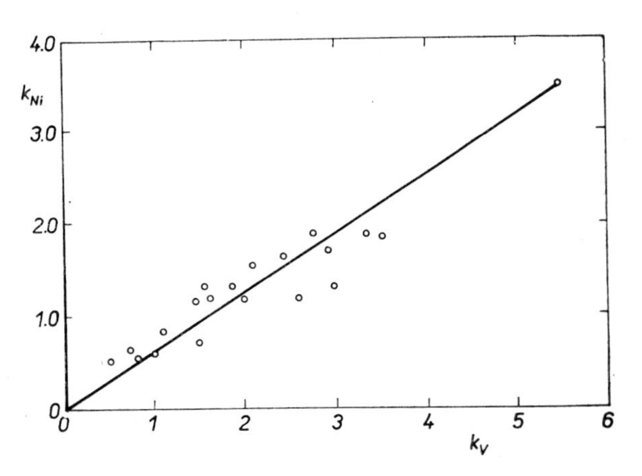

Fig. 94. Comparison of the velocity constants of removal of V and Ni compounds from different raw materials (petroleum residue, malthenic fraction, heavy gas oil) [1201]. k_{Ni} — Pseudo first-order reaction velocity constant of Ni removal. k_V — The same for vanadium.

The major part of the metals is contained in the asphaltenes, so that when these are precipitated, the maltenic fraction contains considerably less metal compounds. Thus the degree of demetallisation rises considerably (Table 84), and the rate of desulphurisation also rises substantially, since the reaction is no longer retarded by asphaltenes [157].

The rate of demetallisation decreases with the rising molecular weight of the material treated. In a broad, non-deasphaltised fraction, differences in the extent of

Table 84.

Difference in the Demetallisation Rate of Vanadium and Nickel Compounds in Hydrodesulphurisation of Petroleum Fractions [1201].

	Feed: 36% atmospheric residue from Kuwait raw petroleum	Hydrogenation product, at LHSV			Feed: Maltenes from Kuwait petroleum tar obtained by visbreaking	Hydrogenation product, at LHSV		
		2	1	0.5		3.8	1.9	1.0
V (ppm)	63.4	15.0	5.2	0.1	27.4	2.2	0.4	0.2
Ni (ppm)	19.4	9.8	6.0	1.1	9.7	1.9	0.9	0.7
Atomic ratio V/Ni	3.77	1.77	1.0	0.105	3.25	1.34	0.51	0.33
Total demetallisation (%)	—	70.0	86.5	98.5	—	88.9	96.5	97.6

removal of V and Ni might be explained by different rates of demetallisation of molecules of different size. Analogous experiments with a petroleum distillate (heavy gas oil) rich in trace metals, however, showed a marked disparity even in this case, where the differences in the molecule sizes of the metal compounds cannot be very large. Fig. 94 is a plot of the apparent rate constant of V and Ni removal assuming that the course of demetallisation is a first-order reaction. It can be seen that with different materials (non-deasphaltised petroleum residue, naphthenes, heavy gas oil) the rate constant for Ni removal shows a maximum of 60% the value for V elimination. Iron compounds are similarly more resistant to demetallisation than vanadium compounds [1201].

The reasons for this different behaviour of metal compounds in hydrodesulphurisation are not yet known, although Larson and Beuther [1201] mention some significant details. It has been found that the major part of the vanadium is present in crude petroleum in the vanadyl V^{+4} form [1802]. On the other hand, it seems that Ni is exclusively present in the form of Ni^{+2}. Only a part of the V and Ni content of petroleum (5 to 44%) was identified as being clearly porphyrin-bound [356, 379]. Yen et al. [2239] showed, however, that a part of the metals is bound in coordination complexes with asphaltene heteroatoms. There are a number of possible vanadium complexes with nitrogen, sulphur and oxygen. It is seen that vanadium petroleum compounds are more polar than Ni or Fe compounds, so that in chromatography on bauxite or in contact with the catalyst [770] vanadium compounds are adsorbed far more strongly [1201].

The rate of deactivation of refining catalysts in hydrogenation of crude petroleum and petroleum residues is greater the lower the hydrogen pressure employed. Small amounts of trace metals (e.g., about 1 ppm V or Ni) [448] do not decrease the catalyst activity substantially, and its lifetime in desulphurisation is satisfactory. Under mild reaction conditions, a good catalyst lifetime may even be expected when hydrogenating crude petroleum [1537]. In the H-Oil process with a moving catalyst, a good catalyst lifetime was also recorded [1618].

An outstanding resistivity to the effect of trace metals was recorded in the case of a special desulphurisation catalyst used in the Gulf-HDS process. For example, after processing of 240 volumes of raw materials (containing 453 ppm V) per 1 volume of the catalyst, 11.5% V was recovered after several regeneration cycles, without a substantial decrease in the desulphurisation activity being observed [1339].

With higher metal concentrations in the feed, considerable amounts of deposits contaminate the catalysts and the activity and selectivity of the catalyst alters (cf. p. 116) [363, 1056, 1347]. Perfect removal of alkalis and vanadium by pretreatment on bauxite will prolong the lifetime of a hydrodesulphurisation catalyst considerably [1663, 1664, 1668].

Whole petroleum and particularly its residues (atmospheric and vacuum residues) are hydrorefined either under mild conditions with no substantial extent of splitting, or desulphurisation is accompanied by hydrocracking. In the main, mild hydrorefining enables the raw material to be desulphurised and decreases its viscosity to some extent. The process is performed at temperatures up to 420 °C and pressures usually up to 70 atm.

The Co–Mo catalyst was mainly found useful in hydrogenating crude petroleum [2a, 717, 1959, 1660, 1664, 2058]. The regeneration periods of the catalyst were 50 to 200 hours [1347, 1959]. In the case of mild hydrogenation of atmospheric residuum

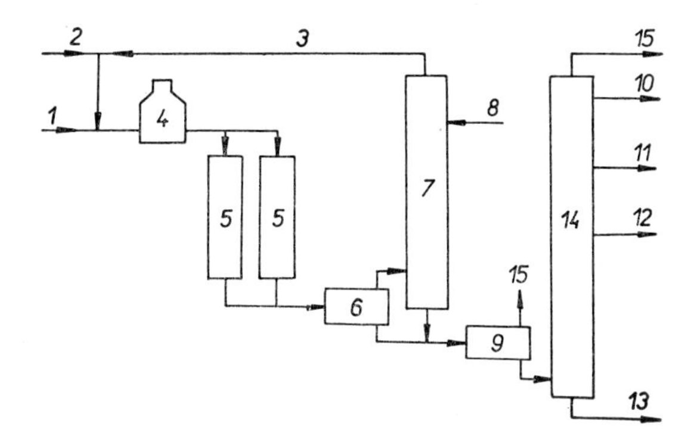

Fig. 95. Simplified scheme of the Gulf-HDS process. 1 — Feed. 2 — Make-up hydrogen. 3 — Recycle hydrogen. 4 — Oil heater. 5 — Reactors. 6 — High-pressure separator. 7 — Absorber. 8 — Inlet of absorption oils. 9 — Low-pressure separator. 10 — Light gasoline. 11 — Heavy gasoline. 12 — Light gas oil. 13 — Residue over 345 °C. 14 — Fractionator. 15 — Hydrogen sulphide and fuel gas.

(400 °C, 21 to 28 atm, LHSV 0.5 to 1) on a Co–Mo catalyst, a high activity of the catalyst was maintained on average for 100 hours [816]. The H-Oil process may also be applied to mild hydrogenation of atmospheric residues (see p. 352). Recently, Gleim and Gatsis [614b] tested mixtures of molybdenum sulphide and Lewis acids ($ZnCl_2$, $AlCl_3$) for use as catalysts in hydrorefining crude oils.

A basic drawback of mild hydrorefining lies in the fact that only partial desulphurisation is achieved. The demand for efficient desulphurisation therefore necessitates more severe reaction conditions, under which hydrorefining is accompanied by hydrocracking. A disadvantage is the higher hydrogen consumption, but on the other hand the degree of desulphurisation may be selected in a wide range.

The Gulf-HDS process (Gulf Research Dev. Co.) [64, 155, 1339] is representative of processes for refining petroleum residues with simultaneous hydrocracking, carried out at temperatures above 420 °C and at medium pressures (up to 70 atm). The development of the Gulf-HDS process started with supported NiO, which was converted to the sulphide in the course of the working cycle and so bound the H_2S formed. H_2S penetrated into the product only when 50 to 62% of the nickel content had been converted to the sulphide. The hydrocracking activity of the commercial catalyst is higher than that of the Ni catalyst, it is very resistant to deactivation and can be repeatedly regenerated. The catalyst activity gradually decreases in the medium-pressure process and regular regeneration is necessary (either with steam or a mixture of combustion gases and air) [155, 1347]. The usual working conditions for the medium-pressure version of the Gulf-HDS process are: temperature 425 to 450 °C, pressure 35 to 70 atm, LHSV 0.5 to 2, H_2: oil ratio 445 to 1780 Nm^3/m^3. A simplified scheme for the process is shown in Fig. 95. Depending on the nature of the raw material and products required, the process can also be performed in two stages [64]. The results of desulphurisation of Kuwait atmospheric residuum are presented in Table 85 [155, 157, 1121].

Table 85.

Desulphurisation of Kuwait Atmospheric Residue by the Gulf-HDS Process [155, 157,1121]

		Raw material	Hydrogenation product			
			Operating pressure (atm)			
			35	70	140	210
Yields (% by vol.)	Gasoline C_4 — 204 °C	—	3	3	3	2
	Fraction b. p. 204—354 °C	—	4	7	15	12
	Fuel oil	—	94	92	84	88
Fuel oil quality	Density (at 15 °C)	0.994	0.959	0.946	0.925	0.916
	Sulphur content (% by wt.)	4.57	1.47	1.03	0.51	0.27
Hydrogen consumption (Nm^3/m^3)		—	92	106	—	—

The high-pressure Gulf-HDS process (performed, e.g., at 140 to 210 atm) allows the basic nitrogen content to be decreased considerably, which is important for refining materials which are to be catalytically cracked [155, 795]. The hydrogen pressure also determines the rate of catalyst contamination with coke [157, 937]. Beuther and Flinn [155] state that the catalyst can be used as long as the amount of carbonaceous deposits on it is not more than 30% (cf. also [579a]).

Table 86.

Desulphurisation of Romashkino Atmospheric Residue [1120, 1121].
(Operating pressure 300 atm, temperature 395 °C, LHSV 0.75)

	Raw material	Hydrogenate
Density (at 20 °C)	0.925	0.908
Fractions up to 350 °C (% by wt.)	13	21.5
Sulphur content (% by wt.)	2.69	1.17
Vanadium content (ppm)	136	88
Asphaltenes content (%)	3.9	3.1
Resins content (%)	30.2	25.6
Carbon residue (%)	7.25	4.7
Ash content (%)	0.07	0.028

A number of experiments were carried out in Czechoslovakia for the purpose of obtaining partly desulphurised fuel oil from petroleum residues [1118]. Atmospheric residue from Romashkino petroleum was hydrodesulphurised on a Co–Mo catalyst at 300 atm, and it was found that only 50 to 60% desulphurisation can be achieved by means of mild hydrocracking. With increasing desulphurisation, the hydrogen consumption rises excessively and, moreover, the catalyst is rapidly coked up. The results of an industrial-scale experiment are shown in Table 86 [1120, 1121]. The petroleum must be carefully desalted in order to maintain a sufficient activity and

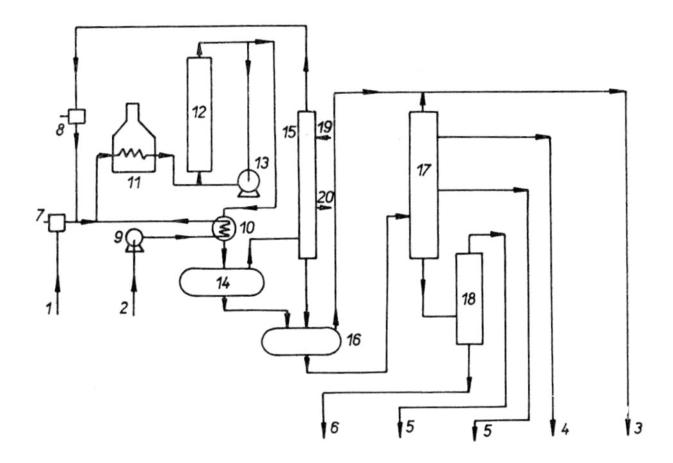

Fig. 96. Simplified scheme of the H-Oil process [64]. 1 — Make-up hydrogen. 2 — Petroleum residue. 3 — Fuel gas. 4 — Gasoline. 5 — Distillate fraction. 6 — Residue. 7 — H_2 compressor. 8 — Hydrogen recycle compressor. 9 — Feed pump. 10 — Exchanger. 11 — Oil heater. 12 — H-Oil reactor. 13 — Circulation pump. 14 — High pressure separator. 15 — Absorber. 16 — Low pressure separator. 17 — Atmospheric column. 18 — Vacuum column. 19 — Lean oil. 20 — Rich oil.

lifetime of the catalyst. In this procedure, the operating cycles are approximately 2,000 hours and the hydrogen consumption is roughly 90 Nm^3/t [1123]. A similar hydrogen consumption is stated in a tender by the Institut Français du Pétrole [874]. Special types of Co–Mo and Ni–Mo catalysts, considerably more resistant to deactivation in desulphurisation of petroleum residues as compared to conventional types, were developed in Czechoslovakia [1124b, 1124c].

The refining process is modified in various ways in order to conserve the activity of the desulphurisation catalysts used to treat atmospheric and vacuum residues. For example, gasoline, diesel fuel and vacuum oil are distilled from crude petroleum; then the vacuum oil is desulphurised [2130] and mixed with excess vacuum residue in order to obtain a fuel oil with the required sulphur content. Another method involves preliminary deasphaltisation of the vacuum residue by means of light hydrocarbons and desulphurisation of the deasphaltisate [86a, 1123, 1124b, 1124c]. In the Combifining process, deasphaltisation is carried out by catalytic means (activated petroleum coke serving as catalyst) at 380–410 °C, hydrogen pressure of 30 atm in a fluidised-bed reactor. Deasphaltisation, as well as partial desulphurisation and demetallisation is achieved [151]. To increase the yield of white products and improve the economy of the process in treating petroleum residues it is recommended that catalytic desulphurisation and visbreaking are combined [1809].

The utilisation of the H-Oil process for the above purpose is also important (for a scheme, see Fig. 96). A fine-grained spherical catalyst (recently the process was improved by the use of microspherical catalyst [579a, 2138a]) is kept as an ebullating bed in a special reactor, and the system guarantees perfect contact between hydrogen and oil. A part of the catalyst is removed for regeneration and replaced by a fresh catalyst portion without interrupting the process, so that a constant mean activity is secured. Part of the outlet product is mixed with the feed and returned to the reactor, which secures perfect temperature control. After expansion the outlet is processed in the conventional manner [64]. This process may be carried out under mild conditions, with substantially limited hydrocracking, as a desulphurisation process [1621]. The catalysts are usually Co and Mo sulphides on Al_2O_3 [301a].

At present, attempts are being made to conduct the H-Oil process under efficient hydrocracking conditions to achieve a high yield of gasoline, diesel fuel and a feed for cracking while limiting the amount of fuel oil produced [1123].

7.2.3.2.7.

Special desulphurisation procedures

Desulphurisation of hydrocarbon mixtures with steam has been tested in the presence of sulphide catalysts. The process was carried out at 250 to 400 °C and medium pressures [427, 1075, 1698]. For example in the presence of NiS, a petroleum distillate was desulphurised at 15 atm and a molar oil: steam ratio of 1 : 2 from 500 to 3 ppm S [1075]. The use of carbon monoxide for desulphurisation has also been tested; a suitable catalyst being found to be MoS_3 at temperatures of about 300 °C and pressures of 700 atm [1279].

Eng and Tiedje [465] describe a special method of gas oil desulphurisation in the presence of the catalyst $CoS–MoS/Al_2O_3$ and utilising the effect of low-boiling olefins, the respective alkyl mercaptans being formed.

7.3.

Cracking and dehydrocracking of hydrocarbon mixtures

In contrast to hydrocracking, sulphide catalysts have been used very rarely in cracking hydrocarbon mixtures, and their technical application in this branch of fuel technology is in the experimental stage only.

Much effort is at present being devoted to the economic preparation of higher n-olefins by selective dehydrocracking of paraffins on non-acid catalysts, which allow a high yield of higher olefins to be obtained while limiting the formation of gases and other side products to a minimum [350, 1414]. Paraffin waxes are a particularly suitable raw material; these being obtained from the refining industry in sufficient amounts. However, the problem of finding a suitable catalyst with sufficient selectivity for the formation of α-olefins is very difficult if the catalyst is also to have a satisfactory lifetime.

Some sulphide catalysts were tested for this purpose (p. 148). Huibers and Waterman [818] found that a mixed Ni–W sulphide catalyst is suitable for dehydrocracking of paraffin waxes; it is sufficiently resistant to deactivation and satisfactorily selective for the formation of higher olefins. Dehydrocracking was performed at temperatures of 500 to 660 °C, with pressure of 0.5 to 2 atm, LHSV 2.1 to 3.25 and ratio by weight of paraffin wax to steam 0.8 to 2.3. The catalyst employed also causes a shift of the double bond. Some degree of deactivation is observed when the catalyst is regenerated with steam at 660 °C, caused most probably by partial sintering of the catalyst surface and also, in part, by oxidation of some of the sulphide with steam.

The possibility of employing sulphide catalysts based on tungsten for dehydrocracking of n-paraffins was later also confirmed by Soviet authors. Sklyar et al. [1908, 1909, 1911] cracked paraffins isolated from diesel fuel by the use of urea (40% of the raw material was C_{16}—C_{18} n-paraffins) in the presence of steam. Among the catalysts tested, the selectivity of the catalysts 5058 and 3076 was best; in the presence of these catalysts, 30 to 31% of a liquid olefin fraction b.p. up to 240 °C being formed, together with roughly 20% gases under optimum conditions (590 °C, LHSV 2 to 3). The dehydrocracking products contained 50 to 70% α-olefins. The catalyst 8376 (sulphidic) exhibited a lower selectivity with respect to the formation of liquid olefins and caused, moreover, a higher degree of isomerisation of α-olefins as well as structural isomerisation. The relative content of α-olefins in the reaction product rises with the decreasing time of contact between the reacting substances and the catalyst.

Oxide catalysts of the types 8376 or 8197 (MoO_3–NiO–Al_2O_3) catalyse more efficiently the unwanted side reaction (isomerisation of α-olefins, aromatisation etc.), their advantage however is their easier regeneration [1910].

Promotion of the sulphide catalyst (NiS/SiO_2) with a basic hydroxide (KOH, LiOH) has been found useful for increasing the selectivity of the catalyst for dehydrocracking n-paraffins (cf. p. 149) [1571, 2264, 2264a, 2265]. The technical literature, however, contains no data relating to the industrial application of sulphide catalysts in dehydrocracking paraffins.

Some additional literature and patent data concerning the application of sulphide catalysts in cracking hydrocarbons help to provide a deeper understanding of the processes which take place in the course of hydrocracking and destructive hydrogenation. A catalyst containing nickel sulphide was tried out [2138] in fluid cracking of hydrocarbons. Some other cracking catalysts contained FeS [1367] and a mixture of Groups VI and VIII sulphides [685]. Splitting of hydrocarbon oils in the presence of a mixture of metal sulphides and sulphates [2167], and cracking of gas oils in the presence of a melt of P, As and Bi sulphides [396] have also been described.

8.

ISOMERISATION IN THE PRESENCE OF SULPHIDE CATALYSTS

Following a period of stagnation, interest in isomerisation of low-boiling n-paraffins was renewed after the year 1950, when there was great demand for high-octane gasolines, and a number of modern processes were developed [64].

Besides low molecular weight paraffins, isomerisation of middle-boiling paraffins, particularly when combined with simultaneous desulphurisation, is also an attractive possibility, since fuels with a low pour point are obtained [1424, 2063]. Attention has also been given to hydroisomerisation of high molecular weight paraffins, resulting in lubricating oils with a high viscosity index [1205]. Nowadays, isomerisation of C_8 aromatic hydrocarbons to *p*-xylene is being developed intensively all over the world [1205].

The technically most important isomerisation catalysts are Friedel-Crafts catalysts on the one hand and bifunctional platinum-based catalysts on the other. Although sulphur catalysts have not yet been used on an industrial scale, isomerisation of all hydrocarbon types has been carried out in their presence and a number of theoretical and pilot-plant research studies have been undertaken.

The thermodynamics of isomerisation reactions, particularly of the lower hydrocarbons, have been thoroughly examined, and the data are analysed in the specialised literature [342, 1650, 1770, 2154].

The composition of the equilibrium mixture varies rapidly with temperature in the case of aliphatic hydrocarbons. At higher temperatures, the content of n-hydrocarbons rises and the concentration of branched ones decreases, especially in the case of disubstituted isomers. With olefins moreover, the double bond shifts away from the terminal position with rising temperature.

With cycloalkanes, the six-membered ring tends to contract as the temperature rises, cyclopentane isomers being formed. The equilibrium in isomerisation of aromatic compounds is also temperature-dependent; the most important case of C_8-aromatic compounds being characterised by a low dependence of the *p*-xylene concentration on temperature [1205].

8.1.

Chemistry and kinetics of isomerisation of hydrocarbons in the presence of sulphide catalysts

Isomerisation proceeds by an ionic mechanism on metal sulphides which are of distinctly polyfunctional character [416, 628]. The ratio between the hydrogenation-dehydrogenation and isomerisation activities of a sulphide catalyst may be adjusted as required by supporting the sulphide on an acid carrier.

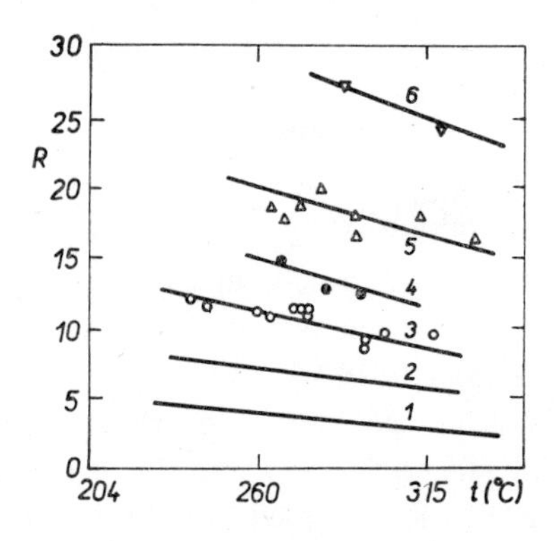

Fig. 97. Hydroisomerisation of 1- and 2-pentenes on the NiS/Al_2O_3–SiO_2 catalyst [569]. R — Molar ratio i-/n-hydrocarbons. 1 — Equilibrium ratio i-C_5H_{12}/n-C_5H_{12}. 2 — Equilibrium ratio i-C_5H_{10}/n-C_5H_{10}. Experimental values for: 3 — 70 atm. 4 — 35 atm. 5 — 17.5 atm. 6 — 7 atm. (Values for 1- and 2-pentene are not distinguished.)

Hydrogen must be present in high-temperature isomerisation on bifunctional catalysts [310, 1315]. In the case of isomerisation of saturated hydrocarbons the reaction is started by dehydrogenation, olefins being formed as reaction intermediates [312]. The olefinic hydrocarbon then reacts by an ionic mechanism.

The hydroisomerisation mechanism of pentenes in the presence of nickel sulphide on an acid carrier was studied by Frye et al. [569]. These authors found that

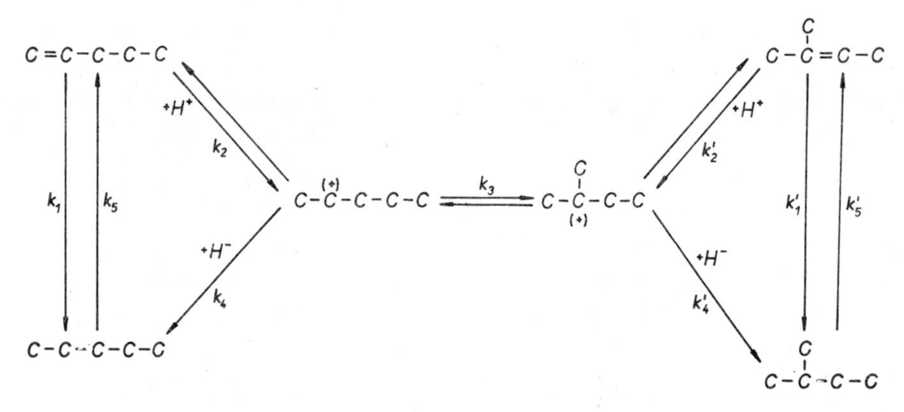

Fig. 98. Hydroisomerisation of 1-pentene in the presence of the bifunctional catalyst NiS/Al_2O_3–SiO_2 [569].

the presence of the nickel compound in the sulphide form is essential to secure a high degree of isomerisation. Partial reduction of the sulphide to the metal considerably decreases the iso- to n-paraffins ratio. Experimental results achieved with the NiS/Al_2O_3–SiO_2 are shown in Table 87. Furthermore, the influence of the reaction conditions on the degree of hydroisomerisation of 1-pentene can be seen in Fig. 97. (the behaviour of 2-pentene is analogous under the reaction conditions employed). Under all conditions, the isopentane/n-pentane ratio is considerably greater than the equilibrium ratio and rises with the decreasing overall pressure. The influence of a molar excess of hydrogen greater than two alters the iso- to n-hydrocarbon ratio only very slightly [569]. The iso- to n-pentane ratio decreases slightly with rising temperature.

Table 87.

Hydroisomerisation of 1-Pentene with Various Sulphur Contents. (Catalyst NiS/Al_2O_3–SiO_2[a]), total pressure 70 atm, 315 °C, molar ratio H_2 : hydrocarbon 6 : 1; sulphur in the feed in the form of CS_2, measurement carried out after five hours of dosage with raw material) [569].

S content in the feed, %	0	0.01	0.1	1.0
Molar ratio isopentane : n-pentane[b])	1.1	2.7	11.1	10.6

[a]) Under the operating conditions, the active hydrogenating component of the catalyst is the subsulphide Ni_3S_2.

[b]) At the temperature employed, the equilibrium ratio is 5.4 for olefins and 3.2 for paraffins.

Frye et al. [569] found furthermore, that under the reaction conditions employed (temperatures up to 340 °C) isomerisation of paraffins on Ni sulphide supported on aluminosilicate proceeds to a limited extent only, probably due to the low hydrogenation-dehydrogenation activity of the catalyst employed.

Fig. 98 shows a scheme for the mechanism of high-temperature hydroisomerisation of paraffins and olefins, using the example of n-pentane and 1-pentene. The catalyst must have a certain hydrogenation-dehydrogenation activity in order to allow an olefin to be formed from the paraffin. Provided the rate of the hydrogenation reaction (k_1) is not substantially greater than the rate of olefin protonisation (k_2) (and this will be the case for a catalyst with a sulphide component which has a hydrogenation activity less than that of the metal), an adequate concentration of secondary carbonium ions will be formed from the olefin. These ions are subject to isomerisation forming the thermodynamically more stable tertiary ions, which predominate at equilibrium. When the rate of isomerisation of the ions (k_3) is high enough, equilibrium is achieved rapidly and a high yield from the isomerisation reaction is assured. The tertiary carbonium ion formed may be stabilised either by abstraction of a proton with formation of an isomeric olefin which is then hydrogenated to the isoparaffin or, by addition of a hydride anion (both types of hydrogen ions are formed from atomic hydrogen which is present in the reaction mixture). Obviously, the temporary secondary carbonium ion may react in the same way. Furthermore both types of carbonium ions may react with an olefin molecule to form a dimer which produces reaction side products on hydrocracking [569].

The use of a catalyst in which platinum sulphide and $AlBr_3$ are supported on Al_2O_3 was tested in connection with the development of new medium-temperature isomerisation catalyst (Friedel-Crafts catalysts supported on high-temperature isomerisation catalysts [1205]). These catalysts were found to be very efficient in isomerising paraffins at temperatures of around 50 °C [1805].

High-temperature isomerisation of cycloalkanes mainly involves conversion of cyclohexanes to cyclopentanes and, in part, hydrogenolysis of the ring or dehydrogenation to aromatic compounds. Here there is a possibility of geminal cyclopentanes being formed, although these isomers are the most difficult ones to form [976]:

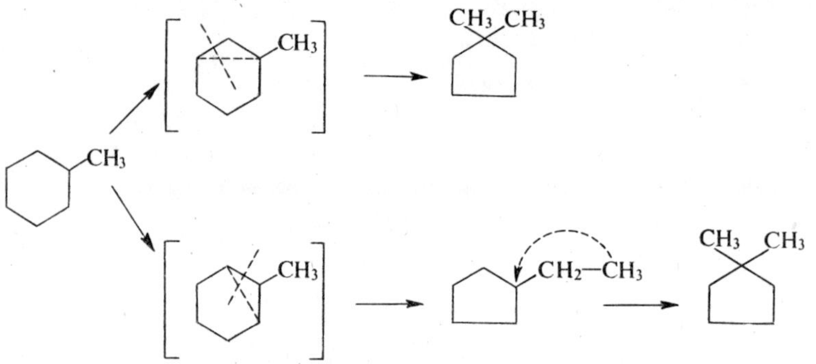

(CXLI)

(CXLII)

Ciapetta [311, 313] believes that geminal cyclopentane is mainly formed by the reaction (CXLI). However, the process is far more complicated [1651], as was proved [2159] for hydroisomerisation of methylcyclohexanes (catalyst WS_2) labelled with ^{14}C carbon in all ring positions.

High-temperature isomerisation of aromatic compounds is characterised by migration of alkyl groups from the side chains into the ring. This is the usual isomerisation reaction under reforming conditions [708], although under isomerisation conditions a reaction involving alteration in the position of substituents on the ring is easier [1205].

It is typical of hydroisomerisation of aromatic compounds that more alkylcyclopentanes are formed than in the hydroisomerisation of the respective hydroaromatic hydrocarbons. This observation applies to the platinum catalyst as well as to a number of sulphide catalysts (MoS_2, WS_2, 8376 etc.) [957, 962]. This may be explained by the fact that it is considerably more difficult to form cyclohexene (as reaction intermediate in hydroisomeriastion under a relatively high hydrogen pressure) by dehydrogenation of cyclohexane than by partial benzene hydrogenation.

Soviet authors have devoted much attention to the kinetics of hydroisomerisation taking place in the presence of sulphide catalysts. In the case of n-butane hydroisomerisation (at 380 to 440 °C, pressure up to 75 atm, molar ratio $H_2 : C_4H_{10} = 1-6$) on tungsten sulphide, the reaction rate increased with the rising partial pressure of the hydrocarbon from 5 to 20 atm. The retarding effect of hydrogen was observed at the same time; the isomerisation rate being decreased 1.4 times with a twofold rise in the hydrogen partial pressure [1312].

Table 88.

Hydroisomerisation of n-Hexane (flow reactor, pressure 100 atm, molar ratio H_2 : hexane 6 : 1) [1318].

Catalyst	Reaction conditions		Ratio	
	Temperature (°C)	Space velocity (l/l/hour)	$\frac{i\text{-}C_6}{n\text{-}C_6}$	$\frac{n\text{-}C_5}{n\text{-}C_6}$
WS_2	360	0.4	0.55	0.0
	400	1.0	1.4	0.03
50% WS_2 on kieselguhr	400	0.4	0.35	0.0
50% WS_2 on alumina	400	0.4	0.45	0.0
42% MoS_2 on alumina	440	0.4	0.70	0.01

A high isomerisation activity is observed particularly with tungsten disulphide, as can be seen from the results of n-hexane isomerisation presented in Table 88. The hydroisomerisation activity decreases when this catalyst is supported on a carrier. The reaction rate is not influenced by water nor by sulphur compounds (CS_2 or H_2S formed from it), nor by naphthenic hydrocarbons [1318]. The observation (when n-hexane with 4% CS_2 was isomerised) that mixed oxide catalysts (Co–Mo/Al_2O_3, Ni–Mo/Al_2O_3) lost their isomerisation activity to a large extent in the presence of sulphur compounds is remarkable. On the other hand, the catalyst MoO_3/Al_2O_3 was resistant and kept its isomerisation activity even in the presence of sulphur compounds. When this catalyst has been activated with fluorine, its isomerisation activity equalled that of WS_2 or was even higher [1313].

A kinetic equation for an irreversible reaction may be applied to isomerisation reactions in which the equilibrium is strongly shifted to the side of one isomer under the reaction conditions employed. In the case of isomerisation of paraffins, however, kinetic equations for a reversible reaction must be used. Maslyanskii and Bursian

[1319] start from an empirical equation, for isomerisation of paraffinic hydrocarbons on tungsten sulphide, in the form of:

$$r = \frac{k_1P_1 - k_2P_2}{(1 + b_1P_1 + b_2P_2)\, P_H^n} \qquad \text{(CXLIII)}$$

where: k_1, k_2 are the velocity constants of isomerisation of the n-paraffin to the isoparaffin and of the reverse reaction, P_1, P_2, P_H are the partial pressures of the n-paraffin, isoparaffin and hydrogen, resp., b_1, b_2 are equilibrium adsorption constants of the n-paraffin and isoparaffin, resp., n is a constant

By integration and modification of the rate equation (CXLIII) the authors [1319] arrive at an expression for the isomerisation velocity constant:

$$k = \frac{P_H^n}{\left(\frac{W}{F}\right)}\left[\left(\frac{1}{P'} + A\right)\ln\frac{x_r}{x_r - x} - Bx\right] \qquad \text{(CXLIV)}$$

where: $A = b_1 + (b_2 - b_1)/1 + (1 : K_p)$, $B = b_2 - b_1$; K_p is the equilibrium constant of the isomerisation reaction, P' is the partial pressure of the n-paraffin on the

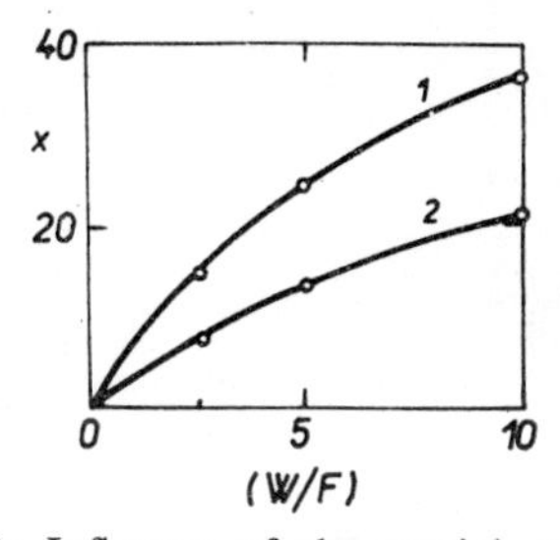

Fig. 99. Influence of the partial nydrogen pressure on the rate of n-hexane isomerisation (catalyst WS_2, temperature 330°C, partial hydrocarbon pressure 4 atm) [1319]. x — Degree of isomerisation (%). (W/F) — Reciprocal values of space velocity (hour^{-1}). 1 — $P_H = 26$ atm. 2 — $P_H = 96$ atm.

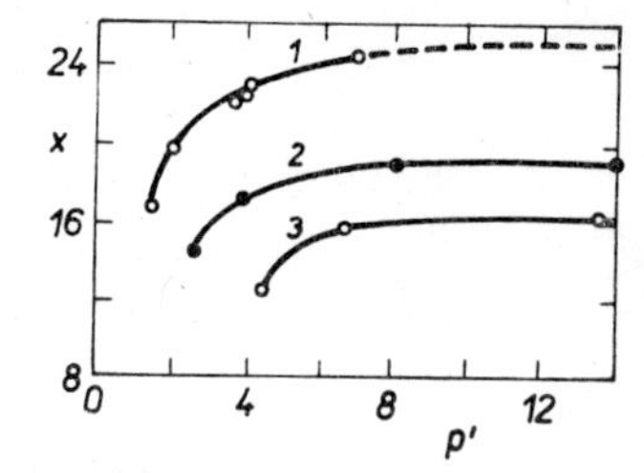

Fig. 100. Influence of partial n-hexane pressure on the degree of its isomerisation (catalyst WS_2, temperature 330 °C) [1319]. x — Degree of conversion (%). P' — Partial n-hexane pressure (atm). Partial hydrogen pressures P_H: 1 — 26 atm. 2 — 56 atm. 3 — 96 atm.

reactor inlet, (F/W) is space velocity, x_r the equilibrium conversion at the given temperature and x is the true conversion value.

With a constant overall pressure and $(P_H/P') = \text{const.}$, the following expression is also constant at a given temperature, i.e.,

$$P_H^n\left[\frac{1}{P'} + b_1 + \frac{b_2 - b_1}{1 + \frac{1}{K_p}}\right] = \text{const.} \qquad \text{(CXLV)}$$

and we obtain for the velocity constant, from the relation (CXLIV)

$$k' = \left(\frac{F}{W}\right) \ln \frac{x_r}{x_r - x} - \left(\frac{F}{W}\right) \beta x \qquad \text{(CXLVI)}$$

which is the form of Frost's equation [76].

Isomerisation experiments have been carried out at 340, 360 and 380 °C and different space velocities with a constant molar ratio of $H_2 : n\text{-}C_6H_{14}$ and total pressure of 50 atm in the presence of WS_2. With a value of the coefficient $\beta = 0.95$ the measured values agreed in a satisfactory way with the kinetic curves derived from equation (CXLVI) [1319]. It follows from equation (CXLIV), that hydrogen retards the isomerisation reaction (although it is essential for the course of the isomerisation process), since at a constant partial pressure of the hydrocarbon (P') the same degree of conversion x is achieved at an elevated partial hydrogen pressure P_H with a lower space velocity (F/W). Furthermore it follows from equation (CXLIV) that the reaction rate is independent of the partial pressure of the hydrocarbon, when this pressure has reached a certain value. This is because as P' increases, the value of $1/P'$ decreases, until it can be neglected in comparison to A after a certain value of P'. The value of the rate constant k (and thus also the reaction rate) will be independent of the partial pressure of the hydrocarbon.

Table 89.

Relative Isomerisation Rates of n-Paraffins in the Presence of WS_2 [1317].

Hydrocarbon	Relative isomerisation rate at a temperature of	
	340 °C	360 °C
n-Pentane	1.0	1.0
n-Hexane	2.1	1.9
n-Heptane	3.1	2.9
n-Octane	4.2	—

These two conclusions have been confirmed by experiment in the case of n-hexane isomerisation on the WS_2 catalyst. Fig. 99 shows the retarding effect of hydrogen. From Fig. 100 it follows that the reaction rate is constant when a partial pressure of n-hexane of 7 to 9 atm has been exceeded. After this pressure is achieved, the catalyst surface is saturated with the hydrocarbon and a further increase in its partial pressure has no effect. Up to this pressure the reaction rate rises with the partial pressure of the hydrocarbon. Similarly in isomerisation of n-butane, the rate of isomerisation rose twofold when the partial pressure of the initial hydrocarbon had risen from 5 to 20 atm. [1912].

The values of the apparent activation energy of n-butane and n-hexane hydro-isomerisation on WS_2 are close to each other (40 to 44 kcal/mole). The rates of iso-merisation rise with the molecular weight of the n-paraffin, as we see in Table 89 [1317]. A detailed analysis of the mixture from n-hexane isomerisation showed that the main products are monomethyl- and di-methylderivatives [1317].

From experimental data for hydroisomerisation of n-hexane on the catalyst WS_2–NiS supported on an acid carrier, the Chinese authors [1241] derived a kinetic equation of the form:

$$\frac{1}{AP_H^{0.5} + B} = \frac{\left(\frac{F}{W}\right)}{P_0'} [1 - (1 - x)^{0.5}] \qquad \text{(CXLVII)}$$

where P_H is the partial pressure of hydrogen, P_0' the partial pressure of the hydro-carbon (both data relate to the reactor inlet), (F/W) is the space velocity of the hydrocarbon, x the degree of conversion, A and B are constants. The above authors found a value of 26.6 kcal/mole for the apparent activation energy of hexane iso-merisation.

Among cycloparaffins, the kinetics of cyclohexane isomerisation were studied in detail. Maslyanskii did not record in an older study a retarding influence of hydrogen on the isomerisation reaction when using the catalyst MoS_2 at 50 atm and 372 to 432 °C [1309]. The problem of cyclohexane isomerisation on WS_2 was studied in detail by Gonikberg [627]. According to these authors, the apparent reaction order is 0.4 with respect to cyclohexane in the 320 to 430 °C range. At 340 °C the reaction rate is considerably decreased by an increased hydrogen partial pressure, the depend-ence of the rate constant on the partial pressure of hydrogen being approximately linear. At 430 °C this relationship is not valid, since the rate of isomerisation is not altered by a change in the hydrogen partial pressure over the wide range of 60 to 172 atm.

A change in the dependence of the cyclohexane isomerisation velocity constant on the hydrogen partial pressure might be caused by a fundamental change in the nature of the reaction, i.e., transition from the kinetic to the transition or diffusion region. Levitskii and Gonikberg [1233] investigated this problem in connection with a de-tailed study of cyclohexane isomerisation in the presence of the catalyst WS_2. These authors confirmed the earlier observation [627] of a linear relationship between the isomerisation rate constant and hydrogen partial pressure

$$\log k_{P_H} = \log k_0 - cP_H \qquad \text{(CXLVIII)}$$

where k_0 is the isomerisation rate constant at a specified temperature, extrapolated to a zero partial pressure of hydrogen, k_{P_H} is the rate constant with a partial pressure of P_H and c is a temperature-dependent constant. The authors set out from the apparent reaction order determined earlier (0.4 with respect to cyclohexane) and derived the following relation for the reaction rate at different temperatures and pressures

$$r = k_0 P^{0.4}_{C_6H_{12}} e^{-mP_H} \quad \text{(CXLIX)}$$

where $m = 2.303c$ [1233].

Fig. 101 illustrates the linear course of the relationship (CXLVIII), determined by experimental means for different partial pressures P_H and different temperatures in the 320 to 430 °C range. The value k_0 is determined by extrapolating to $P_H = 0$ at every temperature, the value m is found from the slope of these linear relationships. The value of m decreases with the temperature, arriving $m \sim 0$ for 430 °C, i.e., at this temperature hydrogen does not retard the isomerisation reaction any more.

Fig. 102 shows the temperature dependence of the extrapolated rate constant k_0. This dependence is linear between 320 and 380 °C, and the value of the apparent

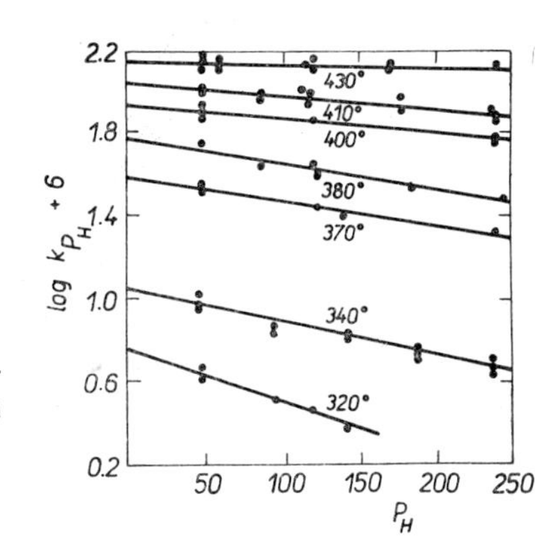

Fig. 101. Dependence of the logarithm of the isomerisation rate constant of cyclohexane on the hydrogen partial pressure (P_H, atm) [1233].

activation energy is roughly 30 kcal/mole. This rather high value indicates that the retarding effect of diffusion does not make itself felt in the 320 to 380 °C range. A sudden change takes place above 380 °C and the activation energy decreases, indicating a transition of the reaction from the kinetic region to the transition or diffusion region: moreover, the apparent reaction order of 0.4 at 430 °C with respect to cyclohexane shows that internal diffusion participates in the process [1233].

This observation holds not only for the value of the extrapolated rate constant k_0, but also for the other individual rate constants k_{P_H}, corresponding to higher partial pressures of hydrogen, i.e., the transition to the transition or diffusion region above 380 °C sets in independently of the value of P_H, and retardation of the isomerisation reaction is characteristic of the kinetic region. The independence of the reaction rate on the size of the catalyst particles may also be taken as evidence for the course of cyclohexane isomerisation on WS_2 in the kinetic region [1233].

In explaining the retarding effect of hydrogen in isomerisation, Levitskii and Gonikberg [1233] make the assumption that dehydrogenation of the original hydrocarbon is the first and rate-controlling step of the isomerisation reaction:

$$\rightleftarrows \quad + H_2 \qquad \text{(CL)}$$

$$\longrightarrow \qquad \text{(CLI)}$$

$$+ H_2 \rightleftarrows \quad . \qquad \text{(CLII)}$$

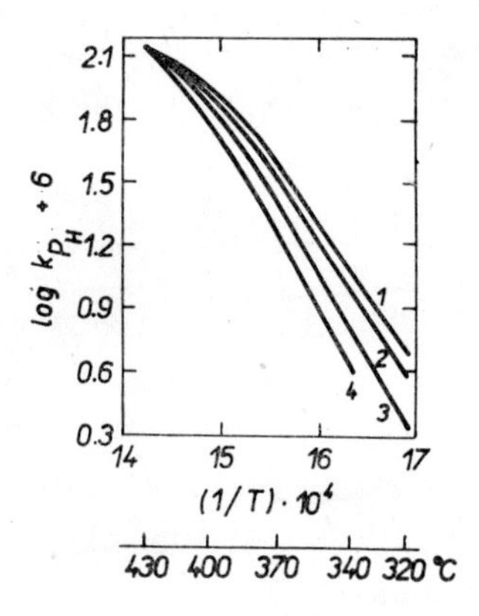

Fig. 102. Dependence of the logarithm of the cyclohexane isomerisation rate constant on temperature [1233]. 1 — Dependence for $k_{PH} = k_0$. Other relationships of log k_{PH} for hydrogen partial pressures of: 2 — 50 atm. 3 — 150 atm. 4 — 250 atm.

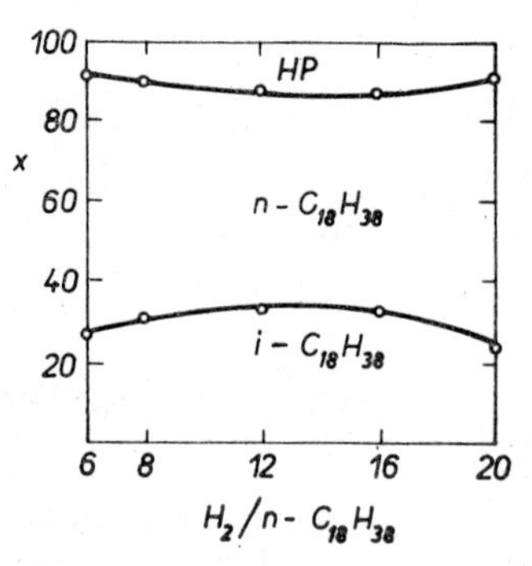

Fig. 103. Influence of the molar ratio H_2: n-$C_{18}H_{38}$ on n-octadecane isomerisation in the presence of the catalyst 8376 ($t = 435$ °C, $P = 20$ atm) [1545]. x — Reaction product composition (% by wt.). *HP* — Hydrocracking products.

The overall rate of the isomerisation reaction will be determined by the slowest step in this consecutive scheme. Increasing the partial hydrogen pressure causes two opposing effects:

a) The overall rate of the isomerisation reaction will decrease with the rising partial pressure of hydrogen, because the equilibrium of the reaction (CL) will shift to the side of the initial cyclohexane.

b) An increase of the partial pressure of hydrogen increases, in the opinion of the authors, the degree of protonisation of the catalyst and thus also the number of active acid isomerisation centres, which are essential to the ionic course of the isomerisation reaction (CLI).

With higher partial pressures of hydrogen, the influence on the dehydrogenation–hydrogenation equilibrium of the reaction (CL) prevails and therefore the retarding effect of hydrogen becomes marked [283]. Under certain reaction conditions, however, the two effects may both participate to comparable degrees, i.e., a maximum should be formed on the isomerisation yield vs. hydrogen partial pressure curve. This maximum was actually confirmed by experimental means, e.g., in the case of n-octadecane hydroisomerisation in the presence of the catalyst 8376 (Fig. 103) [1545] as well as in hydroisomerisation of C_{15} to C_{18} paraffins in the presence of the catalyst WS_2 [1583, 2063]. The maximum was rather indistinct in n-octadecane hydroisomerisation on WS_2 [1546]. In hydroisomerisation of n-octadecane on WS_2, a large amount of hydrocracking products was formed [1546]. Addition of benzene decreases the intensity of the hydrocracking reaction and improves the selectivity of the process [1547].

8.2.

Review of isomerisation of individual hydrocarbons in the presence of sulphide catalysts

8.2.1.

Isomerisation of paraffins

Sulphide catalysts have been used for hydroisomerisation of C_4 to C_{18} n-paraffins; a medium pressure (20 to 70 atm) and a temperature of about 400 °C being recommended as optimum. n-Butane and n-pentane were isomerised in the presence of WS_2 [804, 1314, 1315, 1317]. The WS_2 catalyst is characterised by a long lifetime. When this catalyst was employed in n-hexane isomerisation [1315, 1319], a slight decrease in activity (by only 7%) was recorded after 1500 hours of operation [1315].

n-Hexane [1404, 1460, 1464, 1612], n-heptane [1612, 1711] and n-octane [1361, 1612] were hydroisomerised on molybdenum sulphide. Moreover, sulphides of metals of Group VIII of the Periodic Table have also been employed [648, 847, 1654] and most recently, sulphides of the noble metals rhodium [474] and platinum [1805].

The most important group of catalysts employed in isomerising higher paraffins are those based on tungsten disulphide. n-Dodecane has been isomerised in the presence of the catalyst 8376 [1545], cetane in the presence of WS_2 and WS_2 on aluminosilicate [1288] and n-octadecane in the presence of the catalysts 5058, 8376 and

6434 [1546, 1547, 1583]. These catalysts were also employed for isomerisation of a mixture of $C_{15}-C_{18}$ n-paraffins isolated from petroleum [1582, 2063]. The main products of isomerisation of higher n-paraffins are mono- and disubstituted derivatives, so that the distillation curve changes on isomerisation of a mixture of technical paraffins, and in particular the pour point is lowered [1546, 1583].

8.2.2.

Isomerisation of olefins

When olefins are isomerised in the presence of sulphide catalysts, the double bond is shifted along the carbon chain and this may be accompanied by structural isomerisation. In the presence of hydrogen, the raw material and reaction products are frequently hydrogenated to an extent which depends on the partial pressure of hydrogen, so that the result is a mixture of paraffins, iso-paraffins, iso-olefins and initial olefins.

At relatively low temperatures (50 to 250 °C) the double bond shifts from the terminal position to the centre of the molecule. Nickel sulphide supported on sepiolite is an efficient catalyst [209, 225, 805]: e.g., isomerisation of 1-butene, contained in the C_4-fraction, to 2-butene proceeded with a high-yield at 100 °C, 10.5 atm and raw material space velocity of 0.5 l/l hour in a hydrogen atmosphere [225]. When 4-methyl-1-pentene was isomerised at 105 °C, a mixture of 4-methyl-2-pentene and 2-methyl-2-pentene was obtained [209]. Lebedev et al. [1215a] compared the rate of shifting of the double bond in terminal C_5 to C_7 olefins in the presence of WS_2, WS_2–NiS and WS_2–NiS/Al_2O_3 at 227 to 427 °C.

One very efficient catalyst, which speeds up shifting of the double bond into the centre of the molecule, is mixed WS_2–NiS with an excess of nickel sulphide [1298].

Structural isomerisation as well as hydrogenation takes place when olefins are hydroisomerised at higher temperatures. The acid catalyst carrier employed for this purpose was usually aluminosilicate [569, 1758, 1873, 1943]; NiS, CoS [569, 1653b, 1873] or a mixture of Ni and W sulphides [1758] generally serving as the active dehydrogenation–hydrogenation components. The catalyst WS_2 was found to have a strong isomerising effect, as follows from a study of dehydrocyclisation of 4-methyl--1-pentene [1658a].

Cyclic terpenes undergo isomerisation in the presence of sulphide catalysts very readily. As we have shown earlier from results reported by Kikuchi [1024, 1027], Günther [697] as well as results obtained in our own laboratory [2185] (cf. pp. 121 and 123) intensive isomerisation takes place on hydrogenation of α-pinene, dipentene and other hydrocarbons in the presence of sulphide catalysts, and a complicated reaction mixture is obtained. Thermal isomerisation also takes place, but the reaction is greatly enhanced when a sulphide catalyst is employed. For example in a nitrogen atmosphere in the presence of the catalyst 5058, α-pinene was isomerised to 94% at 220 °C; the main isomerisation products being camphene and *p*-cymene, although tricyclene, β-pinene and limonene were also identified in the reaction product [2185]. Efficient isomerisation of α-pinene also takes place on addition of H_2S, so that the non-reacted hydrocarbon portion cannot be recycled during pinene mercaptan production [2181, 2182].

8.2.3.

Isomerisation of naphthenic and aromatic hydrocarbons

Hydroisomerisation of cyclohexane to methylcyclopentane takes place intensively on sulphide catalysts under hydrogen pressure at temperatures of about 400 °C. The catalysts most often employed are molybdenum disulphide, pure [278, 1309, 1705] or mixed with Co sulphide [1695], or supported on an acid clay [51]. The reverse reaction has also been studied in the presence of MoS_2, i.e., isomerisation of methylcyclopentane [278, 1695]. Isomerisation is accompanied by hydrocracking [278, 1705]. WS_2 is also an efficient catalyst of cyclohexane hydroisomerisation [278, 627, 628, 1233].

In isomerisation of methylcyclohexane on MoS_2, the main products were 1,2- and 1,3-dimethylcyclopentanes as well as a small amount of splitting products (mainly 3-ethylpentane) [1706].

In a number of older studies, Prokopets et al. investigated the hydroisomerisation of some polycyclic napthenes (decalin, perhydroanthracene) in the presence of MoS_2 [1681, 1692] and of some partly hydrogenated aromatic hydrocarbons [1689, 1690].

Sulphide catalysts were also tested for use in isomerising C_8-aromatic hydrocarbons to *p*-xylene. Employed were non-supported WS_2 [1918], W, Mo or Cr sulphides on carriers, e.g., Al_2O_3 [1615] or on a synthetic aluminosilicate [2125]. Nickel sulphide on aluminosilicate was also found useful, e.g., supported on a cracking catalyst [1538] or a used cracking catalyst [450].

8.3.

Isomerisation of hydrocarbon mixtures

Friedel-Crafts catalysts are used in practice for low-temperature isomerisations, while high-temperature isomerisations are carried out with difunctional catalysts [2137]. Pretreated raw materials are used in nearly all cases, so that the high-temperature processes are usually carried out in the presence of efficient isomerisation catalysts in which a metal serves as the hydrogenation-dehydrogenation component. Platinum is used most often, other platinum-group metals being used only rarely. The carrier is usually alumina with added boron trioxide or an acid aluminosilicate, though zeolites have come into use recently [1205]. Platinum is used as the hydrogenation-dehydrogenation component in the industrial processes Butamer, Penex (both UOP), Pentafining (Atlantic Ref. Co.), Iso-Kel (Kellogg) and Octafining (Atlantic Ref. Co.). Palladium is employed in the Linde process (Linde) and in the Sinclair process which is now being developed. The British Petroleum process uses a catalyst in which a noble metal is supported on a promoted carrier; the Isomerate process (Pure Oil) is based on the use of a non-platinum metallic hydrogenation catalyst on an acid carrier of large surface area [1205, 2137].

Sulphide catalysts are not employed in hydroisomerisation processes on the industrial scale. Their use, however, would be desirable in all cases where non-purified materials containing sulphur compounds would be used for the reaction, particularly higher-boiling petroleum fractions. A quantity of literature and patent data indicate the applicability of resistant sulphide catalysts in hydroisomerisation of raw materials in the petroleum industry. For example, intensive isomerisation of a gasoline fraction was achieved on the WS_2 catalyst [1317]. The octane number was increased considerably (from 55 to 71 by the motor method) under conditions similar to those used for isomerisation of pure hydrocarbons (p. 366) with a low degree of splitting (5 to 8%).

The Shell Co. has developed a process for hydroisomerisation of light cracked gasoline (b.p. up to 100 °C) to a high-octane product with very low content of olefins [1878, 1883, 1884]. The reaction was carried out at temperatures starting from 250 °C and gradually rising to compensate for the decreasing catalyst activity, at 40 to 120 atm and a molar hydrogen excess of 6 : 1. NiS was found to be the optimum catalyst (in amounts up to 10%) supported on Al_2O_3–SiO_2 (with more than 60% SiO_2). An adequate catalyst lifetime (1,500 to 1,800 hours) was secured by the addition of a small amount of easily decomposed sulphur compounds (e.g. butyl mercaptan) to the feed. Kelley and Baeder [1013a] also used this catalyst. The gasoline fraction was isomerised in a similar manner in the presence of a sulphurised platinum catalyst [182a].

Studies dealing with hydroisomerisation, and especially isomerisation accompanied by desulphurisation, of higher petroleum fractions such as diesel fuels and paraffin waxes are of great technical significance. In these cases, hydrocracking must be suppressed to a minimum. Therefore, catalysts must be preferred which have minimum splitting properties (or, the splitting properties of conventional sulphide catalysts must be limited, e.g., by the addition of nitrogen compounds when NiS/Al_2O_3–SiO_2 is used) [1871, 1881]. Refining catalysts, particularly 8376,were found useful in isomerising petroleum-derived C_{15} to C_{18} n-paraffins. At 380 to 400 °C and a hydrogen pressure of 20 atm the yield of isoparaffins was up to 35%; the extent of splitting being roughly 10%. The pour point decreased from −27 °C to −42 °C and even $-52\,°C$ [1583, 2063]. The pour point of gas oil was decreased in a similarly efficient manner on the WS_2–NiS catalyst [216].

It is very advantageous to carry out hydroisomerisation and extensive desulphurisation of sulphur-containing diesel fuel in one stage, i.e., selective in order to obtain a minimum amount of gasoline. Co–Mo catalysts supported on weakly acid aluminosilicates [1155, 1294] or Mo, Mo–Ni and Mo–W–Ni catalysts supported on acid alumina activated with fluorine [1054] were found to be very good.

Mráz [1424] studied the problem of isomerisation refining of diesel fuel in great detail. Heavy diesel fuel obtained from paraffinic petroleum containing sulphur was hydrogenated on catalysts with suitable acidity under such conditions that extensive desulphurisation and considerable decrease of the pour point were accompanied by a minimum of hydrocracking. Highly selective isomerisation of higher alkanes with no hydrocracking whatever is impossible, since other components are present besides paraffins (naphthenes, aromatic compounds, sulphur and nitrogen compounds etc.), which have an unfavourable effect on the hydroisomerisation reaction. For this reason the inhibitive effect of an excess of hydrogen, which was clearly observed when pure n-paraffins were hydroisomerised [1319, 1545], was not detected in the isomerisation refining of heavy sulphurous diesel fuel [1424].

Among other sulphide catalysts, the types 5058, 8376 and Co–Mo–Al_2O_3 were tested, and the best was found to be a selective Mo–Ni splitting catalyst supported on a mixed Al_2O_3-kaolin carrier with modified acidity (e.g., by the addition of fluorine). Compared to the raw material, the isoparaffin content of the products of isomerisation refining under optimum reaction conditions (430 °C, 40 to 150 atm, space velocity 0.5 l/l hour, hydrogen to feed ratio 500 to 2,000 Nl/l) is higher. Selective hydrocracking takes place, the mean molecular weight decreases and extensive desulphurisation also occurs. The pour point and temperature of paraffin separation are lower (the pour point decreased by 20 to 25 °C on the average) and the degree of desulphurisation is 97 to 98%. A suitable catalyst acidity is maintained by the addition of a slight amount of a halide derivative to the feed (e.g., $5 \times 10^{-3}\,\%$ dichloroethane) [1424].

Hydroisomerisation of petroleum wax is a promising method of obtaining lubricating oils with a high viscosity index from cheap and easily available sources [1205]. Among catalysts suitable for this purpose, Pt/Al_2O_3, Co–Mo/Al_2O_3 and Mo/Al_2O_3 have been tested, the carriers being modified to the optimum acidity by addition of B_2O_3 or fluorine [629, 1205]. Among sulphide catalysts, WS_2 and the type 8376 were tried out. Although these two catalysts gave a lower degree of isomerisation than given by platinum catalysts [7, 987, 988], sulphur-containing feeds can be processed to good lubricating oils in their presence with a good catalyst lifetime [7]. Oils with pour point as low as $-35\,°C$ and viscosity index up to 145 [216a] were obtained on hydroisomerisation of petroleum wax in the presence of platinum sulphide catalysts [216a, 1806a].

9.

DEHYDROGENATION AND REFORMING IN THE PRESENCE OF SULPHIDE CATALYSTS

Dehydrogenation and reforming of hydrocarbons are among the most important technologies of the petrochemical industry. Although from a technical point of view most important catalysts for these processes are oxides or metals supported on various carriers [313, 709, 796, 1006], sulphide catalysts have been tested in a number of cases for use in dehydrogenation of pure hydrocarbons and reforming hydrocarbon mixtures. Moreover, with regard to the regular presence of sulphur compounds in technical raw materials, it is essential to know the catalytical effect of the sulphides of those metals which are present as the most important components in technical dehydrogenating and reforming catalysts and which may convert to the sulphide component when sulphur-containing feeds are processed. In some cases, dehydrogenation catalysts are deactivated by the influence of sulphur compounds (e.g., the nickel catalysts) [1775], in other cases a metal sulphide was found to posses a high dehydrogenation activity; e.g., MoS_2 acts as an active dehydrocyclisation catalyst [1403].

The thermodynamics, chemistry and kinetics of dehydrogenation and reforming reactions have been fully elucidated in a number of monographs [313, 709, 796, 1006, 1960]. In this chapter, attention will be paid only to literature and patent data which relate to the effect of sulphide catalysts in dehydrogenation reactions and in reforming or, to the influence of sulphur compounds on technically important non-sulphide reforming catalysts. With respect to the reversibility of hydrogenation-dehydrogenation reactions, every hydrogenating sulphide catalyst could, in principle, be used for dehydrogenation, although the process would have to be carried out at a higher temperature. In practice, only selected sulphide hydrogenation catalysts are being used in this way.

9.1.

Dehydrogenation of aliphatic hydrocarbons

Good yields of olefins have been achieved in dehydrogenating C_2 to C_{12} paraffins with sulphide catalysts. For example, propane gave a nearly 64% yield of propylene at 477 °C and 32 atm in the presence of the MoS_2/Al_2O_3 catalyst, which was a better result than that achieved with the analogous oxide catalyst [944].

Similar good yields of olefins were obtained in the dehydrogenation of butanes, pentanes and hexanes in the presence of Mo, W, Cr, Ni, V, etc. sulphides supported on Al_2O_3 [184], dehydrogenation of n-hexane and n-nonane on MoS_2 on MgO or SiO_2–Al_2O_3 [1906], in the hydrogenation of C_4 and C_5 paraffins (and olefins) with sulphur in the presence of the Ni–Mo/Al_2O_3–SiO_2 sulphide catalyst [1834c] (a similar procedure may be used to dehydrogenate ethylbenzene to styrene [1834a]) and in the dehydrogenation of ethane and propane (in the presence of H_2S) under the catalytic effect of Cu_2S on Al_2O_3 [210]. A sulphurised platinum catalyst was likewise found to have a satisfactory dehydrogenation activity [753a]. Besides dehydrogenation to olefin, the WS_2 catalyst was also found to catalyse dehydrocyclisation processes. At 500 °C, atmospheric pressure and a space velocity of 1.5 l/l hour, n-heptane formed equal yields of olefins and aromatic compounds [1863].

Olefins have also been dehydrogenated in the presence of sulphide catalysts to form dienes. For example, butenes and amylenes converted to butadiene and isoprene, resp., in the presence of MoS_2 or potassium sulphomolybdate [434]. The reaction is carried out with advantage in the presence of sulphur oxides [434, 1967]. Adams [4b, 4c] studied the activity of a number of sulphide catalysts (Mo, W, Ni sulphides and mixed Bi–Mo and Bi–W sulphides) in oxidative dehydrogenation of paraffins and olefins in the presence of SO_2. In the case of olefins, the reaction proceeds very well and, in some cases, the yield of dienes is higher than in the reaction with oxygen.

Ni sulphide on Al_2O_3 (besides other catalysts) was tested in the dehydrogenation of ethylnaphthalene to vinylnaphthalene. At 650 °C and slightly less than atmospheric pressure, approximately 39% conversion per cycle was attained with a tenfold molar excess of steam [1458].

Sulphide catalysts are not applied to dehydrogenation of paraffins on an industrial scale. The feasibility was tested using a tungsten—nickel sulphide catalyst in dehydrogenating n-paraffins of b.p. 200 to 300 °C and soft paraffins (b.p. 250 to 350 °C) obtained by dewaxing petroleum fractions with urea, α-olefins being the reaction products. Although the dehydrogenating activity was good, the selectivity of the catalyst was low, and formation of *trans*-olefins of the type $R_1CH{=}CHR_2$ predominated [980]. On the other hand, the dehydrogenation catalyst developed by the Universal Oil Product Co. (Pt-based according to the firm's data) for a new process of α-olefin production is far more selective [183].

9.2.

Dehydrogenation and dehydroisomerisation of cyclic hydrocarbons

It has been already shown in section 6.2. (p. 124) that the distance between the metal atoms in the lattice of basic sulphide catalysts (MoS_2, WS_2) is too large for dehydrogenation of the cyclohexane ring to proceed by a planar sextet mechanism, although a triangular arrangement of active metal centres would be available after partial reduction of the surface to the metal [103, 1570]. Since, however, dehydrogenation of cyclohexane compounds does take place on sulphide catalysts, it must be possible for cyclohexane dehydrogenation to take place, at least in part, on a doublet with gradual rotation of the cyclic molecule, so that the reaction intermediate is cyclohexene [103, 1570]. Dehydroisomerisation of cyclopentanes [1819] and dehydrocyclisation [1403] also takes place on sulphide catalysts.

In one of their first studies [1406], Moldavskii et al. found in respect of cyclohexane dehydrogenation that pure MoS_2 had the greatest dehydrogenation activity of a number of sulphide catalysts (MoS_2, WS_2, CoS, CoS + MoS_2 pure or supported on activated carbon or silica gel were tested). At 410 to 440 °C an apparent activation energy of 27 kcal/mole was found for this catalyst, but its catalytic efficiency is considerably lower than that of e.g. Cr_2O_3. When MoS_2 is supported on a carrier, the system MoS_2/activated carbon is the most efficient one, but this catalyst rapidly loses its activity [905].

Table 90.

Dependence of Dehydrogenation Activity of the Re_2S_7/Al_2O_3 Catalyst on the Rhenium Content. (Dehydrogenation of cyclohexane in a flow apparatus, 500 °C, $P_H = 5$ atm, LHSV 1, ratio cyclohexane : H_2 = 1 : 5) [1375, 1378].

Composition of the reaction product	Rhenium content in the catalyst (%)[a]				
	1	5	10	15	20
Isopentane	—	0.1	0.5	1.4	1.2
n-Pentane	—	0.1	0.3	0.6	1.2
2,2-Dimethylbutane	—	traces	traces	0.2	traces
2-Methylpentane	—	2.0	3.3	5.8	1.2
3-Methylpentane (+ cyclopentane)	—	2.5	2.7	4.5	1.2
n-Hexane	—	2.2	2.1	4.2	1.6
Methylcyclopentane	5.1	16.3	11.4	11.2	0.8
Benzene	2.9	20.2	31.4	34.6	76.1
Toluene	—	—	—	—	9.4
Cyclohexane	92.0	56.3	47.8	35.9	3.5

[a]) The remainder up to 100% is C_1-C_4 hydrocarbons.

With the WS_2 catalyst, the activation energy of cyclohexane dehydrogenation at 380 to 650 °C was found to be about 19 kcal/mole (the measurement was carried out in a flow reactor with programmed temperature rise) [220]. The experiments reported by Balandin et al. [2061] show a somewhat higher activation energy for cyclohexane dehydrogenation in the temperature range 420 to 490 °C (22.6 kcal/mole). The activation energy of cyclohexene dehydrogenation over the same temperature range is 21 on WS_2 and 17.7 kcal/mole on MoS_2. Olefins were found among the dehydrogenation products of cyclohexane, indicating a reaction course by a doublet mechanism.

Rhenium sulphide supported on alumina is an efficient dehydrogenation catalyst. Minachev et al. [1375, 1378, 1785, 1786] found that the benzene yield in cyclohexane dehydrogenation depends to a marked extent on the rhenium content of the catalyst (Table 90). The dehydrogenation is accompanied by isomerisation, splitting and alkylation. A catalyst with low rhenium content has little activity but the dehydrogenation efficiency is considerably improved by the addition of palladium. For example a catalyst containing a mixture of rhenium and palladium sulphides (1% Re, 1% Pd) on Al_2O_3 or aluminosilicate gave a benzene yield of 87% at 480 °C [1785, 1786]. Methylcyclopentane was a by-product of this reaction [1787b].

Unsupported mixed nickel–tungsten sulphide is a very efficient dehydrogenation catalyst. The original German dehydrogenation catalyst IG 5615 contained an excess of tungsten ($NiS : WS_2 \sim 1 : 2$) [756, 1347]. This catalyst was highly selective, since practically no isomerisation or splitting reactions took place in its presence. For example at 420 °C, 14.5 atm (the partial hydrogen pressure was 11.5 atm) and space velocity of 1.7 l/l hour, roughly 96% conversion to benzene was achieved on dehydrogenation of cyclohexane. Similarly good conversion to tetralin and naphthalene was obtained on dehydrogenating decalin [756]. In the presence of hydrogen sulphide (or of sulphur compounds which form H_2S on hydrodesulphurisation) the overall dehydrogenation activity of the catalyst 5615 as well as its selectivity decreases due to the increased acidity of the catalyst surface [757].

This finding is significant, e.g., for the Autofining process (cf. p. 312), where production of high-quality gasoline with efficient desulphurisation is demanded. In the case of sulphur-containing feeds, circulating hydrogen must be made free of H_2S.

A catalyst with the reverse molar ratio of the two components ($NiS : WS_2 \sim 2 : 1$) also has good dehydrogenation properties [56]. The catalyst developed by the Shell Co. was used in World War II for dehydrogenation of cyclohexane and methylcyclohexane to benzene and toluene [1049, 1872]. A catalyst of similar composition was used by Cole [322, 323, 329], Johnson [938, 939, 940] and Peck [1591] for dehydrogenation of naphthenic gasolines with a high content of aromatic hydrocarbons. On the other hand, an inferior dehydrogenating activity was observed in the case of the catalyst 8376 (WS_2–NiS/Al_2O_3) [1819]. A mixed catalyst composed of Mo and Ni sulphides was also used with success for dehydrogenation of alkylcyclohexanes [1889]. On this catalyst, supported on Al_2O_3, octahydroazulene was successfully dehydrogenated to azulene in the presence of CS_2:

$$\text{(octahydroazulene)} + CS_2 \longrightarrow \text{(azulene)} + 2\,H_2S + CH_4 \qquad \text{(CLIII)}$$

60% conversion being attained at 405 °C [1081]. Polynuclear aromatic hydrocarbons from the respective naphthenes, contained in the oil, were obtained by dehydrogenation in the presence of Co, Ni or Mo sulphides [1367b]. Palladium sulphide was a selective catalyst of some dehydrogenation reactions (e.g., dehydrogenation of 3,3′,4,4′-tetrahydro-1,1′-dinaphtyl to benzo[j]fluoranthene) [363a].

One very important reaction is dehydroisomerisation of five-membered naphthenes to aromatic compounds. This reaction is technically significant, particularly in reforming and as an application of sulphide catalysts, e.g., in the HTM process, where it is one of the reactions which improve the quality of the gasoline obtained [1819, 1820] (p. 383).

At the elevated temperatures required for efficient dehydrogenation (above 400 °C) the isomerisation equilibrium is shifted far to the side of the five-membered naphthenes [1770]. When dehydrogenation takes place at the same time, however, equilibrium yields of the corresponding aromatic compounds are considerably higher due to the low thermal stability of the six-membered naphthenes which are temporarily formed. The equilibrium yield of the respective aromatic compound also rises with the increasing molecular weight of the alkylcyclopentane, as we see by comparing the dehydroisomerisation equilibria of methyl- and ethylcyclopentane (Fig. 104).

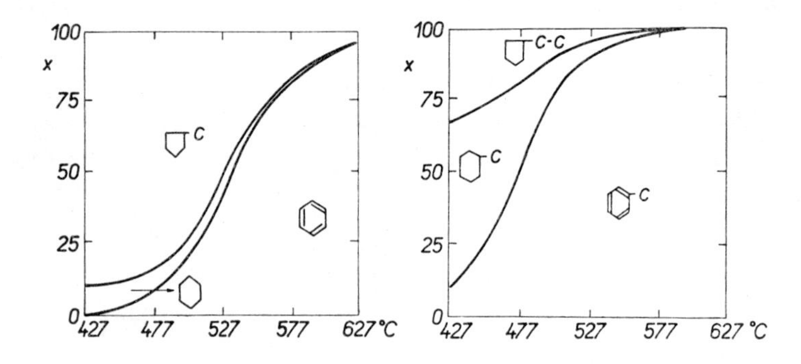

Fig. 104. Temperature relationship of the equilibrium composition of methyl- and ethylcyclopentane dehydroisomerisation products at $P_H = 60$ atm (in the case of ethylcyclopentane, data for this hydrocarbon also include dimethylcyclopentanes) [1819]. x — Mole % of the component in the equilibrium mixture.

Good results in dehydroisomerisating five-membered naphthenes were achieved in the presence of catalysts containing WS_2 or MoS_2 and NiS [2069]. In the presence of rhenium heptasulphide at 500 to 510 °C, methylcyclopentane was likewise efficiently dehydrogenated and isomerised, and a benzene yield in excess of 52% was achieved. The process, however, was accompanied by side reactions, mainly hydrocracking and the formation of paraffins [1375].

As in the case of dehydrogenation, a mixed catalyst composed of Pd and Re sulphides supported on aluminosilicate or Al_2O_3 is more active than Re_2S_7 alone in dehydroisomerisation reactions [1785, 1786].

Schneider et al. [1819, 1820] investigated the dehydroisomerisation efficiency of the catalyst 8376 in connection with a study of the formation of aromatic compounds in the HTM process. At temperatures above 500 °C, this catalyst causes mainly isomerisation, while its dehydrogenating activity is relatively low (differing from the unsupported NiS–WS_2 catalyst). Comparison of results for several alkylcyclopentanes (Table 91) shows that dehydroisomerisation is easier for higher-boiling cyclopentanes; while only slight conversion is obtained with methylcyclopentane, ethylcyclopentane is converted to more than 63% of the equilibrium yield [1820]. Schneider et al. found that the rate-controlling step in dehydroisomerisation in the presence of the catalyst 8376 is dehydrogenation of the six-membered ring formed to the respective aromatic hydrocarbon. The different yield of six-membered hydrocarbons in dehydroisomerisation of methylcyclopentane and higher alkylcyclopentanes is due to thermodynamic factors (the equilibrium content of cyclohexane in methylcyclopentane isomerisation is considerably less than the methylcyclohexane content of an equilibrium mixture with ethylcyclopentane) [1819] as well as to the different isomerisation mechanisms. Schneider assumes that cyclohexane formation from methylcyclopentane proceeds by way of a low-stability primary carbonium ion intermediate:

$$R^{\oplus} + \text{C}_5\text{H}_9\text{—CH—CH}_3 \rightarrow RH + \text{C}_5\text{H}_9\text{—}\overset{\oplus}{\text{C}}\text{—CH}_3$$

$$\rightleftarrows \text{C}_5\text{H}_9\text{—CH—}\overset{\oplus}{\text{CH}_2} \rightleftarrows \text{C}_6\text{ ring—}\overset{\oplus}{\text{CH}} \xrightarrow{RH} \text{C}_6\text{H}_{12} + R^{\oplus}$$

(CLIV)

Table 91.

Dehydroisomerisation of Alkylcyclopentanes in the Presence of the Catalyst 8376.
(580 °C, total pressure 15 atm, LHSV 0.3, volume ratio H_2 : hydrocarbon = 100 : 1) [1820].

Cyclopentane	Methyl-	Ethyl-	Propyl-	Isopropyl-	Butyl
Aromatic substances (% by wt.)	0	22.6	26.7	30.8	29.7
Cyclohexanes (% by wt.)	0.5	17.7	11.5	26.7	24.8

On the other hand, the expansion of the ethylcyclopentane ring does not necessitate the formation of a primary ion, only the far easier production of secondary and tertiary ions:

$$R^{\oplus} + C_5H_9\text{—}CH\text{—}CH_2\text{—}CH_3 \rightarrow RH + C_5H_8\text{—}\overset{\oplus}{C}\text{—}CH_2\text{—}CH_3$$

$$\updownarrow$$

$$C_5H_9\text{—}CH\text{—}\overset{\oplus}{CH}\text{—}CH_3 \rightleftarrows C_6H_{10}\text{—}\overset{\oplus}{C}\text{—}CH_3$$

$$\downarrow RH$$

$$C_6H_{11}\text{—}CH_3 + R^{\oplus}. \qquad \text{(CLV)}$$

The ionic course of the dehydroisomerisation reaction was proved by the decrease in activity of the catalyst 8376 on poisoning of the acid centres of the catalyst, e.g., by basic nitrogen compounds [1820].

Terpenes and turpentine have been dehydroisomerised to aromatic hydrocarbons in the presence of molybdenum sulphide [147, 148].

9.3.

Dehydrocyclisation of aliphatic hydrocarbons

The process of dehydrocyclisation of paraffins and olefins was discovered in 1936 by Moldavskii and Kamusher [1403], who aromatised some hydrocarbons possessing straight chains (n-heptane, n-octane, etc.) in the presence of Cr_2O_3 and MoS_2.

Soon after it had been discovered, this reaction was practically applied in catalytic reforming, where it is one of the partial processes by means of which the octane number of gasoline is increased.

From the thermodynamic point of view, the course of dehydrocyclisation is facilitated by elevated temperatures, lower pressures and higher molecular weight of the initial substance. The influence of the reaction conditions on the equilibrium yield in a typical example of n-heptane dehydrocyclisation is shown in Fig. 105. Although hydrogen has an adverse effect on the equilibrium of this reaction, its presence is essential to maintain a high degree of activity in the catalyst. The reaction is highly endothermic (in the case of n-heptane at about 400 °C, the value of ΔH is about 60 kcal/mole) [313].

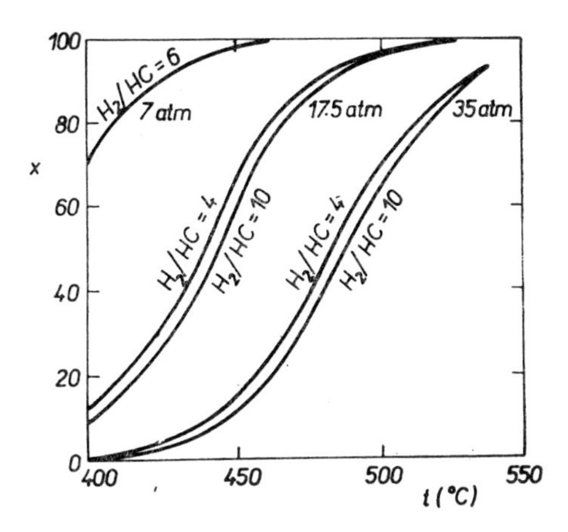

Fig. 105. Equilibrium composition of the mixture n-heptane–toluene–hydrogen [313]. *x* — Mole % toluene.

The mechanism of dehydrocyclisation reactions on sulphide catalysts has not yet been studied in detail. Considering analogous results for the reaction on both the catalysts MoS_2 and Cr_2O_3 [1403, 1960] and with regard to the satisfactory dehydrocyclisation activity of a chromium sulphide-containing catalyst [2080], a similar mechanism to that which operates in the case of the chromium oxide catalyst [768, 1960] may be assumed.

Typical dehydrocyclisation catalysts are transition metal oxides (Cr, Mo, V, Zr) supported usually on Al_2O_3 [1960]. Some sulphides, particularly molybdenum and chromium sulphides, also exhibit a good dehydrocyclisation activity [1410]. In his first studies, Moldavskii [1403, 1407] reported good yields from the dehydrocyclisation of n-heptane to toluene and n-octane to *o*-xylene in the presence of MoS_2. The yield of aromatic hydrocarbons rises when MoS_2 is supported on different carriers in the order: $SiO_2 < Al_2O_3\text{–}SiO_2 < MgO < Al_2O_3$ (experiments at 500 °C) [1655].

A catalyst composed of a mixture of chromium sulphide and oxide, prepared by partial reduction of chromium sulphate, was found to be sufficiently active in n-heptane aromatisation [2050]. Some aromatic hydrocarbons of suitable structure may also be dehydrocyclised in the presence of oxide or sulphide catalysts. For example, *o*-benzyltoluene converts to anthracene [871, 1835].

9.4.

Dehydrogenation and reforming of raw materials in the petroleum industry in the presence of sulphide catalysts

Catalytic reforming, in which the main purpose is conversion of low-quality gasoline to high-octane fuels and aromatic hydrocarbons is one of the most important industrial catalytic reactions [314, 711].

Reforming is a strongly endothermic complex of simultaneous reactions, the most important being reactions of the dehydrogenation type (dehydrogenation of cyclohexanes to aromatic hydrocarbons, dehydroisomerisation of alkylcyclopentanes to aromatic hydrocarbons, dehydrocyclisation of paraffins and olefins to aromatic hydrocarbons) and isomerisation reactions (isomerisation of paraffins to isoparaffins, isomerisation of cyclopentanes to cyclohexanes, hydroisomerisation of olefins and isomerisation of some aromatic compounds). Moreover, intensive hydrocracking reactions are involved (splitting of paraffins and olefins, hydrogenolysis of alkylcyclopentanes to isoparaffins, etc.) and refining reactions (particularly desulphurisation) [313, 709, 1724]. The reaction rates of these individual partial processes vary greatly, since dehydrogenation of naphthenes, isomerisation of paraffins and hydrocracking of the heavier fractions of the raw material are considerably more rapid than dehydrocyclisation of paraffins, hydrocracking of the lighter fractions and dealkylation [711, 2137]. The result of reforming will therefore depend to a large extent on the selectivity of the catalyst employed as well as on the reaction conditions.

9.4.1.

The influence of sulphur compounds on the most important industrial reforming catalysts

Although mainly platinum catalysts are employed in modern reforming processes, the early development of this technological process is closely related to the application of sulphide catalysts. The first technical dehydrogenation unit started to operate in 1930 in Leuna. On the basis of laboratory and pilot-plant experiments with oxide and sulphide catalysts [840, 882], low-quantity gasoline was reformed at 510 °C and 50 atm in the presence of hydrogen on a WS_2–NiS catalyst. Later, in 1938, this process was used in reforming gasoline originating in the liquid-phase hydrogenation of bituminous-coal pitch [998]. During World War II, a tungsten–nickel sulphide catalyst was employed in the USA in the production of aircraft gasoline. After the war this technology was completely converted to producing aromatic hydrocarbons [709, 1524].

The HF process was developed in Leuna in the year 1939. This process involved reforming of straight run gasolines on a MoO_3–ZnO–Al_2O_3 catalyst at 480 to 520 °C and 15 atm [1101]; however, the low hydrogen pressure did not suffice to protect the catalyst from rapid contamination with coke. Therefore, the process was modified and in 1940 the DHD process came into operation, involving a hydrogen pressure of 50 atm and periodically regenerated catalyst containing 10% MoO_3 on Al_2O_3 [1101, 1468]. The analogous Hydroforming process which was developed in the USA at the same time, operates at pressures up to 20 atm with the same fixed-bed catalyst which, however, is regenerated cyclically after 4 to 8 hours of operation [313]. In the Fluid-Hydroforming and Orthoforming processes (M. W. Kellogg Co.) the continuously regenerated fluidised-bed catalyst is used [313]. Two modern processes with oxide catalysts employ a mobile catalyst bed with continuous regeneration. The Thermofor Catalytic Reforming process (Socony Mobil Oil Co.) operates with a catalyst containing 32% Cr_2O_3 on Al_2O_3 which is highly resistant (except to the effect of steam), desulphurises efficiently and is distinguished by a long catalyst life [313]. The Hyperforming process (Union Oil Co. of California) [136] operates with a mobile Co–Mo catalyst supported on alumina which is stabilised with SiO_2.

The advantage common to all these processes, in which the catalyst is in the oxide form when fresh, is the possibility of treating sulphur-containing feeds (sometimes even with a high sulphur content, e.g., the Hyperforming process was used to treat heavy gasoline from coking, containing up to 3.5% S) [373]. Since reforming is accompanied by intensive desulphurisation with formation of H_2S, the oxide catalyst can convert in part to sulphides under the reaction conditions employed. The desulphurisation efficiency of oxide catalysts is high, as nearly 90% of the total sulphur content is converted to H_2S during the reforming process [1724]. The sulphides thus formed are sufficiently active for the main reaction, as will be shown later, and they do not cause deactivation of the oxide reforming catalysts [1724, 2137]. Loss of catalyst activity is mainly due to intensive contamination with carbonaceous deposits [1101].

On the other hand, sulphur compounds and particularly hydrogen sulphide formed from them, cause reversible poisoning of platinum catalysts [313, 715]. The dehydrogenation and dehydrocyclisation activity, as well as the lifetime of the catalyst, decrease under the influence of hydrogen sulphide [190, 313, 769, 1316, 1374, 1724]. Therefore, platinum catalysts demand prerefined feeds (in processes with periodic regeneration, the feed is desulphurised down to 0.01% S and less).

Recent results of American authors indicate, however, that sulphur-containing substances need not always have a deactivating effect on reforming catalysts; under certain conditions, there exists an optimum level of sulphur in the system, at which good activity and life-time of the catalyst are secured as well as the required selectivity of the process [743B, 743e, 1614a, 1656A, 1656b]. Hayes et al. [743B, 743e] found in the case of low-pressure reforming on sulphurised platinum catalysts that the optimum sulphur concentration in the feed (in the form of H_2S or of degradable sulphur compounds containing neither O or N) lies in the 300 to 3000 p. p. m. range (calculated as elementary sulphur for the amount of input hydrocarbons); at the same time, water and oxygenous substances (which might hydrogenate to form water) must be thoroughly removed [743B, 743e, 1656b]. It is likewise advantageous to add to the raw matrial small amounts of halogen-containing compounds [743B].

At present, little is known about the way in which sulphur is bound to the platinum reforming catalyst [1373] (cf. also section 4.2, p. 49). Partial formation of platinum sulphides cannot be excluded and according to some authors, these sulphides act as reforming catalysts [173, 466, 713, 744]. Although in catalytic reforming more than 95% of the sulphur contained in the feed leaves the reactor in the form of H_2S, a certain amount of sulphur still remains bound to the catalyst. Using thiophene labelled with radioactive suphur, Minachev et al. found 0.06 to 0.14% S on the catalyst, depending on the sulphur concentration of the feed [1374]. Up to nearly 0.5% S was found on catalysts of elevated acidity after reforming sulphur containing feeds [715], i.e., 3 to 4 times the stoichiometric value (assuming conversion of all platinum in the catalyst to PtS_2). Sulphur is removed in the form of SO_2 by oxidative regeneration. It is interesting that the original activity of the catalyst is completely renewed when only 60% of the sulphur originally bound by the catalyst has been removed [1724]. In connection with this, Hayes [743c] recommends removal of sulphur from the platinum reforming catalyst before its oxidative regeneration, as otherwise platinum crystallites would grow and the activity of the regenerated catalyst would be lowered. Sulphidation of the regenerated catalyst is carried out as the last operation to secure prolonged activity in case of processing sulphur-containing feeds [743b]. Nowadays, platinum reforming catalysts are frequently sulphided [5a, 743a, 743B, 743b, 743d, 743e, 1656b, 2103e]; this may be related to the positive influence of sulphur on maintenance of the optimum size of Pt crystallites and considerable decrease of the rate of formation of deposits (cf. also [743c]). Buss [240b] states, that another reason for presulphidation of reforming catalysts consists in decreasing the formation of light hydrocarbon gases during startup.

The almost complete reversibility of poisoning of platinum reforming catalysts [313, 715, 1373, 1614a] indicates that strong adsorption of H_2S on hydrogenation-dehydrogenation centres of the bifunctional catalyst is the main cause of deactivation. The degree to which aromatisation reactions are suppressed, and on the other hand the promotion effect of H_2S in the formation of cyclopentanes, is proportional to the sulphur content of the feed. The sensitivity of the catalyst to H_2S rises further with the increasing activity of the catalyst, particularly in respect of aromatisation reactions [715].

9.4.2.

The direct use of sulphide catalysts in reforming

The sulphide catalysts most efficient in gasoline reforming are mixed NiS–WS_2 catalysts which, as was shown earlier, have also been used for industrial-scale dehydrogenation of cyclohexane and of its homologues [756, 1049, 1347, 1872] or for reforming low-quality gasolines [709, 840, 882, 998, 1524].

Combinations of sulphides of Groups VI and VIII metals [1636, 1639], particularly nickel–tungsten sulphidic catalysts, were also used in quite a number of other cases for reforming gasoline. Either gasoline or light oil obtained by coal hydrogenation were reformed [861, 893] or virgin petroleum gasoline [1426, 1491, 1500, 1503, 1591, 1804, 1831]. The working temperatures employed ranged from 450 to 510 °C, pressures from 30 to 70 atm.

Of the other sulphide catalysts, sulphides of Group VI [92, 241, 1346] or mixtures of sulphides and oxides, e.g., NiO–NiS/Al_2O_3 [422] and mixtures of CdS or ZnS with some oxides (MoO_3, ZnO, Al_2O_3) [834, 868, 869] were tried out in reforming processes. The acidity of sulphide reforming catalysts supported on Al_2O_3 [139, 1508], particularly MoS_2/Al_2O_3 (1501, 1954, 1955], was adjusted by the addition of halides, mainly fluorides.

Data relating to the activity of noble metal sulphides in reforming low-octane gasolines are very interesting from the technological point of view. A number of patent data concerning the application of platinum sulphides as components of catalysts in reforming [173, 273a, 466, 712, 713, 744] indicates that partial conversion of platinum to the sulphide is not the main reason for deactivation of these catalysts in reforming of sulphur-containing gasoline, and that the primary reason is deposition of carbonaceous substances and blocking of the hydrogenating-dehydrogenating centres of the catalyst.

Results achieved with sulphide catalysts based on a similar noble metal, palladium, which Minachev describes in a number of studies [1371, 1372, 1894] confirm the high activity of noble metal sulphides in gasoline reforming. At the same time these findings justify the efforts made to develop a highly resistant and efficient catalyst based on noble metal sulphides, which would be suitable for reforming untreated sulphur-containing gasoline. As in the case of platinum (cf. p. 49), no palladium sulphide phase was detected by X-ray examination of the sulphurised palladium catalyst (due to the very low concentration of sulphide in the catalyst). In spite of this, Minachev et al. [1372] ascribe the relatively high activity, and particularly the stability of the activity, of the palladium sulphide catalyst in reforming of sulphur-containing gasoline to the existence of active PdS on the catalyst surface.

The sulphides of other noble or rare metals, e.g., rhodium sulphide [2046] or a catalyst containing 20% rhenium sulphide supported on Al_2O_3 [1376, 1377] and, moreover, sulphurised mixed Pt–Rh [1656a] and Pd–Re [1378a] catalysts have also been tested.

9.5.

Aromatising hydrogenation and the HTM process

Both these hydrogenolytic processes, which have been developed for the purpose of producing high-quality gasoline from tar-derived materials, operated (particularly in respect of temperature) under dehydrogenating conditions, which permit highly aromatic products to be made.

All destructive hydrogenation processes mentioned in section 7.2.1. (pp. 293 and following) were conducted under effective hydrogenating conditions and the catalysts were selected in such a way as to obtain a large amount of paraffins and naphthenes. The requirement of obtaining high-octane gasoline by means of destructive hydrogenation of tar, and thus to lower the hydrogen consumption, meant primarily that the operating conditions must be chosen so as to secure the maximum yield of aromatic hydrocarbons.

Suitable feeds for this purpose were particularly middle oils from bituminous-coal tar. The major problem involved was the development of an optimum catalyst, which had to contain a strong splitting component, capable of splitting the thermodynamically stable aromatic medium oils to gasoline. The hydrogenating component of the catalyst had to guarantee rapid hydrogenation of the split chains in order to avoid any unwanted polymerisation and condensation reactions. At the same time, the hydrogenation activity had to be modified in such a manner as to keep the catalyst from bringing about undue hydrogenation of simple benzene rings, since under the working conditions applied (pressure 300 to 700 atm, needed to keep up a high catalyst activity, temperatures close to 500 °C) equilibrium is shifted to the side of hydroaromatic compounds due to the high pressure of hydrogen. Finally, in contrast to conventional hydrocracking catalysts, the catalyst had to be totally insensitive to nitrogen compounds in the feed as well as in the recycled gas. An industrial catalyst possessing these properties was developed by the former I.G.F. A.G. and served to hydrogenate bituminous-coal middle oil to aviation gasoline by the aromatising hydrogenation process. The catalyst contained 0.6% Mo, 2% Cr, 5% Zn and 5% S on HF-activated alumina [1101]. Mixtures of Fe, Mn and Zn sulphides with Mo, W, V, Co or Ni sulphides were also tested [1637, 1901].

The HTM process (Hochtemperatur-Mitteldruck-Verfahren) was developed at Zeitz (German Democratic Republic) for single-stage hydroprocessing of brown-coal light oil to stable aromatic gasoline. This process, which uses the catalyst 8376, operates at temperatures above 500 °C and medium pressures, i.e., under conditions which enable dehydrogenation and dehydroisomerisation reactions to take place. These reactions are the main source of newly-formed aromatic hydrocarbons and, therefore, of an increase in the octane number [1810, 1819, 1820], as dehydrocyclisation of paraffins and olefins on the catalyst 8376 takes place to a very limited extent under the operating conditions employed [1819]. Isomerisation reactions are retarded by the presence of nitrogen compounds in the raw material [1820]. Since efficient refining of the raw material takes place besides the reactions mentioned above, the HTM process is rather a compromise between hydrorefining and reforming [702].

9.6.

Dehydrogenation of alcohols

ZnS and CdS are the sulphide catalysts which have been employed most often to dehydrogenate alcohols. These were also found to be the most suitable sulphide catalysts for a possible industrial application in dehydrogenating primary and secondary alcohols [833].

Zinc sulphide supported on pumice was employed in Germany for the industrial production of isobutyraldehyde by dehydrogenation of isobutyl alcohol at 370 °C, the product being subjected to reductive amination to isobutylamine [500]. Falkovskii et al. [504] found that butyl alcohol, isobutyl alcohol and isoamyl alcohol are dehydrogenated in the presence of this catalyst by a first-order reaction with an activation energy of roughly 22 kcal/mole. ZnS supported on active carbon was used to study the dehydrogenation of isopropyl alcohol in a fluidised-bed reactor. It was found that the activation energy of the reaction (in the temperature range of 300 to 440 °C) is practically identical when the reaction is carried out in the fixed- or fluidised-bed reactors [1392]. The dehydrogenation activity of a mixed ZnS–CuS catalyst in methanol decomposition was likewise investigated [1139].

Dehydrogenation of alcohols is also catalysed by sulphides of Group VI of the Periodic Table [1167, 2060, 2061]. Here, MoS_2 is considerably more active than WS_2, which primarily catalyses the dehydration reaction [1167, 2061]. The main product of pinacol dehydrogenation at 250 to 300 °C in the presence of molybdenum sulphide was acetone [28].

Rhenium disulphide has also been found to be an efficient catalyst in alcohol dehydrogenation. At temperatures above 400 °C methanol, ethanol and isopropyl alcohol are converted to the respective carbonyl compounds in good yields and with a high degree of selectivity [1653].

In isopropyl alcohol decomposition, NiS, Cr sulphide, ZnS and mixed NiS–ZnS were found to have mainly a dehydrating effect. The fresh catalysts exhibit some degree of dehydrogenating activity, although this decreases gradually. Rubinstein et al. [1776] showed that the composition of the catalysts alters during the reaction (NiS converts to the subsulphide, ZnS hydrolyses in part to ZnO). This alters the structure and interatomic distances of the active catalyst layer and thus also its selectivity. The authors deduced from a number of measurements, that adsorption takes place in the case of dehydrogenation of alcohols and hydrocarbons on oxides and sulphides by preferential bonding of the dissociated hydrogen to free valencies of oxygen or sulphur. The high temperature needed for dehydrogenation on oxides and sulphides, as compared to metallic dehydrogenation catalysts, is essential, because more energy is needed to form a free oxygen or sulphur valency than is required with a metal [1195, 1196].

9.7.

Dehydrogenation of some nitrogen, sulphur and oxygen compounds

Among nitrogen compounds, amines have been dehydrogenated to nitriles on supported zinc sulphide: e.g., isobutyl amine gave a high yield of the respective nitrile at 475 °C [850]. Selective dehydrogenation of saturated nitriles to unsaturated ones also takes place on the WS_2–NiS catalyst. E.g., propionitrile was converted to acrylonitrile in a good yield [2047].

Some sulphur compounds were also dehydrogenated. Nickel sulphide on Al_2O_3 was used to dehydrogenate thiophane to thiophene: at 350 °C, 18% conversion was achieved at atmospheric pressure [2245]. ReS_2 supported on non-acidic Al_2O_3 was used to the same purpose [1303a]. Tall oil has been dehydrogenated in the presence of WS_2 or MoS_2 at 240 °C [692].

One point of interest is the use of As, Sb and Bi sulphides for dehydrogenation of some oxygen substances. For example, isobutyraldehyde converted to methacryl aldehyde in the presence of arsenic sulphide [729a].

10.

OXIDATION IN THE PRESENCE OF SULPHIDE CATALYSTS

In spite of the fact that active metal sulphides are not typical oxidation catalysts, they have been used by a number of authors to oxidise sulphur compounds, carbon monoxide and also hydrocarbons of various molecular weight (from gaseous hydrocarbons through lubricating oils up to hydrocarbons contained in asphalt).

Oxygen alters the activity of sulphide catalysts to a considerable extent (cf. p. 107). Oxidation reactions involve substantial changes of the catalytic surface in the presence of oxygen. For example, the active form of the CoS catalyst is the "basic" cobalt sulphide CoS(OH) in the case of low-temperature carbon monoxide oxidation. Oxidation of CoS to the sulphate takes place as a side reaction when CO is oxidised, the rate of the side reaction depending on the partial O_2 pressure in the mixture with carbon monoxide [418]. When H_2S is oxidised in the presence of nickel sub-sulphide, rhombohedral Ni_3S_2 gradually converts to rhombohedral NiS and Ni_3S_4 and, finally, to cubic NiS_2 which is probably the active oxidation catalyst. Oxidation proceeds in part as far as a sulphate [113, 387].

10.1.

Oxidation of sulphur compounds

Oxidation of hydrogen sulphide and sulphites

Of sulphide catalysts [1296] the most important ones in this reaction are nickel sulphides [113, 387, 664, 774a, 1937, 1938]. Oxidation has been carried out with atmospheric oxygen, with the purpose of removing H_2S from synthesis gas [1296] or air [113]. The reaction is performed at low or medium temperatures, the temperature being kept carefully at the minimum level needed. In some cases, very low temperatures were found sufficient (e.g., below 80 °C), the products being elementary sulphur and water. At higher temperatures (around 200 °C) SO_2 is obtained [387, 390] or a mixture of elementary sulphur and SO_2 [1937, 1938]. Ethylene was found to act as inhibitor in this oxidation reaction [113] as well as in carbon disulphide oxidation [655].

Alkali sulphites were oxidised in the presence of a catalyst, a component of which was CoS [1890b].

Oxidation of organic sulphur compounds

Carbonisation gases contain carbon disulphide, carbonyl sulphide and thiophene as the main sulphur components, as well as smaller amounts of other sulphur compounds (mercaptans, sulphides etc.). Griffith [656, 659] suggested oxidative degradation of sulphur compounds in carbonisation gases as an alternative to hydrodesulphurisation (cf. p. 326). This procedure, however, has not been applied on a wide scale in the gas industry.

Carbon disulphide oxidation proceeds readily even as a homogenous reaction, although it is strongly retarded by the presence of ethylene. It was found that nickel subsulphide is a suitable catalyst for CS_2 and COS oxidation [653]. Kinetic measurements in the presence of nickel subsulphide have shown that oxidative degradation of CS_2 in the temperature range of 200 to 350 °C is a zero-order reaction with respect to the sulphur compound. On the other hand, carbonyl sulphide is more resistant, the apparent reaction order for its oxidation (at the same temperature as with CS_2) is between 0 and 1, and the reaction is retarded by the strong adsorption of SO_2 which is formed in the reaction [655]. Under the same conditions, thiophene is not oxidised. In the practical application of this reaction, the temperature of the catalyst bed was 370 °C, and the catalyst was regenerated by careful oxidation after 3 to 4 months of operation [659].

Efficient degradation of mercaptans, sulphides and disulphides in town gas also takes place in the presence of Ni_3S_2 [595]. The mixed fluidised catalyst WS_2–NiS was tested for oxidation of higher mercaptans to aliphatic carboxylic acids [1940]. Some sulphide catalysts were tested for use in oxidation of organic sulphides to sulphoxides; nickel sulphide was the most active catalyst [1305a, 1305b].

The use of sulphide catalysts for oxidative desulphurisation of petroleum fractions has been described in several patents. PbS was tested as a catalyst of the oxidation of mercaptans in the case of sweetening of petroleum distillates [227, 228, 1610, 2171]. Petroleum distillates were desulphurised oxidatively in the gas phase at 290 to 500 °C in mixture with oxygen on CuS supported on natural clay, alumina or aluminosilicate [792].

10.2.

Oxidation of hydrocarbons and asphalts

The catalytic influence of metal sulphides on the oxidation of hydrocarbons is sometimes of great practical significance. It was found, for example, that microcrystals with a hexagonal configuration catalyse the formation of formic acid in low-temperature oxidation of hydrocarbons (in combustion engines). This acid increases the tendency of the fuel to knocking and its influence must be counteracted by the addition of an antiknock agent. In this connection, the catalytic influence of some metals, together with their oxides and sulphides, on the oxidation of n-heptane and of other hydrocarbons was investigated [1386].

The problem of the influence of MoS_2, which is being used increasingly as a solid lubricant, on the oxidation stability of lubricating oils is most important. Therefore the oxidation of lubricating oils has been studied in the presence of this catalyst at 200 °C. It was found that low-temperature oxidation is catalysed neither by colloidal nor by finely ground MoS_2 (in concentrations of 0.35 and 1%) [371].

Some sulphidic promoters were components of a supported Cu_2O catalyst used for the oxidation of paraffinic C_3 to C_6 hydrocarbons to unsaturated carbonyl compounds, particularly α-methylene aldehydes [292]. Ruthenium sulphide was used for oxidation of olefins to acids [1006a]. Among cyclic hydrocarbons, MoS_2 was tested for use in oxidation of cyclohexane [275]. The oxidation of benzene to phenol in the liquid phase was tried with the use of ZnS and simultaneous irradiation with X-rays or ultraviolet radiation [1411, 1674].

A number of papers describe the use of sulphide catalysts for oxidation of hydrocarbons with steam. Good yields of ketones were obtained by the reaction of a two- to tenfold steam excess with olefins (1-octene, 1-pentene and 2-pentene were tested) at 250 to 400 °C and 10 to 240 atm in the presence of mixed sulphides, particularly WS_2–NiS [518].

Synthesis gas was obtained by the oxidation of gaseous hydrocarbons, e.g., hydrocarbons from cracking processes, mixtures of C_3-hydrocarbons, propane, etc., with steam or an oxygen–steam mixture. Besides oxides some sulphide catalysts, especially Ni and Mo [118, 598, 1435, 1849, 1880, 1886], a sulphurised Co–Mo catalyst alkalised with K_2CO_3 or $Ba(OAc)_2$ [174a] and rhenium sulphide (promoted with Ag) supported on Al_2O_3 [1871a] were also tested. Mo or W sulphides supported on a porous carrier have been used in the fluidised state for the oxidation of liquid hydrocarbons (e.g., oil fractions) to synthesis gas [1472].

Phosphorus sulphides have been used as catalysts in some cases of asphalt blowing [231, 1760, 2231]. The catalyst concentration used was 0.1 to 5% by wt. of the asphalt [231].

11.

POLYMERISATION IN THE PRESENCE OF SULPHIDE CATALYSTS

11.1.

Polymerisation of hydrocarbons

It has been established, that sulphur compounds deactivate polymerisation catalysts which contain transition metal oxides supported on a carrier. It has also been observed that molybdenum sulphide on Al_2O_3 or MoS_2–ZnO–MgO suppress the initiation of ethylene polymerisation [230]. These observations are surprising, particularly in view of the general chemical similarity of transition metal oxides and sulphides; moreover, certain literature and patent data show that some sulphides are efficient catalysts of polymerisation reactions.

After determining the activating effect of sulphates on the polymerising activity of the respective sulphides (Cu, Fe, Ni, Co, Bi, etc.) [2166, 2167, 2168] Wassermann et al. investigated the efficiency of a number of metal sulphides in catalysing the polymerisation of olefins. From experiments with cyclopentadiene dimerisation (Table 92), it follows that black sulphides are active, while white and yellow ones are inactive. Exceptions are MoS_2 and FeS_2. This observation also applies to the reverse reaction, i.e., dicyclopentadiene depolymerisation [1019, 1020]. The polymerisation activity of black sulphides depends on the manner of their preparation, as will be seen from the results achieved with the most highly polymerisation-active catalyst, copper sulphide (Table 93). The catalyst achieves the highest activity when part of the sulphide has been oxidised to sulphate. Only the combination thus formed is active, since the sulphate, oxide or sulphide-oxide mixture do not exhibit any catalytic effect under the same reaction conditions [872].

The CuS + $CuSO_4$ catalyst was also used for polymerisation of isobutylene (at 150 °C, 40% conversion to 2,2,4-trimethylpentenes was achieved, which then converted to iso-octane on hydrogenation) [2164, 2165]. The above catalyst may also be used to polymerise a number of other pure olefins and their mixtures [872]. Differing from data reported by Wassermann, Brown and Saunders [230] also found a certain, though relatively low, polymerisation activity in MoS_2 as well

Table 92.

The Catalytic Influence of Different Sulphides in Cyclopentadiene Dimerisation [872].

Sulphide	CuS	Cu_2S	Ag_2S	SrS	ZnS	CdS	HgS	Tl_2S	SnS_2	SnS
Colour of sulphide	black	black	black	white	white	yellow	black	black	yellow	black
Catalytic activity	+	+	+	—	—	—	+	+	—	+
Sulphide	PbS	As_2S_3	Bi_2S_3	MoS_2	FeS	FeS_2	$KFeS_2$	$FeCuS_2$	NiS	
Colour of sulphide	black	yellow	black	black	black	black	black	black	black	
Catalytic activity	+	—	+	—	+	—	+	+	+	

as WS_2 and V_2S_3. For example, the WS_2 catalyst supported on Al_2O_3 gave a small amount of a solid polymer on ethylene polymerisation at 240 °C and 140 atm in xylene solution, tri-isobutylaluminium serving as activator. When aluminium or cadmium sulphide was used as carrier for MoS_2, low-molecular weight liquid ethylene polymers were obtained. The WS_2/Al_2O_3 catalyst was also employed to polymerise propylene [230]. The catalytic activity of MoS_2 mixed with trialkylaluminium in the case of ethylene polymerisation to give a solid product was also corroborated by Stuart [1978] and Japanese authors [574]. Some metal sulphides served as components for a complex catalyst for polymerising α-olefins [94a].

Table 93.

Dimerisation of Trimethylethylene in the Presence of CuS Activated by Different Procedures (Reaction temperature 100 °C, molar ratio C_5H_{10} : catalyst = 7 : 1) [872].

Catalyst No.	1	2	3	4
Procedure in catalyst preparation	CuS prepared from $CuSO_4$ and Na_2S	No. 1 + 20% S	No. 1 + 20% $CuSO_4 . 5\,H_2O$	CuS prepared from $CuSO_4$ and H_2S in acid solution and exposed to the atmosphere until30% of the sulphide is oxidised to $CuSO_4$
% Dimerisation after 60—64 hours	3	11	41	70

Aluminium sulphide [2092] is one of the substances which have been tested for use as catalysts of isobutylene and propylene polymerisation. The $ZnCl_2$–ZnS catalyst was used in dehydrocyclopolymerisation of C_4 and C_8 olefins to give aromatic compounds [77]. The polymerising effect of FeS and FeS_2 was investigated in connection with a study of the structure of coal, viewed as a polymer of aromatic monomers (e.g., in polymerising anthracene or some oxygen-containing aromatic compounds) [573].

11.2.

Polymerisation of non-hydrocarbon compounds and some condensation reactions

A number of metal sulphides (CuS, CdS, FeS, PbS, MoS_2, CoS, etc.) were tested in low-temperature (40 °C) polymerisation of gaseous formaldehyde [997, 2127].

CuS activated with copper sulphate is also an efficient catalyst of the polymerisation of acetaldehyde to paraldehyde [2163]. Some sulphides of Groups II, VI or VIII catalyse the dimerisation of acrylonitrile to 1,2-dicyanocyclobutane [1930]. Further applications of sulphide catalysts include the use of cadmium sulphide [464, 1145a, 2055a] and zinc sulphide [1554b] in polymerisation of episulphides, the use of ZnS as component of a catalyst for polymerising propylene oxide [806a] and the use of ZnS and CdS for the photopolymerisation of acrylic acid, its salts and amide, and also for polymerisation of vinyl acetate and styrene [1232]. A number of catalysts were tested for use in trimerisation of chlorocyanogen to cyanuric chloride; one of the most efficient catalysts being ZnS [506]. Sb_2S_3 was furthermore used as catalyst in obtaining high-molecular poly(ethylene terephtalate) [273b], zinc sulphide in polymerising cyclic organosiloxanes [206a] and sodium sulphide in polymerising *p*-benzene bis (ω-S-alkylthiosulphate) [1433a] and bis (*p*-chlorophenyl)sulphone [871a].

Antimony sulphide catalyses the condensation of aldehydes and amines to synthetic resins [2226] and MoS_2 supported on activated carbon has been employed in the reductive condensation of aldehydes and lower ketones to higher ketones [1360].

12.

THE USE OF SULPHIDE CATALYSTS IN INORGANIC REACTIONS AND IN SOME SPECIAL PROCEDURES

In a number of cases, sulphide catalysts have been applied to some important inorganic reactions, e.g., hydrogen sulphide synthesis, preparation of metal hydrides or purification of technical gases. Moreover the literature includes data relating to the application of sulphide catalysts in many interesting syntheses, e.g., of metal carbonyls, manufacture of diamond, etc.

12.1.

The use of sulphide catalysts in reactions of hydrogen with elements, reduction of carbon monoxide and sulphur and nitrogen oxides and purification of gases

Reactions of elements with hydrogen

Hydrogen sulphide synthesis from hydrogen and sulphur is a thermodynamically highly advantageous reaction, and virtually total equilibrium conversion is achieved at atmospheric pressure with temperatures as low as 500–530 °C [2256]. When no catalyst is employed, H_2S is formed from sulphur vapour and hydrogen at 600 °C. However, the reaction may be implemented even at 300 °C, although the reaction rate is very low under these conditions [2209]. In practical work, therefore, catalysts, particularly metal sulphides are used in sulphur hydrogenation. Certain literature and patent data show that Mo, W, Fe, Co, V and Zn sulphides are suitable catalysts [851, 863, 1330]. Cobalt and nickel catalysts, the active component of which are the respective sulphides, were used in laboratory-scale preparation of H_2S [123, 604]. A very efficient hydrogen sulphide generator for laboratory use, working with liquid sulphur and hydrogen at atmospheric pressure in the presence of supported MoS_2, is described by Zimmermann [2267]. Mo and W sulphides were used for industrial-scale production of high-concentration H_2S [1331, 1648]. The mixed W–Ni sulphide is also very efficient; enabling practically total conversion of sulphur to H_2S at 360 °C [2260]. The catalytic efficiency of the sulphide catalysts 5058, 8376, MoS_2

and nickel polysulphide was compared in sulphur hydrogenation under pressure; MoS_2 being found to be the most efficient of this group [1171]. The hydrogenation kinetics of sulphur was measured on the Ag_2S catalyst in the temperature range of 350 to 550 °C [1729].

Compared to hydrogen sulphide, the synthesis of hydrogen selenide from the elements is less advantageous in thermodynamic respects. At 350 °C the equilibrium constant of selenium hydrogenation is 2.5 orders of magnitude lower than in the case of sulphur hydrogenation [2256]. In spite of this, selenium conversion is practically quantitative at this temperature when excess hydrogen is present, but the hydrogenation rate in the presence of sulphide catalysts is less than with sulphur [1171]. At the same time, however, the lower thermal stability of hydrogen selenide [1421] sets an upper temperature limit to the possibility of improving the rate of selenium hydrogenation.

The efficient catalytic activity of sulphide catalysts, particularly of MoS_2, has been utilised for hydrogenative refining of selenium to an extremely pure elementary state for use in electrotechnical purposes. It was found possible to selectively hydrogenate selenium at 350 °C, neither tellurium nor arsenic being hydrogenated under the optimum reaction conditions and mercury similarly did not enter the purified material. On oxidation of the H_2Se obtained, low-ash selenium suitable for use in making rectifiers was obtained [1190, 1421].

The hydrogenation of oxygen, involved in purifying hydrogen-containing gases may likewise be catalysed with the aid of metal sulphides. Ni subsulphide promoted with Cu, Cd, Mo, W or Fe compounds (a maximum of 20 atoms of the promoter metal for every 100 Ni atoms) or Co subsulphide promoted with nickel were employed to this purpose. Practically total oxygen removal is achieved at temperatures below 200 °C [1453]. Hammar and Löfgren [722] suggest a mixture of Groups VI and VIII metal sulphides, supported on a suitable carrier (Al_2O_3, SiO_2. aluminosilicate) for the same purpose.

A combination of sulphides of Groups VI and VIII was also tested for use in ammonia synthesis. For example, in the presence of a mixture of 40% Ni sulphide and 60% W sulphide, the reaction rate of ammonia formation was found to be satisfactory at 250 to 260 °C and atmospheric pressure [1931].

Metal sulphides were found to have a significant catalytic effect in the direct reaction of alkali metals with hydrogen to form the respective hydrides [1168, 1169]. The hydrogenation was done in autoclaves at hydrogen pressures above 100 atm and in the presence of small amounts of MoS_2 or WS_2 (of the order of 0.1% of the metal employed). Potassium is easiest to hydrogenate (quantitative hydrogenation to KH takes place at 140 to 155 °C), next comes sodium (220 to 300 °C), while lithium is the most difficult to hydrogenate, its reaction rate being lowest. The hydrides obtained contain no free alkali metal.

The last application of sulphide catalysts in the reaction of elements with hydrogen is the synthesis of silanes. A number of metal sulphides (Cu, Ag, Pt, Ni, Mo, Bi,

Pb, Hg, etc.) were tested in silicon hydrogenation. It has been found that different silanes may be prepared (SiH_4 to Si_4H_{10}) depending on the partial hydrogen pressure employed [2034].

Hydrogenation and conversion of CO and hydrogenation of the oxides of sulphur, nitrogen and conversion of hydrogen cyanide

MoS_2, alone or supported on silica gel, was tried out as catalyst for methanisation of gases containing sulphur compounds besides CO and H_2. This catalyst was less active than metallic methanisation catalysts, but its activity was found to be adequate and its lifetime satisfactory. Higher hydrocarbons as well as CH_4 were formed on hydrogenation of CO [1848]. Hydrogen sulphide was found to cause reversible deactivation of the catalyst [1894a]. When WS_2 was used, a high degree of CO methanisation was achieved in high-pressure hydrogenation (475 °C, 140 to 280 atm) [561a]. WS_2 supported on clay is also a satisfactory methanisation catalyst. At 350 to 480 °C and 150 to 300 atm, a considerable rise in the calorific value was achieved in fuel gases which had originally contained a high percentage of CO [140].

Sulphide catalysts have also been tested for use in selective reduction of carbon monoxide to alcohols, e.g., the catalyst Cu–MnS–CoS [414, 2044] or ZnS with additions of other sulphides [413].

Carbon monoxide conversion with steam is another application of sulphide catalysts. For example, an original gas containing 30.6% CO, 53.5% H_2, 14.6% CH_4 and 1.3% N_2 was converted with steam at 500 °C and 20 atm in the presence of MoS_2 supported on clay, the composition of the resulting crude gas being 12.8% CO_2, 12.9% CO, 49.7% H_2, 22.9% CH_4 and 1.7% N_2 and the calorific value 4,100 kcal/Nm^3 [2230]. MoS_2–CoS/Al_2O_3 is a very efficient conversion catalyst, its activity being similar to that of the conventional iron catalyst without the disadvantage of the sensitivity to sulphur compounds [1732]. Ni_3S_2 was also tested for use as conversion catalyst [1738a].

Sulphide catalysts, particularly MoS_2 and WS_2 [2071] or NiS [1297] were used to purify gases to be used in hydrogenative removal of nitrogen oxides (mainly of NO). Supported hydroxides or sulphides of a number of metals (Ti, Zr, Th, Fe, Mn, Mo, Co, Cr, Ni, etc.) were used in high temperature (200 to 400 °C) hydrolysis of hydrogen cyanide present, together with hydrogen sulphide, in carbonisation gases [1474].

Reduction of sulphur dioxide and nitrogen monoxide

A two-stage process has been suggested for removing SO_2 from gases, in the first stage SO_2 being hydrogenated to H_2S at atmospheric pressure and 300 °C on a FeS/Al_2O_3 catalyst, while unreacted SO_2 is removed in the second stage on active aluminia by reaction with the H_2S formed in the preceding reaction [421]. This latter reaction, however, may also be catalysed by sulphide catalysts, e.g., Ag_2S, Co-thiomolybdate, CoS or MoS_2 [1436, 1716]. Rhenium heptasulphide is a very efficient catalyst of SO_2 reduction to H_2S (at 100 °C and 10 to 30 atm) [1912, 1912a]. The reduction of sulphur dioxide to H_2S with methane has been catalysed by the use of MnS [1767] and CaS [2162]. One of the most efficient catalysts for the reaction between SO_2 and carbon disulphide was found to be titanium sulphide [364]. Ferrous sulphide is an efficient catalyst of the reduction of nitrogen monoxide with hydrogen sulphide [1077]:

$$2\,NO + 2\,H_2S \rightarrow N_2 + 2\,H_2O + 2\,S\,. \qquad \text{(CLVI)}$$

12.2.

Special applications of sulphide catalysts

In some cases, sulphide catalysts have been employed for special synthesis and decomposition reactions, which will be mentioned briefly in the following survey for the sake of completeness.

MoS_2 supported on alumina was used in the synthesis of urea and substituted ureas from CO_2 and NH_3 or amines [1578]. In some cases, metal sulphides have been used as catalysts in the synthesis of nickel carbonyl; the optimum catalysts being iron sulphides [749, 750] or nickel and barium sulphides [1412]. An interesting report deals with the use of iron sulphide in the synthesis of diamond [594] and with the catalytic influence of chromium sulphide in graphitisation of petroleum coke [606a]. The influence of nickel subsulphide on the decomposition of volatile arsenic compounds in tar has been unvestigated in connection with the deactivating influence of arsenic on refining catalysts (cf. p. 114) [1995].

The application of sulphide catalysts in some photochemical or exchange reactions also deserves to be mentioned. Cd, Zn, Ga and Hg sulphides were found useful in the photosynthesis of hydrogen peroxide [677, 1569, 1569a, 1962]. The rate of decomposition of hydrogen peroxide was measured in the presence of CdS [1090]. This catalyst is similarly active in the photoreduction of methylene blue with formaldehyde [1569]. Finally, let us also mention the use of W, Mo and Re sulphides in exchange reactions of deuterated hydrogen and water [2087, 2100], the application of a ZnS catalyst in measuring surface recombination kinetics of atomic hydrogen [1980a] and of the catalyst Cu_2S [1824] or CdS + Cr_2O_3 [221] in conversion of *para*-hydrogen to *ortho*-hydrogen.

BIBLIOGRAPHY

1. Abbott M. D., Liedholm G. E., Sarno D. H.: Petrol. Ref. *34*, No. 6, 118 (1955).
2. Abbott M. D., Liedholm G. E., Sarno D. H.: Oil Gas J. *54*, No. 11, 92 (1955).
2a. Abdou I. K.: *Conf. on the Chemistry and Chem. Processing of Petroleum and Natural Gas*, p. 269. Akadémiai Kiadó, Budapest 1968.
2b. Abdou I. K., Elbadrawy S., Asmy M.: Schmierstoffe Schmierungstech. *1968*, No. 31, 5.
2c. Abo-Lemon F. S., Abdou I. K., Mustafa A.: J. Chem. U.A.R. *11* (2), 253 (1968); C. A. 71, 5121 (1969).
3. Ackermann L.: Magyar Tudemanyos Akad. Kém. Tudemanyok Osztalyanok Közlemenyei *8*, 95 (1956); C. A. 52, 9049 (1958).
4. Ackermann L.: Periodica Polytechnica (Budapest) (14) *3*, 149 (1959).
4a. Adams C. E., House W. T.: Fr. 1569510 (1969).
4b. Adams C. R.: J. Catalysis *11*, 96 (1968).
4c. Adams C. R.: Ind. Eng. Chem. *61*, No. 6, 30 (1969).
5. Adams N. R., Watkins C. H., Stine L. O.: Chem. Eng. Progr. *57*, No. 12, 55 (1961).
5a. Addison G. E., Mitsche R. T.: US 3511773 (1970).
6. Adlington D., Thompson E.: *Third European Symposium on Chem. Reaction Engineering*, Amsterdam 1964.
7. Agafonov A. V., Khavkin V. A., Osipov L. N., Rogov S. P.: Novosti Neft. i Gaz. Tekhn. Neftepererabotka i Neftekhim. *1962* (9), 3; C. A. 60, 10444 (1964).
8. Agafonov A. V., Rysakov M. V., Goldshtein D. L., Gusenkova E. A., Alfimova E. A., Borovaya M. S., Puchkov N. G., Kazanskii V. L., Badyshtova K. M., Altshuler L. A. E., Gerasimenko N. M., Yastrebov G. I., Zhadanovskii V. B.: *6th World Petrol. Congr.*, Sect. III. Paper 23. Frankfurt 1963.
9. Agafonov A. V., Goldshtein D. L., Osipov L. N., Rogov S. P., Rysakov M. V.: *Proceedings, Symposium on the Application of the Hydrogenation Processes in Petroleum Processing.* Index B_1, Moscow 1963.
10. Agafonov A. V., Rysakov M. V., Osipov L. N., Semenova E. S., Perezhigina I. Ya., Khavkin V. A.: *Proceedings, Intern. Conference on Catalysts*, Karlovy Vary 1964.
11. Agafonov V. A., Osipov L. N., Khavkin V. A., Rogov S. P.: USSR 176025 (1962).
11a. Ahuja S. P., Derrien M. L., Le Page J. F.: Ind. Eng. Chem., Prod. Res. Dev. *9*, 272 (1970).
12. Aldridge C. L., Rigney J. A.: US 3267170 (1966).
12a. Aldridge C. L.: Ger. Offen. 1928389 (1970).
13. Alekseeva K. A., Moldavskii B. L.: Khim. Tekhnol. Topliv Masel *4*, No. 1, 43 (1959).
14. Allbright C. S., Schwartz F. G., Ward C. C.: US Bureau of Mines, Rep. Invest. No. 6961 (1967); C. A. 64, 75082 (1967).

15. Allen C. C.: Brit. 464951 (1937).
16. Allen C. C.: Canad. 370264 (1937).
17. Allen C. C.: US 2051807 (1936).
18. Allied Chemical Dye Corp.: Brit. 718615 (1954).
19. Alpert S. B., Johanson E. S., Schuman S. C.: Hydrocarbon Process. *43*, No. 11, 193 (1964).
20. Alsén N.: Geol. För. Förh. *47*, 19 (1925); C. A. 19, 2620 (1925).
21. Altman L. S., Nemtsov M. S.: Acta Physicochim. URSS *1*, 429 (1934).
22. Altman L. S., Nemtsov M. S.: Zhur. Fiz. Khim. *6*, 221 (1935); C. A. 29, 7767 (1935).
23. Altman L. S., Diner I. S., Mitkalev B. A., Nemtsov M. S., Ryskin M. I.: Khim. Tverd. Topliva *7*, 31 (1936); C. A. 31, 1577 (1937).
23a. Altshuler S. A., Shvets A. F., Rozhdestvenskaya M. S.: Neftepererab. Neftekhim. *1970*, No. 6, 11.
24. Alvarado M. A.: Brit. 578124 (1946).
25. Alvarado M. A.: US 2402586 (1946).
26. Amemiya T.: J. Chem. Soc. Japan *63*, 1214 (1942); C. A. *41*, 3756 (1947).
27. Amemiya T.: J. Chem. Soc. Japan *63*, 1219 (1942); C. A. *41*, 3756 (1947).
28. Amemiya T.: J. Chem. Soc. Japan *64*, 1186 (1943); C. A. *41*, 3756 (1947).
29. Amemiya T.: J. Chem. Soc. Japan *65*, 573 (1944); C. A. *41*, 3757 (1947).
30. Anderson J., McAllister S. H., Derr E. L., Peterson W. H.: Ind. Eng. Chem. *40*, 2295 (1948).
31. Anderson W. De F., Manley T. R.: Brit. 953484 (1964).
32. Ando S.: J. Soc. Chem. Ind. Japan *34*, Suppl. Bind., 320 (1931); C. A. *26*, 700 (1932).
33. Ando S.: J. Soc. Chem. Ind. Japan *35*, Suppl. Bind., 455 (1932); C. A. *27*, 710 (1933).
34. Ando S.: J. Soc. Chem. Ind. Japan *38*, Suppl. Bind., 267 (1935); C. A. *39*, 6402 (1935).
35. Ando S.: J. Soc. Chem. Ind. Japan *39*, Suppl. Bind., 278 (1936); C. A. *30*, 8577 (1936).
36. Ando S.: J. Soc. Chem. Ind. Japan *40*, Suppl. Bind., 12 (1937); C. A. *31*, 4796 (1937).
37. Ando S.: J. Soc. Chem. Ind. Japan *40*, Suppl. Bind., 124 (1937); C. A. *31*, 6851 (1937).
38. Ando S.: *Proc. 2nd World Petrol. Congr.*, Vol. 2, p. 237, Paris 1937.
39. Ando S.: J. Fuel Soc. Japan *16*, 21 (1937); C. A. *31*, 5540 (1937).
40. Ando S.: J. Fuel Soc. Japan *17*, 33 (1938); C. A. *32*, 6440 (1938).
41. Ando S.: J. Soc. Chem. Ind. Japan *41*, Suppl. Bind., 191 (1938); C. A. *32*, 8745 (1938).
42. Ando S.: J. Soc. Chem. Ind. Japan *41*, Suppl. Bind., 215 (1938); C. A. *32*, 9448 (1938).
43. Ando S.: J. Soc. Chem. Ind. Japan *41*, Suppl. Bind., 247 (1938); C. A. *32*, 9448 (1938).
44. Ando S.: J. Soc. Chem. Ind. Japan *41*, Suppl. Bind., 292 (1938); C. A. *33*, 1473 (1939).
45. Ando S.: J. Soc. Chem. Ind. Japan *41*, Suppl. Bind., 386 (1938); C. A. *33*, 3564 (1939).
46. Ando S., Usiba T.: J. Soc. Chem. Ind. Japan *41*, Suppl. Bind., 390 (1938); C. A. *33*, 3564 (1939).
47. Ando S.: J. Soc. Chem. Ind. Japan *41*, Suppl. Bind., 413 (1938); C. A. *33*, 6807 (1939).
48. Ando S.: J. Soc. Chem. Ind. Japan *42*, Suppl. Bind., 69 (1939); C. A. *33*, 6030 (1939).
49. Ando S.: J. Soc. Chem. Ind. Japan *42*, Suppl. Bind., 147 (1939); C. A. *33*, 7077 (1939).
50. Ando S.: J. Soc. Chem. Ind. Japan *42*, Suppl. Bind., 268 (1939), C. A. *34*, 872 (1940).
51. Ando S.: J. Soc. Chem. Ind. Japan *42*, Suppl. Bind., 322B (1939); C. A. *34*, 2339 (1940).
52. Ando S.: J. Soc. Chem. Ind. Japan *43*, Suppl. Bind., 328 (1940); C. A. *35*, 1770 (1941).
53. Ando S.: J. Soc. Chem. Ind. Japan *43*, Suppl. Bind., 355 (1940); C. A. *35*, 3980 (1941).
54. Andreevskii I. L., Diner I. S., Mitkalev B. A., Nemtsov M. S., Ryskin M. I.: Khim. Tverd. Topliva *6*, 926 (1935).
54a. Angulo J., Frutos D., Galeote V., Gasca M., Gonzales E., Hernandes-Prieto A., Martinez Cordon J. L.: Ann. Quim. *65*, 1063 (1969); C. A. 72, 102352 (1970).
55. Anisimov S. B., Polozov V. F.: Zh. Prikl. Khim. *11*, 297 (1938).

56. Annable W. G., Haines R. M.: US 2769761 (1956).
57. Annable W. G., Jacobs W. L.: US 2915448 (1959).
58. Anon: Oil Gas J. *54*, No. 46, 137; 160; 162; 163; 164; 165; 166 (1956).
59. Anon: Chem. Weekblad *26*, 68 (1959).
60. Anon: Chem. Eng. News *1961*, No. 43, 56.
61. Anon: Hydrocarbon Process. Petrol. Ref. *41*, No. 9, 153; 192; 195; 199 (1962); Petrol. Ref. *32*, No. 12, 84 (1953).
62. Anon: Oil Gas J. *59*, No. 14, 136 (1961).
63. Anon: Petrol. Ref. *42*, No. 11, 202 (1963).
64. Anon: Hydrocarbon Process. *43*, No. 9, 145; 171; 186; 187; 189; 190; 191; 192; 193; 194; 195; 196 (1964).
65. Anon: Oil Gas J. *62*, No. 24, 74 (1964).
66. Anon: Europa Chemie *1965*, No. 20, 12 .
67. Anon: Chem. Ind. *18*, No. 3, 133.
68. Anon: Erdöl u. Kohle *19*, 466 (1966).
69. Anon: Petro-Chem. Engineer *1966*, February, 68.
70. Anon: Chem. Eng. News *1967*, June 12, 96.
71. Anon: Hydrocarbon Process. *46*, No. 11, 162; 185 (1967).
72. Anon: Oil Gas J. *65*, No. 6, 123 (1967).
73. Anon: Oil Gas J. *65*, No. 10, 58 (1967).
74. Anon: Oil Gas J. *65*, No. 16, 152 (1967).
74a. Anon: Hydrocarbon Process. *49,* No. 9, 163; 168; 172; 205; 210; 211; 215; 220; 221; 223; 224; 225; 226; 230; 231 (1970).
75. Anthes E., v. Füner W., Simon W.: Ger. 725604 (1940).
76. Antipina T. V., Frost A. V.: Usp. Khim. *19*, 342 (1950).
77. Antsus L. I., Petrov A. D., Shchenlina O. I.: Izv. AN SSSR, Ser. Khim. *1964*, 1866.
78. Appleby W. G., Lovell L. L., Love M. P. L.: US 2429575 (1947).
79. Appleby W. G., Sartor A. F.: US 2558508 (1951).
80. Archibald R. C., Greensfelder B. S., Holzman G., Rowe D. H.: Ind. Eng. Chem. *52*, 745 (1960).
81. Archibald R. C.: Brit. 582416 (1946).
82. Archibald R. C.: US 2394739 (1946).
83. Archibald R. C., Trimble R. A.: US 2435380 (1948).
84. Archibald R. C.: US 3012963 (1961).
85. Aretos K., Vialle J.: *Rhenium Papers Symp.*, Chicago 1960, 171 (1962); C. A. *59*, 5145 (1963).
86. Arey W. F., Blackwell N. E., Reichle A. D.: *7th World Petrol. Congr.*, PD 20, Paper 2. Mexico City 1967.
86a. Arey W. F., Mayer F. X.: Amer. Chem. Soc., Div. Petrol. Chem., Preprints 1967, *12* (4), A 25-A 39.
86b. Arey W. F., Hamner G. P.: Brit. 1137639 (1968).
87. Arich G., Cocco A.: *Univ. Studi Trieste*, Fac. Ing., Inst. Chim. Appl. No. 12, 84 (1963); C. A. *60*, 7838 (1964).
88. Arkel van A. E.: Rec. trav. chim. *45*, 437 (1926).
89. Arnold H. R., Lazier W. A.: US 2025032 (1936).
90. Arnold M. H. M.: Brit. 535583 (1940).
91. Arnold M. H. M.: US 2336916 (1943).
92. Arundale E., Guyer W. R. F., Thorn J. P., Mayer M. W.: Fr. 1067976 (1952).
93. Asboth K.: Fr. 1027016 (1950).
93a. Ashley J. H., Mitchell P. C. H.: J. Chem. Soc. *A 1968* (11), 2821.

93b. Ashley J. H., Mitchell P. C. H.: J. Chem. Soc. *A 1969*, 2730.
94. Asselin G. F.: US 3260765 (1966).
94a. Atarashi Y., Fukumoto O.: Japan 3018 ('67); C. A. 67, 22310 (1967).
94b. Audibert F., Marcellin R., Trambouze P.: Fr. 1516733 (1968).
95. Badger E. H., Griffith R. H., Newling W. B. S.: Proc. Roy. Soc. (London) *A 197*, 184 (1949).
96. Badische Anilin u. Soda Fabrik A. G.: Brit. 762811 (1956).
96a. Badische Anilin u. Soda Fabrik A. G.: Fr. 1554369 (1969).
97. Badyshtova K. M., Rogacheva L. M.: Khim. Tekhnol. Topliv Masel *6*, No. 5, 21 (1961).
98. Badyshtova K. M.: Khim. i Tekhnol. Topliv Masel *6*, No. 6, 21 (1961).
99. Badyshtova K. M.: *Proceedings, Symposium on the Application of the Hydrogenation Processes in Petroleum Processing*, Index G_{13}. Moscow 1963.
100. Badyshtova K. M.: Trudy Kuibyshevsk. Gos. NII NP, No. 25, 108 (1964); C. A. 64, 10997 (1966).
101. Bailey Wm. A., Nager M.: *7th World Petrol. Congr.*, PD 20, Paper 3. Mexico City 1967.
102. Baker W., Warburton W. K., Breddy L. J.: J. Chem. Soc. *1953*, 4149.
103. Balandin A. A.: Usp. Khim. *31*, 1265 (1962).
104. Ballard H. D. Jr.: *Advances in Petroleum Chemistry and Refining* (Ed. J. J. McKetta, Jr.), Vol. X., p. 219. Interscience Publ. Corp., New York 1965.
105. Ballard W. P., Dickens S. P., McKinley J. K., Smith B. F.: US 2888393 (1959).
106. Ballard W. P., Dickens S. P., McKinley J. K.: US 2983773 (1961).
107. Ballou E. V., Ross S.: J. Phys. Chem. *57*, 653 (1953).
108. Ballou E. V.: J. Am. Chem. Soc. *76*, 1199 (1954).
109. Belon P., Derrien M.: Fr. 1465372 (1967).
110. Bapseres P.: Chim. et ind. *90*, No. 4, 358 (1963).
110a. Baral W. J., Hendricks G. E. W., Damskey L. R.: Petrol. Ref. *37*, No. 10, 133 (1958).
110A. Baral W. J., Huffman H. C.: 8th World Petrol. Congress, PD 12, Paper 1. Moscow 1971.
110b. Barger B. D. Jr., Hengstebeck R. J., Moore T. M., Russum L. W.: US 3506567 (1970).
111. Baroni A.: It. 391304 (1941).
112. Barr F. T., Keyes D. B.: Ind. Eng. Chem. *26*, 1111 (1934).
113. Barret P., Lefebvre J., Watelle-Marion G.: Compt. rend. *249*, 2204 (1959).
114. Barry A. W.: US 2511453 (1950).
115. Bartovská L., Černý Č., Kochanovská A.: Collection Czech. Chem. Commun. *31*, 1439 (1966).
116. Basak N. G., Bose S. K., Ghosh S., Lahiri A.: J. Sci. Ind. Research (India) *14B*, 284 (1955); C. A. *50*, 2951 (1956).
117. Buss W. C.: US 3265615 (1966).
117a. Bat I. I., Chistyakova G. A., Ovchinnikov P. N., Guseva L. G., Rebrova V. V.: Tr. Gos. Inst. Prikl. Khim. No. *62*, 69 (1969); C. A. 73, 34928 (1970).
118. Batchelder H. R.: Am. Gas Assoc. Monthly *29*, 222 (1947); C. A 41, 5371 (1947).
119. Batchelder H. R.: *Advances in Petroleum Chemistry and Refining* (Ed. J. J. McKetta, Jr.), Vol. 5, p. 3. Interscience Publ., New York 1962.
120. Bauer K.: US 2116182 (1938).
121. Baumgarten P. K., Hoffmann E. J., Wadley E. F.: US 2694671 (1954).
122. Baumgarten P. K., Hoffmann E. J., Wadley E. F.: US 2717861 (1955).
123. Baxter J. P., Burrage L. J., Tanner C. C.: J. Soc. Chem. Ind., Trans. *53*, 410 (1934).
124. Beach L. K., Barnett A. E.: US 2667515 (1954).
125. Bell R. E., Herfert R. E.: J. Am. Chem. Soc. *79*, 3351 (1957).
126. Bell R. T.: US 2592646 (1952).

127. Bell R. T.: US 2647151 (1953).
128. Bell R. T.: US 2685605 (1954).
129. Bell R. T.: US 2808441 (1957).
130. Bell R. T.: US 2874129 (1959).
131. Belchetz A.: US 2487039 (1949).
131a. Bennerr R. N., Parker R. N., Burbidge B. W.: Ger. Offen. 1914457 (1969).
132. Benzie R. J.: Brit. 832939 (1960).
133. Bercik P. G., Henke A. M.: US 2953519 (1960).
133a. Bercik P. G., Henke A. M.: US 3392112 (1968).
134. Berents A. D., Vol'Epshtein A. B., Krichko A. A., Mukhina T. N., Pokorskii V. N., Yablokhkina M. N.: Novye Sposoby Poluchenia Khim. Prod. Osn. Gor. Iskop., AN SSSR, Inst. Gor. Iskop. 1966, 91; C. A. 66, 20761 (1967).
134a. Berents A. D., Borisova L. V., Vol'Epshtein A. B., Krichko A. A., Puchkov V. A.: *Tr. Inst. Gor. Iskop.* Moscow *23* (4), 146 (1968).
134b. Berents A. D., Vol'Epshtein A. B., Krichko A. A., Malyavinskii L. V., Mukhina T. N., Puchkov V. A.: *Tr. Inst. Gor. Iskop.* Moscow *24* (2), 131 (1968).
135. Berg C., Bradley W. E., Stirton R. I., Fairfield R. G., Leffert C. B., Ballard J. H.: Chem. Eng. Progr. *43*, No. 1, 1 (1947).
136. Berg C.: Petrol. Ref. *33*, No. 10, 153 (1954).
137. Berg L., Green K. J., Munro B. L., Hartwig J. R., Silver F. C., Hooper H. C., Harris A. N., Jacobson R. L., Westby A. J.,: Am. Chem. Soc., Div. Petrol. Chem., Gen. Papers Preprints 2, No. 1, 375 (1957).
137a. Berg L., Isaacson W. B.: US 3312750 (1967).
138. Berger C. V.: US 2905625 (1959).
139. Berger C. V.: US 2939837 (1960).
140. Berger W., Schulze—Bentrop R.: Ger. 975067 (1953).
141. Bergius F.: Z. angew. Chem. 34, 341 (1921).
142. Bergius F.: Z. angew. Chem. 38, 502 (1925).
143. Bergius F.: VDI.—Z. *69*, 1313; 1359 (1925); C. A. 20, 654 1926).
144. Bergius F.: Chemical Reactions Under High Pressure. *Nobel Lectures, Chemistry 1922 to 1941*, p. 245. Elsevier Publ. Co., Amsterdam—London—New York 1966.
145. Bergius F., Billwiller J.: Ger. 301231 (1919).
146. Bergius F., Billwiller J.: Ger. 303893 (1919).
147. Bergström H. O. V., Cederquist K. N., Trobeck K. G.: Ger. 625994 (1936).
148. Bergström H. O. V., Cederquist K. N., Trobeck K. G.: Swed. 85933 (1935).
149. Bernardini F., Brill R.: Chem. Berichte *96*, 2340 (1963).
149a. Bernardini F., Collepardi M., Armisi I.: Chim. Ind. *49* (4), 366 (1967).
150. Berthelot M.: Bull. Soc. Chim. *11*, 278 (1869).
151. Berti V., Padovani C., Todesca F.: *6th World Petrol. Congr.*, Sect. III, Paper 8, Frankfurt 1963.
152. Bethea S. R.: US 2425506 (1947).
153. Beuther H., Flinn R. A., McKinley J. B.: Preprint. Am. Chem. Soc., Div. Petrol. Chem. *3*, No. 3, 35 (1958).
154. Beuther H., Flinn R. A., McKinley J. B.: Ind. Eng. Chem. *51*, 1349 (1959).
155. Beuther H., Flinn R. A.: Petroleum. Ref. *39*, No. 4, 143 (1960).
155a. Beuther H., McKinley J. B., Flinn R. A.: Amer. Chem. Soc., Div. Petrol. Chem. Preprints 1961, *6*, No. 3, A 75.
156. Beuther H., Flinn R. A.: Ind. Eng. Chem., Prod. Res. Dev. *2*, 53 (1963).
157. Beuther H., Schmid B. K.: *6th World Petrol. Congr.*, Sect. III, PD 7, Paper 20. Frankfurt 1963.

158. Beuther H., Donaldson R. E., Henke A. M.: Ind. Eng. Chem., Product Research Dev. *3*, 174 (1964).
159. Beuther H. Larson O. A.: Ind. Eng. Chem., Process Design Dev. *4*, 177 (1965).
160. Beuther H., Mansfield R. F., Stauffer H. C.: Hydrocarbon Process. *45*, No. 5, 149 (1966).
161. Beuther H., Flinn R. A., Henke A. M., McKinley J. B.: Fr. 1228378 (1960).
162. Beuther H., McKinley J. B., Rice T., Schmid B. K., Stewart M. M., Sutphin E. M.: Fr. 1388593 (1965).
163. Beuther H., Flinn R. A., Henke A. M., McKinley J. B.: Ger. 1115870 (1959).
164. Beuther H., Flinn R. A., Henke A. M., McKinley J. B.: Ger. 1136785 (1962).
165. Beuther H., Henke A. M., McKinley J. B.: US 2960458 (1960).
166. Beuther H., Flinn R. A., McKinley J. B.: US 2967204 (1961).
167. Beuther H., Henke A. M., Offutt W. C.: US 2998377 (1958).
168. Beuther H., Hills P., Odioso R. C., Schmid B. K., Zabor R. C.: US 3006970 (1961).
169. Beuther H., Flinn R. A., Henke A. M., McKinley J. B.: US 3078221 (1963).
170. Beuther H., Flinn R. A., Henke A. M., McKinley J. B.: US 3078238 (1963).
171. Beuther H., Flinn R. A.: US 3153627 (1964).
171a. Beuther H., Schmid B. K.: US 3354076 (1967).
171b. Beuther H., Schmid B. K.: US 3383301 (1968).
171c. Beuther H., Schmid B. K.: US 3392109 (1968).
172. Beyer H.: *Lehrbuch der organischen Chemie*, p. 372. Hirzel Verlag, Leipzig 1954.
173. Bilisoli J. P., Polack J. A., Segura M. A.: US 2767147 (1956).
174. Bille R.: IVA *24*, 124 (1953); C.A. *47*, 10203 (1953).
174a. Billings W. G.: US 3506418 (1970).
175. Billon A., Derrien M., Lavergne J. C., Nonnenmacher H., Oettinger W., Reitz O.: Hydrocarbon Process. *45*, No. 3, 129 (1966).
176. Biltz W., Jusa R.: Z. anorg. Chem. *190*, 161 (1930).
177. Biltz W.: Z. anorg. Chem. *228*, 275 (1936).
178. Biltz W., Köcher A.: Z. anorg. Chem. *248*, 172 (1941).
179. Birthler R., Szkibik Ch.: Freiberger Forschungs-H. *A 36*, 42 (1955).
180. Birthler R., Karolyi J.: Freiberger Forschungs-H. *A 131*, 31 (1960).
181. Birthler R., Deuthoff E., Szkibik Ch.: GDR 12392 (1956).
182. Bitepazh Yu. A., Maslyanskii G. N., Kamusher G. D., Bursian R. N.: USSR 108257 (1952).
182a. Bittner C. W.: Brit. 1110170 (1968).
183. Bloch H. S.: European Chemical News, Normal Paraffins Suppl., December 2, 1966, p. 46.
184. Bloch H. S., Kvetinskas B.: US 2886614 (1959).
185. Blom R. H., Kollonitsch V., Kline C. H.: Ind. Eng. Chem., Int. Ed. *54*, No. 4, 16 (1962).
186. Blom R. H., Kline C. H.: Petrol. Refiner *42*, No. 10, 132 (1963).
187. Blonskaya A. I., Lozovoi A. V., Muselevich D. L., Ravikovich T. M., Titova T. A.: *Trudy Instit. Gor. Iskop.*, Tom IX, p. 5. Moscow 1959.
188. Blonskaya A. I., Lozovoi A. V., Gavrilova A. E., Gonikberg M. G., Kazanskii B. A.: *Trudy Inst. Gor. Iskop.*, Tom IX, p. 50, Moscow 1959.
189. Blonskaya A. I., Lozovoi A. V.: *Trudy Inst. Gor. Iskop.* Tom XVII, p. 187, Moscow 1962.
189a. Blue E. M., Spurlock B.: Chem. Eng. Progr. *56*, 54 (1960).
189b. Blue E. M.: US 3291723 (1966).
190. Blume H., Szkibik Ch., Pfeiffer F., Klotzsche H., Strich E. R., Becker K., Weidenbach G.: *Symposium über Hydrokatalytische Processe in der Erdölverarbeitung und Petrolchemie*, Band 2, p. 189. Leuna 1966.
191. Blume H., Becker K., Welker J., Strich E. R., Naundorf W.: *Symposium über Hydrokatalytische Processe in der Erdölverarbeiung und Petrolchemie*, Band 3, p. 119. Leuna 1966.

191a. Blume H., Strich E. R., Naundorf W., Becker K., Welker J.: Chem. Technik *18*, 623 (1966).
191b. Blume H., Strich E. R., Welker J.: Chem. Technik *21*, 342 (1969).
192. Bobyshev V. I., Dyakova M. K., Lozovoi A. V.: Zh. Prikl. Khim. *13*, 942 (1940).
193. Boelhouwer C., Hoolboom M. A., Perquin J. N. J., Waterman H. I.: Chem. Eng. Sci. *2*. 69 (1953).
194. Boente L.: Reichsamt Wirtschaftsausbau, Prüf. - Nr. 37 (PB 52005) 77 (1940); C. A. *41*, 6558 (1947).
195. Böeseken J., v. de Linde N.: Rec. trav. chim. Pays-Bas *54* ((4), 16), 739 (1935).
196. Bogdanova T. A., Sidorov R. I.: Zh. Vses. Khim. Obshch. Im. D. I. Mendeleeva *10*, No. 4, 479 (1965).
197. Bogdanova T. A., Lipovich T. V., Kalechits I. V.: Neftekhimiya *6*, 27 (1966).
198. Bogdanova T. A., Kalechits I. V.: Nefteper. i Neftekhim. *1966*, No. *12*, 27.
199. Bognar A.: Magyar Chem. Folyoirat *40*, 105 (1934).
200. Boivin J. L., Mc Donald R.: Can. J. Chem. *33*, 1281 (1955).
201. Bonnetain L., Duval X., Letort M., Souny P.: Comt. rend. *242*, 1979 (1956).
202. Boreskov G. K., Berger I. I., Klimenko M. Ya., Raeva V. S., Afanaseev M. M., Lisovskii P. V.: Tr. Nauchno- Issled. Inst. Sintet. Spirtov i Organ. Produktov *1960* (2), 213; C. A. 59, 9370 (1963).
203. Borglin J. N., Ott E.: US 2076875 (1937).
204. Bose S. K., Ganguli A. K., Basak N. G., Lahiri A.: J. Inst. Petrol. *45*, No. 428, 252 (1959).
205. Bose S. K., Kini K. A., Basak N. G., Lahiri A.: Chem. Age India *12*, 126 (1961).
206. Bose S. K., Basak N. G., Ganguli A. K., Lahiri A.: Ind. 65861 (1961).
206a. Bostick E. E.: US 3415777 (1968).
207. Boswell F. W. C.: *Proc. European Regional Conf. Electron Microscopy*. Delft 1, 409 (1962).
208. Boucher H. G., Leeds T. F.: US 2426483 (1947).
209. Bourne K. H., Holmes P. D.: Brit. 1002394 (1965).
210. Box E. O.: US 3275705 (1966).
211. Boyd J. H. Jr.: US 2440671 (1948).
211a. Boyer A. F., Payen P.: Khim. Tverd. Topliva *1968* (2), 125.
212. Braae B.: Acta Polytechn., Chem. Met. Ser. *3*, No. 9, 76 pp. (1954); C. A. *49*, 601 (1955).
213. Braae B.: Swed. 129091 (1950).
214. Bratton A. C., Bailey J. R.: J. Am. Chem. Soc. *59*, 175 (1937).
214a. Bray B. G., Kozlowski R. H.: US 3420768 (1969).
215. Bray W. B., Pollock R. C., Merrill D. R.: US 1981305 (1929).
216. Bredenberg J. B-son., Gustafsson O., Harva O.: Kem. Teollissuus *22* (6), 419 (1965); C. A. 63, 16101 (1965).
216a. Breimer F., Waterman H. I., Weber A. B. R.: J. Inst. Petrol. *43*, 297 (1957).
217. Brennan H. M., Den Herder M. J.: US 3214366 (1965).
218. Brewer M. B., Cheavens T. H.: Oil Gas J. *64* (17), 176, 181 (1966).
219. Brewer M. B., Cheavens T. H.: Hydr. Process. Petrol. Ref. *45*, No. 4, 203 (1966).
220. Bridges J. M., Houghton G.: J. Am. Chem. Soc. *81*, 1334 (1959).
221. Briones A. G.: Rev. Real Acad. Cienc. Exact., Fis. Nat. Madrid *58* (4), 431 (1964); C. A. 62, 13895 (1965).
222. British Petrol. Co. Ltd.: Belg. 647083 (1964).
223. British Petrol. Co. Ltd.: Belg. 659679 (1965).
224. British Petrol. Co. Ltd.: Fr. 1424344 (1966).
225. British Petrol. Co. Ltd.: Neth. Appl. 6613369 (1967).
226. Broadbent H. S., Slaugh J. H., Jarvis N. L.: J. Am. Chem. Soc. *76*, 1519 (1954).
226a. Broadbent H. S.: Ann. N. Y. Acad. Sci. *145* (1), 58 (1967).

226b. Bronovitskii V. E., Salyamova F., Volochkovich M. A.: Uzb. Khim. Zh. *11* (4), 68 (1967).
226c. Bronovitskii V. E., Volochkovich M. A., Kalinskaya L. L., Nam A.: Uzb. Khim. Zh. *12* (3), 71 (1968).
227. Brooks F. W., Sharpless H. R.: US 2879227 (1959).
228. Brooks F. W.: US 3050460 (1962).
229. Brown C. L.: US 2432087 (1947).
230. Brown C. P., Saunders J.: J. Polymer Sci. *43*, 580 (1960).
231. Brown E. K., Dobbins E. V., Burge R. A.: US 2886506 (1959).
232. Brown O. W., Raines E. D.: J. Phys. Chem. *43*, 383 (1939).
233. Bube R. H.: Phys. Rev. *80*, 655 (1950).
234. Buchmann F. J., Mertzweiller J. K., Frasce E. V.: US 2796436 (1957).
235. Buckley J. W. A.: Brit. 719159 (1954).
236. Bull F. W.: Virginia J. Sci. (N. S.) *1*, 63; Bull Virginia Polytech. Inst. Eng. Expt. Sta. Ser. No. 74 (1950); C. A. *44*, 8089 (1950).
237. Bullock M. W., Hand J. J., Stokstad E. L. R.: J. Am. Chem. Soc. *79*, 1978 (1957).
238. Bundel A. A., Rusanova A. I.: Izv. AN SSSR, Ser. Fiz. *13*, 166 (1949).
239. Bunte K., Lorenz F.: Gas u. Wasserfach *75*, 787 (1932).
240. Bureau of Mines: Rep. of Investigation No. 5506. US State Dep. Interior, 1959.
240a. Buss W. C.: US 3385782 (1968).
240b. Buss W. C.: S. African 69 05858 (1970).
241. Butkov N. A., Rabinovich E. L.: Neft. Khoz. *1936*, No. 6, 48.
242. Byrne P. J., Gohr E. J., Haslam R. T.: Ind. Eng. Chem. *24*, 1129 (1932).
243. Byrns A. C., Bradley W. E., Lee M. W.: Ind. Eng. Chem. *35*, 1160 (1943).
244. Byrns A. C.: Canad. 413254 (1943).
245. Byrns A. C.: US 2325033 (1943).
246. Byrns A. C.: US 2325034 (1944).
247. Byrns A. C.: US 2369432 (1945).
248. Caglioti V., Roberti G.: Gazz. chim. ital. *62*, 19 (1932); C. A. 26, 3175 (1932).
248a. California Research Corp.: Brit. 891096 (1962).
249. California Research Corp.: Brit. 928794 (1960).
250. California Research Corp.: Brit. 962347 (1964).
251. California Research Corp.: Brit. 962778 (1964).
252. Campaigne E., Diedrich J. L.: J. Am. Chem. Soc. *73*, 5240 (1951).
253. Campbell R. W.: US 3345381 (1967).
254. Candea C., Marschal A.: Congr. chim. ind. Bruxelles, *15*, I, 542 (1935); Chem. Zentr. *107*, II, 1825 (1936).
255. Candea C., Marschall A.: Chim. et ind. *36*, 463 (1936).
256. Candea C., Sauciuc L.: Petroleum Z. *34*, No. 42, 3 (1938).
257. Candea C., Sauciuc L.: Petroleum Z. *34*, No. 43, 1 (1938).
258. Candea C., Sauciuc L.: Petroleum Z. *34*, No. 50, 1 (1938).
259. Candea C., Kühn J.: Bull. Sci. école polytechn. Timisoara, *8*, 72 (1938); C. A. *32*, 7251 (1938).
260. Candea C., Sauciuc L.: Petroleum Z. *35*, 361 (1939).
261. Candea C., Sauciuc L., Fridlovschi A.: Bull. Inst. Natl. Cercetari Technol. *2*, 72 (1947); C. A. *42*, 9117 (1948).
262. Candea C., Fridlovschi A., Gropsianu Z.: Bull. études et recherches tech. (Bucharest) *1*, 159 (1949); C. A. 44, 9133 (1950).
263. Cannon P.: J. Phys. Chem. *64*, 858 (1960).
264. Cannon P.: J. Phys. Chem. *64*, 1285 (1960).
265. Cannon P.: J. chim. phys. *58*, 126 (1961).

265a. Cannon P.: General Electric Research Lab. Report No. 59-RL-2293 C.
266. Carl P. R.: US 3222274 (1965).
267. Carlile J. H., Cawley C. M., Hall C. C.: J. Soc. Chem. Ind. *57*, 240 (1938).
268. Carlile J. H., Cawley C. M.: J. Soc. Chem. Ind. *57*, 347 (1938).
269. Carlsmith L. E., Haig R. R.: Petrol. Ref. *36*, No. 9, 233 (1957).
270. Carpenter C. C.: J. Gas Lighting *122*, 1010 (1913).
271. Carpenter C. C.: J. Gas Lighting *123*, 30 (1913).
272. Carpenter H. C., Cottingham P. L.: US Bur. Mines, Rept. Invest. No. 5533, 29 pp. (1959); C. A. 54, 5057 (1960).
273. Carpenter H. C., Cottingham P. L.: US Dep. Interior, Inf. Circ. 8156, Bureau of Mines (1962).
273a. Carter J. L., Sinfelt J. H.: US 3424669 (1969).
273b. Carter M. E., Price J. A.: US 3425995 (1969).
274. Casagrande R. M., Meerbott W. K., Sartor A. F., Trainer R. P.: Ind. Eng. Chem. *47*, 744 (1955).
275. Cates H. L., Punderson J. O., Wheateroft R. W., Stiles A. B.: US 2851496 (1958).
276. Cawley C. M., Hall C. C.: J. Soc. Chem. Ind. *62*, 116 (1943).
277. Cawley C. M., Kingman F. E. T.: Fuel *23*, 4 (1944).
278. Cawley C. M., Hall C. C.: J. Soc. Chem. Ind. *63*, 33 (1944).
279. Cawley C. M., Carlile J. H. G., Newall H. E., Kingman F. E. T.: J. Inst. Petrol. *32*, 660 (1946).
280. Cawley C. M.: Research (London) *1*, 553 (1948).
281. Cawley C. M.: *Proc. 3rd World Petrol. Congr.*, Sect. IV, 294. The Hague 1951.
282. Černý Č., Habeš M., Zelená M., Erdös E.: Collection Czech. Chem. Commun. 24, 3836 (1959).
283. Chang Fu-Liang et al.: Acta Foculio-Chim. Sinica *6* (3), 187 (1965); C. A. *65*, 18379 (1966).
284. Chang K. H., Itabashi K.: Kogyo Kagaku Zasshi *69* (1), 63 (1966); C. A. *65*, 5349 (1966).
285. Chang K. H.: Kogyo Kagaku Zasshi *69* (6), 1160 (1966); C. A. *65*, 20049 (1966).
286. Chang K. H.: Bull. Chem. Soc. Jap. *40* (7), 166 (1967); C. A. *67*, 99700 (1967).
286a. Chang K. H.: Kogyo Kagaku Zasshi *70* (9), 1512 (1967).
286b. Chang K. H., Itabashi K.: Kogyo Kagaku Zasshi *71*, 142 (1968).
286c. Chapell S. F.: Belg. 634763 (1964).
287. Charbonnages de France: Fr. 1260709 (1961).
288. Charbonnages de France: Fr. 1359910 (1964).
288a. Charbonnages de France: Ger. 1272921 (1968).
289. Chemokomplex Vegyipari Gep-es Buenderes Export-Import Vallalat: Fr. 1474773 (1967).
290. Chen K. Ch., Chang F. L.: Acta Focalia Sinica *3*, No. 1, 34 (1958); Chem. Zentr. *131*, 8062 (1960).
291. Cheney L. C.: Brit. 608969 (1948).
292. Cheney H. A., Breier I.: US 2807647 (1957).
293. Cheng L. P., Liang Ch.: Jan Liao Hsüeh Pao *2*, 136 (1957); C. A. *52*, 16022 (1958).
294. Cheng L. P., Liang Ch., Hsi T. W., Hsü C. T.: Jan Liao Hsüeh Pao *4*, No. 2, 119 (1959); C. A. *54*, 8042 (1960).
295. Cheng L. P., Liang Ch., Hsi T. W., Hsü C. T.: Jan Liao Hsüeh Pao *4*, No. 2, 113 (1959); C. A. *54*, 7313 (1960).
296. Chenicek J. A.: US 2531767 (1950).
297. Cheradame R., Letort M.: CERCHAR Rappt. *1960*, 103 (Pub. 1961); C. A. 61, 14433 (1964).
298. Chereau J.: Fr. 1473900 (1967).
299a. Chernozhukov N. I., Kuliev R. Sh., Sadykhova B. A., Rasulova A. M.: Issled. Primen.

Gidrogenizatsionnykh Protsessov Neftepererab. Neftekhim. Prom. *1968*, 261; C. A. 72, 13485 (1970).
300. Chervenak M. C., Johnson C. A., Schuman S. C.: Petrol. Ref. *39*, No. 10, 151 (1960).
301. Chervenak M. C., Feigelmann S., Wolk R., Byrd C. R., Hellwig L. R., van Driesen R. P.: Chem. Eng. Progr. *59*, No. 2, 53 (1963)
301a. Chervenak M. C., Schuman S. C.: US 3418234 (1968).
301A. Chevron Research Co.: Brit. 1064776 (1967).
302. Chevron Research Co.: Brit. 1073364 (1967).
303. Chevron Research Co.: Brit. 1075165 (1967).
304. Chiang P. N., Kalechits I. V., Wei S. P.: Khim. i Tekhnol. Topliv i Masel *4*, No. 10, 16 (1959).
305. Chiang P. N., Lin L. Y., Chou F. L., Kalechits I. V.: Trudy Vost. Sibir. Filiala AN SSSR, Ser. Khim. *18*, 107 (1959).
306. Chiang P. N., Wei S. P., Chou F. L.: Tr. Vost. Sibir. Filiala AN SSSR, Ser. Khim. *26*, 128 (1959).
307. Chiang P. N., Wei S. P., Kuan C., Kalechits I. V.: Khim. i Tekhnol. Topliv i Masel *6*, No. 2, 21 (1961).
308. Chiang P. N., Wei S. P., Lin L. Y., Kuan C., Chou F. L., Kalechits I. V.: Izv. Sibir. Otd. AN SSSR *1959*, No 2, 81—96.
308a. Chibnik S., Foster H. M., Glick L. A., Kaufman H. A.: Fr. 1482037 (1967).
309. Chuchla J.: *Thesis*. Institute of Chemical Technology, Prague 1959.
309a. Chujo K., Tomihisa N.: Jap. Chem. Quart. 5 (3), 40 (1969); C. A. 71, 62749 (1969).
310. Ciapetta F. G., Hunter J. B.: Ind. Eng. Chem. *45*, 147 (1953).
311. Ciapetta F. G.: Ind. Eng. Chem. *45*, 159 (1953).
312. Ciapetta F. G.: Ind. Eng. Chem. *45*, 162 (1953).
313. Ciapetta F. G., Dobres R. M., Baker R. W.: *Catalysis* (Ed. P. H. Emmett), Vol. 6, p. 495. Reinhold Publ. Corp., New York 1958.
314. Ciapetta F. G.: *7th World Petrol. Congr.*, RP 11. Mexico City 1967.
315. Ciborowski S.: Pol. 41482 (1958).
316. Cinque J. J., Grove H. D., Hoot W. E., Jarboe J. A.: US 2822400 (1958), 2797191 (1955), 2822401 (1955); Fr. 1161066 (1956).
317. CIOS-Rep., File No. XXXII-107, 115—124 (1945).
317a. Cír J.: Ropa a uhlie *11*, 7 (1969).
318. Clough H.: Ind. Eng. Chem. *49*, 673 (1957).
319. Cnu C., Hsiao K.: K'o Hsueh T'ung Pao *1963* (9), 52; C. A. *60*, 11414 (1964).
320. Cockram C., Sawyer E. W.: Ind. Chem. *35*, 221 (1959).
321. Cole R. M., Davidson D. D.: Ind. Eng. Chem. *41*, 2711 (1949).
322. Cole R. M.: Brit. 574800 (1946).
323. Cole R. M.: US 2373626 (1945).
324. Cole R. M.: US 2392579 (1946).
325. Cole R. M.: US 2394751 (1946).
326. Cole R. M.: US 2398175 (1946).
327. Cole R. M.: US 2406200 (1946).
328. Cole R. M.: US 2413312 (1946).
329. Cole R. M.: US 2423176 (1947).
330. Cole R. M.: US 2431920 (1947).
331. Cole E. L., Skelton Wm. E.: US 2779713 (1957).
332. Cole E. L., Skelton Wm. E., Meyers R. K.: US 2879223 (1959).
333. Cole E. L.: US 2967144 (1961).
334. Colson J. C., Barret P.: C. R. Acad. Sci., Paris, Ser. *C 265* (5), 303 (1967).

334a. Colson J. C., Delafosse D., Barret P.: Bull. soc. chim. France *1968* (1), 146.
335. Columbian Carbon Co.: Brit. 1023499 (1966).
336. Compagnie Française des essences synthetiques: Fr. 977656 (1951).
337. Compagnie Française des essences synthetiques: Fr. 981468 (1951).
338. Compagnie Française de Raffinage: Fr. 866322 (1940).
339. Compagnie Française de Raffinage: Fr. 1379062 (1964).
340. Condit P. C.: Ind. Eng. Chem. *41*, 1704 (1949).
341. Condit P. C.: US 2560555 (1951).
342. Condon F. E.: *Catalysis* (Ed. P. H. Emmett), Vol. VI, p. 43. Reinhold Publ. Corp., New York 1958.
343. Conrad C., Hermann R.: Erdöl u. Kohle *17*, 897 (1964).
344. Constabaris G., Lindquist R. H.: US 3235486 (1966).
345. Constabaris G., Unverferth J. W.: US 3256205 (1966).
346. Constantinescu M., Constantinescu T.: Petrol si Gaze *10*, 298 (1959).
347. Constantinescu M., Constantinescu T., Fedin T.: Petrol si Gaze *12*, 33 (1961).
348. Constantinescu M., Constantinescu T., Fedin T.: Rom. 44309 (1966).
349. Coonradt H. L., Garwood W. E.: Ind. Eng. Chem., Process Design Dev. *3*, 38 (1964).
349a. Coonradt H. L., Leaman W. K., Miale J. N.: Amer. Chem. Soc., Div. Petrol. Chem., Preprints 1964, *9* (1), 59.
350. Coons E. A.: US 2611789 (1952).
351. Cope A. C., Farkas E.: J. Org. Chem. *19*, 385 (1954).
352. Copenhaver J. W., Bigelow M. H.: *Acetylene and Carbon Monoxide Chemistry*, p. 282. Reinhold Publ. Corp. New York 1949.
353. Corson B. B., Monroe G. S.: US 2298346 (1943).
354. Corson B. B., Monroe G. S.: US 2298347 (1943).
355. Corson B. B., Detrick R. S.: US 2438148 (1948).
356. Costantinides G., Arich G.: *Proc. 6th World Petrol Congr.*, Sect. 5, Paper 11. Frankfurt 1963.
356a. Cosyns B., Derrien M. L., Le Page J., Vidal S.: Ropa a uhlie *10*, 121 (1968).
357. Cottingham P. L., Carpenter H. C.: Ind. Eng. Chem., Process Design Dev. *6*, 212 (1967).
358. Coulson A. E., Handley R., Holt E. C., Stonestreet D. A.: J. Soc. Chem. Ind. *65*, 396 (1946).
359. Council of Scientific and Industrial Research (India): Ind. 79211 (1964).
360. Coussemant F., Jungers J. C.: Bull. Soc. Chim. Belges *59*, 295 (1950).
361. Cox K. E., Berg L.: Chem. Eng. Progr. *56*, No. 12, 54 (1962).
362. Craig R. G., White E. A., Henke A. M., Kwolek S. J.: Hydrocarbon Process. *45*, No. 5, 159 (1966).
363. Crawford J. P., Miller R. M.: *Proc. Am. Petrol. Inst.*, Sect. III, 43, 106 (1963).
363a. Crawford M., Supanekar V. R.: J. Chem. Soc. *C 1969*, 832.
364. Crawley B., Griffith R. H.: Trans. Farad. Soc. *32*, 1623 (1936).
365. Crawley B., Griffith R. H.: J. Chem. Soc. *1938*, 717.
366. Crawley B., Griffith R. H.: J. Chem. Soc. *1938*, 720.
367. Crawley B., Griffith R. H.: J. Chem. Soc. *1938*, 2034.
368. Čůta F., Karlík M.: Collection Czech. Chem. Commun. *29*, 2151 (1964).
369. Dahlke B., Günther G.: Chem. Technik *14*, 150 (1962).
369a. Daido Steel Co. Ltd.: Brit. 1103442 (1968).
370. Dare H. F., Demeester J.: Petrol Ref. *39*, No. 11, 251 (1960).
371. Datka M.: Freiberger Forschungs-H. *A 196*, 255 (1960).
372. Daussat R. L., Steele C. T.: US 2595772 (1952).
372a. Davenport W. H., Kollonitsch V., Kline C. H.: Ind. Eng. Chem. *60*, No. 11, 10 (1968).

373. Davidson R. L.: Petrol. Processing *10*, No. 8, 1170 (1955).
374. Davis G. H. B.: US 1960204 (1934).
374a. Davis J. R. Jr., Tice J. D.: US 3481862 (1969).
374b. Davis J. R., Tice J. D., Benner R. I.: US 3516926 (1970).
375. Davtyan N. A., Dyakova M. K.: Izv. AN SSSR, Otd. Metal. Toplivo *1959*, 250.
376. Davtyan N. A., Dyakova M. K.: *Trudy Inst. Gor. Iskop.*, Tom. IX, p. 26, Moscow 1959.
377. Day T. D.: US 826089 (1906).
378. Dazeley G. H., Gall D., Hall C. C.: J. Inst. Petroleum *34*, 647 (1948).
379. Dean R. A., Whitehead E. V.: *Proc. 6th World Petrol. Congr.*, Sect. 5, Paper 9. Frankfurt 1963.
380. Debie N. C.: *Chemické zpracování ropy* (*Petrochemical Technology*) p. 33. SNTL, Prague 1966.
381. Debucquet L., Velluz L.: Bull. soc. chim. France (4) *51*, 1571 (1932).
382. Decrue J., Susz B.: Helv. Chim. Acta *39*, 619 (1956).
383. Decrue J.: Helv. Chim. Acta *39*, 812 (1956).
384. Dedinas J., Starnes W. C., Stewart M. M.: US 3244616 (1966).
385. Deger T. E., Buchholz B., Goshorn R. H.: Brit. 931585 (1963).
386. Deger T. E., Buchholz B., Goshorn R. H.: US 3035097 (1962).
387. Delafosse D., Barret P., Abon M., Lavier A.: Compt. rend. *252*, 3250 (1961).
388. Delafosse D., Barret P., Joffrin J. D.: Compt. rend. *256*, 1531 (1963).
389. Delafosse D., Budelot J. P.: Compt. rend. *260* (19), 5037 (1965).
390. Delafosse D., Budelot J. P., Coquillion S.: Bull. soc. chim. France *1965* (2), 341.
391. Demann W., Krebs E., Borchers H.: Tech. Mitt. Krupp Tech. *6*, 59 (1938); C. A. *33*, 6257 (1939).
392. Demeester J.: Fr. 1355789 (1964).
393. Deringer M. L., Hare C. R.: US 3340181 (1967).
394. De Rosset A. J., Riedl F. J., Czajkowski G. J.: Am. Chem. Soc., Div. Petrol. Chem., General Papers No. 33, 137 (1955).
395. De Rosset A. J., Watkins Ch. H.: US 2671754 (1954).
396. De Rosset A. J.: US 2719113 (1955).
397. Deryagina E. N.: *Thesis*. Institute of Petrochemical Syntheses, Academy of Sciences USSR. Moscow 1966.
398. Desikan P., Amberg C. H.: Canad. J. Chem. *41*, 1966 (1963).
399. Desikan P., Amberg C. H.: Canad. J. Chem. *42*, 843 (1964).
400. Dets M. M., Agafonov A. V.: Nefteper. i Neftekhim., Nauchno-Tekhn. Sb. *1965* (5), 6.
400a. Dets M. M., Agafonov A. V.: Neftepererab. Neftekhim. *1965*, No. 6, 29.
401. Deutsche Gold- u. Silber-Scheideanstalt: Brit. 453859 (1936).
402. Deutsche Gold- u. Silber-Scheideanstalt: Fr. 711046 (1931).
403. Deutsche Gold- u. Silber-Scheideanstalt: Fr. 711520 (1931).
404. Deutsche Gold- u. Silber-Scheideanstalt: Fr. 714557 (1927).
405. Deutsche Gold- u. Silber-Scheideanstalt: Fr. 804112 (1936).
406. Deutsche Gold- u. Silber-Scheideanstalt: Ger. 665370 (1938).
406a. Develotte J., Mazza M., Payen P.: *7th International Conference on Coal Science*, Papers, Section 2. Prague 1968.
407. Dickinson R. G., Pauling L.: J. Am. Chem. Soc. *45*, 1466 (1923).
408. Dierichs A., Kubička R.: *Fenoly a zásady z uhlí* (*Phenols and Bases From Coal*). SNTL, Prague 1956.
409. Dierichs A., Müller E.: Freiberger Forschungs-H. *A 87*, 46 (1958).
410. Directie van de Staatsmijnen in Limburg: Neth. 70816 (1952).

411. Dirksen H. A., Linden H. R., Pettyjohn E. S.: Inst. Gas Technol., Research Bull. No. 4, 27 pp. (1953); C. A. 47, 11700 (1953).
411a. Doane E. P.: US 3378482 (1968).
412. Docksey P., Porter F. W. B.: Brit. 670619 (1952).
412a. Doelp L. C., Kreider E. R., Macarus D. P.: US 3498907 (1970).
413. Dolgov B. N., Karpinskii M. H., Silina N. P.: Khim. Tverd. Topliva *5*, 470 (1934).
414. Dolgov B. N.: *Katalýza v organickej chémii* (*Catalysis in Organic Chemistry*) p. 61. SVTL-SNTL, Prague 1962.
415. Dombart R. C., Eld A. C., Lehrian W. R.: Fr. 1348259 (1964).
416. Donath E. E.: *Advances in Catalysis*, Vol. VIII, p. 239. Academic Press Inc., New York 1956.
417. Donath E., Nonnenmacher H.: Ger. 802398 (1951).
418. Dönges E.: Z. anorg. Chem. *254*, 133 (1947).
419. Dontsova V. A., Barabanshchikova N. F., Lyudkovskaya V. G., Volynkina A. Ya., Polazhchenko M. A., Osipova V. S.: Khim. Prom. *43* (7), 515 (1967).
419a. Dorer F. H.: J. Catalysis *13*, 65 (1969).
420. Dorogochinskii A. Z., Kupriyanov V. A., Melnikova N. P.: *Proceedings, Symposium on the Application of the Hydrogenation Processes in Petroleum Processing*, Index D_5, Moscow 1963.
421. Doumani T. F., Deery R. F., Bradley W. E.: Ind. Eng. Chem. *36*, 329 (1944).
422. Doumani T. F.: US 2780584 (1957).
423. Doumani T. F.: US 2816146 (1957).
424. Doumani T. F.: US 2820831 (1958).
425. Doumani T. F.: US 2829171 (1958).
426. Douwes C. T., t'Hart M.: Erdöl u. Kohle *21*, 202 (1968).
427. Douwes C. T., van Weeren P. A., de Ruiter H.: Brit. 907770 (1962).
428. Douwes C. T., van Weeren P. A., de Ruiter H.: Neth. 104874 (1963).
429. Dovell F. S., Greenfield H.: J. Org. Chem. *29*, 1265 (1964).
430. Dovell F. S., Greenfield H.: J. Am. Chem. Soc. *87*, 2767 (1965).
430a. Dovell F. S., Ferguson W. E., Greenfield H.: Ind. Eng. Chem., Prod. Res. Dev. *9*, 224 (1970).
430b. Dovell F. S., Greenfield H.: US 3336386 (1967).
430c. Dovell F. S., Greenfield H.: US 3350450 (1967).
431. Drabkina I. E., Zyryanov B. F., Orechkin D. B.: Khim. i Tekhnol. Topliv i Masel *6*, No. 10, 12 (1961).
432. Drabkina I. E., Zyryanov B. F., Orechkin D. B., Popova T. S.: Nefteper. i Neftekh., Nauchno-Tekhn. Sb. *1965* (4), 14.
433. Drescher K., Hummel H., Rossmanith G., Sander H.: GDR 17295 (1959).
434. Duck E. W., Timms D. G.: Brit. 984901 (1965).
435. Dufek L., Goppoldová M.: *Proceedings, Intern. Conference on Catalysts*, p. 66. Karlovy Vary 1964.
435a. Dufek L., Goppoldová M.: Czech. 126974 (1968).
436. Dudkin L. D.: *Vysokotemperaturnye Metalokeram. Materialy*, (*High-temperature Metaloceramic Materials*), AN Ukr. SSR, Inst. Metalokeram. i Spets. Splavov 1962, p. 87; C. A. 58, 6300 (1963).
437. Dutta A. K.: Indian J. Phys. *18*, 249 (1944); C. A. *39*, 3983 (1945).
438. Dutta A. K.: Nature *156*, 240 (1945).
439. Duyverman C. J., Vlugter J. C., van de Weerdt W. J.: *Proc. Intern. Congr. Catalysis*, 3rd, Amsterdam, 1964, 2, 1416; discuss. 1427 (Pub. 1965).
440. Dyakova M. K., Lozovoi A. V.: Khim. Tverd. Topliva *5*, 719 (1934).

441. Dyakova M. K., Lozovoi A. V.: Doklady Akad. Nauk SSSR *1935*, II, 254.
442. Dyakova M. K., Vol'Epshtein A. B., Aleksi E. A., Vasilchikova E. I.: Khim. i Tekhnol. Topliv *1956*, No. 9, 44.
443. Dyakova M. K., Vol'Epshtein A. B., Aleksi E. A., Vasilchikova E. I.: Zh. Prikl. Khim. *30*, 1056 (1957).
444. Dyakova M. K., Vol'Epshtein A. B., Zharova M. N., Zasukhina J. A.: Zh. Prikl. Khim. *32*, 2120 (1959).
445. Dyakova M. K., Davtyan N. A., Zharova M. N., Avramenko V. I., Karandasheva V. M.: Koks i Khim. *1962*, No. 10, 40.
446. Eagle S., Rudy Ch. E. Jr.: Ind. Eng. Chem. *42*, 1294 (1950).
447. Eastwood A. H.: Brit. 605838 (1948).
448. Eberline C. R., Wilson R. T., Larson L. G.: Ing. Eng. Chem. *49*, 661 (1957).
449. Egan C. J., Langlois G. E., White R. J.: J. Am. Chem. Soc. *84*, 1204 (1962).
450. Egan C. J.: US 3120569 (1964).
451. Egan C. J.: US 3340183 (1967).
452. Eggertsen F. T., Roberts R. M.: J. Phys. Chem. *63*, 1981 (1959).
453. Ehrlich P.: Z. anorg. Chemie *257*, 247 (1948).
454. Ehrmann K.: Fr. 1336648 (1963).
455. E. I. du Pont de Nemours and Co: Brit. 574936 (1946).
456. E. I. du Pont de Nemours and Co: Brit. 577279 (1946).
457. E. I. du Pont de Nemours and Co: Brit. 646408 (1950).
457a. Eigenson A. S., Krichko A. A., Kozik B. L., Pestrikov S. V., Sadykova S. R., Ivanova L. S., Skundina L. Ya., Ayazyan G. N.: Khim. Tekhnol. Topliv Masel *13*, No. 9, 7 (1968).
458. Elbert E. I., Katsobashvili Ya. R., Revva M.: Podz. Gazif. Uglei Kuzbasse No. 1, 95 (1966); C. A. 68, 23376 (1968).
458a. El Kady F. Y., Hassan H. A., Abdou I. K.: J. Inst. Petrol. *55*, 338 (1969).
459. Elliott G. E., Moore R. J.: Belg. 640048 (1964).
460. Elliott G. E., Mc Kinley J. B., Stewart M. M.: US 3017368 (1962).
461. Elliott M. A., Kandiner H. J., Kallenberger R. H., Hiteshue R. W., Storch H. H.: Ind. Eng. Chem. 42, 83 (1950).
462. Elliott N.: J. Chem. Phys. *33*, 903 (1960).
463. Eltekov Yu. A., Samoilov S. M.: Izv. AN SSSR, Otd. Khim. Nauk *1960*, 794.
464. Endo K., Kojima H.: Jap. 21496 ('63); C. A. *60*, 3122 (1964).
465. Eng J., Tiedje J. L.: US 3340184 (1967).
465a. Eng. J., Thomson G. H., Vanderlinden R. C., Lewis E. H., Tripp R. G.: Fr. 1542017 (1968).
465b. Eng J.: Fr. 1566324 (1969).
465c. Eng J., Guggisberg W. R.: US 3480540 (1969).
466. Engel W. F., van Weeren H. J.: Ger. 1086375 (1960).
467. Engel W. F.: Neth. 72052 (1953).
468. Engel W. F.: Neth. 72677 (1953).
469. Engel W. F.: Neth. 77643 (1955).
470. Engel W. F., Hoog H.: US 2650906 (1953).
471. Engel W. F.: Krijger P.: US 2662860 (1953).
472. Engel W. F., Hoog H.: US 2690433 (1954).
473. Engel W. F., Hoog H.: US 2697683 (1954).
474. Engelhard Industries, Inc.: Brit. 822670 (1959).
475. Engelhard Industries, Inc.: Brit. 1066579 (1967).
476. Engelhard Industries, Inc.: Neth. 6608097 (1966).

477. Engelhard P., Trambouze Y.: Bull. soc. chim. France 1959, 195.
478. Engelhard P. A.: Chim. mod. *4*, 61 (1959).
479. Epstein M. B., Pitzer K. S., Rossini F. D.: J. Research Nat. Bur. Standards *42*, 379 (1949).
480. Erbelding W. F.: US 3236765 (1966).
481. Erickson H., Johnson M. F. L., Keith C. D.: US 3132111 (1964).
482. Erickson H., Sanford R. A.: US 3223652 (1965).
483. Eru I. I.: Zh. Prikl. Khim. *7*, 145 (1934).
484. Eru I. I.: Khim. Tverd. Topliva *6*, 831 (1935).
485. Eru I. I., Sakhnovskaya E. M., Pychko V. A.: Zh. Obshch. Khim. 8, 1563 (1938).
486. Eru I. I., Volkov Yu. M., Lange A. A.: Koks i Khim. *1959*, No. 2, 35.
487. Eru I. I., Volkov Yu. M., Lange A. A.: Sb. Nauch. Tr. Ukr. Nauch.-Issled. Uglekhim. Inst. *1960*, No. 10, 164; C. A. 57, 14091 (1962).
488. Eru I. I., Lange A. A., Timoshenko V. A., Kiryakova E. T.: Koks i Khim. *1961*, No. 5, 44.
489. Eru I. I., Lange A. A., Zeidlin E. M., Strelnikova V. P.: Koks i Khim. *1962*, No. 10, 46.
490. Esso Research and Eng. Co.: Brit. 748154 (1952).
491. Esso Research and Eng. Co.: Brit. 802133 (1958).
492. Esso Research and Eng. Co.: Brit. 804308 (1958).
493. Esso Research and Eng. Co.: Brit. 882958 (1960).
494. Esso Research and Eng. Co.: Brit. 933663 (1963).
495. Esso Research and Eng. Co.: Fr. 1180418 (1959).
496. Esso Research and Eng. Co.: Fr. 1482144 (1967).
496a. Esso Research and Eng. Co.: Fr. 1486689 (1967).
496b. Esso Research and Eng. Co.: Fr. 1491344 (1967).
497. Esso Research and Eng. Co.: Ger. 1050946 (1959).
498. Esso Research and Eng. Co.: Neth. Appl. 6409702 (1965).
499. Ettel V.: *Organická technologie I* (*Organic Technology, Vol. I*), p. 21. SNTL Prague, 1955.
500. Ettel V.: *Organická technologie II* (*Organic Technology,* Vol. *II*), p. 226, 278. SNTL Prague, 1956.
501. Evans B. L., Young P. A.: Proc. Roy. Soc. (London) *A 284,* 402 (1965).
502. Evans E. V.: J. Soc. Chem. Ind. *34,* 9 (1915).
503. Evans E. V., Stanier H.: Proc. Roy. Soc. (London) *A 105,* 626 (1924).
504. Falkovskii V. B., Lvov S. V., Starkov A. V., Konareva Z. P.: Kinetika i Kataliz, AN SSSR, Sb. Statei *1960*, p. 120.
505. Farbenfabriken Bayer A. G.: Brit. 880300 (1961).
505a. Farbenfabriken Bayer A. G.: Fr. Demande 2008330 (1970).
505b. Farbenfabriken Bayer A. G.: Fr. Demande 2008331 (1970).
506. Farbenfabriken Bayer A. G.: Ger. 1179213 (1964).
507. Farlow W. M., Hunt M., Langkammer G. M., Lazier W. A., Peppel W. J., Signaigo F. K.: J. Am. Chem. Soc. *70*, 1392 (1948).
508. Farlow M. W., Lazier W. A., Signaigo F. K.: Ind. Eng. Chem. *42*, 2547 (1950).
509. Farlow M. W., Signaigo F. K.: US 2402613 (1946).
510. Farlow M. W., Signaigo F. K.: US 2402614 (1946).
511. Farlow M. W., Signaigo F. K.: US 2402615 (1946).
512. Fear J. V. D.: US 2908638 (1957).
513. Fear J. V. D.: US 2928787 (1960).
514. Fear J. V. D.: US 2952628 (1960).
515. Feofilov E. E., Garnovskaya G. N.: Trudy Vsesoyuz. Nauchno-Issled. Instituta po Pererab. Slantsev *1958*, No. 6, 183; C. A. 53, 14481 (1959).
516. Fetterly L. C.: US 2897245 (1959).

517. Finke M., Heinze G.: Erdöl u. Kohle *19*, 17 (1966).
518. Finch H. D., Furman K. E.: US 2635119 (1953).
519. Fischer F., Bahr T., Petrick A. J.: Brennstoff-Chem. *13*, 45 (1932).
520. Fischer F., Koch H.: Brennstoff-Chem. *19*, 245 (1938).
521. Fischer P. E., Buss W. C., Goldsmith E. A.: US 3320181 (1967).
522. Fischmeister H. F.: Acta Chem. Scand. *13*, 852 (1959).
523. Fisher C. H., Eisner A.: Ind. Eng. Chem. *29*, 939 (1937).
524. Fisher C. H., Eisner A.: Ind. Eng. Chem. *29*, 1371 (1937).
524a. Fivaz R.: Helv. Phys. Acta *39* (3), 247 (1966).
524b. Fivaz R., Mooser E.: Phys. Rev. *163*, 743 (1967).
525. Fleck R. N.: US 2608534 (1952).
526. Fleck R. N.: US 2945824 (1960).
527. Fleming H. W., Gutmann W. R.: US 2959627 (1960).
528. Fleming H. W., Gutmann W. R.: US 3003008 (1962).
529. Fleming H. W., Gutmann W. R.: US 3050571 (1962).
530. Fleming H. W., Gutmann W. R.: US 3152193 (1964).
531. Fleming H. W.: US 3155739 (1964).
532. Flinn R. A., Larson O. A., Beuther H.: Ind. Eng. Chem. *52*, 153 (1960).
533. Flinn R. A., Beuther H., Schmid B. K.: Petrol. Ref. *40*, No. 4, 139 (1961).
534. Flinn R. A., Larson O. A., Beuther H.: Hydrocarbon Process. Petrol. Ref. *42*, No. 9, 129 (1963).
535. Flinn R. A., Mc Kinley J. B.: US 2880171 (1956).
536. Flinn R. A., Larson O. A.: US 3260666 (1966).
536a. Flinn R. A., Tarhan M. O.: US 3503872 (1970).
537. Flockhart E. A., Henke A. M., Kwolek S. J., Hornaday G. F.: Hydrocarbon Process. Petrol. Ref. *44*, No. 12, 99 (1965).
538. Folkins H. O., Miller E., Hennig H.: Ind. Eng. Chem. *42*, 2202 (1950).
539. Folkins H. O., Miller E. L.: *Proc. Am. Petrol. Inst.*, Sect. III, *42*, 188 (1962).
540. Folkins H. O., Miller E. L.: Ind. Eng. Chem., Process Design Dev. *1*, 271 (1962).
541. Folkins H. O., Miller E. L.: *Lecture on 140th Congress of Am. Chem. Soc.*, Chicago, 3—8 Sept. 1961; Erdöl u. Kohle *15*, 557 (1962).
542. Folkins H. O.: *Kirk-Othmer*, *Encyclopedia of Chem. Technol.*, Vol. 4, p. 376. Interscience Publ., New York 1964.
543. Folkins H. O., Kempf A., Miller E. L.: Canad. 552791 (1958).
544. Folkins H. O., Kempf A., Miller E. L.: Canad. 552792 (1958).
545. Folkins H. O., Kempf A., Miller E. L.: Canad. 552793 (1958).
546. Folkins H. O.: US 2786079 (1957).
547. Folkins H. O., Miller E. L.: US 2820060 (1958).
548. Folkins H. O., Miller E. L.: Kempf A.: US 2820061 (1958).
549. Folkins H. O., Miller E. L., Kempf A.: US 2820062 (1958).
550. Folkins H. O., Miller E. L.: US 2820063 (1958).
551. Folkins H. O., Miller E. L.: US 2951872 (1960).
552. Folkins H. O., Miller E. L.: US 2951873 (1960).
553. Folkins H. O., Kempf A.: US 2976322 (1961).
554. Folkins H. O., Miller E. L.: US 2976323 (1961).
555. Ford F. A.: US 3045053 (1962).
556. Forney R. C., Smith J. M.: Ind. Eng. Chem. *43*, 1841 (1951).
556a. Foster H. M.: US 3381018 (1968).
557. Fox A. L.: Brit. 580366 (1946).
558. Fox A. L.: US 2407266 (1946).

559. Franceway J. A.: US 1904218 (1933).
559a. Frese E.: Status of Recent Research Work on Hydrogenation With a Fixed Catalyst at 700 Atm. Hydrogenation of Middle Oil. Bureau of Mines Special Comm., November 1946.
560. Fridshtein I. L., Kamlyuk L. M., Dmitriev I. I.: Neftekhimiya *4* (6), 824 (1964).
561. Friedman L. D., Rambo M. L., Estes J. H.: US 3025231 (1962).
561a. Friedman S., Hiteshue R. W.: US 3429679 (1969).
562. Friedrich E.: Z. Physik. *31*, 813 (1925).
563. Frindt R. F., Yoffe A. D.: Proc. Roy. Soc. (London) *A 273*, 69 (1963).
564. Friz H.: Z. Elektrochem. *54*, 538 (1950).
565. Frolich P. K.: US 2035121 (1936).
566. Frolich P. K., Wiezevich P. J.: US 2045766 (1936).
567. Frost A. V.: Vestnik Mosk. Gos. Univ., No. 3—4, 111 (1946).
568. Frye C. G., Barger B. D., Brennan H. M., Coley J. R., Gutberlet L. C.: *Hydroisomerization of Olefins*. Preprints, Div. Petrol. Chem., Am. Chem. Soc. *7*, No, 4. A-21 (1962).
569. Frye C. G., Barger B. D., Brennan H. M., Coley J. R., Gutberlet L. C.: Ind. Eng. Chem., Product Research Dev. *2*, 40 (1963).
570. Frye C. G., Mosby J. F.: Chem. Eng. Progress *63*, No. 9, 66 (1967).
571. Fuchs O., Brendlein H.: Z. angew. Chem. *52*, 49 (1939).
572. Fuchs W.: *Die Chemie der Kohle*, p. 445. Springer Verlag, Berlin 1931.
572a. Fucki K., Kudomi H., Ishida H., Mukaibo T.: J. Electrochem. Soc. Japan *37* (3), 137 (1969); C A. *72*, 104402 (1970).
573. Fugassi P., Trammel R.: Chem. and Ind. *1959*, 654.
574. Fukui K., Yuasa Y., Hirooka M.: Jap. 3484 ('62); C. A. *59*, *11685* (1963).
575. v. Füner W., Simon W.: Ger. 801395 (1951).
576. v. Füner W., Simon W.: Ger. 814138 (1951).
577. v. Füner W.: Ger. 816853 (1951).
578. Furukawa T., Iijima K., Kotani T., Izume Y.: Bull. Japan Petrol. Inst. 6, 1 (1964); C. A. 62, 2649 (1965).
579. Gagnon P. E., Boivin J. L., Watson D. C.: Canad. J. Chem. *37*, 1846 (1959).
579a. Galbreath R. B., van Driesen R. P.: 8th World Petrol. Congress, PD 12, Paper 2, Moscow 1971.
580. Gall D. D.: J. Soc. Chem. Ind. *65*, 185 (1946).
580a. Gallagher J. P., Rausch M. K.: US 3494854 (1970).
581. Galle E., Michelitsch W.: Petrol. Z. *31* (8), 1 (1935); Montan. Rundschau *27*, No. 5, 1 (1935); C. A. *29*, 4251 (1935).
582. Ganguli A. K., Basu A. N., Basak N. G., Lahiri A.: Chem. Age India *17* (3), 225 (1966). 606A. C. A. 65, 2034 (1966).
583. Ganguli N. C., Kini K. A., Basak N. G., Lahiri A.: Chem. Age India *17* (12), 1013 (1966); C. A. 66, 77978 (1967).
583a. Ganguli N. C., Bose S. K., Kim K. A., Basak N. G., Lahiri A.: Chem. Age India *12*, 126 (1961).
584. Ganz E.: Ann. chim. *19*, 202 (1944).
585. Garbo P. W.: US 2774717 (1956).
586. Gardner L. E., Hogan R. J.: US 2946738 (1960).
587. Gardner L. E.: US 2970102 (1961).
587a. Garwood W. E.: Ger. 1301410 (1969).
588. Garwood W. E.: US 2687983 (1954).
589. Gáti G., Meisel A.: Z. Chem. 4, 171 (1964).
590. Gatsis J. G.: US 3173860 (1965); C. A. 62, 11607 (1965).
591. Gatsis J. G.: US 3180822 (1965); C. A. 63, 4079 (1965).

592. Gatsis J. G.: US 3262874 (1966).
592a. Gatsis J. G.: US 3269958 (1966).
592b. Gatsis J. G.: US 3331769 (1967).
593. Gelms I. E.: *Proceedings, Symposium on the Application of Hydrogenation Processes in Petroleum Processing*, Index D_1, Moscow 1963.
593a. Genas M., Rull T.: Fr. 1518809 (1968).
594. Gen. Electric Co.: Brit. 830743 (1960).
595. Georgius F.: Gas World *138*, 1306 (1953).
596. Gerald C. F.: US 2790751 (1957).
597. Gerhold C. G.: US 2773007 (1956).
597a. Gerhold C. G.: Ger. 1269757 (1968).
598. Gerhold M.: Austrian 205154 (1959).
599. Gerlach J.: Metall *17*, 32 (1963).
600. Germain J. E.: *Heterogenní katalysa* (*Heterogenous Catalysis*) p. 120. SNTL Prague, 1962.
601. Ghosal S. R., Ghosh S. C., Majumdar M. M., Roy D., Dutta B. K.: Technology (Sindri) *2*, 211 (1965).
602. Ghosal S. R., Ghosh S. C., Majumdar M. M., Dutta B. K.: Technology (Sindri) *3*, 3 (1966).
603. Ghosal S. R., Ghosh S. C., Majumbar M. M., Dutta B. K.: Technology (Sindri) *3*, 126 (1966).
603a. Ghosal S. R., Majumdar M. M., Ghosh S. C., Dutta B. K.: Technology *4* (3), 63 (1967).
604. Giauque W. F., Blue R. W.: J. Am. Chem. Soc. *58*, 831 (1936).
605. Gilbert J. B., Kartzmark R., Sproule L. W.: J. Inst. Petrol. *53*, 317 (1967).
606. Gilbert J. B., Kartzmark R.: *7th World Petrol. Congr.*, PD 20, Paper 4. Mexico City 1967.
606A. Gilbert J. B., Walker J.: 8th World Petrol Congress, PD 12, Paper 4. Moscow 1971.
606a. Gillot J., Lux B., Cornuault P., Du Chaffaut F.: J. Chim. Phys. Physicochim. Biol. *1969* (No. Spec.), 172; C. A. 71, 5084 (1969).
607. Ginsberg H. H., Lewis P. S., Anderson R. B., Hiteshue R. W.: US Dep. Interior, Bur. Mines, Rep. Invest. 1960, No. 5674, 1—57. Pittsburgh, Pa., Bur. Mines. Chem. Zentr. 133, 8472 (1962).
608. Giona A. R., Vlugter J. C., van de Weerdt W. J.: Calore *31*, 663 (1960); C. A. 59, 2613 (1963).
608a. Girotti P. L., Floris T., Pecci G.: Ger. Offen. 1907495 (1969).
609. Gladrow E. M., Parker P. T.: Brit. 913923 (1962).
610. Gleim W. K. T.: US 3161584 (1964).
611. Gleim W. K. T., Gatsis J. G.: US 3161585 (1964).
612. Gleim W. K. T., Gatsis J. G.: US 3165463 (1965).
613. Gleim W. K. T., Arrigo J. T.: US 3196104 (1965).
614. Gleim W. K. T., Gatsis J. G.: US 3252895 (1966).
614a. Gleim W. K. T.: US 3282828 (1966).
614b. Gleim W. K. T., Gatsis J. G.: US 3483118 (1969).
615. Glemser O., Sauer H., König P.: Z. anorg. Chem. *257*, 241 (1948).
616. Glushenkova E. V.: Trudy Vsesoyuz. Nauchno-Issled. Instituta po Pererab. Slantsev 1958, No. 6, 163.
617. Glushenkova E. V., Semenov S. S.: Trudy Vsesoyuz. Nauchno-Issled. Inst. Pererab. Slantsev 1958, No. 6, 206.
618. *Gmelins Handbuch der anorg. Chem.*, System Nr. 53, Teil Molybdän, p. 183. Verlag Chemie G. m. b. H., Weinheim/Bergstr. u. Berlin 1935.
619. *Gmelins Handbuch der anorg. Chem.*, System Nr. 54, p. 185, Berlin 1935.

620. Go Si-Sjan et al.: *Problemy Kinetiki i Kataliza* (*Problems of Kinetics and Catalysis*), Vol. 10. Izd. AN SSSR, Moscow 1960.
620a. Goheen D. W.: *Lignin Structure and Reactions*, p. 205. Advances in Chemistry Series 59, Am. Chem. Soc., Washington 1966.
621. Gohr E. J., Barr F. T., Roetheli B. E.: US 2415817 (1947).
622. Gohr E. J.: US 2421608 (1947).
623. Goldsmith E. A.: US 2928784 (1960).
624. Goldsmith E. A.: US 3151175 (1964).
625. Goldshtein D. L., Osipov L. N., Agafonov A. V.: *Khim. Sera- i Azotorg. Soedinenii Soderzhashchikhsya v Nefti i Nefteprodukt.*, Akad. Nauk SSSR, Bashkirsk. Filial *3*, 389 (1960).
626. Golubzova V. A.: Petroleum *16*, 220 (1953); Chem. Zentr. 125, 4064 (1954).
626a. Goncharenko A. D., Martynenko A. G., Volkov A. I., Vovk L. M.: Neftepererab. Neftekhim. *1970*, No. 3, 36.
627. Gonikberg M. G., Levitskii I. I., Kazanskii B. A.: Izv. AN SSSR, Otdel. Khim. Nauk *1959*, 611.
628. Gonikberg M. G., Levitskii I. I.: Izv. AN SSSR, Otd. Khim. Nauk *1960*, 1170 (1960).
629. Good G. M., de Nooijer C. M. J.: Neth. 82266 (1956).
630. Goodson L. B.: US 2541237 (1951).
631. Goppoldová M., Dufek L.: Ropa a uhlie *7*, 74 (1965).
632. Goppoldová M.: Ropa a uhlie *10*, 232 (1968).
632a. Goppoldová M.: Issled. Primen. Gidrogenizatsionnykh Protsessov Nefteper. Neftekhim. Prom. *1968*, 208; C. A. 71, 103771 (1969).
633. Goppoldová M., Dufek L.: Czech. 121929 (1967).
634. Gordon K.: J. Inst. Fuel *9*, 69 (1935).
635. Gordon K.: J. Inst. Fuel *20*, 42 (1946).
636. Gordon K.: J. Inst. Fuel *21*, 53 (1947).
637. Gordon K.: Petroleum (London) *11*, 35 (1948).
638. Goshorn R. H., Buchholz B., Deger T. E.: US 3036133 (1962).
639. Gossedin A.: Rev. inst. franc. pétrole et Ann. combustibles liquides *1*, 145 (1946); C. A. 44, 5080 (1950).
640. Gould G. D., Paterson N. J., Laity D. S., Bray B. G.: Hydrocarbon Process. Petrol. Ref. *43*, No. 5, 117 (1964).
641. Gould G. D.: US 3267021 (1966).
642. Grace W. R. and Co.: Neth. Appl. 6413872 (1965).
643. Graham J. I., Skinner D. G.: J. Soc. Chem. Ind. *48*, 129 T (1929).
643a. Craig R. G., Doelp L. C. Jr., Friedman L.: US 3487120 (1969).
644. Greber W.: J. prakt. Chem., Reihe *1* (4), 98 (1954).
645. Greber W., Brücks B., Kranig L.: Erdöl u. Kohle *20*, 404 (1967).
646. Greensfelder B. S., Voge H. H., Good G. M.: Ind. Eng. Chem. *41*, 2573 (1949).
647. Greensfelder B. S., Moore R. J.: Brit. 603103 (1948).
648. Greensfelder B. S.: US 2393041 (1946).
649. Greensfelder B. S., Peterson W. H.: US 2402493 (1946).
650. Greenfield H., Dovell F. S.: J. Org. Chem. *31*, 3053 (1966).
651. Greenfield H., Dovell F. S.: J. Org. Chem. *32*, 3670 (1967).
651a. Greenfield H.: Ann. N. Y. Acad. Sci. *145* (1), 108 (1967).
652. Gribb G. S., Marsh J. D. F.: Gas Council (Gt. Brit.) Research Commun. GC 66, 32 pp. (1959); C. A. 54, 6087 (1960).
653. Griffith R. H.: Inst. Gas Eng. 1937, Publ. 175, 48.
654. Griffith R. H., Hill S. G.: J. Chem. Soc. *1938*, 717.
655. Griffith R. H., Hill S. G.: J. Chem. Soc. *1938*, 2037.

656. Griffith R. H., Plant J. H. G.: Gas J. *244*, 48 (1944).
657. Griffith R. H.: *Advances in Catalysis*, Vol. I., p. 91. Academic Press Inc., New York 1948.
658. Griffith R. H., Marsh J. D. F., Newling W. B. S.: Proc. Roy. Soc. (London) *A 197*, 194 (1949).
659. Griffith R. H.: Ind. Eng. Chem. *44*, 1011 (1952).
660. Griffith R. H., Marsh J. D. F.: *Contact Catalysis*. Oxford Univ. Press 1957.
661. Griffith R. H.: Brit. 370909 (1932).
662. Griffith R. H., Geoffrey J. H.: Brit. 529711 (1939).
663. Griffith R. H.: Brit. 577816 (1946).
664. Griffith R. H., Morcon A. R., Newling W. B. S.: Brit. 600118 (1948).
665. Griffith R. H., Newling W. B. S., Plant J. H. G.: Brit. 600787 (1948).
666. Griffith R. H., Plant J. H. G.: US 2295653 (1943).
667. Griffiths D. J., James J. L., Luntz D. M.: Erdöl u. Kohle *21*, 83 (1968).
668. Griffitts F. A., Brown O. W.: J. Phys. Chem. *41*, 477 (1937).
669. Grigoryan L. A., Novoselova A. V.: Dokl. AN SSSR *144*, 795 (1962).
670. Grimmeiss H. G., Rabenau A., Hahn H., Ness P.: Z. Elektrochem. *65*, 776 (1961).
671. Gring J. L.: US 3152091 (1964).
672. Griswold C. R., van Driesen R. P.: Hydrocarbon Process. *45*, No. 5, 153 (1966).
673. Groenvold F., Thurmann-Moe T., Westrum E. F., Chang E.: J. Chem. Phys. *35*, 1665 (1961).
674. Groggins P. H.: *Základní pochody organické synthesy* (*Unit Processes in Organic Synthesis*). SNTL, Prague 1958.
675. Grosse A. V.: US 2029100 (1936).
676. Grosse A. V.: US 2037781 (1936).
677. Grossweiner L. I.: J. Phys. Chem. *59*, 742 (1955).
678. Groszek A. J.: Nature *204*, 680 (1964).
679. Groszek A. J.: ASLE Trans. *9* (1), 67; discussion 74 (1966).
680. Grote H. W., Watkins C. H., Poll H. F., Hendricks G. W.: Oil Gas J. *52*, No. 50, 211, (1954).
681. Grzechowiak J., Tomasik Z.: Chem. Stosowana *4*, 101 (1960).
682. Grzechowiak J., Tomasik Z.: Chem. Stosowana, Ser. A, No. 9 (2), 231 (1965).
683. Guichard M.: Bull. soc. chim. *51*, 563 (1932).
684. Gulf Research and Dev. Co.: Brit. 996780 (1965).
685. Gulf Research and Dev. Co.: Fr. Addn. 79175 (1962).
686. Gulf Research and Dev. Co.: Neth. 300693 (1965).
687. Gulf Research and Dev. Co.: Neth. Appl. 6410024 (1965).
688. Gulf Research and Dev. Co.: Neth. Appl. 6510949 (1966).
689. Gulf Research and Dev. Co.: Neth. Appl. 6608758 (1966).
689a. Gulf Research and Dev. Co.. Neth. Appl. 6516982 (1966).
690. See 460.
691. Gully A. J., Ballard W. P.: *Advances in Petroleum Chemistry and Refining* (Ed. J. J. Mc Ketta Jr.), Vol. VII, p. 240—282. Interscience Publ., Wiley, New York, London 1963.
692. Gumlich W., Kränzlein P.: Ger. 1041037 (1958).
693. Günther G.: Freiberger Forschungs-H. A, Bergbau No. 17, 38 (1953).
694. Günther G., Kühnhanss G., Hüttig E.: Chem. Technik *7*, 656 (1955).
695. Günther G.: J. Prakt. Chem. (4), *3*, 241 (1956).
696. Günther G.: Chem. Technik *9*, 410 (1957).
697. Günther G.: Chem. Technik *12*, 181 (1960).
698. Günther G.: Chem. Technik *13*, 427 (1961).

699. Günther G.: Chem. Technik *13*, 718 (1961).
700. Günther G.: Chem. Technik *13*, 720 (1961).
701. Günther G.: *Paper presented at Symposium on Catalysts.* Marianské Lázně 1961.
702. Günther G.: *Proceedings, Intern. Conference on Catalysts*, p. 76. Karlovy Vary 1964.
703. Günther G.: Chem. Technik *18*, 358 (1966).
704. Guthke F. W.: US 1897798 (1933).
705. Guthke F. W.: US 1932369 (1934).
706. Gwin G. T.: Proc. API *39* (III), 193 (1959).
707. Haas R. H., Tulleners A. J.: US 3119763 (1964).
708. Haensel V., Donaldson G. R.: Ind. Eng. Chem. *43*, 2102 (1951).
709. Haensel V., Berger C. V.: *Advances in Petroleum Chemistry and Refining* (Ed. J. J. Mc Ketta, Jr.), Vol. I, p. 387. Interscience Publ. Inc., New York 1958.
710. Haensel V., Pollitzer E. L., Watkins C. H.: *6th World Petrol. Congr.*, Sect. III, Paper 17. Frankfurt 1963.
711. Haensel V., Addison G. E.: *7th World Petrol. Congr.*, PD-17, Paper 4. Mexico City 1967.
712. Haensel V.: US 2479109 (1949).
713. Haensel V.: US 2566521 (1948).
714. Hagemeyer H. J. Jr.: US 2691045 (1954).
715. Hähner E., Kaufmann H., Leibnitz E.: Freiberger Forschungs-H. A 329, 67 (1964).
716. Hájek B.: Naturwissenschaften *46*, 424 (1959).
717. Hale J. H., Simmons M. C., Whisenhunt F. P.: Ind. Eng. Chem. *41*, 2702 (1949).
718. Halik R. R., Ireland H. R., Smilski M. T.: US 3284340 (1966).
719. Hall C. C.: Fuel *12*, 76 (1933).
720. Hall C. C., Cawley C. M.: J. Soc. Chem. Ind. *58*, 7 (1939).
721. Hammar C. G. B.: *Proc. 3rd World Petrol. Congress*, Sect. IV, 295. The Hague 1951.
722. Hammar C. G. B., Löfgren A. W.: Swed. 151523 (1955).
722a. Hamner G. P., Mason R. B.: Fr. 1580689 (1969).
722b. Hamner G. P.: US 3511771 (1970).
722c. Haney S. C., Rausch M. K., Divijak J. M. Jr.: US 3528910 (1970).
723. Hanisch H., Korn F.: Ger. 938613 (1956).
724. Hannay N. B.: *Semiconductors.* Reinhold Publ. Co., New York 1959.
725. Hass R. H., Reeg C. P., Riddick F. C.: US 3256178 (1966).
726. Hansen C. A. Jr.: US 2963425 (1960).
727. Hansford R. C.: US 2885349 (1959).
727a. Hansford R. C.: US 3499835 (1970).
727b. Hanson F. V., Snyder P. W. Jr.: US 3523887 (1970).
728. Harcourt G. A.: Am. Mineralogist *27*, 91 (1942).
729. Haresnape J. N., Morris J. E.: Brit. 701217 (1953).
729a. Hargis C. W., Young H. S.: US 3370087 (1968).
730. Harnsberger H. F., Mulaskey B. F., Hickson D. A.: US 3242100 (1966).
731. Harris E. E., Saeman J. F., Bergstrom C. B.: Ind. Eng. Chem. *41*, 2063 (1949).
732. Hartley F. L., Inwood T. V.: US 2883337 (1959).
733. Hartung G. K., Jewell D. M., Larson O. A., Flinn R. A.: J. Chem. Eng. Data *6*, 477 (1961).
733a. Hartung S., Freese H., Wedemeyer K. F., Dierichs H., Hagedorn F.: Ger. Offen. 1903968 (1970).
734. Harvey P. D., Robbers J. A., Scott J. W., Mason H. F.: Fr. 1351998 (1963).
735. Harvey P. D., Robbers J. A., Scott J. W., Mason H. F.: Fr. 1359378 (1964).
736. Harvey P. D., Robbers J. A., Scott J. W., Mason H. F.: US 3166491 (1965).
737. Harvey P. D., Unverferth J. W.: US 3259588 (1966).

738. Hashimoto H., Naiki T.: J. Electronmicroscopy (Tokyo) *13* (3), 178 (1964); C. A. *63*, 8091 (1965).
739. Hassel O.: Z. Kristallogr., Mineralog., Petrogr. *61*, 92 (1925).
740. Hatch G. B.: US 2648675 (1953).
741. Hauffe K., Flindt H. G.: Z. phys. Chem. *200*, 199 (1952).
742. Hauffe K.: *Advances in Catalysis*, Vol. VII, p. 213. Academic Press Inc., New York 1955.
742a. Hauffe K.: Angew. Chem. *67*, 189 (1955).
742b. Hawk C. O., Schlesinger M. D., Ginsberg H. H., Hiteshue R. W.: US Bureau of Mines, Rep. Invest. No. 6548 (1964).
742c. Hawk C. O., Schlesinger M. D., Dobransky P., Hiteshue R. W.: US Bureau of Mines, Rep. Invest. No. 6655 (1965).
743. Hayashi H., Hayashi S., Takizawa N., Morita T., Hirai K.: Sekiyu Gakkai Shi *8* (4), 254 (1965); C. A. 65, 15119 (1966).
743a. Hayes J. C.: US 3379641 (1968).
743b. Hayes J. C.: US 3442796 (1969).
743B. Hayes J. C.: US 3448036 (1969).
743c. Hayes J. C.: US 3481861 (1969).
743d. Hayes J. C., Mitsche R. T.: US 3499836 (1970).
743e. Hayes J. C.: US 3515665 (1970).
744. Heard L., Den Herder M. J.: US 2659701 (1953).
745. Heckelsberg L. F., Bailey G. C., Clark A.: J. Am. Chem. Soc. *77*, 1373 (1955).
745a. Heckelsberg L. F.: Fr. 1562396 (1969).
745b. Heckelsberg L. F.: S. African 68 01707 (1968).
745c. Heckelsberg L. F.: S. African 68 01878 (1968).
745d. Heckelsberg L. F.: US 3340322 (1967).
745e. Hedvall J. A.: Angew. Chem. *54*, 505 (1941).
746. Heinemann H., Kirsch F. W., Burtis T. A.: Erdöl u. Kohle *10*, 225 (1957).
747. Heinemann H., Milliken T. H.: US 2985579 (1961).
748. Heinemann H.: US 2985580 (1961).
749. Heinicke G.: Chem. Technik *15*, 197 (1963).
750. Heinicke G., Harenz H.: Z. anorg. Chem. *324*, 185 (1963).
751. Hellwig K. C., Johanson E. S., Johnson C. A., Schuman S. C., Stotler H. H.: Hydrocarbon Process *45*, No. 5, 165 (1966).
752. Hellwig L. R., van Driesen R. P., Byrd R. C.: Oil Gas J. *61*, 227 (Oct. 14, 1963).
753. Hendel F. J.: US 2697078 (1954).
753a. Henderson D. W., Raley J. H.: US 3439061 (1969).
754. Hendricks G. W., Huffman H. C., Parker R. L., Stirton R. I.: *Catalytic Desulphurization of Petroleum Distillates*. Petroleum Division Preprints, A. C. S. Meeting, April 1946.
755. Hendricks G. W.: US 2687370 (1954).
756. Hendriks W. J., Vlugter J. C., Waterman H. I., van de Weerdt W. J.: Brennstoff-Chem. *42*, 185 (1961).
757. Hendriks W. J., Vlugter J. C., Waterman H. I., van de Weerdt W. J.: Brennstoff-Chem. *42*, 215 (1961).
758. Henke A. M., Peterson R. E.: US 3046218 (1962).
759. Henke A. M., Mc Kinley J. B.: US 3053760 (1962).
760. Henke A. M., Schmid B. K., Strom J. R.: US 3349025 (1967).
760a. Henke A. M., Horne W. A., Schmid B. K.: Fr. 1521459 (1968).
760b. Henke A. M., Horne W. A., Schmid B. K.: US 3493493 (1970).
761. Henke C. O., Vaughen J. V.: US 2198249 (1940).
762. Hennig H.: US 2727853 (1955).

763. Hennig H., Tierney J. W.: US 2807649 (1957).
764. Hennig H.: US 2819313 (1958).
765. Haraldsen H., Klemm W.: Z. anorg. Chem. *223*, 409 (1935).
766. Haraldsen H.: Z. anorg. Chem. *224*, 85 (1935).
767. Herbst W., Dorschner O., Eisenlohr K. H.: Ger. 956539 (1957).
768. Herington E. F. G., Rideal E. K.: Proc. Roy. Soc. *A 184*, 434; 447 (1945).
769. Hettinger W. P., Keith C. D., Gring J. L., Teter J. W.: Ind. Eng. Chem. *47*, 719 (1955).
770. Hiemenz H.: *Comments at 6th World Petrol. Congr.*, Sect. III, p. 307. Frankfurt 1963.
771. Hilberath F., Meisenburg E.: Ger. 1221751 (1966).
772. Hill B. N.: US 2436170 (1948).
773. Hlavica B.: Paliva a topení *11*, 150; 161 (1929).
773a. Hodgson R. L.: S. African 68 07725 (1969).
773b. Hodgson R. L.: S. African 69 05181 (1970).
774. Hoeffelman J. M., Berkenbosch R.: US 2352435 (1944).
774a. Hoekstra J.: S. African 67 07622 (1968).
775. Hoffert W. H., Wendtner K.: J. Inst. Petrol. *35*, 171 (1949).
776. Hoffmann E. J., Lewis E. W., Wadley E. F.: Ind. Eng. Chem. *49*, 656 (1957).
776a. Hoffmann E. J., Lewis E. W., Wadley E. F.: Petrol. Ref. *36*, No. 6, 179 (1957).
777. Hoffmann E. J., Wadley E. F.: US 2638438 (1953).
778. Hogan R. J., Mills K. L., Lanning W. C.: J. Chem. Eng. Data *7*, 148 (1962).
779. Hogan R. J.: US 2926130 (1960).
780. Holden D. L.: US 3016348 (1962).
781. Hollings H., Evans E. V.: Inst. Gas. Engrs. Publ. No. 146, 10 pp. (1936).
782. Holmes P. D., Bourne K. H.: Belg. 623777 (1963).
783. Holmes P. D., Pitkethly R. C.: Belg. 625486 (1963).
783a. Holmes P. D., Pitkethly R. C.: Brit. 1044333 (1966).
784. Holroyd R.: Bureau of Mines, Inf. Circ. 7370 (1946).
785. Hoog H.: J. Inst. Petrol. *36*, 738 (1950).
786. Hoog H.: Rec. trav. chim. *69*. 1289 (1950).
787. Hoog H., Reman G. H., Brezesinska-Smithuysen W. C.: *Proc. 3rd World Petrol. Congr.*, Sect. IV, 282. The Hague 1951.
788. Hoog H., Klinkert H. G., Schaafsma A.: Petrol. Ref. *32*, No. 5, 137 (1953).
789. Hoog H., Klinkert H. G., Schaafsma A.: Proc. Am. Petrol. Inst. *33*, III, 71 (1953).
790. Hoog H.: Neth. 71308 (1952).
791. Hoog H., Koome J.: Neth. 71604 (1953).
792. Hoover Ch. O.: US 2640010 (1953).
793. Höring M.: *Die Chemie der Braunkohle* (Ed. A. Lissner, A. Thau), Vol. 2, p. 238. W. Knapp Verlag, Halle 1953.
794. Höring M., Oettinger W., Reitz O.: Erdöl u. Kohle *16*, 361 (1963).
795. Horne W. A., McAfee J.: *Advances in Petroleum Chemistry and Refining* (Ed. J. J. Mc Ketta, Jr.), Vol. III, p. 193. Interscience Publ., New York 1960.
796. Hornaday G. F., Ferrell F. M., Mills G. A.: *Advances in Petroleum Chemistry and Refining* (Ed. J. J. Mc Ketta, Jr.), Vol. IV, p. 451. Interscience Publ., Inc., New York 1961.
797. Hotelling E. B., Windgassen R. J., Previc E. P., Neuworth M. B.: J. Org. Chem. *24*, 1598 (1959).
798. Hotelling E. B., Neuworth M. B., Previc E. P.: US 2882319 (1959).
799. Hougen O. A., Watson K. M.: *Chemical Process Principles*, Part. III. Wiley, New York 1948.
800. House W. T., Watts R. N.: US 3203998 (1965).
801. Howard F. A.: US 1702899 (1929).

802. Howes D. A.: Brit. 682309 (1952).
803. Howes D. A., Fawcett E. W. M.: US 2357741 (1944).
804. Howes D. A., Fawcett E. W. M.: US 2399927 (1946).
805. Howman E. J., Lacey R. N., Turney L.: Belg. 634702 (1964).
806. Hrubesch A.: Ger. 1088049 (1960).
806a. Hsieh H. L.: US 3383333 (1968).
807. Huang Y. M.: *Thesis*. Inst. Fiz. Khim. AN SSSR, 1960.
808. Huang Y. M., Keier N. P., Roginskii S. Z.: Dokl. AN SSSR *133*, 413 (1960).
809. Huang Y. M., Keier N. P., Roginskii S. Z.: Dokl. AN SSSR *133*, 641 (1960).
810. Hubáček J., Kessler F. M., Ludmila J., Tejnický B.: *Chemie uhlí* (*Chemistry of Coal*). SNTL, Prague 1962.
811. Hübenett F., Schnack N.: Ger. 1016261 (1959).
812. Hübenett F., Schnack N.: US 3062891 (1962).
813. Hudlický M., Trojánek J.: *Preparativní reakce v organické chemii I* (*Preparative Reactions in Organic Chemistry*, Vol. I), p. 51. Nakl.ČSAV, Prague 1953.
814. Hudson N.: Gas J. *243*, 498 (1944).
815. Huffman H. C.: US 2437533 (1948).
816. Hughes E. C., Stine H. M., Faris R. B.: Ind. Eng. Chem. *42*, 1879 (1950).
817. Hughes E. C., Shaw W. G., Strecker H. A.: US 3329603 (1967).
817a. Hughes T. R.: US 3364150 (1968).
818. Huibers D. T. A., Waterman H. I.: Brennstoff-Chem. *41*, 361 (1960).
819. Hülsmann O., Biltz W., Meisel K.: Z. anorg. Chem. *224*, 73 (1935).
820. Hung S. F.: Jan Liao Hsüeh Pao *4*, 69 (1959); C. A. 55, 3045 (1962).
821. Hyde J. W., Porter F. W. B.: *Proc. 4th World Petrol. Congr.*, Sect. III/C, Preprint 2, Rome 1955.
822. I. G. Farbenindustrie A. G.: Brennstoff-Chem. *14*, 378 (1933).
823. I. G. Farbenindustrie A. G.: Brennstoff-Chem. *14*, 379 (1933).
824. I. G. Farbenindustrie A. G.: F. I. A. T. Final Repts., Vol. 1, p. 649.
825. I. G. Farbenindustrie A. G.: Austral. 17448 (1929).
826. I. G. Farbenindustrie A. G.: Brit. 247582 (1925).
827. I. G. Farbenindustrie A. G., Brit. 247583 (1926).
828. I. G. Farbenindustrie A. G.: Brit. 247584 (1926).
829. I. G. Farbenindustrie A. G.: Brit. 247585 (1926).
830. I. G. Farbenindustrie A. G.: Brit. 247586 (1926).
831. I. G. Farbenindustrie A. G.: Brit. 249493 (1925).
832. I. G. Farbenindustrie A. G.: Brit. 251264 (1926).
833. I. G. Farbenindustrie A. G.: Brit. 262120 (1927).
834. I. G. Farbenindustrie A. G.: Brit. 293887 (1927).
835. I. G. Farbenindustrie A. G.: Brit. 315439 (1928).
836. I. G. Farbenindustrie A. G.: Brit. 326580 (1928).
837. I. G. Farbenindustrie A. G.: Brit. 332635 (1930).
838. I. G. Farbenindustrie A. G.: Brit. 348690 (1930).
839. I. G. Farbenindustrie A. G.: Brit. 379335 (1932).
840. I. G. Farbenindustrie A. G.: Brit. 423001 (1935).
841. I. G. Farbenindustrie A. G.: Brit. 429410 (1935).
842. I. G. Farbenindustrie A. G.: Brit. 434141 (1935).
843. I. G. Farbenindustrie A. G.: Brit. 444689 (1936).
844. I. G. Farbenindustrie A. G.: Brit. 454668 (1936).
845. I. G. Farbenindustrie A. G.: Brit. 458699 (1936).
846. I. G. Farbenindustrie A. G.: Brit. 488651 (1938).

847. I. G. Farbenindustrie A. G.: Brit. 527767 (1939).
848. —
849. I. G. Farbenindustrie A. G.: Brit. 675493 (1930).
850. I. G. Farbenindustrie A. G.: Fr. 624890 (1926).
851. I. G. Farbenindustrie A. G.: Fr. 636963 (1927).
852. I. G. Farbenindustrie A. G.: Fr. 655230 (1928).
853. I. G. Farbenindustrie A. G.: Fr. 685564 (1929).
854. I. G. Farbenindustrie A. G.: Fr. 728913 (1932).
855. I. G. Farbenindustrie A. G.: Fr. 45949 (1936); addn. to Fr. 728913.
856. I. G. Farbenindustrie A. G.: Fr. 735295 (1932).
857. I. G. Farbenindustrie A. G.: Fr. 746496 (1933).
858. I. G. Farbenindustrie A. G.: Fr. 777427 (1935).
859. I. G. Farbenindustrie A. G.: Fr. 797606 (1936).
860. I. G. Farbenindustrie A. G.: Fr. 889485 (1942).
861. I. G. Farbenindustrie A. G.: Fr. 893406 (1943).
862. I. G. Farbenindustrie A. G.: Fr. 893492 (1943).
863. I. G. Farbenindustrie A. G.: Ger. 558432 (1932).
864. I. G. Farbenindustrie A. G.: Ger. 685371 (1939).
865. I. G. Farbenindustrie A. G.: Ger. 685595 (1939).
866. I. G. Farbenindustrie A. G.: Swiss 160436 (1933).
867. see 704.
868. I. G. Farbenindustrie A. G.: US 1913940 (1933).
869. I. G. Farbenindustrie A. G.: US 1913941 (1933).
870. Imperial Chem. Ind., Ltd.: Fr. 756724 (1933).
871. Imperial Chem. Ind., Ltd.: Ind. 58161 (1957).
871a. Imperial Chemical Industries, Ltd.: Neth. Appl. 6613475 (1967).
871b. Indyukov N. M., Danielyan M. K.: Khim. Prom. *46*, 226 (1970).
872. Ingold E. H., Wassermann A.: Trans. Farad. Soc. *35*, 1022; 1052 (1939).
873. Inoue S., Tabata A., Chiba Y., Kawakami S.: Aromatikkusu *19* (3), 101 (1967); C. A. 67, 8358 (1967).
874. Institute Française du Pétrole: Manufacturer's catalogue, Paris 1964.
875. Institute of Synthetic Org. Chem. Research: Jap. 153805 (1942); C. A. *43*, 3454 (1949).
876. Intern. Hydrog. Patents Co., Ltd.: Brit. 379587 (1932).
877. Intern. Hydrog. Patents Co., Ltd.: Brit. 424531 (1935).
878. Intern. Hydrog. Patents Co., Ltd.: Brit. 442573 (1936).
879. Intern. Hydrog. Patents Co., Ltd.: Brit. 477944 (1938).
880. Intern. Hydrog. Patents Co., Ltd.: Fr. 728287 (1932).
881. Intern. Hydrog. Patents Co., Ltd.: Fr. 746374 (1933).
882. Intern. Hydrog. Patents Co., Ltd.: Fr. 772811 (1934).
883. Intern. Hydrog. Patents Co., Ltd.: Fr. 774343 (1934).
884. Intern. Hydrog. Patents Co., Ltd.: Fr. 793227 (1936).
885. Intern. Hydrog. Patents Co., Ltd.: Fr. 793436 (1936).
886. Intern. Hydrog. Patents Co., Ltd.: Fr. 794936 (1936).
887. Intern. Hydrog. Patents Co., Ltd.: Fr. 795349 (1936).
888. Intern. Hydrog. Patents Co., Ltd.: Fr. 796443 (1936).
889. Intern. Hydrog. Patents Co., Ltd.: Fr. 800229 (1936).
890. Intern. Hydrog. Patents Co., Ltd.: Fr. 819132 (1937).
891. Intern. Hydrog. Patents Co., Ltd.: Fr. 822306 (1937).
892. Intern. Hydrog. Patents Co., Ltd.: Fr. 826236 (1938).
893. Intern. Hydrog. Patents Co., Ltd.: Fr. 826360 (1938).

894. Intern. Hydrog. Patents Co., Ltd.: Fr. 830135 (1938).
895. Intern. Hydrog. Patents Co., Ltd.: Fr. 851335 (1939).
896. Intern. Hydrog. Patents Co., Ltd.: Fr. 859123 (1940).
897. Intern. Hydrog. Patents Co., Ltd.: Ital. 334639 (1935).
898. Intern. Hydrog. Patents Co., Ltd.: Neth. 50935 (1941).
899. Intern. Hydrog. Patents Co., Ltd.: Pol. 24179 (1932).
900. Ioffe A. V., Ioffe A. F.: Zhur. Exper. i Teoret. Fiz. *9*, 1451 (1939).
901. Ionov I. F., Lavrovskaya E. V., Lozovoi A. V., Malkova S. S., Muselevich D. L.: Nefteper. Neftekhim. *1967*, No. 8, 3.
902. Ipatieff V. N., Pines H., Friedman B. S.: J. Am. Chem. Soc. *60*, 2731 (1938).
903. Ipatieff V. N., Friedman B. S.: J. Am. Chem. Soc. *61*, 71 (1939).
904. Ipatiev V. V. Jr., Levina M. S., Karblom A. I.: Usp. Khim. *8*, 481 (1939).
904a. Ishiguro T., Okagami A., Matsuoka S.: Fr. 1466905 (1967).
904b. Ishii T., Sanada Y., Takeya G.: Kogyo Kagaku Zasshi *71*, 1783 (1968); C. A. 70, 70044 (1969).
905. Isobe H., Tanaka K., Yabuki S.: Bull. Inst. Phys. Chem. Research (Tokio) *21*, 190 (1942); C. A. 43, 7912 (1949).
906. Itabashi K.: Yuki Gosei Kagaku Kyokai Shi *17*, 287 (1959); C. A. 53, 14918 (1959).
907. Itabashi K.: Yuki Gosei Kagaku Kyokai Shi *17*, 330 (1959); C. A. 53, 17978 (1959).
908. Itabashi K.: Yuki Gosei Kagaku Kyokai Shi *17*, 464 (1959); C. A. 53, 21763 (1959).
909. Itabashi K.: Yuki Gosei Kagaku Kyokai Shi *17*, 611 (1959); C. A. 54, 2153 (1960).
910. Itabashi K.: Yuki Gosei Kagaku Kyokai Shi *18*, 48 (1960); Chem. Zentr. 131, 16726 (1960).
911. Itabashi K.: Yuki Gosei Kagaku Kyokai Shi *18*, 347 (1960); C. A. 54, 19466 (1960).
911a. Itabashi K.: Yuki Gosei Kagaku Kyokai Shi *18*, 475 (1960); C. A. 54, 19466 (1960).
912. Itabashi K.: Yuki Gosei Kagaku Kyokai Shi *18*, 480 (1960); C. A. 54, 19466 (1960).
913. Itabashi K.: Yuki Gosei Kagaku Kyokai Shi *19*, 266 (1961); C. A. 55, 11344 (1961).
914. Itabashi K.: Yuki Gosei Kagaku Kyokai Shi *19*, 271 (1961); C. A. 55, 11345 (1961).
915. Itabashi K.: Yuki Gosei Kagaku Kyokai Shi 19, 597 (1961); C. A. 55, 23412 (1961).
916. Itabashi K.: Yuki Gosei Kagaku Kyokai Shi *21*, 531 (1963); C. A. 59, 6242 (1963).
917. Itabashi K.: Yuki Gosei Kagaku Kyokai Shi *23*, 1023 (1965); C. A. 64, 3387 (1966).
918. Ivankiv L. I., Kavich I. V., Savitskii I. V.: Kinetika i Kataliz *9*, 191 (1968).
919. Ivanovskii F. P., Kalvarskaya R. S., Beskova G. S., Sokolova N. P.: Zhur. Fiz. Khim. *30*, 1860 (1956).
920. Ivanovskii F. P., Kalvarskaya R. S., Beskova G. S., Sokolova N. P.: Zhur. Fiz. Khim. *30*, 2353 (1956).
921. Ivanovskii F. P., Kalvarskaya R. S., Beskova G. S., Sokolova N. P.: Zhur. Fiz. Khim. *30*, 2555 (1956).
922. Ivanovskii F. P., Dontsova V. A., Beskova G. S.: Zh. Fiz. Khim. *33*, 2569 (1959).
923. Iwaki T., Egashira S., Okagami A.: Fr. 1468374 (1967).
924. Ismailzade I. G.: Tr. Vses. Sov. po Khim. i Pererab. Nefti, p. 90, Baku, Sept. 1951. AN Azerb. SSR 1953; C. A. *49*, 10028 (1955).
925. Jäckh W.: Erdöl u. Kohle *11*, 625 (1958).
926. Jacobson R. L., Kozlowski R. H.: US 3114701 (1963).
927. Jacobson R. L.: US 3145160 (1964).
928. Jacobson R. L., Kozlowski R. H.: US 3227646 (1966).
928a. Jaffe J.: Ger. Offen. 1938978 (1970).
928b. Jaffe J.: Ger. Offen. 1939468 (1970).
928c. Jaffe J.: US 3434965 (1969).
928d. Janardanarao M., Salvapati G. S., Srikanthareddy P., Vaidyeswaran R.: Erdöl u. Kohle *23*, 20 (1970).

928e. Janardanarao M., Srikanthareddy P., Salvapati G. S., Vaidyeswaran R.: Petrol. Hydrocarbons *5* (2), 11 (1970), Publ. in Chem. Age India *21* (7); C. A. 73, 100833 (1970).
928f. Janardanarao M., Salvapati G. S., Srikanthareddy P., Vaidyeswaran R.: Petrol. Hydrocarbons *5* (2), 19 (1970). Publ. in Chem. Age India *21* (7); C. A. 73, 100831 (1970).
929. Janeček F.: Chem. zvesti *13*, 377 (1959).
930. Janeček F.: Czech. 91133 (1959).
930a. Janeček F., Lupert S., Zdych J.: Czech. 132596 (1969).
930b. Japan Oil Co., Ltd.: Brit. 1137584 (1968).
931. Jarkovský L., Pašek J., Richter P., Růžićka V.: *Scientific Papers of the Institute of Chemical Technology*, Prague C 8, 5 (1966).
932. Jasaitis Z. V., Davidson D. D.: US 2414951 (1947).
933. Jelínek J.: Chem. průmysl *6*, 3 (1956).
934. Jelínek J.: Czech. 95413 (1960).
935. Jellinek F., Brauer G., Müller H.: Nature (London) *185*, 376 (1960).
936. Jellinek F.: Nature (London) *192*, 1065 (1961).
937. Jiříček B., Kostková H., Kubička R., Mráz V., Novák V., Kvapil Z., Vepřek J., Krafft O., Švajgl O., Cír J.: Czech. 121048 (1966).
937a. Jiříček B.: Ropa a uhlie *12*, 30 (1970).
938. Johnson A. J.: Brit. 579766 (1946).
939. Johnson A. J.: Canad. 489867 (1943).
940. Johnson A. J.: US 2528693 (1950).
941. Johnson A. R., Rapp L. M.: Hydrocarbon Process. Petrol. Ref. *43*, No. 5, 165 (1964).
942. Johnson H. L., Stuart A. P.: US 2649419 (1953).
943. Johnson H. L., Stuart A. P.: US 2686763 (1954).
944. Johnson M. M., Hepp H. J.: US 3280210 (1966).
945. Jones S. O., Reid E. E.: J. Am. Chem. Soc. *60*, 2452 (1938).
946. Jones W. H.: US 2112292 (1938).
947. Jungers J. C., Coussemant F.: J. Chim. Phys. *47*, 139 (1950).
948. Kafka Z.: *Thesis*. Institute of Chemical Technology, Prague 1967.
948a. Kainuma Y., Uyeda R.: Proc. Phys. Mat. Soc. Japan *18*, 563 (1936).
948b. Kalchenko V. M., Derkach V. L., Kholodenko N. A., Lebedev E. V., Anistratenko G. A., Manza I. A.: Neft. Gazov. Prom. *1970*, No. 3, 45.
949. Kalechits I. V., Strakhova K. A., Skvortsov Yu. M.: Tr. Vost.- Sibir. Filiala AN SSSR, Ser. Khim. *3*, 88 (1955).
950. Kalechits I. V., Katkova L. M.: Tr. Vost.-Sibir. Filiala AN SSSR, Ser. Khim. *3*, 99 (1955).
951. Kalechits I. V., Salimgareeva F. G.: Tr. Vost.-Sibir. Filiala AN SSSR, Ser. Khim. *4*, 12 (1956).
952. Kalechits I. V., Pavlova K. A.: Tr. Vost.-Sibir. Filiala AN SSSR, Ser. Khim. *4*, 81 (1956).
953. Kalechits I. V., Pavlova K. A., Samoilov S. M.: Tr. Vost.-Sibir. Filiala AN SSSR, Ser. Khim. *4*, 123 (1956).
954. Kalechits I. V., Nikolaeva D. Kh.: Tr. Vost.-Sibir. Filiala AN SSSR, Ser. Khim. *4*, 130 (1956).
955. Kalechits I. V., Strakhova K. A., Katkova L. M.: *Khim. Pererab. Topliva* (*Chemical Processing of Fuels*), p. 206. Izd. AN SSSR, Moscow 1957.
956. Kalechits I. V., Chen L. B., Syui Ch. D.: Izv. Sibir. Otdel. AN SSSR *1958*, No. 10, 3.
957. Kalechits I. V., Chen L. B., Syui Ch. D.: Chuagun Sjuebao (Chemical Industry and Engineering), No. 1, 51 (1958).
958. Kalechits I. V., Pavlova K. A., Samoilov S. M.: Tr. Vost.-Sibir. Filiala AN SSSR, Ser. Khim. *18*, 81 (1959).

959. Kalechits I. V., Sung Y. S., Chu K., Wang K. Y.: Tr. Vost.-Sibir. Filiala AN SSSR, Ser. Khim. *26*, 5 (1959).
960. Kalechits I. V., Sung Y. S., Chu K., Wang K. Y.: Tr. Vost.-Sibir. Filiala AN SSSR, Ser. Khim. *26*, 13 (1959).
961. Kalechits I. V., Sung Y. S., Chu K., Wang K. Y.: Tr. Vost.-Sibir. Filiala AN SSSR, Ser. Khim. *26*, 26 (1959).
962. Kalechits I. V., Sung Y. S., Chu K., Wang K. Y.: Tr. Vost.-Sibir. Filiala AN SSSR *26*, 33 (1959).
963. Kalechits I. V., Si Czu-Vej, Salimgareeva F. G.: Tr. Vost.-Sibir. Filiala AN SSSR Ser., Khim. *26*, 45 (1959).
964. Kalechits I. V., Ying Y. K.: Tr. Vost.-Sibir. Filiala AN SSSR, Ser. Khim. *26*, 108 (1959).
965. Kalechits I. V.: *Problemy Kinetiki i Kataliza* (*Problems of Kinetics and Catalysis*), Vol. 10, p. 121. Izd. AN SSSR, Moscow 1960.
966. Kalechits I. V., Sung Y. S., Chu K., Wang K. Y.: Jan Liao Hsüeh Pao *5*, 1 (1960); C. A. 54, 19131 (1960).
967. Kalechits I. V., Sung Y. S., Chu K., Wang K. Y.: Jan Liao Hsüeh Pao *5*, 7 (1960); C. A. 54, 19132 (1960).
968. Kalechits I. V., Lipovich V. G., Vykhovanets V. V.: Dokl. AN SSSR *138*, 381 (1961).
969. Kalechits I. V., Lipovich V. G., Vykhovanets V. V.: Tr. Vost.-Sibir. Filiala AN SSSR, Ser. Khim. *38*, 5 (1961).
970. Kalechits I. V., Pavlova K. A., Kaliberdo L. M., Skvortsova G. G., Bogdanova T. A., Sidorov R. I., Trotsenko Z. P.: Tr. Vost.-Sibir. Filiala AN SSSR, Ser. Khim. *38*, 31 (1961).
971. Kalechits I. V., In Yuen-ken: Zh. Fiz. Khim. *35*, 501 (1961).
972. Kalechits I. V., Lipovich V. G., Vykhovanets V. V., Petrova V. N.: Kinetika i Kataliz *2*, 748 (1961).
973. Kalechits I. V., Deryagina E. N., Lipovich V. G.: Sb. *Primenenie Mechennych Atomov dlya Izucheniya Neftekhim. Protsessov* (Collection of papers: *Application of Labelled Atoms in the Investigation of Petrochemical Processes*), p. 59. Min. Nefteper. i Neftekhim. Prom. SSSR, Moscow 1965.
974. Kalechits I. V., Deryagina E. N., Pavlova K. A., Lipovich V. G.: *Symposium über Hydrokatalytische Processe in der Erdölverarbeitung und Petrolchemie*, Band 2, p. 76. Leuna 1966.
975. Kalechits I. V., Deryagina E. N.: Kinetika i Kataliz *8*, 604 (1967).
976. Kalechits I. V.: Kinetika i Kataliz, *8*, 1114 (1967).
977. Kalechits I. V., Bogdanova T. A.: Neftepererab. i Neftekhim. *1967*, No. 1, 22.
977a. Kalechits I. V., Kazantseva V. M., Lipovich V. G.: Neftepererab. Neftekhim. 1969, No. 6, 31.
978. Kaliberdo L. M.: *Thesis*. Institute of Organic Chemistry, Academy of Sciences USSR, Moscow 1955.
979. Kaliberdo L. M., Kalechits I. V.: Tr. Vost.-Sibir. Filiala AN SSSR, Ser. Khim. *38*, 25 (1961).
980. Kaliko M. A., Pelevina R. S., Pervushina M. N., Fedotova T.V.: Neftekhimiya *5*, 24 (1965).
981. Kamigaichi T., Masumoto K., Hihara T.: J. Phys. Soc. Japan *15*, 1355 (1960); C. A. *57*, 10621 (1962).
982. Kamm E. D.: Brit. 401724 (1933).
983. Kandiner H. J., Hiteshue R. W., Clark E. L.: Chem. Eng. Progr. *47*, 392 (1951).
984. Karzhev V. I., Kasatkin D. F., Orochko D. I.: Chem. průmysl *8*, 571 (1958).
985. Karzhev V. I., Kasatkin D. F., Orochko D. I.: Khim. i Tekhnol. Topliv i Masel *3*, No. 12, 3 (1958).
986. Karzhev V. I., Kasatkin D. F., Bulekova J. A.: Koks i Khim. *1961*, No. 5, 50.

987. Karzhev V. I., Silchenko E. I., Zherdeva L. G.: *Proceedings, Symposium on the Application of the Hydrogenation Processes in Petroleum Processing.* Index Ž$_1$. Moscow 1963.
988. Karzhev V. I., Silchenko E. I., Robozheva E. V., Lebedeva A. M.: Khim. i Tekhnol. Topliv i Masel *10*, No. 11, 4 (1965).
988a. Karzhev V. I., Shavolina N. V., Kasatkin D. F., Khailov V. S., Druzhinina O. A., Fisenko N. P., Sharko M. S.: Tr. Vses. Nauch.-Issled. Inst. Pererab. Nefti *1970*, No. 13, 136.
989. Kashima H., Tabata H., Watanabe H., Kubo T.: Kami-pa Gikyoshi *16* (11), 901 (1962); C. A. *61*, 12186 (1964).
990. Kato R.: J. Fuel Soc. Japan *34*, 40 (1955); C. A. 49, 6584 (1955).
991. Kato R.: J. Fuel Soc. Japan *34*, 82 (1955); C. A. 49, 8591 (1955).
992. Katsobashvili Ya. R., Kurkova N. S.: Trudy Inst. Nefti, Alad. Nauk SSSR *6*, 100 (1955).
993. Katsobashvili Ya. R., Kurkova N. S.: Zhur. Prikl. Khim. *31*, 874 (1958).
994. Katsobashvili Ya. R., Golosov S. A.: Khim. i Tekhnol. Topliv i Masel *5*, No. 1, 8 (1960)
995. Katsobashvili Ya. R., Golosov S. A.: Zhur. Prikl. Khim. *33*, 1369 (1960).
996. Katsobashvili Ya. R., Mikheev G. M.: Khim. Tekhnol. Topliv i Masel *9*, No. 12, 28 (1964).
996a. Katsobashvili Ya. R., Mikheev G. M.: Khim. Seraorg. Soed. Soderzh. Neftyakh Nefteprod. *8*, 264 (1968).
997. Kaufhold R., Ratz L., Kloss R.: GDR 44108 (1965).
998. Kaufmann H., v. Sahr E.: Chem. Technik *11*, 403 (1959).
999. Kaufmann H., Welker J., Bätz R.: Acta Chimica Acad. Sci. Hung. *36*, 131 (1963).
1000. Kaufmann H., Welker J., Blume H., Onderka E., Risse F.: Brit. 1082334 (1967).
1001. Kaufmann H.: GDR 12808 (1957).
1002. Kaufmann H., Welker J., Bätz R., Münzing J.: USSR 168661 (1965).
1003. Kawa W., Hiteshue R. W.: US Bur. Mines, Rept. Invest. No. 6179, 17 pp. (1963).
1003a. Kay N. L., Pullen E. A., Reeg C. P.: US 3365385 (1968).
1003b. Kay N. L.: US 3513086 (1970).
1004. Kazanskii V. L., Badyshtova K. M., Denisenko K. K.: Khim. i Tekhnol. Topliv i Masel *11*, No. 10, 19 (1966).
1004a. Kazantseva V. **M.,** Neudachina V. I., Kalechits I. V.: Neftepererab. Neftekhim. *1968*, No. 8, 23.
1004b. Kazantseva V. M., Kalechits I. V.: Neftepererab. Neftekhim. *1968*, No. 11—12, 24.
1005. Kazeev S. A.: *Kineticheskie Osnovy Metalurgicheskikh Protsessov* (*Fundamental Kinetics of Metallurgic Processes*). Oborongiz, Moscow 1946.
1006. Kearby K. K.: *Catalysis* (Ed. P. H. Emmett), Vol. III, p. 453. Reinhold Publ. Corp., New York 1955.
1006a. Keblys K. A., Dubeck M.: US 3409649 (1968).
1007. Ketslakh M. M., Rudkovskii D. M.: Khim. Tekhnol. Topliv i Masel *2*, No. 3, 1 (1957).
1008. Keil G., Roth H., Wenzel B.: Chem. Technik *16*, 545 (1964).
1009. Keil G., Roth H., Wenzel B.: Ropa a uhlie *9*, 12 (1967).
1010. Keith C. D., Bair D. L.: US 3275567 (1966).
1011. Kelley A. E., Deering R. F.: US 2952625 (1960).
1012. Kelley A. E., Deering R. F.: US 2952626 (1960).
1013. Kelley A. E., Barnet W. I.: US 3132086 (1964).
1013a. Kelley J. M., Baeder D. L.: Fr. 1488430 (1967).
1014. Kelley K. K.: US Bur. Mines Bull. No. 406 (1937).
1015. Kelley K. K.: *Thermodynamic Properties of Molybdenum Compounds.* Bull. Cdb-2, Climax Molybdenum Co., 1954.
1016. Key A., Eastwood A. H.: Gas Research Board, Copyright Pub. No. 14/4, 42 pp. (1946); C. A. *43*, 3169 (1949).

1017. Key A.: Brit. 561679 (1944).
1018. Key A.: Brit. 563350 (1944).
1019. Khambata B. S., Wassermann A.: J. Chem. Soc. *1946*, 1090.
1020. Khambata B. S., Rubin W., Wassermann A.: *Proc. 11th Intern. Congr. Pure and Applied Chem. (London)* 2, 155—9 (1947); C. A. 45, 7024 (1951).
1021. Kieran P., Kemball C.: J. Catalysis *3*, 426 (1964).
1022. Kieran P., Kemball C.: J. Catalysis *4*, 380 (1965).
1023. Kieran P., Kemball C.: J. Catalysis *4*, 394 (1965).
1023a. Kiezel L., Rutkowski M., Tomasik Z.: Pol. 57045 (1969).
1023b. Kukuchi E., Morita Y., Yamamoto K.: Bull. Jap. Petrol. Inst. *11*, 34 (1969); C. A. 71, 51912 (1969).
1024. Kikuchi M.: Kogyo Kagaku Zasshi *61*, 1865 (1959); C. A. 58, 4602 (1963).
1025. Kikuchi M.: Kogyo Kagaku Zasshi *63*, 1414 (1960); C. A. 57, 7314 (1962).
1026. Kikuchi M.: Kogyo Kagaku Zasshi *63*, 2174 (1960); C. A. 58, 2473 (1963).
1027. Kikuchi M.: Kogyo Kagaku Zasshi *64*, 334 (1961); C. A. 57, 3485 (1962).
1028. Kikuchi M.: Kogyo Kagaku Zasshi *65*, 1239 (1962); C. A. 59, 3961 (1963).
1029. Kikuchi M.: Kogyo Kagaku Zasshi *66*, 461 (1963); C. A. 59, 11569 (1963).
1030. Kikuchi M.: Kogyo Kagaku Zasshi *66*, 464 (1963); C. A. 59, 11569 (1963).
1031. King J. G., Matthews M. A.: Gas J. *200*, 213 (1932).
1032. King J. G., Matthews M. A.: J. Inst. Fuel *6*, No. 25, 33 (1932).
1033. King J. G.: *The Science of Petroleum* (Ed. A. E. Dunstan), Vol. III, p. 2149. Oxford Univ. Press, London 1938.
1034. King J. G.: *The Science of Petroleum* (Ed. A. E. Dunstan), Vol. III, p. 2156. Oxford Univ. Press, London 1938.
1035. King J. G., Cawley C. M.: *IIe Congr. mondial pétrole 2*, Sect. 2, *Phys., chim., raffinage*, 249 (1937); C. A. *33*, 346 (1939).
1036. King J. G.: Gas Research Board Publ. No. 8; Gas J. *1944*, 623, 660, 689, 691; Gas World *120*, 6245; Gas World *121*, 18.
1037. Kingman F. E. T., Rideal E. K.: Nature *137*, 529 (1936).
1038. Kingman F. E. T.: Trans. Faraday Soc. *33*, 784 (1937).
1039. Kirkpatrick W. J.: *Advances in Catalysis*. Vol. III, p. 329. Academic Press, New York 1951.
1040. Kirsch F. W., Shalit H., Heinemann H.: Ind. Eng. Chem. *51*, 1379 (1959).
1041. Kirsch F. W., Shull S. E.: Ind. Eng. Chem., Product Research Dev. *2*, 48 (1963).
1042. Kirsch F. W., Shull S. E.: Fr. 1253588 (1961).
1043. Kirsch F. W.: US 3123626 (1964).
1044. Klemm W., Schüth W.: Z. anorg. Chem. *210*, 33 (1933).
1045. Klever H. W.: Ger. 301773 (1921).
1046. Klier K.: Chem. listy *58*, 621 (1964).
1047. Klimov B. K., Bogdanov I. F.: Trudy Inst. Gor. Iskop. AN SSSR *3*, 140 (1954).
1048. Klimov B. K., Bogdanov I. F.: Trudy Inst. Gor. Iskop, AN SSSR *3*, 151 (1954).
1049. Kline C. H., Kollonitsch V.: Ind. Eng. Chem. *57*, No. 7, 53 (1965).
1050. Kline R. E., Mc Kinley J. B.: US 3213012 (1965).
1051. Kling A., Florentin D.: Compt. rend. *193*, 1023 (1931).
1052. Kling A., Florentin D.: Compt. rend. *193*, 1198 (1931).
1053. Klotzsche H.: Freiberger Forschungs-H. *A 352*, 47 (1965).
1054. Klotzsche H.: Freiberger Forschungs-H. *A 387*, 69 (1966).
1055. Klotzsche H.: *Symposium über Hydrokatalytische Processe in der Erdölverarbeitung und Petrolchemie*, Band 1, p. 191. Leuna 1966.
1055a. Kluck H., Staab W., Staffehl J., Peper F. K.: GDR 61584 (1968).

1056. Klvaňa D.: *Thesis.* Institute of Chemical Technology, Prague 1967.
1057. Klyukvin N. A., Polozov V. F., Lobus I. I.: Khim. Tverd. Topliva *5*, 357 (1934).
1058. Knights D. L., Winsor J.: Brit. 897238 (1962).
1059. Knights D. L., Winsor J.: Ger. 1238144 (1967).
1060. Knyunyants I. J., Fokin A. V.: Usp. Khim. *19*, 545 (1950).
1061. Kobozev N. I.: Usp. Khim. *25*, 545 (1956).
1062. Kobozev N. I., Danchevskaya M. N.: *Proceedings of the I. International Catalysis Conference* (Ministry of Technical and Higher Education) 1958, No. 1, pt. 1, 313 (1962); C. A. 59, 1134 (1963).
1063. Kobozev N. I., Krylova I. V., Shashkov A. S.: *Proc. Intern. Congr. Catalysis, 3rd.* Amsterdam 1964, No. 1, 614 (Published 1965).
1063b. Koestler D.: *Roentgenspektren Chem. Bindung,* Vortr. Int. Symp., Leipzig 1965, 175; C. A. 68, 72720 (1968).
1064. Köhler K.: *Proceedings, Intern. Conference on Catalysts,* Pt. I., p. 203. Karlovy Vary 1964.
1065. Kolbanovskii Yu. A., Kustanovich I. M., Polak L. S., Shcherbakova A. S.: Dokl. AN SSSR 129, 145 (1959).
1066. Kolboe S., Amberg C. H.: Canad. J. Chem. *44*, 2623 (1966).
1066a. Kolboe S.: Canad. J. Chem. *47*, 352 (1969).
1067. Kolkmeijer H. H., Moesveld A. L. T.: Z. Krist. *A 80*, 91 (1931).
1068. Komarewsky V. I., Knaggs E. A.: Ind. Eng. Chem. *43*, 1414 (1951).
1069. Kono K., Ogoro H., Inaba T.: Kogyo Kagaku Zasshi *64*, 525 (1961); C.A. *57*, 3720 (1962).
1070. Kono K., Ogoro H., Inaba T.: Kogyo Kagaku Zasshi *64*, 532 (1961); C.A. *57*, 3720 (1962).
1071. Kono K., Ogoro H., Inaba T.: Kogyo Kagaku Zasshi *64*, 631 (1961); C.A. *57*, 3720 (1962).
1072. Kono K., Ogoro H., Inaba T.: Kogyo Kagaku Zasshi *64*, 888 (1961); C. A. *57*, 3720 (1962).
1073. Koome J., Stijntjes G. J. F., Meerbott W. K.: US 2939836 (1960).
1074. Koome J., Stijntjes G. J. F.: US 2947685 (1960).
1075. Heinrich Koppers G. m. b. H.: Fr. 1457058 (1966).
1075a. Koros R. M., Bank S., Hofmann J. E., Kay M. I.: Amer. Chem. Soc., Div. Petrol. Chem., Preprints 1967, *12* (4), B 165-B 174.
1076. Korsh M. P., Ivanovskii F. P.: Zh. Prikl. Khim. *29*, 1561 (1956).
1077. Korsh M. P., Ivanovskii F. P.: Zh. Prikl. Khim. *31*, 980 (1958).
1078. Kosaka Y., Kayamori H., Taniyama T.: J. Chem. Soc. Ind. Japan *46*, 663 (1943); C. A. 42, 6332 (1948).
1079. Kosiba J. K., Rice T.: US 3360456 (1967).
1080. Kostková H., Kubička R., Jiříček B.: Czech. 102828 (1961).
1080a. Kovach S. M., Rogers E. S.: US 3394077 (1968).
1080b. Kovach S. M., Rogers E. S.: US 3422002 (1969).
1080c. Kovach S. M., Rogers E. S.: US 3437588 (1969).
1081. Kovats E., Günthard Hs. H., Plattner Pl. A.: Helvetica Chimica Acta *37*, 2123 (1954).
1082. Kozlovskii A. L., Shlyakova K. S.: Tr. Vses. Nauch.-Issled. Inst. Autogen. Obrab. Metal. *1960*, No. 6, 136; C. A. *55*, 16062 (1961).
1083. Kozlowski R. H., Mason H. F., Scott J. W.: Ind. Eng. Chem., Process Design Dev. *1*, 276 (1962).
1084. Kozlowski R. H., Mason H. F., Scott J. W.: US 3092567 (1963).
1085. Kozlowski R. H., Jacobson R. L.: US 3189540 (1965).
1086. —

1087. Kozlowski R. H., Jacobson R. L.: US 3228993 (1966).
1088. Krafft O.: *Sborník prací z výzkumu chemického využití uhlí, dehtu a ropy* (*Collection of Papers on Chemical Utilisation of Coal, Tar and Petroleum*), Vol. 2, 156 (1962).
1089. Krafft O., Kubička R., Mráz V., Hála S.: Czech. 122575 (1967).
1089a. Kramer W. E., Kimble R. C.: US 3280175 (1966).
1090. Kranz M.: Studia Muzealne *3*, 141—55 (1957).
1091. Krauch C., Pier M.: Z. angew. Chem. *44*, 953 (1931).
1092. Krauch C., Pier M.: Austral. 1217/26 (1926).
1093. Kravtsov S. L.: USSR 173203 (1965).
1094. Krebs R. W.: Fr. 1048905 (1953).
1095. —
1096. Kreuz K. L.: US 2557664 (1951).
1097. Kreuz K. L.: US 2557665 (1951).
1098. Kreuz K. L.: US 2557666 (1951).
1099. Krichko A. A., Lozovoi A. V., Pchelina D. P.: Trudy Inst. Gor. Iskop., Vol. IX, p. 37. Moscow 1959.
1100. Krichko A. A., Konyashina P. A., Lozovoi A. V.: Trudy Inst. Gor. Iskop., Vol. IX, p. 68. Moscow 1959.
1100a. Krichko A. A., Skvortsov D. V., Sovetova L. S., Filippov B. S.: Tr. Inst. Gor. Iskop. *23* (4), 172 (1968).
1100b. Krichko A. A., Skvortsov D. V., Titova T. A., Filippov B. S., Dogadkina N. E.: Khim. Tekhnol. Topliv Masel 14, No. 1, 14 (1969).
1101. Krönig W.: *Die katalytische Druckhydrierung von Kohlen, Teeren und Mineralölen.* Springer-Verlag, Berlin 1950.
1102. Krönig W.: Erdöl u. Kohle *18*, 432 (1965).
1103. Kruglova V. G., Sidorenko G. A., Polupanova L. I.: Tr. Mineralog. Muzeya, AN SSSR, No. 16, 233 (1965).
1104. Krüss G.: Ann. *225*, 29 (1884).
1105. Krylov O. V., Fokina E. A.: Izv. AN SSSR, Otd. Khim. Nauk *1958*, 266.
1106. Krylov O. V., Roginskii S. Z.: Dokl. AN SSSR *118*, 523 (1958).
1107. Krylov O. V., Roginskii S. Z.: Izv. AN SSSR, Otd. Khim. Nauk *1959*, 17.
1108. Krylova I. V., Danchevskaya M. N., Kobozev N. I.: Zh. Fiz. Khim. *29*, 1684 (1955).
1109. Krylova I. V., Kobozev N. I.: Zh. Fiz. Khim. *35*, 911 (1961).
1110. Krylova I. V., Shashkov A. S., Kobozev N. I.: Zh. Fiz. Khim. *35*, 2657 (1961).
1111. Krylova I. V., Shashkov A. S., Kobozev N. I.: Optika i Spektroskopiya *12*, 635 (1962).
1112. Kubička R., Kvapil Z.: *Sborník prací z výzkumu chemického využití uhlí, dehtu a ropy* (*Collection of Research Papers on Chemical Utilisation of Coal, Tar and Petroleum*), Vol. 1, 87 (1960).
1113. Kubička R., Kvapil Z.: Freiberger Forschungs- H. *A 131*, 9 (1960).
1114. Kubička R.: Chem. průmysl *11*, 449 (1961).
1115. Kubička R.: *Scientific Papers of the Institute of Chemical Technology*, Prague 1961, Technology of Fuels *5*, 159.
1116. Kubička R., Kvapil Z.: Freiberger Forschungs- H. *A 340*, 73 (1964).
1117. Kubička R., Mráz V.: Ropa a uhlie *6*, 35 (1964).
1118. Kubička R., Vepřek J.: Ropa a uhlie *6*, 116 (1964).
1119. Kubička R., Kvapil Z.: *Sborník prací z výzkumu chemického využití uhlí, dehtu a ropy* (*Collection of Research Papers on Chemical Utilisation of Coal, Tar and Petroleum*), Vol. 3, 4 (1964).
1120. Kubička R., Vepřek J.: Ropa a uhlie *7*, 48 (1965).
1121. Kubička R., Vepřek J.: *Sborník prací z výzkumu chemického využití uhlí, dehtu a ropy*

(*Collection of Research Papers on Chemical Utilisation of Coal, Tar and Petroleum*), Vol. 5, 46 (1965).

1122. Kubička R., Kvapil Z.: *Sborník prací z výzkumu chemického využití uhlí, dehtu a ropy* (*Collection of Research Papers on Chemical Utilisation of Coal, Tar and Petroleum*), Vol. 6, 3 (1966).
1123. Kubička R., Vepřek J., Novák V., Mráz V.: Ibid., Vol. *6*, 69 (1966).
1124. Kubička R., Novák V.: Ibid., Vol. 8, 3 (1968).
1124a. Kubička R., Cír J., Novák V., Vepřek J.: Brennstoff-Chem. 49, 268 (1968).
1124b. Kubička R., Cír J., Novák V., Vepřek J.: Brennstoff-Chem. 49, 308 (1968).
1124c. Kubička R., Cír J., Novák V., Vepřek J.: *Sborník prací z výzkumu chem. využití uhlí, dehtu a ropy* (*Collection of Research Papers on Chemical Utilisation of Coal, Tar and Petroleum*), Vol. *9*, 19 (1969).
1124d. Kubička R., Kvapil Z.: *Sborník prací z výzkumu chem. využití uhlí, dehtu a ropy* (*Collection of Research Papers on Chemical Utilisation of Coal, Tar and Petroleum*) Vol. 10, 3 (1970).
1125. Kubička R., Jelínek J., Zogala J.: Czech. 92329 (1959).
1126. Kubička R.: Czech. 103744 (1962).
1127. Kubička R., Kvapil Z., Zogala J.: Czech. 108502 (1963).
1127a. Kubička R., Kvapil Z.: Czech. 129864 (1968).
1128. Kuliev R. Sh., Sadykhova B. A.: Khim. i Tekhnol. Topliv i Masel *7*, No. 5, 32 (1962).
1129. Kuliev R. Sh., Sadykhova B. A.: Azerb. Neft. Khoz. *41*, No. 11, 32 (1962).
1130. Kuliev R. Sh., Sadykhova B. A., Musaev G. T.: Khim. i Tekhnol. Topliv i Masel *8*, No. 7, 6 (1963).
1131. Kuliev R. Sh., Sadykhova B. A., Musaev G. T.: *Proceedings, Symposium on the Application of the Hydrogenation Processes in Petroleum Processing*, Index G_6, Moscow 1963.
1132. Kuliev R. Sh., Sadykhova B. A.: Ibid, index G_7, Moscow 1963.
1133. Kuliev R. Sh., Sadykhova B. A.: Ibid, index G_8, Moscow 1963.
1133a. Kuliev R. Sh., Sadykhova B. A.: Issled. Primen. Gidrogenizatsionnykh Protsessov Neftepererab. Neftekhim. Prom. *1968*, 257; C. A. 71, 62756 (1969).
1134. Kullerud G., Yund R. A.: J. Petrol. *3*, 126 (1962).
1135. Küntscher W., Klopfer O., Legutke G., Hähnel G.: GDR 8979 (1955).
1136. Kupriyanov V. A., Dorogochinskii A. Z., Melnikova N. P.: Tr. Grozn. Neft. Nauchn-. Issled. Inst. *1961*, No. 11, 181; C. A. 57, 6209 (1962).
1137. Kupriyanov V. A., Dorogochinskii A. Z.: *Proceedings, Symposium on the Application of the Hydrogenation Processes in Petroleum Processing*, Index E_2, Moscow 1963.
1138. Kurihara K., Yoshioka A.: J. Soc. Chem. Ind. Japan *44*, Suppl. Bind. 250 (1941); C.A. *44*, 7511 (1950).
1139. Kurin N. P.: Khim. Referat. Zhur. No. *4*, 100 (1941); C.A. *37*, 5306 (1943).
1140. Kurokawa M.: J. Soc. Chem. Ind. Japan *45*, 1030 (1942); C. A. 43, 8640 (1949).
1141. Kurokawa M.: J. Soc. Chem. Ind. Japan *45*, 1033 (1942); C. A. 43, 8640 (1949).
1142. Kurokawa M.: J. Soc. Chem. Ind. Japan *46*, 535 (1943); C. A. 43, 8640 (1949).
1143. Kurokawa M.: J. Soc. Chem. Ind. Japan *46*, 538 (1943); C. A. 43, 8640 (1949).
1144. Kurokawa S., Fujihara M.: J. Soc. Chem. Ind. Japan *48*, 40 (1945); C. A. 42, 6080 (1948).
1145. Kuss E., Moos J., Stegemeyer H.: Naturwissenschaften *48*, 73 (1961).
1145a. Kutch E. L., Osborn S. W., Wells T. F.: US 3337513 (1967).
1146. Kuznetsov V. G., Eliseev A. A., Shpak Z. S., Palkina K. K., Sokolova M. A., Dmitriev A. V.: *Voprosy Met. i Fiz. Poluprovod., AN SSSR, Trudy 4-go Soveshaniya* (*Proceedings of the 4th Conference on Met. and Physical Problems of Semiconductors*), p. 159, Moscow 1961; C. A. 56, 5444 (1962).
1147. Kuyl A.: *Proc. 3rd World Petrol Congr.*, Sect. IV, 307. The Hague 1951.

1148. Kwan T., Okada K., Matsushita S.: Bull. Inst. Phys. Chem. Research (Tokyo), Chem. Ed. *23*, 173 (1944); C.A. *43*, 7802 (1949).
1149. Kwan T.: Bull. Inst. Phys. Chem. Research (Tokyo), Chem. Ed. *23*, 205 (1944); C.A. *43*, 7802 (1949).
1150. Lackowicz S., Ridez J.: Fr. 1147441 (1957).
1151. Laffitte M.: Bull. soc. chim. France *1959*, 1211 (1959).
1152. Laffitte M.: Rev. nickel *25*, 79; 109 (1959).
1153. Lagrenaudie J.: J. Physique Radium *13*, 311 (1952).
1154. Lagrenaudie J.: J. Physique Radium *15*, 299 (1954).
1155. Lainé B., Vernet Ch.: US 3125509 (1964).
1156. Lalet P.: Fr. 1272087 (1962).
1156a. Lambertin M., Colson J. C., Delafosse D.: C. r. Acad. Sci., Ser. C, *270* (11), 974 (1970).
1157. Landa S., Landová M.: Petroleum *28*, No. 51, 10 (1932).
1158. Landa S.: Paliva *35*, 315 (1955).
1159. Landa S., Mostecký J.: Chem. listy *49*, 67 (1955); Collection Czech. Chem. Commun. *20*, 430 (1955).
1160. Landa S.: *Paliva a jejich použití* (*Fuels and Their Utilization*). SNTL, Prague 1956.
1161. Landa S.: *Trans. 5th World Power Conf.*, Vienna 1956, Div. 2, Sect. D, Paper 137D/4·
1162. Landa S., Mostecký J.: Chem. listy *50*, 565 (1956).
1163. Landa S., Weisser O.: Chem. listy *50*, 569 (1956); Collection Czech. Chem. Commun. *22*, 93 (1957).
1164. Landa S., Macák J.: Chem. listy *51*, 1851 (1957).
1165. Landa S., Weisser O., Mostecký J.: Chem. listy *51*, 452 (1957); Collection Czech. Chem. Commun. *22*, 1006 (1957).
1166. Landa S., Mostecký J.: Chem. průmysl *7*, 393 (1957).
1167. Landa S., Weisser O., Mostecký J.: Chem. listy *52*, 60 (1958); Collection Czech. Chem. Commun. *24*, 1036 (1959).
1168. Landa S., Petrů F., Mostecký J., Vít J., Procházka V.: Chem. listy *52*, 1357 (1958).
1169. Landa S., Petrů F., Vít J., Procházka V., Mostecký J.: *Scientific Papers of the Institute of Chemical Technology*, Prague, Inorg. and Org. Technol. *1958*, 495.
1170. Landa S., Weisser O., Kaplan E. P., Gao C. L., Petrov A. D.: Izv. AN SSSR, Otd. Khim. Nauk *1959*, 1425; Ropa a uhlie *1*, 263 (1959).
1171. Landa S., Weisser O., Mostecký J.: Collection Czech. Chem. Commun. *24*, 2197 (1959).
1172. Landa S.: *Syntetická paliva* (*Synthetic Fuels*). SNTL, Prague 1960.
1173. Landa S., Macák J.: Collection Czech. Chem. Commun. *25*, 761 (1960).
1174. Landa S., Macák J.: Collection Czech. Chem. Commun. *25*, 766 (1960).
1175. Landa S., Mostecký J., Weisser O.: Collection Czech. Chem. Commun. *25*, 1165 (1960).
1176. Landa S., Szebenyi I., Weisser O., Mostecký J.: Ropa a uhlie *2*, 260 (1960); Acta Chimica Acad. Sci. Hung. *29*, Fasc. 2, 237 (1961).
1177. Landa S., Chuchla J.: *Scientific Papers of the Institute of Chem. Technol.*, Prague 1961, Fuel Technology *5*, 35.
1178. Landa S., Andrzejak A., Weisser O.: Collection Czech. Chem. Commun. *27*, 979 (1962).
1179. Landa S., Rábl V.: Ropa a uhlie *4*, 168 (1962).
1180. Landa S., Rábl V.: Brennstoff-Chem. *43*, 50 (1962).
1181. Landa S.: Ropa a uhlie *4*, 107 (1962).
1182. Landa S., Eyem J.: Fuel *42*, 265 (1963).
1183. Landa S., Urban M.: Brennstoff-Chem. *44*, 377 (1963).
1184. Landa S., Mrnková A.: Collection Czech. Chem. Commun. *31*, 2202 (1966).
1185. Landa S., Mrnková A.: *Scientific Papers of the Institute of Chemical Technology*, Prague, *D 11*, 5 (1966).

1186. Landa S., Mrnková A., Hrušková A.: *Scientific Papers of the Institute of Chemical Technology*, Prague, *D 13*, 151 (1967).
1187. Landa S., Mrnková A., Bártová N.: *Scientific Papers of the Institute of Chemical Technology*, Prague, *D 16*, 159 (1969).
1187a. Landa S., Mrnková A., Hercíková J.: *Scientific Papers of the Institute of Chemical Technology*, Prague *D 16*, 149 (1969).
1188. Landa S., Kafka Z., Galík V., Šafář M.: Collection Czech. Chem. Commun. *34*, 3588 (1969).
1188a. Landa S., Kafka Z., Galík V., Šafář M.: Collection Czech. Chem. Commun. *34*, 3967 (1969).
1189. Landa S.: Brit. 792919 (1958).
1190. Landa S., Mostecký J., Weisser O.: Czech. 90577 (1959).
1191. Landa S., Weisser O., Mostecký J.: Czech. 90578 (1959).
1192. Landa S., Mostecký J.: Czech. 107495 (1963).
1193. Landa S., Weisser O., Mostecký J.: Czech. 108422 (1963).
1194. Landa S., Maťáš M., Mostecký J.: Czech. 117097 (1965).
1195. Landau M. A., Shchekin V. V.: Izv. AN SSSR, Otd. Khim. Nauk *1960*, 946.
1196. Landau M. A., Shchekin V. V.: Tr. Inst. Nefti, AN SSSR *14*, 150 (1960).
1197. Langer A. W., Stewart J., Thompson C. E., White H. T., Hill R. M.: Ind. Eng. Chem., Intern. Ed. *53*, 27 (1961).
1198. Langlois G. E., Sullivan R. F., Egan C. J.: Amer. Chem. Soc. Div., Petrol. Chem., Preprints *10* (4), B 127 (1965).
1199. Langlois G. E., Sullivan R. F., Egan C. J.: J. Phys. Chem. *70*, 3666 (1966).
1200. La Porte W. N.: US 2654696 (1953).
1201. Larson O. A., Beuther H.: *Processing Aspects of Vanadium and Nickel in Crude Oils*. Presented at the Petrol. Div., ACS Meeting, Pittsburgh, Penn., March 1966.
1201a. Larson O. A.: Amer. Chem. Soc., Div. Petrol. Chem., Preprints 1967, 12(4), B123—B138.
1202. Laughlin K. C.: US 2514300 (1950).
1203. Lavergne J. C., Clement C., Dutriau R.: Fr. 1388624 (1965).
1204. Lavronskii K. P., Makarov D. V., Nazarova L. M.: Trudy Inst. Nefti, AN SSSR *8*, 145 (1956).
1205. Lawrance P. A., Rawlings A. A.: *7th World Petrol. Congr.*, PD 17, Paper 6. Mexico City 1967.
1206. Lazier W. A., Vaughen J. V.: US 2105665 (1935).
1207. Lazier W. A., Signaigo F. K.: US 2402639 (1946).
1208. Lazier W. A., Signaigo F. K.: US 2402640 (1946).
1209. Lazier W. A., Signaigo F. K.: US 2402641 (1946).
1210. Lazier W. A., Signaigo F. K.: US 2402642 (1946).
1211. Lazier W. A., Signaigo F. K., Werntz J. H.: US 2402643 (1946).
1212. Lazier W. A., Signaigo F. K., Wise L. G.: US 2402645 (1946).
1213. Leak R. J., Throckmorton M. C.: US 3111494 (1963).
1214. Leak R. J., LeBleu H. J.: US 3240698 (1966).
1215. Lebedev E. V., Zakupra V. A., Manza I. A.: *Symposium über Hydrokatalytische Processe in der Erdölverarbeitung und Petrolchemie*, 2. Band, p. 270, Leuna 1966.
1215a. Lebedev E. V., Manza I. A., Zakupra V. A.: Neftepererab. Neftekhim. *1967*, No. 2, 156.
1215b. Lebedev E. V., Manza I. A.: Neftepererab. Neftekhim. (Kiev) *1969*, No. 3, 121.
1215c. Lebedev E. V., Manza I. A., Pliev T. N.: Neftekhim. *9*, 529 (1969).
1216. Lefrançois P. A., McMahon J. F., Solomon E.: US 3169106 (1965).
1217. Lefrançois P. A.: US 3269938 (1966).

1218. Lehmann P., Pindur E., Blume H.: GDR 18499 (1960).
1219. Lehmann P., Blume H., Pindur E.: GDR 20413 (1960).
1219a. Lempert F. L., Solomon E., Schwarzenbek E. F.: US 3492220 (1970).
1219b. Leonard J. D.: Neth. Appl. 6607964 (1966).
1219c. Le Page J. F., Derrien M., Haddad A. C.: World Petrol. Congress, Proc. 7th 1967, 4, 216.
1219d. Le Page J. F., Derrien M.: Ger. Offen. 1804768 (1969).
1220. Lesiak T.,: Roczniki Chem. *39* (5), 757 (1965).
1221. Lester R.: Petrol. Ref. *40*, No. 9, 175 (1961).
1222. Letort M.: Petrol. Ref. *41*, No. 7, 83 (1962).
1223. Letort M.: Chem. et ind. *87*, 371 (1962).
1224. Levin S. Z., Diner I. S., Karpov A. Z.: Khim. i Tekhnol. Topliv i Masel *1*, No. 8, 8 (1956).
1225. Levin S. Z., Karpov A. Z., Diner I. S., Gribova I. G.: Khim. i Tekhnol. Topliv i Masel *3*, No. 5, 51 (1958).
1226. Levin S. Z., Karpov A. Z.: Izvest. Vysshikh. Ucheb. Zavedenii, Khim. i Khim. Tekhnol. *3*, 364 (1960).
1227. Levin S. Z., Diner I. S.: Tr. Vses. Nauch.-Issled. Inst. Per. i Ispolz. Topliva *1960*, No. 9, 65; C. A. 55, 15901 (1961).
1228. Levin S. Z., Karpov A. Z.: Vestn. Tekhn. i Ekon. Inform. Nauchno-Issled. Inst. Tekhn.-Ekon. Issled. Gos. Kom. Khim. i Neft. Prom. pri Gosplane SSSR *1963* (6), 12—15; C. A. 62, 3927 (1965).
1228a. Levin S. Z., Gurevich G. S., Sedova I. G., Batenina A. D.: Protsesy Katal. Gidrirovaniya Proizvod. Monomerov Poluprod., Vses. Nauch.-Issled. Inst. Neftekhim. Protsessov *1966*, 17; C. A, 67, 63655 (1967).
1229. Levin S. Z., Gurevich G. S., Sedova I. G., Batenina A. D., Oparina E. I.: Protsessy Katal. Gidrirovaniya Proizvod. Monomerov Poluprod., Vses. Nauch.-Issled. Inst. Neftekhim. Protsessov *1966*, 41; C. A. *67*, 53324 (1967).
1230. Levin S. Z., Gurevich G. S.: USSR 168278 (1965).
1231. Levina R. Ya., Shusherina N. P.: Usp. Khim. *24*, 181 (1955).
1232. Levinos S.: US 3065160 (1962).
1233. Levitskii I. I., Gonikberg M. G.: Izv. AN SSSR, Otd. Khim. Nauk *1960*, 996.
1234. Levitskii I. I., Gonikberg M. G.: Dokl. AN SSSR *137*, 609 (1961).
1235. Levitskii I. I., Udalťsova E. A., Gonikberg M. G.: Zh. Prikl. Khim. *35*, 205 (1962).
1235a. Levitskii I. I., Minachev Kh. M., Bogomolov V. I., Udaltsova E. A.: Izv. Akad. Nauk SSSR, Ser. Khim. *1968* (9), 1993.
1235b. Lewis E. H.: US 3388056 (1968).
1236. Lyashenko V. I.: Trudy Inst. Fiz. Akad. Nauk Ukrain. SSR *4*, 33 (1953); C. A. 49, 5923 (1955).
1237. Lieber E., Rosen R.: US 1976806 (1934).
1238. Lyman A. L., Nichols H. B., Mithoff R. C.: US 2143078 (1939).
1239. Limido J., Miguel J., Thonon C.: Brit. 842377 (1960).
1240. Limido J., Miquel J., Thonon C.: US 3006844 (1961).
1241. Lin L. W., et al.: Acta Foculio- Chim. Sinica *6* (1), 55 (1965); C. A. *64*, 19264 (1966).
1241a. Linden T. M.: Ger. Offen. 1805664 (1969).
1242. Lindquist R. H.: US 3325396 (1967).
1243. Lindqvist M., Lundqvist D., Westgren A.: Svensk. Kem. Tidskr. *48*, 156 (1936).
1243a. Lipsch J. M. J. G.: *Thesis*, University of Technology, Eindhoven 1968.
1243b. Lipsch J. M. J. G., Schuit G. C. A.: J. Catalysis *15*, 163 (1969).
1243c. Lipsch J. M. J. G., Schuit G. C. A.: J. Catalysis *15*, 174 (1969).
1243d. Lipsch J. M. J. G., Schuit G. C. A.: J. Catalysis *15*, 179 (1969).
1244. Lobus I. I., Polozov V. F., Feofilov E. E.: USSR 42982 (1935).

1245. Loginov G. M.: Zhur. Neorg. Khim. *6*, 261 (1961).
1246. Logwinuk A. K., Friedman L., Weiss A. H.: Erdöl u. Kohle *17*, 532 (1964).
1246a. Lopez Gonzales J. de D., Barea Cuesta E.: An. Quim. *64*, 687 (1968); C. A. 70, 6795 (1969).
1246b. Lopez Gonzales J. de D., Barea Cuesta E., Banares Munoz M. A.: An. Quim. *64*, 699 (1968); C. A. 69, 110132 (1968).
1247. Lorne H. T., Porter F. W. B.: US 2574449 (1951).
1248. Love R. M., Tannich R. A.: US 2731506 (1956).
1249. Lozovoi A. V., Dyakova M. K.: Zh. Obshch. Khim. *7*, 2964 (1937).
1250. Lozovoi A. V., Dyakova M. K.: Zh. Obshch. Khim. *8*, 105 (1938).
1251. Lozovoi A. V., Dyakova M. K.: Zh. Obshch. Khim. *10*, 1 (1940).
1252. Lozovoi A. V., Senyavin S. A.: Zh. Prikl. Khim. *14*, 96 (1941).
1253. Lozovoi A. V., Senyavin S. A.: *Sbornik Statei Obshchei Khim.*, Akad. Nauk SSSR, Vol. I, 254 (1953).
1254. Lozovoi A. V., Senyavin S. A.: *Sbornik Statei Obshch. Khim.*, Akad. Nauk SSSR, Vol. II, 1035 (1953).
1255. Lozovoi A. V., Senyavin S. A.: Zh. Obshch. Khim. *24*, 1803 (1954).
1256. Lozovoi A. V., Senyavin S. A., Vol'Epshtein A. B.: Zh. Prikl. Khim. *28*, 175 (1955).
1257. Lozovoi A. V., Vol'Epshtein A. B., Senyavin S. A.: Trudy Inst. Gor. Iskop., AN SSSR *6*, 16 (1955).
1258. Lozovoi A. V., Senyavin S. A.: *Khim. Pererabotka Topliv* (*Chemical Processing of Fuels*), p. 180. Izd. AN SSSR, Moscow 1957.
1259. Lozovoi A. V., Senyavin S. A., Sovetova L. S.: *Trudy Inst. Gor. Iskop.*, Tom. IX, 122. Moscow 1959.
1260. Lozovoi A. V.: *Trudy Inst. Gor. Iskop.*, Tom IX, 148. Moscow 1959.
1261. Lozovoi A. V., Senyavin S. A., Sovetova L. S.: Zh. Prikl. Khim. *33*, 947 (1960).
1262. Lozovoi A. V., Muselevich D. L., Ravikovich T. M., Senyavin S. A., Cherkasova V. F.: Zh. Prikl. Khim. *34*, 1200 (1961).
1263. Lozovoi A. V., Muselevich D. L., Ravikovich T. M., Senyavin S. A., Titova T. A., Cherkasova V. F.: Tr. Inst. Gor. Iskop. AN SSSR. Tom XVII, 199. Moscow 1962.
1264. Lucke H.: Ger. 837994 (1952).
1265. Lüder H., Drescher K.: Chem. Technik *12*, 16 (1960).
1265a. Lukens H. R., Meisenheimer R. C., Wilson J. N.: J. Phys. Chem. *66*, 469 (1962).
1266. Lundqvist D., Westgren A.: Z. anorg. Chem. *239*, 85 (1938).
1267. Lurgi Gesellschaften, Frankfurt a. M.: *Lurgi Handbuch*, p. 74; 112 (1960).
1268. Luzarreta E., Garand J.: *Compt. rend. 78e, Congr. Socs. Savantes Paris et dépt.*, Sect. sci. 1953, 365; C. A. *49*, 9239 (1955).
1269. Lyman A. L., Mithoff R. C., Nichols H. B.: US 2236216 (1941).
1270. Lyon J. P., Crouch W. W.: US 2527948 (1950).
1271. Mach Yu. M., Khadshinov V. N.: Ukr. Khim. Zhur. *7*, 32 (1932).
1272. Mahugh R. A.: *Thesis*. Montana State College, March 1960.
1273. Mailhe A.: Bull. soc. chim. *15*, 327 (1914).
1274. Makolkin I. A.: Zhur. Fiz. Khim. *14*, 110 (1940).
1275. Makolkin I. A.: Acta Physicochim. USSR *13*, 361 (1940).
1276. Makray v. I.: Chemiker Ztg. *57*, 9 (1933).
1277. Makray v. I.: Chemiker Ztg. *57*, 205 (1933).
1278. Malinowski S., Kobyliński T.: J. prakt. Chem. *14*, 34 (1961).
1279. Malishev B. V.: J. Inst. Petroleum Tech. *22*, 341 (1936).
1280. Mandi T., Gáti G., Meszáros J.: *Symposium über Hydrokatalytische Processe in der Erdölverarbeitung und Petrolchemie*, Band I., p. 214. Leuna 1966.

1281. Manley R. E.: US 2967146 (1961).
1282. Mann R. S.: J. Chem. Soc. *1964*, 1531.
1283. Mann R. S.: Indian J. Technol. *3*, 53 (1965).
1284. Mansfield R., Salam S. A.: Proc. Phys. Soc. *66 B*, 377 (1953).
1285. Marcéaux P.: *Oil Shale and Cannel Coal Conf.*, Inst. of Petroleum, London 2, 673 (1951); C. A. 47, 1364 (1953).
1286. Mardanov M. A., Sultanov S. A.: Azerb. Khim. Zhur. *1960*, No. 2, 47.
1287. Mardanov M. A., Sultanov S. A., Naroditskaya S. G.: Azerb. Khim. Zhur. *1960*, No. 5, 3.
1288. Mardanov M. A., Sultanov S. A.: Azerb. Neft. Khoz. *40*, No. 8, 37 (1961).
1289. Mardanov M. A., Sultanov S. A., Naroditskaya S. G.: Azerb. Neft. Khoz. *40*, No. 9, 31 (1961).
1290. Mardanov M. A., Kuliev R. Sh., Markhaseva S. M., Sadykhova B. A., Alekperova N. G.: Azerb. Khim. Zhur. *1962*, No. 2, 25.
1291. Mardanov M. A., Sultanov S. A., Naroditskaya S. G.: *Proceedings, Symposium on the Application of the Hydrogenation Processes in Petroleum Processing*, Index B_7, Moscow 1963.
1292. Mardanov M. A., Sultanov S. A. et al.: Ibid., Index B_8, Moscow 1963.
1293. Mardanov M. A., Sultanov S. A., Naroditskaya S. G.: Azerb. Khim. Zhur. *1965* (3), 16.
1293a. Mardanov M. A., Sultanov S. A., Kulieva N. N.: Azerb. Khim. Zh. *1969*, No 5, 45.
1294. Marechal J. E. M.: *6th World Petrol. Congr.*, Sect. III, PD 7, Paper 1. Frankfurt 1963.
1295. Mariich L. I.: Khim. Prom., Inform. Nauch.-Tekhn. Zh. *1964* (4), 13.
1296. Marsh J. D. F., Newling W. B. S.: Brit. 867853 (1961).
1297. Marsh J. D. F., Rich J.: Brit. 894451 (1962).
1298. Marsh N. H.: US 2405440 (1946).
1298a. Martin D. J.: J. Org. Chem. *34*, 473 (1969).
1299. Martin H. Z.: US 3109804 (1963).
1300. Martin M. W., Harris G. H., Olivier K. R.: Canad. 552641 (1958).
1301. Martin M. W., Harris G. H., Olivier K. R.: US 2796438 (1957).
1302. Mashkina A. V., Sukhareva T. S.: Kinetika i Kataliz *5*, 751 (1964).
1303. Mashkina A. V., Sukhareva T. S., Chernov V. I.: Neftekhimiya *6*, 619 (1966).
1303a. Mashkina A. V., Vakurova E. M.: Dokl. AN SSSR *168*, 821 (1966).
1304. Mashkina A. V., Sukhareva T. S., Chernov V. I.: Neftekhimiya *7*, 110 (1967).
1305. Mashkina A. V., Sukhareva T. S., Chernov V. I.: Neftekhimiya *7*, 301 (1967).
1305a. Mashkina A. V., Avdeeva L. B.: Neftekhim. *8*, 414 (1968).
1305b. Mashkina A. V.: Khim. Seraorg. Soed., Soderzh. v Neftyakh i Nefteprod. *8*, 125 (1968).
1306. Mashkina A. V., Sukhareva T. S., Chernov V. I., Vakurova E. M.: USSR 175979 (1965).
1307. Maslyanskii G. N., Nemtsov M. S.: Dokl. AN SSSR *11*, 404 (1935).
1308. Maslyanskii G. N., Shenderovich F. S.: Zh. Fiz. Khim. *14*, 1301 (1940).
1309. Maslyanskii G. N.: Zh. Obshch. Khim. *13*, 540 (1943).
1310. Maslyanskii G. N.: Dokl. AN SSSR *45*, 25 (1944).
1311. Maslyanskii G. N., Meshebovskaya E. I.: Zh. Obshch. Khim. *16*, 701 (1946).
1312. Maslyanskii G. N., Kobelev V. A.: *Khim. Pererabotka Neftyanykh Uglevodorodov*, p. 499. Izd. AN SSSR, Moscow 1956.
1313. Maslyanskii G. N., Bursian N. R.: *Sb. Referatov Issled. Rabot LenNII za 1950—1957 gg.*, p. 99. Gostoptekhizdat 1958.
1314. Maslyanskii G. N., Kobelev V. A., Bursian N. R., Ryskin M. I.: *Proceedings, 8th Mendeleev Congress on General and Applied Chemistry, Section on Fuel Chemistry and Technology*, p. 7. Izd. AN SSSR, Moscow 1959.
1315. Maslyanskii G. N., Bursian N. R., Barkan S. A., Kobelev V. A., Telegin V. G.: Izv. Vysshikh Uchebnykh Zavedenii Khim. i Tekhnol. *3*, 359 (1960).

1316. Maslyanskii G. N., Bursian N. R., Kamusher G. D., Barkan S. A., Shuvaev E. S.: Khim. i Tekhnol. Topliv i Masel *5*, No. 9, 1 (1960).
1317. Maslyanskii G. N., Bursian N. R., Kobelev V. A., Ryskin M. I.: Khim. i Tekhnol. Topliv i Masel *6*, No. 1, 11 (1961).
1318. Maslyanskii G. N., Bursian N. R.: Zh. Prikl. Khim. *35*, 816 (1962).
1319. Maslyanskii G. N., Bursian N. R.: Zh. Prikl. Khim. *36* (7), 1549 (1963).
1320. Maslyanskii G. N.: USSR 53575 (1938); C. A. 34, 5258 (1940).
1320a. Masologites G. P., Jacobs H. E., White P. J.: US 3477943 (1969).
1321. Mason H. F., Taylor J. H., Buss W. C.: US 3172838 (1965).
1322. Mason H. F., Bray B. G.: US 3243367 (1966).
1323. Mason R. B.: US 2402423 (1946).
1324. Mason R. B.: US 2497176 (1950).
1325. Mason R. B., Nicholson W. S.: US 2813911 (1957).
1326. Mason R. B., Nicholson W. S.: US 2976254 (1961).
1327. Massagutov R. M., Berg G. A., Kulinich G. M., Kirillov T. S.: *7th World Petrol. Congr.*, PD 20, Paper 7. Mexico City 1967.
1327a. Massagutov R. M., Berg G. A., Kulinich G. M., Sultanov S. A., Safaev A. S., Antipin M. K.: Tr. Bashkir. Nauch.-Issled. Inst. Pererab. Nefti *1968*, No. 8, 126; C. A. 69, 60473 (1968).
1327b. Massoth F. E., Bidlack D. L.: *Referat an 157. Tagung der American Chemical Society*, Mineapolis, Minnesota, 13.—18. 4. 1969; Erdöl u. Kohle *23*, 46 (1970).
1327c. Massoth F. E., Bidlack D. L.: J. Catalysis *16*, 303 (1970).
1327d. Mathiron C., Colson J. C.: C. R. Acad. Sci., Paris, Ser. *C 1968*, 267 (7), 521.
1328. Matic D., Mijatovic I., Kolombo M.: Tehnika (Belgrade) *17*, 1761 (1962).
1329. Matic D., Mijatovic I.: Chem. Eng. Progr. Symp. Ser. *61* (54), 68 (1965).
1330. Maude A. H., Sweeney J. D.: US 2214859 (1940).
1331. Maude A. H., Mac Fadyen D. E.: US 2863725 (1958).
1332. Max N.: Neth. 70867 (1952).
1333. Max N.: US 2644834 (1950).
1334. Maxted E. B., Marsden A.: J. Soc. Chem. Ind. *65*, 51 (1946).
1335. Maxted E. B., Priestley J. J.: Gas J. *247*, 471; 477; 481; 515; 556; 593 (1946).
1336. Maxted E. B.: Brit. 435192 (1935).
1337. Maxted E. B.: Brit. 490775 (1938).
1338. Mayo F. R., Walling C.: Chem. Review *27*, 351 (1940).
1338a. Mayumi O., Takahashi M., Gomi S., Takahasi R., Kawai Y., Suzuki S., Sagamihara K., Saito T., Yamaki K.: Ger. Offen. 1908299 (1969).
1339. Mc Afee J., Montgomery C. W., Hirsch J. N., Horne W. A., Summers C. R. Jr.: Petrol. Ref. *34*, No. 5, 156 (1955).
1340. Mc Allister S. H., Greensfelder B. S., Peterson W. H.: Brit. 589915 (1947).
1341. Mc Cabe C. L.: J. Metals *7*, 61 (1955).
1342. Mc Clellan A. L.: US 3222271 (1965).
1343. Mc Comb W. M.: Brit. 160907 (1920).
1344. Mc Cullough J. P., Hubbard W. N., Frow F. R., Hossenlopp J. A., Waddington G.: J. Am. Chem. Soc. *79*, 561 (1957).
1345. Mc Grath H. G., Knaus J. A.: US 2894903 (1959).
1346. Mc Grath H. G., Smith M. R.: US 2787583 (1957).
1346a. McKinney J. D., Offutt W. C.: Fr. 1498938 (1967).
1346b. McKinney J. D., Sutphin E. M.: US 3509040 (1970).
1347. Mc Kinley J. B.: *Catalysis* (Ed. P. H. Emmett), Vol. 5, p. 405, Reinhold, New York (1957).

1348. Meerbott W. K., Hinds G. P. Jr.: Ind. Eng. Chem. *47*, 749 (1955).
1349. Meharg V. E., Coons K. W.: US 2295672 (1943).
1350. Meharg V. E., Coons K. W.: US 2295673 (1943).
1351. Meharg V. E., Coons K. W.: US 2295674 (1943).
1352. Meharg V. E., Coons K. W.: US 2295675 (1943).
1352a. Menzel R. L., Marks M. E., Crowther R. H.: US 3376218 (1968).
1353. Mering J., Levialdi A.: Compt. rend. *213*, 798 (1941).
1354. Mertes T. S.: US 3080311 (1963).
1355. Mertzweiller J. K., Tenney H. M.: Fr. 1315275 (1963).
1356. Mertzweiller J. K., Watts R. N.: Ger. 1114469 (1960).
1356a. Mertzweiller J. K., Tenney H. M.: Ger. Offen. 1801935 (1969).
1357. Mertzweiller J. K.: US 2671119 (1954).
1358. Mertzweiller J. K.: US 2709714 (1955).
1359. Mertzweiller J. K.: US 2808443 (1957).
1360. Mertzweiller J. K., Watts R. N.: US 3316303 (1967).
1361. Meshcheryakov A. P., Kaplan E. P.: Izv. AN SSSR, Ser. Khim. *1938*, 1055.
1362. Mestern A.: Brennstoff-Chem. *14*, 225 (1933).
1363. Mestern A.: Brennstoff-Chem. *14*, 379 (1933).
1363a. Metcalfe T. B.: Chim. Ind., Genie Chim. *102* (9), 1300 (1969).
1364. Metzner A.: Chem. Industrie *17*, 602 (1965).
1365. Middleton A. W., Ward A. M.: J. Chem. Soc. *1935*, 1459.
1366. Miller R. J.: Ind. Eng. Chem. *43*, 1410 (1951).
1367. Milliken T. M., Oblad A. G.: Fr. 1043097 (1951).
1367a. Mills G. A.: Ind. Eng. Chem. *61*, No. 7, 6 (1969).
1367b. Mills I. W., Dimeler G. R.: Ger. Offen. 1926173 (1970).
1368. Mills I. W., Johnson H. L.: US 2635081 (1953).
1369. Mills I. W., Johnson H. L.: US 2735877 (1956).
1370. Mills K. L. Jr.: US 3256176 (1966).
1371. Minachev Kh. M., Shuikin N. I., Ryashentseva M. A., Kononov N. F.: Izv. AN SSSR, Otd. Khim. Nauk *1958*, 719.
1372. Minachev Kh. M., Ryashentseva M. A., Rubinshtein A. M.: Izv. AN SSSR, Otd. Khim. Nauk *1959*, 819.
1373. Minachev Kh. M., Kondratev D. A.: Izv. AN SSSR, Otd. Khim. Nauk *1960*, 300.
1374. Minachev Kh. M., Isagulyants G. V., Kondratev D. A.: Izv. AN SSSR., Otd. Khim. Nauk *1960*, 902.
1375. Minachev Kh. M., Ryashentseva M. A.: Izv. AN SSSR, Otd. Khim. Nauk *1961*, 103.
1376. Minachev Kh. M., Ryashentseva M. A.: Izv. AN SSSR, Otd. Khim. Nauk *1961*, 107.
1377. Minachev Kh. M., Ryashentseva M. A., Petukhov V. A.: Izv. AN SSSR, Otd. Khim. Nauk *1961*, 1307.
1378. Minachev Kh. M., Ryashentseva M. A., Afanaseva Yu. A.: Izv. AN SSSR, Otd. Khim. Nauk, *1961*, 1673.
1378a. Minachev Kh. M., Ryashentseva M. A., Garanin V. I., Afanaseva Yu. A.: Neftekhim. *5*, 498 (1965).
1379. Mirza A., Masood M. A., Ramaswamy A. V., Vaidyeswaran R.: Brennstoff-Chem. *46*, 355 (1965).
1380. Mirza A., Ramaswamy A. V., Masood M. A., Qader S. A., Vaidyeswaran R.: Indian J. Technol. *3* (2), 57 (1965); C. A. 62, 15955 (1965).
1381. Mirza A., Masood M. A., Mallikarjunan M. M., Vaidyeswaran R.: Chem. Age *17* (3), 240 (1966); C. A. 65, 2019 (1966).

1382. Mirza A., Masood M. A., Mallikarjunan M. M., Vaidyeswaran R.: Brennstoff-Chem. *48*, 310 (1967).

1382a. Mirza M. A., Masood M. A., Mallikarjunan M. M., Vaidyeswaran R.: Petrol. Hydrocarbons *3*, No 1, 13 (1968); C. A. 69, 29012 (1968).

1382b. Mirza M. A., Masood M. A., Mallikarjunan M. M., Vaidyeswaran R.: Petrol. Hydrocarbons *3*, No. 2, 61 (1968); C. A. 69, 60497 (1968).

1383. Mistrík E. J., Polievka M.: Chem. průmysl *12*, 123 (1962).

1384. Mistrík E. J., Polievka M.: Chem. průmysl *13*, 129 (1963).

1385. Mistrík E. J., Polievka M.: Czech. 107326 (1963).

1385a. Mitchell P. C. H.: *The Chemistry of Some Hydrodesulphurisation Catalysts Containing Molybdenum.* The Climax Molybdenum Co., Ltd., London 1967.

1386. Miyanishi M.: Mem. Fac. Eng., Hiroshima Univ. *1*, 309 (1961); C. A. *55*, 20400 (1961).

1387. Mizera S.: Chem. Technik *9*, 286 (1957).

1388. Mizera S.: Angew. Chem. *71*, 372 (1959).

1389. Mobil Oil Corp.: Brit. 1056301 (1967).

1390. Mobil Oil Corp.: Fr. 1457382 (1966).

1391. Mobil Oil Corp.: Neth. Appl. 6511448 (1967).

1392. Mohr K. H., Runge F., Wolf F.: Chem. Technik *16*, 457 (1964).

1393. Moldavskii B. L., Prokopchuk N. J.: Zh. Prikl. Khim. *5*, 619 (1932).

1394. Moldavskii B. L., Livshits S. E.: Zh. Obshch. Khim. *3*, 603 (1933).

1395. Moldavskii B. L., Kumari Z. I.: Zh. Obshch. Khim. *4*, 298 (1934).

1396. Moldavskii B. L., Kumari Z. I.: Zh. Obshch. Khim. *4*, 307 (1934).

1397. Moldavskii B. L., Livshits S. E.: Zh. Obshch. Khim. *4*, 948 (1934).

1398. Moldavskii B. L., Livshits S. E.: Khim. Tverd. Topliva *5*, 91 (1934).

1399. Moldavskii B. L.: Destructive Hydrog. of Fuels ONTI, Goschimtechizdat (Leningrad) 1, 255 (1934); C. A. *29*, 1961 (1935).

1400. Moldavskii B. L., Vainstein S. M.: Khim. Tverdovo Topliva *6*, 656 (1935).

1401. Moldavskii B. L., Pokorskii V. N.: Khim. Tverdovo Topliva *6*, 943 (1935).

1402. Moldavskii B. L., Pokorskii V. N., Andreevskii I. L.: Neft. Khoz. *28*, No. 2, 52 (1935).

1403. Moldavskii B. L., Kamusher H. D.: Doklady AN SSSR *10*, 355 (1936).

1404. Moldavskii B. L., Kobylskaya M. V., Livshits S. E.: Zh. Obshch. Khim. *6*, 616 (1936).

1405. Moldavskii B. L., Pokorskii V. N.: Neft. Khoz. *1936*, No. 1, 48.

1406. Moldavskii B. L., Kamusher H. D., Livshits S. E.: Zh. Obshch. Khim. *7*, 131 (1937).

1407. Moldavskii B. L., Kamusher H. D., Kobylskaya M. V.: Zh. Obshch. Khim. *7*, 1835 (1937).

1408. Moldavskii B. L., Nizovkina T. V.: Zh. Obshch. Khim. *10*, 653 (1940).

1409. Moldavskii B. L., Turetskaya L. E.: Zh. Obshch. Khim. *16*, 445 (1946).

1410. Moldavskii B. L., Kamusher H. D.: USSR 53573 (1938).

1411. Montarnal R., Preve J.: Brit. 904697 (1962).

1412. Montecatini Societa generale per l'industria mineraria e chimica: Ital. 598102 (1959).

1413. Montecatini Societa generale per l'industria mineraria e chimica: Ital. 601436 (1960).

1413a. Montgomery D. P.: Ind. Eng. Chem., Product Research Dev. *7*, 274 (1968).

1414. Moore O. K.: US 2768225 (1956).

1415. Moore R. J., Greensfelder B. S.: J. Am. Chem. Soc. *69*, 2008 (1947).

1416. Moore R. J., Trimble R. A., Greensfelder B. S.: J. Am. Chem. Soc. *74*, 373 (1952).

1417. Moore R. J., Trimble R. A.: US 2471077 (1949).

1417a. Morikawa K., Okamura T., Abe R.: J. Soc. Chem. Ind. Japan *41*, Suppl. Bind., 431 (1938).

1418. Morikawa K., Takagi T.: Jap. 3521 ('58) (1958).

1418a. Morin F. J.: J. Appl. Phys. *32*, 2195 (1961).

1418b. Morita S., Inoue T., Eto H., Yoshimitsu K.: Fr. 1487588 (1967).

1419. Morita Y., Kikuchi E., Yamamoto K.: Kogyo Kagaku Zasshi *70* (5), 670 (1967); C. A. 67, 101617 (1967).
1419a. Morozov V. F., Parakonov V. B., Popov V. D.: Neftepererab. Neftekhim. *1970*, No. 8, 1.
1420. Morris J. E.: Brit. 719640 (1954).
1421. Mostecký J., Weisser O., Landa S., Ruprych M.: Chem. průmysl *11*, 2 (1961).
1422. Mostecký J., Pecka K.: *Sborník prací z výzkumu chemického využití uhlí, dehtu a ropy* (*Collection of Papers on Chemical Utilisation of Coal, Tar and Petroleum*), Vol. 2, 97 (1962).
1423. Mráz V., Krafft O., Kubička R.: Acta Chim. Acad. Sci. Hung., *36*, 269 (1963).
1424. Mráz V.: *Thesis*. Institute of Chemical Technology, Prague 1965.
1425. Mráz V.: Ropa a uhlie *7*, 296 (1965).
1426. Mukhin I. N., Kovalskii E. V.: Tr. Khark. Politekhn. Inst. *39*, 36 (1962); C. A. 60, 7845 (1964).
1426a. Mulaskey B. F.: US 3399132 (1968).
1426b. Mulaskey B. F.: US 3453204 (1969).
1427. Munday J. C.: US 2620356 (1952).
1428. Münzing E., Blume H., Rost H.: Z. Chem. *1*, 257 (1961).
1429. Münzing E., Blume H., Pindur E.: Z. für Chemie *2*, 76 (1962).
1430. Münzing E.: Acta Chim. Acad. Sci. Hung., *36*, 279 (1963).
1431. Münzing E., Blume H.: *Proceedings, Symposium on the Application of the Hydrogenation Processes in Petroleum Processing*, Moscow 1963.
1432. Münzing E.: Freiberger Forschungs-H. *A 264*, 5 (1963).
1433. Münzing E.: *Proceedings, Intern. Conference on Catalysts*, p. 263. Karlovy Vary 1964.
1433a. Murayama K., Kato Y., Morimoto S.: Bull. Chem. Soc. Japan *40* (11), 2645 (1967); C. A. *68*, 50168 (1968).
1434. Murphree E. V.: US 2414889 (1947).
1435. Murphy E. J.: Gas World *125*, 259 (1946).
1436. Murthy A. R. V., Rao B. S.: Proc. Indian Acad. Sci. *34A*, 283 (1951); C. A. *47*, 5227 (1953).
1436a. Muscheid E.: Ann. Physik (6) *13*, 305 (1953).
1436b. Muths R., Taillardat J.: Fr. 1540296 (1968).
1437. Myers C. G., Garwood W. E., Rope B. W., Wandlinger R. L., Hawthorne W. P.: J. Chem. Engng. Data *7*, 257 (1962).
1437a. Myers C. G., Rope B. W., Garwood W. E.: US 3384572 (1968).
1438. Nádeník O.: Chem. průmysl *6*, 441 (1956).
1439. Nagiev M. F., Vechkhaizer I. V., Sadykhova S. A.: Dokl. AN Azerb. SSR *17*, 681 (1961).
1440. Nagiev M. F., Vechkhaizer I. V., Sadykhova S. A.: Azerb. Khim. Zh. *1961*, No. 4, 61.
1441. Nahin P. G., Huffman H. C.: US 2486361 (1949).
1442. Nahin P. G., Huffman H. C.: US 2510189 (1950).
1442a. Nakamura H., Satori H., Tabuchi K., Nishizaki S., Kurita S., Watanabe T., Sakuma J.: Sekiyu Gakkai Shi *12*, 778 (1969); C. A. 72, 68885 (1970).
1442b. Nakano K., Itabashi K.: Kogyo Kagaku Zasshi *72*, 1029 (1969); C. A. 71, 60603 (1969).
1443. Nametkin S. S., Sanin P. I., Makover S. V., Zyba A. N.: Zh. Prikl. Khim. *6*, 494 (1933).
1444. Nametkin S. S., Sanin P. I. et al.: Gor. Slantsii *4*, No. 1, 44 (1934); C. A. 28, 6557 (1934).
1445. Nametkin S. S., Sanin P. I., Tzuiba A. N.: Gor. Slantsii *4*, No. 2, 40 (1934); C. A. 28, 6557 (1934).
1446. Nametkin S. S., Sanin P. I., Rudakova E. F.: Khim. Tverdogo Topliva *4*, 332 (1933).
1446a. Nastasi I., Welther E., Cioc D.: Petrol Gaze *18*, 271 (1967); C. A. 68, 61304 (1968).
1447. Neiser J.: Koks, smola, gaz *10*, 86 (1965).
1448. Neiser J.: Ropa a uhlie *9*, 191 (1967).

1449. Nelson W. L.: *Petroleum Rafinery Engineering*. McGraw-Hill Book Co., Inc., New York 1958.
1450. Nelson W. L.: Oil Gas J. *65* (12), 170 (1967).
1451. Nemtsov M. S.: Destr. Gidrog. Topliv ONTI, Goskhimtekhizdat (Leningrad) 1, 119 (1934); C. A. *29*, 1962 (1935).
1452. Nemtsov M. S., Sipovskii G. V.: Dokl. AN SSSR *1*, 494 (1934).
1453. Newling W. B. S.: Brit. 587070 (1947).
1454. Newling W. B. S.: Brit. 844093 (1960).
1455. Newling W. B. S., Marsh J. D. F.: Brit. 909973 (1962).
1456. Nicholson D. E.: Anal. Chem. *32*, 1365 (1960).
1457. Nicholson D. E.: Anal. Chem. *34*, 370 (1962).
1458. Nickels J. E., Corson B. B.: Ind. Eng. Chem. *43*, 1685 (1951).
1459. Nikolaeva A. F., Puchkov P. V.: Dokl. AN SSSR *24*, 345 (1939).
1460. Nikolaeva A. F., Puchkov P. V.: Zh. Obshch. Khim. *9*, 277 (1939).
1461. Nikolaeva A. F., Puchkov P. V.: Zh. Obshch. Khim. *9*, 2153 (1939).
1462. Nikolaeva A. F.: Zh. Obshch. Khim. *12*, 240 (1942).
1463. Nikolaeva A. F., Frost A. V.: Zh. Obshch. Khim. *12*, 646 (1942).
1464. Nikolaeva A. F.: Zh. Obshch. Khim. *16*, 1819 (1946).
1465. Nikolaeva D. Kh.: Tr. Vost.-Sibir. Filiala AN SSSR, Ser. Khim. *18*, 78 (1959).
1466. Nishio A., Higuchi S., Ozaki S., Yasuda H., Iguchi M., Yoshida K.: J. Fuel Soc. Japan *39*, 648 (1960); Chem. Zentr. *133*, 4397 (1962).
1467. Niwa K., Yokokawa T., Isoya T.: Bull. Chem. Soc. Japan *35*, 1543 (1962); C. A. 57, 14479 (1962).
1468. Nonnenmacher H.: Brennstoff-Chem. *31*, 138 (1950).
1469. Nonnenmacher H., Reitz O., Schmidt P.: Erdöl u. Kohle *8*, 407 (1955).
1470. Nonnenmacher H., Oettinger W.: Ger. 851493 (1952).
1471. Nonnenmacher H., v. Füner W.: Ger. 893389 (1953).
1472. Nonnenmacher H., Meyer H.: Ger. 1124927 (1962).
1473. Nonnenmacher H., v. Füner W.: US 2705733 (1955).
1474. North Thames Gas Board: Ger. 955318 (1957).
1475. Novák V.: *Sborník prací z výzkumu chemického využití uhlí, dehtu a ropy* (*Collection of Research Papers on Chemical Utilisation of Coal, Tar and Petroleum*), Vol. 2, 264 (1962).
1476. Novák V., Večerka M., Žídek J.: *Proceedings. International Conference on Catalysts*, p. 285, Karlovy Vary 1964.
1477. Novák V., Večerka M., Žídek J.: Ropa a uhlie *7*, 204 (1965).
1478. Novák V., Čech J.: *Sborník prací z výzkumu chemického využití uhlí, dehtu a ropy* (*Collection of Research Papers on Chemical Utilisation of Coal, Tar and Petroleum*), Vol. 5, 77 (1965).
1479. Novák V., Mráz V., Večerka M.: Freiberger Forschungs-H. *A 367*, 71 (1965).
1479a. Novák V.: *Sborník prací z výzkumu chemického využití uhlí, dehtu a ropy* (*Collection of Research Papers on Chemical Utilization of Coal, Tar and Petroleum*) Vol. 6, 102 (1966).
1480. Novák V., Kubička R.: Brenntoff-Chem. *48*, 258 (1967).
1480a. Novák V., Kubička R., Jiříček B.: Ropa a uhlie *10*, 362 (1968).
1481. Nozaki K.: US 2744052 (1956).
1482. Nozaki K.: US 2813835 (1957).
1483. Nozdrina E. V., Sergienko S. R.: Trudy Inst. Nefti, AN SSSR, *13*, 127 (1959).
1484. N. V. de Bataafsche Petroleum Maatschappij: Brit. 332944 (1929).
1485. N. V. de Bataafsche Petroleum Maatschappij: Brit. 349470 (1931).
1486. N. V. de Bataafsche Petroleum Maatschappij: Brit. 358180 (1931).
1487. N. V. de Bataafsche Petroleum Maatschappij: Brit. 532676 (1941).

1488. N. V. de Bataafsche Petroleum Maatschappij: Brit. 660215 (1951).
1489. N. V. de Bataafsche Petroleum Maatschappij: Brit. 668592 (1952).
1490. N. V. de Bataafsche Petroleum Maatschappij: Brit. 693407 (1953).
1491. N. V. de Bataafsche Petroleum Maatschappij: Brit. 711670 (1951).
1492. N. V. de Bataafsche Petroleum Maatschappij: Brit. 766498 (1957).
1493. N. V. de Bataafsche Petroleum Maatschappij: Fr. 694192 (1930).
1494. N. V. de Bataafsche Petroleum Maatschappij: Fr. 745074 (1933).
1495. N. V. de Bataafsche Petroleum Maatschappij: Fr. 745468 (1933).
1496. N. V. de Bataafsche Petroleum Maatschappij: Fr. 754664 (1933).
1497. N. V. de Bataafsche Petroleum Maatschappij: Fr. 804482 (1936).
1498. N. V. de Bataafsche Petroleum Maatschappij: Fr. 804483 (1936).
1499. N. V. de Bataafsche Petroleum Maatschappij: Fr. 1036720 (1953).
1500. N. V. de Bataafsche Petroleum Maatschappij: Fr. 1058230 (1954).
1501. N. V. de Bataafsche Petroleum Maatschappij: F. 1075730 (1954).
1502. N. V. de Bataafsche Petroleum Maatschappij: Ger. 681078 (1939).
1503. N. V. de Bataafsche Petroleum Maatschappij: Neth. 61913 (1948).
1504. N. V. de Bataafsche Petroleum Maatschappij: Neth. 67093 (1951).
1505. N. V. de Bataafsche Petroleum Maatschappij: Neth. 68159 (1951).
1506. N. V. de Bataafsche Petroleum Maatschappij: Neth. 68368 (1951).
1507. N. V. de Bataafsche Petroleum Maatschappij: Neth. 69617 (1952).
1508. N. V. de Bataafsche Petroleum Maatschappij: Neth. 83987 (1957).
1509. N. V. Internationale Hydrogeneeringsoctrooien Maatschappij: Fr. 839358 (1939).
1510. N. V. Internationale Hydrogeneeringsoctrooien Maatschappij: Neth. 52727 (1942).
1511. N. V. Internationale Hydrogeneeringsoctrooien Maatschappij: Neth. 61702 (1948).
1512. Oblad A. G., Milliken T. H. Jr., Mills G. A.: *Advances in Catalysis*, Vol. III, p. 199. Academic Press, New York 1951.
1513. Obolentsev R. D., Mashkina A. V.: Dokl. AN SSSR *119*, 1187 (1958).
1514. Obolentsev R. D., Mashkina A. V.: *Khim. Seraorganicheskikh Soedinenii Soderzh. v Neftyakh i Nefteprod.* (*Collection of Papers on the Chemistry of Organic Sulphur Compounds Contained in Petroleum and its Products*), Vol. II, p. 228. Izd. AN SSSR, 1959.
1515. Obolentsev R. D., Mashkina A. V., Kuzyev A. R., Gribkova G. P.: *5. Nauch. Sessiya po Khim. Seraorg. Soedinenii Soderzh. v Neft. i Nefteprod.* (*Proceedings, 5th Meeting on the Chemistry of Sulphur-Containing Organic Compoundsin Petroleum and its Products*), p. 49. Izd. Bash. Fil. AN SSSR, Ufa 1959.
1516. Obolentsev R. D., Mashkina A. V.: Dokl. AN SSSR *131*, 1092 (1960).
1517. Obolentsev R. D., Mashkina A. V.: *Khim. Seraorg. Soedinenii Soderzh. v Neft. i Nefteproduktakh* (*Chemistry of Organic Sulphur Compounds Contained in Petroleum and its Products*), Vol. III, p. 295. Izd. Bash. Fil. AN SSSR, Ufa 1960.
1518. Obolentsev R. D., Mashkina A. V.: *Gidrogenoliz Seraorg. Soedinenii Nefti* (*Hydrogenolysis of Sulphur-Containing Organic Compounds from Petroleum*). Gos. Nauch.-Tekhn. Izd. Neft. i Gorno-Toplivoi Lit., Moscow 1961.
1519. Obolentsev R. D., Kuzyev A. R., Mikheev G. M.: *Khim. Seraorg. Soed., Soderzhashch. v Neft. i Nefteprod.* (*Chemistry of Organic Sulphur Compounds Contained in Petroleum and its Products*), Vol. VI, p. 331. Izd. Bash. Fil. AN SSSR, 1964.
1520. Obryadchikov S. N.: *Tekhnol. Nefti* (*Petroleum Technology*), Tom. II. GTTI, Moscow 1952.
1521. Obryadchikov S. N., Saskind D. M.: Neft. Khoz. *33*, No. 6, 72 (1955).
1522. Odell W. W., Morrell C. E., Scheeline H. W.: US 2663622 (1953).
1523. Odioso R. C., Schmid B. K., Zabor R. C.: US 3129253 (1964).
1524. O'Donell J. P.: Oil Gas J. 42, No. 32, 43 (1943).

1525. Oettinger W., Reitz O.: Erdöl u. Kohle 18, 267 (1965).
1526. Oettinger W., Reitz O., Faltings V., Brandl F., Knopf H. J., Letort M.: Belg. 638806 (1964).
1527. Oettinger W., Reitz O., Faltings V., Brandl F., Knopf H. J.: Belg. 638807 (1964).
1528. Oettinger W., v. Füner W.: Ger. 945646 (1956).
1529. Oettinger W.: US 2839450 (1958).
1529a. Oettinger W., Wodtcke F., Seubert R.: Ger. 1271868 (1968).
1530. Oganesyan V. Kh., Rud B. M.: Poroshkovaya Met., AN SSSR *5* (12), 54 (1965); C. A. 64, 13495 (1966).
1531. Oganesyan V. Kh.: Izv. AN Arm. SSR, Ser. Tekhn. Nauk *19* (1), 30 (1966)
1531a. Oganesyan V. Kh.: Izv. AN Arm. SSR, Fiz. *1967*, 2 (1), 55; C. A. 69, 55302 (1968).
1532. Ogata E.: Nenryo Kyokaishi *46* (4), 226 (1967); C. A. 68, 4606 (1968).
1533. Ogata E.: Nenryo Kyokaishi *46* (483), 528 (1967); C. A. 68, 51586 (1968).
1534. Ogata I.: Koru Taru *14*, 57 (1962); C. A. 57, 11903 (1962).
1534a. O'Grady M. A., Parsons B. I.: Can., Dep. Energy, Mines Resour., Mines Br., Res. Rep. *R 194*, 53 pp. (1967); C. A. 68, 88776 (1968).
1535. O'Hara M. J.: US 3016347 (1962).
1536. Ohtsuka T., Shimizu S., Nagata T., Nakamura N., Shiba T.: Bull. Japan Petrol. Inst. *2*, 13 (1960).
1537. Ohtsuka T., Hasegawa Y., Koizumi M., Ono T.: Bull. Japan Petrol. Inst. *9*, 1 (1967); C. A. 67, 55863 (1967).
1537a. Ohtsuka T., Hasegawa Y., Takanari N.: Bull. Jap. Petrol. Inst. *10*, 14 (1968).
1538. Oldenburg Ch. C.: US 3088984 (1963).
1539. Olin J. F., Buchholz B., Loev B., Goshorn R. H.: US 3070632 (1960).
1540. Opalovskii A. A., Fedorov V. E.: Usp. Khim. *35*, 427 (1966).
1540a. Opalovskii A. A., Fedorov V. E.: Khalkogenidy, Mater. Seminara, 1 st, Kiev 1965, 79; C. A. *68*, 65276 (1968).
1540b. Opalovskii A. A., Fedorov V. E.: Izv. AN SSSR, Neorg. Mater. *4* (2), 293 (1968).
1541. Orchin M., Storch H. H.: Ind. Eng. Chem. *40*, 1385 (1948).
1542. Orechkin D. B.: Tr. Vost.-Sibir. Filiala AN SSSR, Ser. Khim. *3*, 105 (1955).
1543. Orechkin D. B., Ovsyanikov L. F., Bogdanova T. A.: Tr. Vost.-Sibir. Filiala AN SSSR, Ser. Khim. *26*, 71 (1959).
1544. Orechkin D. B.: Tr. Vost.-Sibir. Filiala AN SSSR, Ser. Khim. *26*, 141 (1959).
1544a. Orkin B. A., Braid M.: US 3530061 (1970).
1545. Orlov Kh. Ya., Martynov A. A., Bulychev V. P.: Izv. AN SSSR, Otd. Khim. Nauk *1963*, 1636.
1546. Orlov Kh. Ya., Martynov A. A., Bulychev V. P.: Izv. AN SSSR, Ser. Khim. *1965*, 792.
1547. Orlov Kh. Ya., Martynov A. A.: Izv. AN SSSR, Ser. Khim. *1965*, 796.
1548. Orlov N. A.: Ber. *64*, 2631 (1931).
1549. Orlov N. A., Broun A. S.: Khim. Tverd. Topliva *3*, 817 (1932).
1550. Orlov N. A.: Dokl. AN SSSR *4*, 286 (1934).
1551. Orlov N. A., Tishchenko V. V., Tarasenkova E. M.: Zh. Prikl. Khim. *8*, 501 (1935).
1552. Orlov N. A., Glinskikh S. A., Ignatovich N. I.: Zh. Prikl. Khim. *8*, 1170 (1935).
1553. Orlov N. A., Radchenko O. A.: Zh. Prikl. Khim. *9*, 249 (1936).
1554. Orochko D. I., Shavolina N. V.: Khim. Tekhnol. Topliv Masel *4*, No. 9, 48 (1959).
1554a. Orochko D. I., Perezhigina I. Ya,. Rogov S. P., Rysakov M. V., Chernakova G. N.: Khim. Tekhnol. Topliv Masel *15*, No. 8, 2 (1970).
1554b. Osborn S. W., Wells T. F., Kutch E. L.: US 3359248 (1967).
1554c. Oshima M., Kashima K., Kubo T., Tabata H., Watanabe H.: Bull. Chem. Soc. Jap. *39* (12), 2750 (1966); C. A. 66, 56775 (1967).

1554d. Oshima M., Kashima K., Tabata H., Watanabe H., Kubo T.: Bull. Chem. Soc. Jap. *39* (12), 2763 (1966); C. A. 66, 56778 (1967).
1555. Oshima M., Maeda Y., Kashima K.: Ger. 1115737 (1961).
1556. Oshima M., Maeda Y., Kajima H.: Jap. 13864 ('63).
1557. Oshima M., Maeda Y., Kajima H.: Jap. 17018 ('63).
1558. Oshima M., Maeda Y., Kashima K.: US 3223698 (1965).
1559. Osipov L. N., Goldshtein D. L.: Khim. i Tekhnol. Topliv i Masel *4*, No. 8, 22 (1959).
1560. Osipov L. N., Antipin M. K., Khavkin V. A.: Neftepererab. i Neftekhim., Nauchn.-Tekhn. Sb. *1965* (7), 7.
1561. Osipov L. N., Khavkin V. A., Semenova E. S., Agafonov A. V., Rogov S. P.: Khim. i Tekhnol. Topliv i Masel *11*, No. 5, 5 (1966).
1561a. Ovchinnikov P. N., Bat I. I., Chistyakova G. A., Rebrova V. V., Guseva L. G., Ardemasova M. A., Mironova G. A., Grachev A. M.: USSR 189441 (1966).
1562. Ovsyanikov L. F., Orechkin D. B.: Tr. Vost.-Sibir. Filiala AN SSSR, Ser. Khim. *26*, 63 (1959).
1563. Owen J. J.: US 2402439 (1946).
1564. Owen J. J.: US 2402440 (1946).
1565. Owens P. J., Amberg C. H.: Advances in Chem. Series, No. 33, 182 (1961).
1566. Owens P. J., Amberg C. H.: Canad. J. Chem. *40*, 941 (1962).
1567. Owens P. J., Amberg C. H.: Canad. J. Chem. *40*, 947 (1962).
1568. Padovani C. et al.: *5th World Petrol. Congr.*, Sect. III, Pap. 7. New York 1959.
1569. Pamfilov A. V., Mazurkevich Ya. S.: Ukr. Khim. Zh. *28* (9), 1014 (1962).
1569a. Pamfilov A. V., Mazurkevich Ya. S.: Katal., AN Ukr. SSR, Respubl. Mezhvedom. Sb. No. 3, 129 (1967); C. A. *68*, 91041 (1968).
1570. Panchenkov G. M., Lebedev B. P.: *Chemická kinetika a katalýza* (*Chemical Kinetics and Catalysis*), p. 377—385. SNTL, Prague 1964.
1571. Panchenkov G. M., Kuznetsov O. I., Zhorov Yu. M.: Mosk. Inst. Naftekhim. Gazov. Prom. No. *69*, 98 (1967); C. A. 68, 51592 (1968).
1572. Parravano N., Malquori G.: Atti acad. Lincei (6), *7*, 19 (1928).
1573. Parravano N., Malquori G.: Atti acad. Lincei (6), *7*, 109 (1928).
1574. Parravano N., Malquori G.: Atti acad. Lincei (6), *7*, 189 (1928).
1575. Pashley D. W., Presland A. E. B.: *Proc. European Regional Conf. Electron Microscopy*, Delft, 1, 417 (1960).
1576. Paterson N. J.: US 3242068 (1966).
1577. Paterson N. J.: US 3338818 (1967).
1577a. Paterson N. J.: US 3404539 (1968).
1578. Patterson J. A.: US 2957911 (1960).
1579. Pauling L.: *The Nature of Chemical Bond.* Cornell Univ. Press 1944.
1580. Paull P. L.: US 3074783 (1963).
1581. —
1582. Paushkin Ya. M., Orlov Kh. Ya., Katsobashvili Ya. R.: Izv. Vysshikh. Ucheb. Zavedenii, Neft i Gaz *1959*, No. 9, 57.
1583. Paushkin Ya. M., Orlov Kh. Ya.: Izv. AN SSSR, Otd. Khim. Nauk *1961*, 657.
1584. Pavlic A. A., Peppel W. J.: US 2397689 (1946).
1585. Pavlic A. A.: US 2467222 (1949).
1586. Pavlova K. A.: *Thesis.* Institute of Organic Chemistry, Academy of Sciences USSR, Moscow 1955.
1587. Pavlova K. A., Kalechits I. V.: Tr. Vost.-Sibir. Filiala AN SSSR, Ser. Khim. *38*, 19 (1961).
1588. Pavlova K. A., Kalechits I. V.: Tr. Vost.: Sibir. Filiala AN SSSR, Ser. Khim. *38*, 61 (1961).

1589. Pavlova K. A., Panteleeva B. D., Deryagina E. N., Kalechits I. V.: Kinetika i Kataliz *6*, 493 (1965).
1590. Pease R. N., Keighton W. B. Jr.: Ind. Eng. Chem. *25*, 1012 (1933).
1591. Peck E. B.: US 2409382 (1946).
1591a. Peck R. A., Franz W. F., Messing D. A.: US 3294673 (1966).
1592. Peck R. A., Messing D. A., Child E. T.: US 3316169 (1967).
1593. Peck R. A., Child E. T., Messing D. A.: US 3360457 (1967).
1594. Pecka K., Mostecký J., Hejda Z., Valdauf B.: Czech. 125351 (1967).
1595. Pedigo C. L., Alexander B. T.: US 3213040 (1965).
1596. Pelipetz M. G., Wolfson M. L., Ginsberg H. H., Clark E. L.: Chem. Eng. Progr. *48*, 353 (1952).
1597. Pelipetz M. G., Salmon J. R., Bayer J., Clark E. L.: Ind. Eng. Chem. *45*, 806 (1953).
1598. Pelipetz M. G., Frank L. V., Ginsberg H. H., Wolfson M. L., Clark E. L.: Chem. Eng. Progr. *50*, 626 (1954).
1599. Peppel W. J., Signaigo F. K.: US 2402665 (1946).
1600. Peppel W. J., Moss P. H.: *Advances in Petroleum Chemistry and Refining* (Ed. J. J. Mc Ketta, Jr.), Vol. IV, p. 369. Interscience Publ., New York—London 1961.
1601. Peralta B., Reeg C. P., Vaell R. P., Hansford R. C.: Chem. Eng. Progr. *58*, No. 4, 41 (1962).
1602. Perchenko V. N., Sergienko S. R.: *Izbiratelnoe Kataliticheskoe Gidrirovanie Seraorganicheskikh Soedinenii.* (*Selective Catalytic Hydrogenation of Organic Sulphur Compounds*). Izd. AN Turkmenskoi SSR, Ashkhabad 1962.
1603. Perna F., Doležalík V.: Paliva *32*, 145 (1952).
1604. Perna F., Pelčík J.: Paliva *33*, 126 (1953).
1605. Perna F., Riedl R.: *Plynárenství* (*The Gas Industry*), Vol. II. SNTL, Prague 1957.
1606. Pertierra J. M.: Anales soc. espan. fis. quim., *31*, 289 (1933).
1607. Pestrikov S. V., Kozik B. L., Mamina S. R., Bnatova L. A.: Neftepererab. i Neftekhim. *1967*, No. 4, 1.
1608. Peterson R. E., Sutphin E. M.: US 3310594 (1967).
1609. Petrik G. K., Zabramnyi D. T.: Doklady Akad. Nauk Uzbek. SSR *1954*, No. 9, 29; Referat. Zh. Khim. *1956*, Abst. No. 40674.
1610. Petrolite Corp.: Fr. 1377392 (1964).
1611. Petrov A. D., Poshiltseva E. A.: Khim. Tverd. Topliva *5*, 273 (1934).
1612. Petrov A. D., Meshcheryakov A. P., Andreev D. N.: Ber. *68 B*, 1 (1935).
1613. Petrov A. D.: *Khim. Motornych Topliv* (*The Chemistry of Motor Fuels*), p. 35. Izd. AN SSSR, Moscow 1953.
1614. Pevere E. F., Riordan M. D., Carter N. D.: US 2973313 (1961).
1614a. Pfefferle W. C.: Oil Gas J. 68, No. 15, 97 (1970).
1615. Pfennig R. F.: US 2632779 (1953).
1616. Pfirrmann T. W.: Ger. 860947 (1952).
1617. Phillips Petrol. Co.: Belg. 668580 (1966).
1618. Pichler H., Chervenak M. C., Johnson C. A., Sze M. C., Campagnolo J. F.: Petrol. Ref. *36*, No. 9, 201 (1957).
1619. Pichler H.: Trans. Inst. Chem. Engrs. *38*, 225 (1960).
1620. Pichler H., Schulz H., Reitemeyer H.-O.: Erdöl u. Kohle *22*, 617 (1969).
1621. Pichler H.: Brennstoff-Chem. *44*, 110 (1963).
1622. Pichler H.: Erdöl u. Kohle *21*, 66 (1968).
1623. Picon M.: Compt. rend. *189*, 96 (1929).
1624. Pictet A.: Ber. *38*, 1946 (1905).
1625. Pier M.: Chemiker Ztg. *57*, 205 (1933).
1626. Pier M.: Z. Elektrochem. *53*, 291 (1949).

1627. Pier M.: Brennstoff-Chem. *32*, 129 (1951).
1628. Pier M.: *Progress in Hydrogenation of Petroleum Oils, Proc. 3rd World Petrol. Congr.*, Sect. 4, Leiden 1951.
1629. Pier M.: Z. Elektrochem. *57*, 456 (1953).
1630. Pier M., Jacob P., Simon W.: Brit. 444779 (1936).
1631. Pier M., Krönig W.: Ger. 670717 (1939).
1632. Pier M., Donath E.: Ger. 711470 (1941).
1633. Pier M., Krönig W.: Ger. 713208 (1941).
1634. Pier M., Krönig W.: Ger. 714810 (1942).
1635. Pier M., Krönig W.: Ger. 715624 (1942).
1636. Pier M., Krönig W., Simon W., Kaufmann H.: Ger. 733239 (1943).
1637. Pier M.: Ger. 734721 (1943).
1638. Pier M., v. Füner W., Simon W.: Ger. 736093 (1943).
1639. Pier M., Simon W., Grassl G.: Ger. 742066 (1943).
1640. Pier M., Peters K.: Ger. 863195 (1953).
1641. Pier M., Donath E.: Ger. 885700 (1953).
1642. Pier M., Urban W., Schmidt P., Oettinger W.: Ger. 933648 (1955).
1643. Pier M., Urban W., Schmidt P., Oettinger W.: Ger. 933826 (1955).
1644. Pier M., Ringer F., Simon W.: US 1932186 (1934).
1645. Pierce R. B., Trusty A. W.: US 2060112 (1937).
1646. Pinchuk L. V., Ovsyanikov L. F., Orechkin D. B., Kalechits I. V.: Tr. Vost.-Sibir. Filiala AN SSSR, Ser. Khim. *4*, 130 (1956).
1647. Pinchuk L. V., Ovsyanikov L. F., Orechkin D. B., Kalechits I. V.: Tr. Vost.-Sibir. Fil. AN SSSR, Ser. Khim. *4*, 137 (1956).
1648. Pindur E.: *Ingenieurarbeit*, Leuna 1957.
1648a. Pine L. A., Ellert H. G.: US 3287401 (1966).
1648b. Pine L. A.: US 3299110 (1967).
1648c. Pine L. A., Ellert H. G., Drushel H. V.: US 3329826 (1967).
1649. Pine L. A.: US 3342879 (1967).
1649a. Pine L. A., Ellert H. G.: US 3361832 (1968).
1650. Pines H.: *Advances in Catalysis*, Vol. I, p. 201. Academic Press Inc., New York 1948.
1651. Pines H.. Hoffman N. E.: *Advances in Petroleum Chemistry and Refining* (Ed. J. J. Mc Ketta Jr.), Vol. 3, p. 127. Interscience Publ. Inc., New York 1960.
1651a. Plank C. J., Rossinski E.: US 3391088 (1968).
1652. Plant J. H. G., Newling W. B. S.: Inst. Gas Engrs., Commun. No. 344; Gas World *129*, 807, 922; Gas J. *256*, 654, 657, 659 (1948).
1653. Platonov M. S.: Zh. Obshch. Khim. *11*, 683 (1941).
1653a. Platonova A. V., Ryashentseva M. A., Ananchenko S. N., Torgov I. V.: Izv. AN SSSR, Ser. Khim. *1966*, 1256.
1653b. Platteeuw J. C., de Ruiter H., van Zoonen D., Kouwenhoven H. W.: Amer. Chem. Soc., Div. Petrol. Chem., Preprints 1966, *11* (4), A 133—A 134.
1654. Platteeuw J. C., Choufoer J. H.: Neth. 102260 (1962).
1655. Polataiko R. I., Kruglikova N. S., Frolova V. S., Galikh P. N., Skarchenko V. K.: Neft. i Gaz. Prom., Inform. Nauch.-Issled. Tekhn. Sb. *1965* (2), 53.
1656. Poll H. F.: US 3306845 (1967).
1656A. Pollitzer E. L., Haensel V., Hayes J. C.: 8th World Petrol. Congress, PD13, Paper 3. Moscow 1971.
1656a. Pollitzer E. L.: Ger. Offen. 1919698 (1970).
1656b. Pollitzer E. L., Hayes J. C.: US 3502573 (1970).
1657. Polozov V. F.: Khim. Tverd. Topliva *6*, 78 (1932).

1658. Polozov V., Feofilov E.: Novosti Tekhniki *1936*, No. 36, 20; C. A. 31, 2757 (1937).
1658a. Popl M., Pecka K., Weisser O., Mostecký J.: Neftekhimiya *9*, 506 (1969).
1659. Porter F. W. B.: J. Inst. Petrol *40*, 18 (1954).
1660. Porter F. W. B., Northcott R. P.: Brit. 710342 (1954).
1661. Porter F. W. B.: Brit. 713832 (1954).
1662. Porter F. W. B., Howes D. A., Morris J. E.: Brit. 719627 (1954).
1663. Porter F. W. B., Isitt J. S.: Brit. 729303 (1955).
1664. Porter F. W. B., Northcott R. P.: Brit. 736782 (1955).
1665. Porter F. W. B., Green F. R. G.: US 2574445 (1951).
1666. Porter F. W. B., Haresnape J. N.: US 2656302 (1953).
1667. Porter F. W. B., Northcott R. P., Rowland J.: US 2672433 (1954).
1668. Porter F. W. B., Northcott R. P.: US 2687985 (1954).
1669. Pospíšil J.: Chem. listy 55, 1210 (1961).
1670. Pospíšil J., Rada A., Vodička L.: Chem. průmysl *12*, 117 (1962).
1671. Pospíšil J., Weisser O., Landa S.: Collection Czech. Chem. Commun. *29*, 1387 (1964).
1671a. Pospíšil J., Prusíková M., Weisser O.: Chem. průmysl *19*, 500 (1969).
1672. Pospíšil J.: Czech. 97599 (1959).
1673. Pospíšil J., Weisser O., Landa S.: Czech. 115968 (1965).
1674. Preve J., Montarnal R.: Compt. rend. *249*, 1667 (1959).
1675. Priestley J. J.: Gas Times *91*, 640 (1957).
1676. Prilezhaeva E. N., Shostakovskii M. F.: Usp. Khim. *32*, 897 (1963).
1677. Procházka J., Jelínek J., Dufek L.: *Research Report*. Institute for Chemical Processing of Coal, Záluží 1962.
1678. Prokofeva V. P., Rozovskii A. Ya., Shchekin V. V.: Tr. Inst. Nefti *10*, 293 (1957).
1679. Prokopets E. I., Eru I. I.: Koks i Khim. *1*, 35 (1932).
1680. Prokopets E. I., Eru I. I.: Ukrain. Khim. Zh. *6*, 244; Chem. Zentr. 104, I, 611 (1933).
1681. Prokopets E. I.: Zh. Prikl. Khim. *7*, 159 (1934).
1682. Prokopets E. I., Eru I. I.: Destr. Gidrog. Topliv *1*, 275 (1934); C. A. 32, 6034 (1938).
1683. Prokopets E. I.: Khim. Tverd. Topliva *5*, 832 (1934).
1684. Prokopets E. I., Eru I. I.: Khim. Tverd. Topliva *6*, 67 (1935).
1685. Prokopets E. I., Khadshinov V. N.: Khim. Tverd. Topliva *6*, 347 (1935).
1686. Prokopets E. I.: Zh. Prikl. Khim. *10*, 126 (1937).
1687. Prokopets E. I., Gavrilova G. E., Klimova L. A.: Zh. Prikl. Khim. *11*, 823 (1938).
1688. Prokopets E. I.: Zh. Prikl. Khim. *11*, 835 (1938).
1689. Prokopets E. I., Pavlenko A. V., Boguslavskaya S. M.: Zh. Prikl. Khim. *11*, 840 (1938).
1690. Prokopets E. I., Boguslavskaya S. M.: Zh. Prikl. Khim. *11*, 847 (1938).
1691. Prokopets E. I., Boguslavskaya S. M.: Zh. Prikl. Khim. *11*, 850 (1938).
1692. Prokopets E. I., Boguslavskaya S. M.: Zh. Prikl. Khim. *11*, 1471 (1938); Foreign Petroleum Technique 7, 169 (1939).
1693. Prokopets E. I., Filaretov A. N., Boguslavskaya S. M.: Zh. Prikl. Khim. *11*, 1475 (1938).
1694. Prokopets E. I., Filaretov A. N., Pychko V. A.: Zh. Prikl. Khim. *11*, 1626 (1938).
1695. Prokopets E. I., Filaretov A. N.: Zh. Prikl. Khim. *11*, 1631 (1938).
1696. Proshkin A. A., Galenko N. P., Khemeris T. A., Kovalenko N. A., Golubchenko I. T.: Gazovaya Prom. *5*, No. 12, 46 (1960).
1697. Prückner H.: Erdöl u. Kohle *16*, 188 (1963).
1697a. Pruiss C. E.: US 3519557 (1970).
1698. Pschorr R.: Ger. 380059 (1923).
1699. Puchkov P. V.: Neft. Khoz. *3*, 51 (1935).
1700. Puchkov P. V.: Khim. Tverdovo Topliva *6*, 852 (1935); Chem. Zentr. 108, I, 482 (1937).
1701. Puchkov P. V.: Petroleum *32*, No. 15, 1 (1936).

1702. Puchkov P. V., Nikolaeva A. F.: Zh. Prikl. Khim. *10*, 327 (1937).
1703. Puchkov P. V., Nikolaeva A. F.: Izv. AN SSSR, Ser. Khim. *1937*, 171.
1704. Puchkov P. V.: Neft. Khoz. *1937*, No. 6, 44; Chem. Zentr. 108, II, 3843 (1937).
1705. Puchkov P. V., Nikolaeva A. F.: Zh. Obshch. Khim. *8*, 1153 (1938).
1706. Puchkov P. V., Nikolaeva A. F.: Zh. Obshch. Khim. *8*, 1159 (1938).
1707. Puchkov P. V.: Zh. Obshch. Khim. *8*, 1677 (1938).
1708. Puchkov P. V., Nikolaeva A. F.: Zh. Obshch. Khim. *8*, 1756 (1938).
1709. Puchkov P. V., Nikolaeva A. F.: Zh. Obshch. Khim. *8*, 1939 (1938).
1710. Puchkov P. V., Nikolaeva A. F.: Zh. Obshch. Khim. *9*, 280 (1939).
1711. Puchkov P. V., Nikolaeva A. F.: Dokl. AN SSSR *24*, 345 (1939).
1712. Puchkov P. V.: Zh. Fiz. Khim. *14*, 1319 (1940).
1712a. Puchkov V. A., Vol'Epshtein A. B., Krichko A. A.: Neftepererab. Neftekhim. *1967*, No. 5, 23.
1713. Qader S. A., Krishna M. G.: Indian J. Technol. *3*, 125 (1965).
1713a. Qader S. A., Wiser W. H., Hill G. R.: Ind. Eng. Chem., Process Design Dev. *7*, 390 (1968).
1713b. Qader S. A., Wiser W. H., Hill G. R.: Amer. Chem. Soc., Div. Fuel Chem., Preprints 1968, *12* (2), 28.
1713c. Qader S. A., Hill G. R.: Ind. Eng. Chem., Process Design Dev. *8*, 450 (1969).
1713d. Qader S. A., Hill G. R.: Ind. Eng. Chem., Process Design Dev. *8*, 456 (1969).
1713e. Qader S. A., Hill G. R.: Ind. Eng. Chem., Process Design Dev. *8*, 462 (1969).
1713f. Qader S. A., Hill G. R.: Hydrocarbon Process. *48*, No. 3, 141 (1969).
1713g. Qader S. A., Hill G. R.: Brennstoff-Chem. *50*, 167 (1969).
1714. Radomyski B., Tomasik Z.: Chem. Stosowana *4*, 501 (1960).
1715. Rálek M., Jírů P., Grubner O.: *Molekulová síta* (*Molecular Sieves*). SNTL, Prague 1966.
1716. Rao B. S.: Current Sci. *12*, 323 (1943).
1717. Rao M. J., Krishna M. G.: Chem. Age India *17* (3), 233 (1966).
1718. Rapoport I. B., Mintschenkov M. P., Konov V. P.: Khim. Tverd. Topliva *6*, 146 (1935).
1719. Rapoport I. B.: Zh. Prikl. Khim. *9*, 1456 (1936).
1720. Rapoport I. B., Masina M. P.: Khim. Tverd. Topliva *7*, 694 (1936).
1721. Rapoport I. B.: Zh. Fiz. Khim. *14*, 1321 (1940).
1722. Rapoport I. B.: *Iskusstvennoe Zhidkoe Toplivo* (*Synthetic Liquid Fuel*), Vol. I. Gos. Nauch.-Tekhn. Izd. Neft. i Gorno-Topl. Liter., Moscow 1955.
1723. Rapoport I. B.: *Iskusstvennoe Zhidkoe Toplivo* (*Synthetic Liquid Fuel*), Vol. II. Gos. Nauch.-Tekhn. Izd. Neft. i Gorno-Topl. Liter., Moscow 1955.
1724. Raseev S. D., Ionescu C. D.: *Katalytisches Reformieren*. VEB Deutscher Verlag für Grundstoffindustrie, Leipzig 1966.
1725. Rasulov A. M., Chernozhukov N. I., Kuliev R. Sh., Sadykhova B. A.: Khim. i Tekhnol. Topliv i Masel *9*, No. 12, 11 (1964).
1725a. Rausch M. K., Isaacson H. V.: US 3427177 (1969).
1726. Ray G. C.: US Appl. 32541. Official Gaz. *652*, 891 (1951); C. A. *47*, 1725 (1953).
1727. Reid E. E.: *Organic Chemistry of Bivalent Sulphur*, Vol. I, p. 165. Chem. Publ. Corp. Inc., New York 1958.
1728. Reid E. E.: *Organic Chemistry of Bivalent Sulphur*, Vol. II, p. 78. Chem. Publ. Corp. Inc., New York 1960.
1729. Reinhold H., Appel W., Frisch P.: Z. physik Chem. *A 184*, 273 (1939).
1729a. Reitemeyer H. O.: *Thesis*. Universität Karlsruhe 1970.
1730. Reitz O.: Chem. Ing. Tech. *21*, 413 (1949).
1731. Reitz O., Kohrt H. U.: Erdöl u. Kohle *11*, 18 (1958).
1732. Reitz O., Lorenz E.: Brit. 940960 (1963).
1733. Reppe W., Nicolai F.: Ger. 625660 (1936).

1734. Reppe W., Nicolai F.: Swed. 82699 (1935).
1735. Reppe W., Nicolai F.: US 2156095 (1939).
1736. Retailliau E. R.: Ger. 1040723 (1958).
1736a. Retallick W. B.: *Kirk-Othmer Encyclopedia of Chemical Technology*, Vol. 11, p. 418. Second Ed., Interscience Publ., John Wiley, New York—London—Sidney 1966.
1737. Rhodes E. O.: Bureau of Mines, Inf. Circ. 7490 (1949).
1738. Rice T., Goldthwait R. G., Kwolek S. J.: Belg. 637762 (1964).
1738a. Rich J., Newling W. B. S.: Brit. 1051626 (1966).
1739. Richardson F. D., Jeffes J. H. E.: J. Iron Steel Inst. (London) *171*, 165 (1952).
1740. Richardson J. T.: Ind. Eng. Chem., Fundamentals *3*, 154 (1964).
1740a. Rickert H.: Festkörperprobleme *6*, 85 (1967).
1741. Rieche A., Redinger L., Lindenhayn K.: Brennstoff-Chem. *47*, 326 (1966).
1742. Ries H. E.: *Advances in Catalysis*, Vol. IV, p. 87. Academic Press Inc., New York 1952.
1743. Riesz C. H., Morritz F. L.: *Advances on Petroleum Chemistry and Refining* (Ed. J. J. Mc Ketta, Jr.), Vol. IV, p. 279. Interscience Publ., Wiley, New York—London 1961.
1744. Riesz C. H., Schwoegler E. J.: US 2920116 (1960).
1745. Rigney J. A., Mason R. B., Hamner G. P.: US 3337447 (1967).
1746. Rittmeister W.: Ger. 611922 (1935).
1747. Robbers J. A., Paterson N. J., Lane W. T.: Hydrocarbon Process. Petrol. Ref. *40*, No. 6, 147 (1961).
1748. Robbins L. V. Jr., Mertzweiller J. K.: US 3118954 (1964).
1749. Roberti G.: Mem. Acad. Italia (Classe sci. fis. mat. e nat.) *1*, Chim. No. 2 (1930); C. A. *25*, 1969 (1931).
1750. Roberti G.: Ann. chim. applicata *21*, 217 (1931).
1751. Roberti G.: Mem. accad. Italia chim. *2*, No. 5, 5 (1931).
1752. Roberti G.: Atti accad. Lincei *13*, 527 (1931).
1753. Roberti G.: Ann. chim. applicata *22*, 3 (1932).
1754. Roberti G.: Giorn. chim. ind. applicata *14*, 437 (1932).
1755. Roberti G.: *World Petroleum Congr.*, London 1933, Proc. 2, 326; C.A. *28*, 3869 (1934).
1756. Roberti G.: Chaleur et ind. *15*, No. 167, 446 (1934).
1757. Rode E. Ya., Lebedev B. A.: Zh. Neorg. Khim. *6*, 1189 (1961).
1757a. Rode E. Ya., Lebedev B. A.: Zh. Neorg. Khim. 6, 1198 (1961).
1758. Rodgers T. A., Nager M.: Belg. 626504 (1963).
1759. Roebuck A. K., Evering B. L.: Ind. Eng. Chem. *50*, 1135 (1958).
1760. Roediger J. C., Morris K. G.: US 2906687 (1959).
1760a. Rogers E. S., Kovach S. M.: US 3383306 (1968).
1761. Roginskii S. Z.: Izv. AN SSSR, Ser. Fiz. *21*, 163 (1957).
1762. Rogov S. P., Goldshtein D. L., Osipov L. N., Agafonov A. V.: Khim. i Tekhnol. Topliv i Masel *6*, No. 8, 13 (1961).
1763. Rohländer W.: *Thesis*. Martin-Luther Universität, Halle/Wittenberg, Math.-Nat.-Fak. 1963.
1764. Rohländer W.: *Proceedings, Int. Conference on Catalysts*, Part II, p. 315. Karlovy Vary 1964.
1765. Romanowski W.: Roczniki Chem. *37*, 1077 (1963).
1765a. Rosenheimer M. O., Kiovsky J. R.: Amer. Chem. Soc., Div. Petrol. Chem., Preprints 1967, *12* (4), B 147—B 164.
1766. Rosenqvist T.: J. Iron Steel Inst. (London) *176*, 37 (1954).
1767. Rosenstein L.: US 1967264 (1934).
1768. Rossi W. J.: Hydrocarbon Proc. Petrol. Ref. *44*, No. 12, 109 (1965).

1769. Rossini F. D., Wagman D. D., Evans W. H., Levine S., Jaffe I.: *Selected Values of Chemical Thermodynamic properties*, Circular 500, Washington D. C., Nat. Bur. Standards, 1952.

1770. Rossini F. D., Pitzer K. S., Arnutt R. L., Braun R. M., Pimentel G. S.: *Selected Values of Physical and Thermodynamic Properties of Hydrocarbons and Related Compounds*. Pittsburgh 1953.

1771. Rossmanith G., Sander H., Sander W.: GDR 12202 (1956).

1772. Rossmanith G., Sander H., Sander W.: GDR 13369 (1957).

1772a. Rostrup-Nielsen J. R.: J. Catalysis *11*, 220 (1968).

1773. Rottig W.: Fr. 1180508 (1959).

1774. Rozovskii A. Ya., Shchekin V. V.: Tr. Inst. Nefti *10*, 286 (1957).

1775. Rubinshtein A. M.: *Problemy Kinetiky i Kataliza AN SSSR, Geterogennyi Kataliz* (*Heterogenous Catalysis*), p. 127. Izd. AN SSSR, Moscow 1949.

1776. Rubinshtein A. M., Kulikov S. G., Zakharov B. A.: Izv. AN SSSR, Otd. Khim. Nauk *1956*, 587.

1777. Rubinshtein A. M., Dulov A. A., Kulikov S. G., Pribytkova N. A.: Izvest. Akad. Nauk SSSR, Otdel. Khim. Nauk *1956*, 596.

1778. Rubinshtein A. M.: Kinetika i Kataliz *8*, 1094 (1967).

1779. Rudkovskii D. M., Trifel A. G., Alekseeva K. A.: Khim. i Tekhnol. Topliv i Masel *3*, No. 6, 17 (1958).

1780. Rudkovskii D. M.: Khim. i Tekhnol. Topliv i Masel *4*, No. 5, 1 (1959).

1781. Ruhrchemie A. G.: Fr. 70335 (1959).

1782. Russel R. P.: *The Science of Petroleum* (Ed. A. E. Dunstan), Vol. III, p. 2139. Oxford Univ. Press, London 1938.

1783. Rutkowski M.: Zeszyty Nauk. Polytech. Wroclaw, Chem. No. *12*, 91 (1965).

1784. Růžička V., Zbirovský M.: *Organická technologie* (*Organic Technology*), Vol. I, p. 22. SNTL, Prague 1967.

1785. Ryashentseva M. A., Minachev Kh. M., Afanaseva Yu. A.: Neftekhimiya *2*, 37 (1962).

1786. Ryashentseva M. A., Minachev Kh. M.: *Renii* (*Rhenium*), AN SSSR, Inst. Met., Tr. 2-go Vses. Soveshch. Moscow 1962, 226 (publ. 1964); C. A. 62, 11707 (1965).

1787. Ryashentseva M. A., Minachev Kh. M., Kalinovskii O. A., Goldfarb Ya. L.: Zh. Organ. Khim. *1* (6), 1104 (1965).

1787a. Ryashentseva M. A., Kalinovskii O. A., Minachev Kh. M., Goldfarb Ya. L.: Khim. Geterotsikl. Soed. *1966* (5), 694.

1787b. Ryashentseva M. A., Minachev Kh. M., Kolesnikov I. M., Panchenkov G. M.: Kinet. Kataliz *8*, 917 (1967).

1787c. Ryashentseva M. A., Minachev Kh. M., Geidysh L. S.: Izv. AN SSSR, Ser. Khim. *1968*, 1601.

1787d. Ryashentseva M. A., Minachev Kh. M.: Uspekhi Khim. *38*, 2050 (1969).

1787e. Ryashentseva M. A., Kalinovskii O. A., Minachev Kh. M., Fedorov B. P., Goldfarb Ya. L.: USSR 165424 (1964).

1787f. Ryashentseva M. A., Minachev Kh. M., Geidysh L. S.: USSR 170998 (1965).

1787g. Ryashentseva M. A., Kalinovskii O. A., Minachev Kh. M.: USSR 172832 (1965).

1787h. Ryashentseva M. A., Kalinovskii O. A., Minachev Kh. M., Goldfarb Ya. L.: USSR 172834 (1965).

1788. Sabatier P., Mailhe A.: Compt. rend. *150*, 1217.

1789. Sabatier P., Mailhe A.: Compt. rend. *150*, 1569.

1790. Sabavin V. I.: Khim. Tverd. Topliva *3*, 587 (1932).

1791. Sakaguchi M., Shibuki K.: Nippon Kagaku Zasshi *86*, 1123 (1965); C. A. *65*, 9774 (1966).

1792. Salimgareeva F. G., Davidovich B. V., Ivanova M. F.: Tr. Vost.- Sibir. Filiala AN SSSR, Ser. Khim. *18*, 95 (1959).
1793. Salvi G.: Riv. combustibili *3*, 369 (1949); C. A. *44*, 4659 (1950).
1794. Samoilov S. M., Rubinshtein A. M.: Izv. Akad. Nauk SSSR, Otdel. Khim. Nauk *1957*, 1158.
1795. Samoilov S. M., Rubinshtein A. M.: Izv. AN SSSR, Otd. Khim. Nauk *1958*, 550.
1796. Samoilov S. M., Rubinshtein A. M.: Izv. AN SSSR, Otd. Khim. Nauk *1958*, 557.
1797. Samoilov S. M., Rubinshtein A. M.: Izv. AN SSSR, Otd. Khim. Nauk *1959*, 1905.
1798. Samoilov S. M., Rubinshtein A. M.: Izv. AN SSSR, Otd. Khim. Nauk *1960*, 427.
1799. Samoilov S. M.: Izv. AN SSSR, Otd. Khim. Nauk *1961*, 1416.
1800. Samoilov S. M.: Izv. AN SSSR, Otd. Khim. Nauk *1961*, 1559.
1800a. Samsonov G. V., Oganesyan V. Kh.: Izv. Akad. Nauk SSSR, Neorg. Mater. *2* (10), 1757 (1966).
1801. Sanches R., Jurado J. R.: *Proc. World Petrol. Congr., 4th*, Rome 1955, Sect. III, 445.
1802. Saraceno A. J., Fanale D. T., Coggeshall N. D.: Anal. Chem. *33*, 500 (1961).
1803. Sartor A. F.: US 2512570 (1950).
1803a. Satterfield Ch. N., Roberts G. W.: A. I. Ch. E. J. *14* (1), 159 (1968).
1804. Schaafsma Y.: Neth. 77118 (1955).
1805. Schwartz D., Eberly P. E. Jr., Gladrow E. M.: US 2985699 (1961).
1806. Schenck R., von der Forst P.: Z. anorg. Chem. *249*, 76 (1942).
1806a. Schenk P., Vervoorn A. B. H., Waterman H. I., Weber A. B. R.: J. Inst. Petrol. *42*, 210 (1956).
1807. Schmerling L.: US 2948755 (1960).
1808. Schmid B. K., Beuther H.: Div. Petrol. Chem., Am. Chem. Soc. Preprint, March 23—26, 11 (2), B 51 (1966).
1809. Schmid B. K., Beuther H.: Ind. Eng. Chem., Process Design Dev. *6*, 207 (1967).
1810. Schmidt R., Günther G.: Chem. Technik *7*, 316 (1955).
1811. Schnabel B.: Chem. průmysl *9*, 10 (1959).
1812. Schnabel B., Večerka M.: Ropa a uhlie *2*, 230 (1960).
1813. Schnabel B., Večerka M.: *Sborník prací z výzkumu chemického využití uhlí, dehtu a ropy* (*Collection of Papers on the Chemical Utilisation of Coal, Tar and Petroleum*), Vol. 2, 165 (1962).
1814. Schnabel B., Večerka M., Žídek J.: Ropa a uhlie *6*, 198 (1964).
1815. Schnabel B., Večerka M., Žídek J.: Freiberger Forschungs-H. *A 367*, 79 (1965).
1816. Schnabel B., Kubička R.: Czech. 85992 (1956).
1817. Schnabel B., Smutný F., Večerka M., Žídek J.: Czech. 113457 (1965).
1818. Schneíder G. G., Bock H., Häusser H.: Ber. *70*, 425 (1937).
1819. Schneider W., Teubel J., Mahrwald R.: Chem. Technik *13*, 139 (1961).
1820. Schneider W., Mahrwald R., Teubel J.: Chem. Technik *13*, 423 (1961).
1821. Schneider W., Teubel J., Schmiedel R.: Chem. Technik *17*, 467 (1965).
1822. Schneider W., Teubel J., Schmiedel R.: Chem. Technik *17*, 577 (1965).
1823. Schöbert A., Wagner A.: *Methoden org. Chemie* (Houben-Weil), Vol. IX, p. 118. Georg Thieme Verlag, Stuttgart 1955.
1824. Schoennagel H. J.: Ber. Bunsenges. Phys. Chem. *71* (9—10), 1060 (1967).
1825. Scholven-Chemie Akt. Ges., Gelsenkirchen und H. Koppers G. m. b. H., Essen: Brit. 863711 (1961).
1826. Schöngut S., Vybíhal J.: Czech. 87760 (1958).
1826a. Schöngut S.: Paliva *34*, 237 (1954).
1827. v. Schuckmann G., Schnell H.: Austrian 208846 (1960).
1828. v. Schuckmann G., Schnell H.: Fr. 1218316 (1959).

1829. v. Schuckmann G., Schnell H.: Ger. 1105428 (1961).
1830. v. Schuckmann G.: Ger. 1176620 (1964).
1831. Schuit G. C. A., t'Hart M. L.: Neth. 84701 (1957).
1831a. Schulman B. L.: S. African 68 03066 (1968).
1832. Schulze W. A., Lyon J. P., Short G. H.: Ind. Eng. Chem. *40*, 2308 (1948).
1832a. Schuman S. C., Shalit H.: Catal. Rev. *4* (2), 245 (1970).
1833. Schuman S. C.: US 3291721 (1966).
1834. Schuman S. C., Wolk R. H., Chervenak M. C.: US 3321393 (1967).
1834a. Schuman S. C.: US 3344201 (1967).
1834b. Schuman S. C., Wolk R. H.: US 3363009 (1968).
1834c. Schuman S. C.: US 3387054 (1968).
1834d. Schutt H. U.: Ger. Offen. 1921693 (1969).
1835. Scott G., Tate F. E. G.: Brit. 798167 (1958).
1836. Scott J. W., Robbers J. A., Paterson N. J., Lavender H. M.: Petrol. Ref. *39*, No. 5, 161 (1960).
1837. Scott J. W., Robbers J. A., Mason H. F., Paterson N. J., Kozlowski R. H.: *6th World Petrol. Congr.*, Sect. III, Paper 18. Frankfurt 1963.
1838. Scott J. W., Paterson N. J.: *7th World Petrol. Congress*, PD 17, Paper 3. Mexico City 1967.
1839. Scott J. W.: Canad. 627923 (1961).
1840. Scott J. W.: US 2944005 (1960).
1841. Scott J. W.: US 2944006 (1960).
1842. Scott J. W.: US 2944089 (1960).
1843. Scott J. W., Mason H. F.: US 3117075 (1964).
1844. Scott J. W.: US 3127339 (1964).
1845. Scott J. W., Jacobson R. L.: US 3268438 (1966).
1846. Seaman W.: US 1993287 (1935).
1847. Seaman W., Huffman J. R.: US 2066189 (1937).
1848. Sebastian J. J. S.: Carnegie Inst. Tech. Coal Res. Lab. Contrib., *35*, 8 pp. (1936); C.A. *30*, 3968 (1936).
1849. Sebastian J. J. S., Riesz C. H.: Ind. Eng. Chem. *43*, 860 (1951).
1850. Seelig H. S., Coley J. R., Brennan H. M., Den Herder M. J.: Belg. 621681 (1962).
1851. Selwood P. W.: *Advances in Catalysis*, Vol. III., p. 27. Academic Press Inc., New York 1951.
1852. Selwood P. W.: *Catalysis* (Ed. P. H. Emmett), Vol. I, p. 353. Reinhold Publ. Co., New York 1954.
1853. Semenov S. S., Glushenkova E. V.: Trudy Vshesoyuz. Nauch. Issled. Inst. Pererabotki i Ispolzovan. Topliva *1960*, No. 9, 99.
1854. Semenova E. S.: Nauch. Osnovy Podbora i Proiz. Katalizatorov, AN SSSR, Sibir. Otd. *1964*, 401.
1855. Samedova-Beibutova Z. A.: Tr. Azerb. Politekh. Inst. *1963* (9), 44; C. A. 61, 9004 (1964).
1856. Semiletov S. A.: Kristallografiya *6* (4), 536 (1961).
1857. Sergienko S. R., Nozhkina I. A., Galpern G. D., Tsedilina A. I., Kolbanovskii Yu. A.: *Kataliticheskoe Gidrirovanie i Okislenie, Trudy Konf.* (*Proceedings of Conference on Catalytical Hydrogenation and Oxidation*), p. 179. Akad. Nauk Kazakh. SSR, 1955; C. A. 50, 12856 (1956).
1858. Sergienko S. R., Nozdrina E. V., Nozhkina I. A.: Tr. Inst. Nefti AN SSSR *12*, 136 (1958).
1859. Sergienko S. R., Nozhkina I. A., Nozdrina E. V.: Trudy Inst. Nefti AN SSSR *12*, 168 (1958).

1860. Sergienko S. R., Perchenko V. N.: Dokl. AN SSSR *128*, 103 (1959).
1861. Sergienko S. R., Nozhkina I. A., Nozdrina E. V.: *Khim. Seraorganicheskikh Soedinenii Soderzh. v Neft. i Nefteprod.* (*Chemistry of Organic Sulphur Compounds Contained in Petroleum and its Products*), p. 85. Izd. AN SSSR, Moscow 1959.
1862. Sergienko S. R., Perchenko V. N.: Izv. AN Turk. SSR, Ser. Fiz.-Tekhn., Khim. i Geol. *1960*, No. 1, 65.
1863. Sergienko S. R., Medvedeva V. D., Garbalinskii V. A.: Izv. AN Turkm. SSR, Ser. Fiz.-Tekhn., Khim. i Geol. Nauk *1964* (3), 25.
1864. Shashkov A. S., Krylova I. V.: Vestnik Mosk. Univ. Ser. II, *21* (5), 45 (1966).
1865. Shavolina N. V., Orochko D. I., Silchenko D. I.: *Khim. i Tekhnol. Pererabotki Nefti.* Trudy VNII NP, *Vyp. 8*, p. 4. Gostoptekhizdat, Moscow 1959.
1866. Shavolina N. V., Karzhev V. I., Orochko D. I.: *Khim. i Tekhnol. Pererabotki Nefti,* Trudy VNII NP, *Vyp. 8*, p. 20. Gostoptekhizdat, Moscow 1959.
1867. Shavolina N. V., Kasatkin D. F., Karzhev V. I.: Koks i Khim. *1959*, No. 11, 45.
1868. Shchukarev S. A., Morozova M. P., Damen Ch.: Zhur. Obshch. Khim. *30*, 2102 (1960).
1869. Shekhter L. N., Myasnikov I. A., Pshezhetskii S. Ya.: Dokl. AN SSSR *109*, 1163 (1956).
1870. Shell Internationale Research Maatschappij N. V.: Belg. 609376 (1962).
1871. Shell Internationale Research Maatschappij N. V.: Belg. 639291 (1964).
1871a. Shell Internationale Research Maatschappij N. V.: Belg. 648528 (1964).
1872. Shell Development Co.: Brit. 596829 (1948).
1873. Shell Internationale Research Maatschappij N. V.: Brit. 878035 (1960).
1874. Shell Internationale Research Maatschappij N. V.: Brit. 891382 (1962).
1875. Shell Internationale Research Maatschappij N. V.: Brit. 897240 (1962).
1876. Shell Internationale Research Maatschappij N. V.: Brit. 908208 (1962).
1877. Shell Internationale Research Maatschappij N. V.: Brit. 911813 (1962).
1878. Shell Internationale Research Maatschappij N. V.: Brit. 917091 (1963).
1879. Shell Internationale Research Maatschappij N. V.: Brit. 945914 (1964).
1880. Shell Internationale Research Maatschappij N. V.: Brit. 1011696 (1965).
1881. Shell Internationale Research Maatschappij N. V.: Fr. 1379249 (1964).
1882. See 428.
1883. Shell Internationale Research Maatschappij N. V.: Neth. 284006 (1964).
1884. Shell Internationale Research Maatschappij N. V.: Neth. 298817 (1965).
1885. Shell Internationale Research Maatschappij N. V.: Neth. Appl. 6404461 (1965).
1886. Shell Internationale Research Maatschappij N. V.: Neth. Appl. 6404815 (1965).
1887. Shell Internationale Research Maatschappij N. V.: Neth. Appl. 6413450 (1966).
1887a. Shell Internationale Research Maatschappij N. V.: Neth. Appl. 6501529 (1966).
1888. Shell Internationale Research Maatschappij N. V.: Neth. Appl. 6506348 (1966).
1888a. Shell Internationale Research Maatschappij N. V.: Neth. Appl. 6601027 (1967).
1888b. Shell Internationale Research Maatschappij N. V.: Neth. Appl. 6609887 (1968).
1888d. Shell Internationale Research Maatschappij N. V.: Neth. Appl. 6702572 (1968).
1888e. Shell Internationale Research Maatschappij N. V.: Neth. Appl. 6707302 (1968).
1888f. Shell Internationale Research Maatschappij N. V.: Neth. Appl. 6709241 (1969).
1889. Shepardson R. M.: US 2404104 (1946).
1890. Sherwood P. W.: Erdöl u. Kohle *6*, 616 (1953).
1890a. Shimizu Y., Inoue K., Nishikata H., Koenuma Y., Takemura Y., Aizawa R., Kobayashi S., Egi K., Matsumoto K., Wakao N.: Kagaku Kogaku *4*, 193 (1970); Bull. Jap. Petrol. Inst. *12*, 10 (1970).
1890b. Shin Y. K.: Ger. 1275036 (1968).
1891. Shono S., Itabashi K., Yamada M., Kikuchi M.: Kogyo Kagaku Zasshi *64*, 1357 (1961); C. A. 58, 4390 (1963).

1892. Shono S., Yamada M.: Koru Taru *15*, 405 (1963); C.A. 61, 6823 (1964).
1893. Shtraler F. E., Moiseenko V. M.: Khim. Tverd. Topliva *6*, 161 (1935).
1894. Shuikin N. I., Minachev Kh. M., Ryashentseva M. A.: Dokl. AN SSSR *101*, 107 (1955).
1894a. Shultz J. F., Karn F. S., Anderson R. B.: US Bureau of Mines, Rep. Invest. No. 6974 (1967).
1895. Signaigo F. K.: US 2230390 (1941).
1896. Signaigo F. K.: US 2402456 (1946).
1897. Signaigo F. K.: US 2402686 (1946).
1898. Signaigo F. K.: US 2402683 (1946).
1899. Signaigo F. K.: US 2406410 (1946).
1900. Signaigo F. K.: US 2454099 (1948).
1900a. Signal Oil and Gas Co.: Fr. 1493227 (1967).
1900b. Šilhánek J., Zbirovský M., Šmídová J.: *Scientific Papers of the Institute of Chemical Technology*, Prague *C 11*, 25 (1967).
1900c. Silverman M. S.: Inorg. Chem. *6* (5), 1063 (1967).
1900d. Silverman M. S.: US 3399962 (1968).
1901. Simon W., Donath E.: Ger. 741224 (1943).
1902. Simon W., Waldman H. J., Süssenguth H.: Ger. 1004605 (1957).
1903. Sinnatt F. S.: Gas J. *208*, 433 (1934).
1904. Sinclair Ref. Co.: Brit. 827298 (1960).
1905. Skach F., Kaška R., Rosenthal J.: Czech. 102447 (1961).
1906. Skarchenko V. K.: Neftekhim., AN Ukr. SSR, Inst. Khim. Vysokomolek. Soed. 1964, 5; C. A. 62, 8906 (1965).
1907. Skinner L. C., Donath E. E., Schappert H., Frese E.: Petroleum Refiner *29*, No. 7, 83 (1950).
1908. Sklyar V. T., Lebedev E. V., Zakupra V. A., Lizogub A. P., Klimenko P. L.: Neftekhim. AN Turkm. SSR *1963*, 36; C. A. 60, 10442 (1964).
1909. Sklyar V. T., Lebedev E. V., Zakupra V. A.: Neftekhimiya *4*, 200 (1964).
1910. Sklyar V. T., Lebedev E. V., Zakupra V. A.: Neftekhimiya *4*, 209 (1964).
1911. Sklyar V. T., Lebedev E. V., Zakupra V. A.: *Nauch. Osnovy Podbora i Proiz. Katalizatorov* (*Scientific Fundamentals of Catalyst Production*), AN SSSR, Sibir. Otd. 1964, p. 316.
1912. Slaugh L. H.: Inorg. Chem. *3* (6), 920 (1964).
1912a. Slaugh L. H.: US 3321274 (1967).
1913. Smeykal K., Moll K. K.: Chem. Technik *19*, 92 (1967).
1914. Smeykal K., Moll K. K.: Brit. 1039856 (1966).
1915. Smeykal K., Pallutz H.: GDR 18230 (1960).
1916. Smeykal K., Pritzkow W., Knopel H.: GDR 19252 (1960).
1917. Smeykal K., Pallutz H.: Ger. 1080538 (1959).
1918. Smeykal K., Pritzkow W., Knopel H., Unger S., Moll K.: US 3119886 (1964).
1919. Smith H. A., Pennekamp E. F. H.: J. Am. Chem. Soc. *67*, 279 (1945).
1920. Smith H. A.: *Catalysis* (Ed. P. H. Emmett), Vol. 5, p. 175, Reinhold Publ. Corp., New York 1957.
1921. Smith D. F., Brown D., Dworkin A. S., Sasmor D. J., van Artsdalen E. R.: J. Am. Chem. Soc. *78*, 1533 (1956).
1921a. Smith J. H., Stover L. E.: S. African 67 05622 (1968).
1922. Smith W. M., Landrum T. G., Phillips G. E.: Ind. Eng. Chem. *44*, 586 (1952).
1923. Smith W. M.: US 2692226 (1954).
1923a. Sodi J., Elliott J. F.: Trans. Met. Soc. AIME *242* (10), 2143 (1968); C. A. 70, 51345 (1969).
1924. Soemantri R. M., Waterman H. I.: Chim. et ind. *82*, 181 (1959).
1925. Sokol L.: Chem. listy *50*, 711 (1956).

1926. Sokol L., Karas V.: Czech. 101009 (1961).
1927. Sokol L., Saleta L.: *Sborník prací z výzkumu chemického využití uhlí, dehtu a ropy* (*Collection of Papers on Chemical Utilisation of Coal, Tar and Petroleum*), Vol. 5, 135 (1965).
1927a. Sokolova N. P., Balandin A. A., Maksimova M. P., Skulskaya Z. M.: USSR 188982 (1966).
1928. Solomon E.: US 3167496 (1965).
1929. Solomon E.: US 3167497 (1965).
1930. Sennewald K., Goetz A., Kallrath G.: Ger. 1103330 (1960).
1931. Souby A. M., Schiller J. C.: US 2550389 (1951).
1932. Sowerwine E. O.: US 2921022 (1960).
1932a. Spackman J. W. C.: Nature *198*, 1266 (1963); Abstr. BD2-53, 19th Ann. Conf. I.U.P.A.C., London 1963.
1933. Spars B. G., Mc Call P. P.: US 3022245 (1962).
1934. Spars B. G., Mason H. F.: US 3288703 (1966).
1934a. Spars B. G., Kozlowski R. H.: US 3365389 (1968).
1935. Spector M. L., Christopher G. L. M., Carpenter E. L.: US 2784236 (1954).
1936. Spengler G.: Erdöl u. Kohle *7*, 156 (1954).
1937. Spillane P. X.: Brit. 642726 (1950).
1938. Spillane P. X.: US 2559325 (1951).
1939. Spitsyn V. I., Pikaeva V. I., Mikhailenko I. E.: Dokl. AN SSSR *159* (5), 1102 (1964).
1939a. Spitsyn V. I., Pikaeva V. I., Mikhailenko I. E.: Dokl. AN SSSR *177*, 1390 (1967).
1940. Sprowls R.: US 2782214 (1957).
1940a. Srikanthareddy P., Salvapati G. S., Janardanarao M., Vaidyeswaran R.: Indian J. Technol. *6* (11), 325 (1968); C. A. 70, 49187 (1969).
1941. Ssu, Wei-Chan: Acta Focalia Sinica *4*, 369 (1959); Chem. Zentr. 132, 14454 (1961).
1942. Standard Oil Dev. Co.: Brit. 477015 (1938).
1943. Standard Oil Dev. Co.: Brit. 534151 (1941).
1944. Standard Oil Dev. Co.: Brit. 587168 (1947).
1945. Standard Oil Dev. Co.: Brit. 598103 (1948).
1946. Standard Oil Dev. Co.: Brit. 599252 (1948).
1947. Standard Oil Dev. Co.: Brit. 602880 (1948).
1948. Standard Oil Dev. Co.: Brit. 605868 (1948).
1949. Standard Oil Dev. Co.: Brit. 660155 (1951).
1950. Standard Oil Dev. Co.: Brit. 664974 (1952).
1951. Standard Oil Dev. Co.: Brit. 667617 (1952).
1952. Standard Oil Dev. Co.: Brit. 672349 (1952).
1953. Standard Oil Dev. Co.: Brit. 699717 (1953).
1954. Standard Oil Dev. Co.: Brit. 709929 (1954).
1955. Standard Oil Dev. Co.: Brit. 714061 (1954).
1956. Standard Oil Dev. Co.: Brit. 718946 (1954).
1957. Standard I. G. Co.: Fr. 854992 (1939).
1958. Standard I. G. Co.: Ger. 703736 (1941).
1959. Štaud M., Radoš M., Šír E., Škarvada A.: Ropa a uhlie *5*, 205 (1963).
1959a. Steenberg L. R., Bicek E. J.: S. African 68 00921 (1968).
1959b. Steffgen F. W.: US 3463829 (1969).
1960. Steiner H.: *Catalysis* (Ed. P. H. Emmett), Vol. IV, p. 529. Reinhold Publ. Corp., New York 1956.
1961. Stengler W., Welker J., Leibnitz E.: Freiberger Forschungs-H. *A 329*, 51 (1964).
1962. Stephens R. E., Ke B., Trivich D.: J. Phys. Chem. *59*, 966 (1955).
1963. Štěpina V.: *Ropa — zdroj energie* (*Petroleum As a Source of Energy*), Vol. II, p. 283—284. Ústav pro technické a ekonomické informace, Prague 1962.

1964. Stevenson D. H., Heinemann H.: Ind. Eng. Chem. *49*, 664 (1957).
1965. Stiles A. B.: US 2620362 (1952).
1966. Stiles V. E., Inwood T. V.: US 2793170 (1957).
1966a. Stolfa F.: US 3502571 (1970).
1967. Stone C. M.: US 2418374 (1947).
1968. Stone F. S.: Trabajos Reunion Intern. Reactividad Sólidos, 3°, Madrid *1956*, 641; C. A. *52*, 1721 (1958).
1969. Storch H. H., Fisher C. H., Eisner A., Clarke L.: Ind. Eng. Chem. *32*, 346 (1940).
1970. Storch H. H.: *12th Report of the Comm. on Catalysis*, p. 86. Wiley, New York 1940.
1971. Storch H. H., Hirst L. L., Fisher C. H., Work H. K., Wagner F. W.: Ind. Eng. Chem. *33*, 264 (1941).
1972. Storch H. H.: Chem. Rev. *29*, 483 (1941).
1973. Stormont D. H.: Oil Gas J. *61*, No. 50, 96 (1963).
1973a. Straschil H. K., Cohn J. G.: *Kirk-Othmer Encyclopedia of Chemical Technology*, Vol. 15, p. 861. Interscience Publ., Div. J. Wiley, New York—London—Sydney—Toronto 1968.
1974. Straus F., Grindel H.: Ann. *439*, 298 (1924).
1975. Strecker H. A.: US 3158563 (1964).
1975a. Strich E. R., Blume H.: Khim. Prom. *44* (9), 705 (1968).
1976. Strunz H.: *Mineralogische Tabellen*, p. 93. III. Ausg., Greest u. Portig, Leipzig 1958.
1977. Stuart A. P.: US 2736689 (1956).
1978. Stuart A. P.: US 2935497 (1960).
1979. Stuart A. P.: US 2966455 (1960).
1980. Stubbles J. R., Richardson F. D.: Trans. Farad. Soc. *56*, 1460 (1960).
1980a. Styrov V. V.: Kinet. Katal. *9*, 124 (1968).
1981. Sukhanov V. P.: *Katal. Protsesy v Neftepererabotke* (*Catalytic Processes in Petroleum Technology*). Gostoptekhizdat, Moscow 1963.
1982. Suchet J. P.: *Chemical Physics of Semiconductors*. D. van Nostrand Co. Ltd., London 1965.
1983. Sudo K.: Science Repts. Research Insts. Tokoku Univ., Ser. *A 2*, 507 (1950); C.A. *46*, 4347 (1952).
1984. Sudo K.: Science Repts. Research Inst. Tokoku Univ., Ser. *A 4*, 182 (1952); C.A. *47*, 6321 (1953).
1984a. Sukhareva T. S., Mashkina A. V.: Kinet. Katal. *11*, 665 (1970).
1985. Sulimov A. D., Lobeev M. V., Kozhina I. N., Piguzova L. I., Papko T. S.: Khim. i Tekhnol. Topliv i Masel *3*, No. 12, 32 (1958).
1986. Sullivan R. F., Egan C. J., Langlois G. E., Sieg R. P.: J. Am. Chem. Soc. *83*, 1156 (1961).
1987. Sullivan R. F., Egan C. J., Langlois G. E.: J. Catalysis *3*, 183 (1964).
1988. Sultanov S. A., Mardanov M. A., Naroditskaya S. G.: Azerb. Khim. Zhur. *1960*, No. 3, 25.
1989. Sultanov S. A., Naroditskaya L. G., Mardanov M. A., Oserova Yu. F.: Azerb. Neft. Khoz. *41*, No. 1, 39 (1962); Chem. Zentr. 133, 15820 (1962).
1989a. Surrena H. J., Kirby A. M. Jr., Reif H. E.: S. African 67 05763 (1968).
1990. Sustmann H., Weinrotter F.: Brennstoff-Chem. *22*, 229 (1941).
1991. Švajgl O.: Collection Czech. Chem. Commun. *24*, 3829 (1959).
1992. Švajgl O.: Chem. průmysl *9*, 230 (1959).
1993. Švajgl O.: Collection Czech. Chem. Commun. *25*, 1883 (1960).
1994. Švajgl O.: *Thesis*. Institute of Chemical Technology, Prague 1962.
1995. Švajgl O.: Chem. průmysl *12*, 473 (1962).
1996. Švajgl O.: Collection Czech. Chem. Commun. *28*, 11 (1963).
1997. Švajgl O.: Freiberger Forschungs-H. *A 340*, 89 (1964).

1998. Švajgl O.: *Sborník prací z výzkumu chem. využití uhlí, dehtu a ropy* (*Collection of Research Papers on Chemical Utilisation of Coal, Tar and Petroleum*), Vol. 3, 105 (1964).
1999. Švajgl O.: *Sborník prací z výzkumu chemického využití uhlí, dehtu a ropy* (*Collection of Research Papers on Chemical Utilisation of Coal, Tar and Petroleum*), Vol. 4, 100 (1964).
2000. Švajgl O.: *Proceedings, International Conference on Catalysts*, Part II, p. 375. Karlovy Vary 1964.
2001. Švajgl O.: *Scientific Papers of the Institute of Chemical Technology Prague*, Technology of Fuels *7*, 187 (1965).
2002. Švajgl O.: Chem. průmysl *15*, 137 (1965).
2003. Švajgl O.: Ropa a uhlie *8*, 35 (1966).
2004. Švajgl O.: Chem. průmysl *16*, 209 (1966).
2005. Švajgl O.: *Sborník prací z výzkumu chemického využití uhlí, dehtu a ropy* (*Collection of Research Papers on Chemical Utilisation of Coal, Tar and Petroleum*), Vol. 6, 31 (1966).
2006. Švajgl O.: *Sborník prací z výzkumu chemického využití uhlí, dehtu a ropy* (*Collection of Research Papers on Chemical Utilisation of Coal, Tar and Petroleum*), Vol. 6, 129 (1966).
2007. Švajgl O.: *Studie hydrokrakování těžkých ropných destilátů na motorová paliva* (*Study of Hydrocracking Heavy Petroleum Distillates to Motor Fuels*). Research Report. Chemical Works-Záluží, 1966.
2008. Švajgl O.: Chem. průmysl *19*, 103 (1969).
2009. Švajgl O.: *Scientific Papers of the Institute of Chemical Technology*, Prague *D 13*, 163 (1967).
2010. Švajgl O.: *Sborník prací z výzkumu chemického využití uhlí, dehtu a ropy* (*Collection of Research Papers on Chemical Utilisation of Coal, Tar and Petroleum*), Vol. 8, 83 (1968).
2011. Švajgl O., Smrž Z.: Chem. průmysl *18*, 294 (1968).
2012. Švajgl O.: *Sborník prací z výzkumu chem. využití uhlí, dehtu a ropy* (*Collection of Papers on Chemical Utilization of Coal, Tar and Petroleum*) Vol. 9, 49 (1969).
2012a. Švajgl O., Smrž Z.: *Sborník prací z výzkumu chemického využití uhlí, dehtu a ropy* (*Collection of Papers on Chemical Utilization of Coal, Tar and Petroleum*) Vol. 9, 63 (1969).
2012b. Švajgl O.: Ropa a uhlie *12*, 300 (1970).
2012c. Švajgl O.: Chem. listy *64*, 591 (1970).
2012d. Švajgl O.: *Sborník prací z výzkumu chemického využití uhlí, dehtu a ropy* (*Collection of Papers on Chemical Utilization of Coal, Tar and Petroleum*) Vol. 10, 27 (1970).
2013. Švajgl O., Kopelent Z., Baláš Z.: Czech. 90194 (1959).
2014. Švajgl O., Kubička R.: Czech. 91962 (1959).
2015. Švajgl O.: Czech. 99097 (1961).
2016. Švajgl O.: Czech. 102086 (1961).
2017. Švajgl O.: Czech. 103170 (1962).
2018. Švajgl O.: Czech. 114646 (1965).
2019. Swanson H., Gilfrich N. T., Cook M. I.: Nat. Bur. Stand. Circ. 539, *Vol. 8*, 65 (1958).
2020. Sweeney W. J.: US 2167339 (1939).
2021. Sweetser S. B., Bronson S. O., Morbeck R. C.: US 2740747 (1956).
2022. Sweetser S. B., Bronson S. O., Weikart J.: US 2761816 (1956).
2023. Sweetser S. B., Bronson S. O.: US 2761817 (1956).
2024. Sweetser S. B., Weikart J.: US 2769754 (1956).
2025. Szayna A.: US 2273297 (1942).
2026. Szayna A.: US 2337358 (1943).
2027. Szebenyi I.: Magyar Kém. Folyóirat *62*, 73 (1956).
2028. Szebenyi I.: Magyar Tud. Akad. Kém. Tud. Osztál. Közleményei *10*, 13 (1958).
2029. Szebenyi I.: Magyar Tud. Akad. Kém. Tud. Osztál. Közlem. *10*, 59 (1958).

2030. Szebenyi I.: Periodica Polytech. *2*, 39 (1958).
2031. Szeszich v. L., Hupe R.: Brennstoff-Chem. *14*, 221 (1933).
2032. Szeszich v. L., Hupe R.: Brennstoff-Chem. *14*, 379 (1933).
2033. Szucs M.: Magyar Tudomanyos Akad. Kém. Tudomanyaok Osztalyának Közleményei *6*, 375 (1955).
2034. Tachiki K., Yamashita Y.: Japan 21507 ('61).
2035. Takenaka Y.: J. Soc. Chem. Ind. Japan *46*, 658 (1943); C.A. *43*, 2415 (1949).
2035a. Takte D. G., Rooney J. J.: Chem. Communications *1969*, 612.
2036. Talibi M. A., Abdullaev G. B.: Doklady AN Azerb. SSR *17*, No 2, 97 (1961).
2037. Tanner H. G.: US 2402694 (1946).
2038. Tarama K., Teranishi S., Hattori K., Higashi M.: J. Fuel Soc. Japan *42*, 99 (1963); C. A. *62*, 13898 (1965).
2039. Tarama K., Teranishi S., Hattori K., Nishida S.: J. Fuel Soc. Japan *42*, 156 (1963); C. A. *62*, 13898 (1965).
2040. Tatarskii S. V., Papok K. K., Semenido E. G.: Neft. Khoz. *24*, No. 2, 52 (1946).
2041. Taylor H. S.: Refiner Nat. Gasol. Manuf. *9*, No. 12, 83 (1930).
2042. Taylor J. H., Mason H. F.: US 3291722 (1966).
2043. Taylor J. H., Jacobson R. L.: US 3336216 (1967).
2044. Taylor R.: J. Chem. Soc. *1934*, 1429.
2045. Tatarskii S. V., Papok K. K., Semenido E. G.: Neft. Khoz. *24*, No. 2, 52 (1946).
2046. Teter J. W., Keith C. D., Rosenblatt E. P.: Fr. 1078528 (1954).
2047. Teter J. W.: US 2452505 (1948).
2048. Teubel J., Schneider W., Schmiedel R., Kluek H.: GDR 32622 (1964).
2049. Texaco Dev. Corp.: Brit. 627247 (1949).
2050. Texaco Dev. Corp.: Brit. 970357 (1964).
2050a. Texaco Dev. Corp.: Brit. 1098659 (1968).
2050b. Tecaco Dev. Corp.: Brit. 1110359 (1968).
2051. Thacker C. M., Miller E.: Ind. Eng. Chem. *36*, 182 (1944).
2052. Thacker C. M.: US 2411236 (1946).
2053. Theilheimer W.: *Synthetic Methods of Organic Chemistry*, Vol. 10, p. 97. S. Karger, Basel, New York 1956.
2054. Thibaut L.: Chimie et industrie *59*, 548 (1948).
2055. Thiele E. W.: Ind. Eng. Chem. *31*, 916 (1939).
2055a. Thiokol Chem. Corp.: Brit. 1059491 (1967).
2056. Thomas W. J., Strickland-Constable R. F.: Trans. Farad. Soc. *53*, 972 (1957).
2057. Thomas W. J., Ullah U.: J. Catalysis *9*, 278 (1967).
2057a. Thomas W. J., Ullah U.: J. Catalysis *15*, 342 (1969).
2057b. Thompson S. L., Olenzak A. T.: US 3511772 (1970).
2058. Thonon C., Limido J.: Brit. 905809 (1962).
2059. Thorn J. P., Guyer W. R. F., Arundale E.: US 2739945 (1956).
2060. Tolstopyatova A. A., Balandin A. A.: *Problemy Kinetiki i Kataliza* (*Problems in Kinetics and Catalysis*), AN SSSR 1960, No. 10, 351.
2061. Tolstopyatova A. A., Balandin A. A., Matyushenko V. Kh., Petrov Yu. I.: Izv. AN SSSR, Otd. Khim. Nauk *1961*, 583.
2062. Tomkinson M. G.: J. Chem. Soc. *125*, 2264 (1924).
2063. Topchiev A. V., Orlov Kh. Ya., Paushkin Ya. M.: Doklady AN SSSR *127*, No. 6, 1235 (1959).
2064. Trambouze Y., de Mourgues L., Perrin M.: J. chim. phys. *51*, 723 (1954).
2065. Traore K., Brenet J.: Compt. rend. *249*, 280 (1959).
2066. Travers A., Marecaux P.: 14me Congr. Chim. ind., Paris, Oct. 1934.

2067. Trdlička V., Mostecký J.: *Scientific Papers of the Institute of Chemical Technology*, Prague *D 9*, 55 (1966).
2068. Trdlička V., Schuman A., Mostecký J.: *Scientific Papers of the Institute of Chemical Technology*, Prague *D 13*, 141 (1967).
2069. Trimble R. A.: US 2414620 (1947).
2069a. Trippler S., Plate A. F., Danilova T. A., Ryashentseva M. A.: Neftekhim. *8*, 783 (1968).
2070. Tropsch H., Hlavica B., Weinstein O.: Brennstoff-Chem. *11*, 449 (1930).
2071. Tropsch H., Kassler R.: Brennstoff-Chem. *12*, 345 (1931).
2072. Tropsch H.: *Proc. Inter. Conf. Bituminous Coal, 3rd Conf.*, 2, 35 (1931).
2073. Tropsch H.: Fuel *11*, 61 (1932).
2074. Tropsch H., Weinstein O.: Mitt. Kohlenforschungsinst. Prague, *No. 4*, 171 (1932).
2075. Truitt P., Holst E. H., Sammons G.: J. Org. Chem. *22*, 1107 (1957).
2076. Trzhtsinskaya B. V., Kalechits I. V.: Kinetika i Kataliz *6*, 346 (1965).
2077. Tubandt C., Haedicke M.: Z. anorg. Chem. *160*, 297 (1927).
2078. Tulleners A. J., Reeg C. P., Price F. C.: US 3256177 (1966).
2079. Tupholme C. H. S.: Ind. Eng. Chem., New. Ed. *11*, 15 (1933).
2080. Turkewich J.: US 2307715 (1943).
2081. Tyson Ch. W.: US 2759769 (1956).
2081a. Tyutyunikov B. N., Bogdan I. V.: Izv. Vyssh. Ucheb. Zaved. Pishch. Tekhnol. *1968* (6), 38.
2082. Uchida A., Tanaka M., Matsuda S.: Kogyo Kagaku Zasshi *65*, 577 (1962); C. A. 58, 4451 (1963).
2083. Ungarisch-Deutsche Varga-Studiengesellschaft, Budapest: *The Varga Hydrocracking Process* (Catalogue).
2084. Unger S., Lehmann P.: Chem. Technik *11*, 252 (1959).
2085. Union Oil Co. of California: Fr. 1398827 (1965).
2085a. Uniroyal, Inc.: Neth. Appl. 6611430 (1967).
2086. Uniroyal, Inc.: Neth. Appl. 6611772 (1967).
2087. United Kingdom Atomic Energy Authority: Brit. 800730 (1958).
2088. United States Rubber Co.: Belg. 643911 (1964).
2089. United States Rubber Co.: Neth. Appl. 6402424 (1964).
2090. Universal Oil Prod. Co.: Brit. 826305 (1959).
2091. Universal Oil Prod. Co.: Brit. 939200 (1960).
2092. Universal Oil Prod. Co.: Brit. 993008 (1965).
2093. Universal Oil Prod. Co.: Fr. 1373253 (1964).
2093a. Universal Oil Prod. Co.: Fr. 1486367 (1967).
2094. Universal Oil Prod. Co.: Ital. 386026 (1940).
2095. Universal Oil Prod. Co.: Neth. Appl. 297593 (1965).
2096. Unverferth J. W.: Ger. 1238883 (1967).
2097. Unverferth J. W.: US 3211642 (1965).
2098. Urban W.: Erdöl u. Kohle *4*, 279 (1951).
2099. Urban W.: Erdöl u. Kohle *8*, 780 (1955).
2100. Urey H. C., Grosse A. V.: US 2690379 (1954).
2101. Vacek J., Patera E.: *Dehet a jeho chemické zpracování* (*Tar and its Chemical Processing*). SNTL, Prague 1960.
2102. Vaculík P.: *Chemie monomerů* (*The Chemistry of Monomers*), Vol. 1, pp. 683, 697. Nakl. ČSAV, Prague 1956.
2102a. Vaell R. P.: US 3505208 (1970).
2103. Vaiselberg K. B.: Khim. Tverd. Topliva *8*, 232 (1937).

2103a. Vakurova E. M., Sukhareva T. S., Mashkina A. V.: Kinetika Kataliz *9*, 649 (1968).
2103b. Valitov N. Kh., Panchenkov G. M., Balandina K. L.: Neftepererab. Neftekhim. *1970*, No. 3, 8.
2103c. Valitov N. Kh., Panchenkov G. M., Zakharov M. A., Poteryazhin V. A., Tanatarov M. A.: Neftepererab. Neftekhim. *1970*, No. 5, 12.
2103d. Van Dyke R. E., Crecelius R. L., Schutt H. U.: Ger. Offen. 1814973 (1969).
2103e. Van Geusau F. E. A., Wiedijk C., Aben P. C.: Brit. 1134144 (1968).
2103E. Van Giesen J. A. van der C.: Ger. Offen. 2010533 (1970).
2103f. Van Venrooy J. J.: US 3477963 (1969).
2103g. Van Zijll Langhout W. C.: Ger. Offen. 1950348 (1970).
2104. Varga J.: Brennstoff-Chem. *9*, 277 (1928).
2105. Varga J.: Math. naturw. Anz. ungar. Akad. Wiss. *50*, 408 (1934).
2106. Varga J., Makray v. I.: Brennstoff-Chem. *17*, 81 (1936).
2107. Varga J.: Magyar Tudemányos Akad. Kém. Tudományok Osztályának Közleményei *5*, 167; discussion 183 (1954); C.A. 50, 13416 (1956).
2108. Varga J., Rabó G., Zalai A.: Brennstoff-Chem. *37*, 244 (1956).
2109. Varga J., Szebenyi I.: Paliva *36*, 235 (1956).
2110. Varga J., Szebenyi I., Koksis E.: Acta Chim. Acad. Sci. Hung. *14*, 133 (1958).
2111. Varga J., Szebenyi I.: Acta Chim. Acad. Sci. Hung. *16*, 193 (1958).
2112. Varga J., Birthler R., Karolyi J., Steingaszner P., Zalai A.: Khim. i Tekhnol. Topliv i Masel *5*, No. 10, 11 (1960).
2113. Varga J., Karolyi J., Steingaszner P., Zalai A., Birthler R., Rabó G.: Petrol. Ref. *39*, No. 4, 182 (1960).
2114. Varga J.: Brit. 313505 (1930).
2115. Varga J.: Fr. 676464 (1929).
2116. Varga J.: Fr. 683069 (1929).
2117. Varga J.: Fr. 683070 (1929).
2118. See 402
2119. See 403
2120. Varga J.: US 1852988 (1928).
2121. Varga J.: US 1894924 (1933).
2122. Varga J.: US 1894926 (1933).
2123. Vartanyan S. A., Karapetyan N. G., Pirenyan S. K. et al.: Izv. AN Arm. SSR, Khim. Nauki *14*, 565 (1961).
2124. Vavon G., Berton A. L.: Bull. soc. chim. *37*, 296 (1925).
2125. VEB Leuna Werke "Walter Ulbricht", Leuna: Austrian 218498 (1961).
2126. VEB Leuna Werke "Walter Ulbricht": Belg. 612177 (1962).
2127. VEB Leuna Werke "Walter Ulbricht": Fr. 1407021 (1965).
2128. VEB Leuna Werke "Walter Ulbricht" (Schade H.): GDR 16279 (1959).
2129. Vepřek J.: *Sborník prací z výzkumu chemického využití uhlí, dehtu a ropy* (*Collection of Papers on Chemical Utilisation of Coal, Tar and Petroleum*), Vol. 5, 67 (1965).
2130. Vepřek J., Novák J.: Ibid. 8, 40 (1968).
2131. Vernon L. W., Richardson J. T.: *Abstracts of the 142nd Meeting of the Am. Chem. Soc., Div. of Colloid and Surface Chem.*, p. 71. Atlantic City, N. Y. 1962.
2132. Vernon L. W., Richardson J. T.: US 3000816 (1959).
2133. Veselov V. V., Orechkin D. B., Popova N. V., Shepoťko O. F.: Tr. Vost.-Sibir. Filiala AN SSSR, Ser. Khim. *26*, 135 (1959).
2134. Veselov V. V., Orechkin D. B., Popova N. V., Shepoťko O. F.: Khim. i Tekhnol. Topliv i Masel *5*, No. 8, 11 (1960).
2135. Veselý V.: *Kapalná paliva* (*Liquid Fuels*). SNTL, Prague 1956.

2136. Veselý V.: *Chemia a technologia ropy* (*Petroleum Chemistry and Technology*), Vol. I. SVTL, Bratislava 1963.

2137. Veselý V.: *Chemie a technologie ropy* (*Petroleum Chemistry and Technology*), Vol. II. SVTL, Bratislava 1967.

2138. Viles P. S.: US 2471131 (1949).

2138a. Vlugter J. C., van't Spijker P.: 8th World Petrol. Congress, PD 12, Paper 5. Moscow 1971.

2139. Voge H. H.: *Catalysis* (Ed. P. H. Emmett), Vol. VI, p. 407. Interscience Publ., New York 1958.

2140. Vol'Epshtein A. B., Dyakova M. K., Surovtseva V. V.: Izv. AN SSSR, Otd. Khim. Nauk *1960*, 2230.

2141. Vol'Epshtein A. B., Zharova M. N., Surovtseva V. V.: Trudy Inst. Gor. Iskop. AN SSSR, Vol. XVII, p. 262, Moscow 1962.

2142. Vol'Epshtein A. B.: *Khim. i Pererab. Topliv* (*Khim. i Tekhnol.*), p. 107. Akad. Nauk SSSR, Inst. Goryuch. Iskop. (1965).

2143. Vol'Epshtein A. B.: Ger. 1219035 (1966).

2144. Vol'Epshtein A. B., Dyakova M. K., Surovtseva V. V.: USSR 132641 (1960).

2145. Voorhies A., Smith W. M., Hemminger C. E.: Ind. Eng. Chem. *39*, 1104 (1947).

2146. Voorhies A. Jr., Smith W. M., Mac Laren D. D.: *6th World Petrol. Congr.*, Sect. III, Paper 19, Frankfurt 1963.

2147. Voorhies A., Metrailer W. J., Kimberlin C. N., Jahnig C. E.: *Intern. Conf. on the Mechanism of Corrosion by Fuel Impurities*, May 1963, Session III, Paper 13, Marchwood Eng. Laboratories.

2148. Voorhies A., Smith W. M.: *Advances in Petroleum Chemistry and Refining* (Ed. J. J. Mc Ketta, Jr.), Vol. VIII, p. 169. Interscience Publ., New York 1964.

2149. Voorhies A.: US 2455713 (1948).

2150. Voorhies A.: US 2554282 (1951).

2151. Voorhies A., Kimberlin C. N.: US 3238123 (1966).

2152. Vos J. M., de Vries R.: US 3059007 (1962).

2153. Vosser J. L., Ford G. W.: Brit. 1006508 (1965).

2154. Vvedenskii V. A.: *Thermodynamické výpočty petrochemických pochodů* (*Thermodynamic Calculations of Petrochemical Processes*). SNTL, Prague 1963.

2155. Vybíhal J., Schöngut S.: Chem. průmysl *6*, 358 (1956).

2156. Vybíhal J.: *Sborník prací z výzkumu chemického využití uhlí, dehtu a ropy* (*Collection of Papers on Chemical Utilisation of Coal, Tar and Petroleum*), Vol. 6, 151 (1966).

2157. Vybíhal J.: Czech. 102446 (1961).

2158. Vybíhal J.: Czech. 111030 (1964).

2159. Vykhovanets V. V., Kalechits I. V.: *Sb. Primenenie Mechennykh Atomov dlya Izucheniya Neftekhimicheskikh Protsessov* (*Utilisation of Labelled Atoms for Research of the Petrochemical Processes*), p. 59 .VNIIOENG, Moscow 1965.

2160. Wadley E. F., Bennett R. B.: Fr. 982298 (1951).

2161. Wainwright H. W.: US Dep. Interior, Bur. Mines, Inform. Circ. *1961*, Nr. 8054, 1—85. Morgantown, W. Va., Bur. of Mines, Morgantown Coal Res. Center.

2162. Walker S. W.: Ind. Eng. Chem. *38*, 906 (1946).

2162a. Wardlaw R. S.: Gas J. *329* (5396), 283 (1967).

2162b. Warthen J. L., Briggs W. S., Ciapetta F. G.: Ger. Offen. 1807013 (1969).

2163. Wassermann A.: Nature *147*, 391 (1941).

2164. Wassermann A., Weller W. T.: Nature *149*, 669 (1942).

2165. Wassermann A., Weller W. T.: J. Chem. Soc. *1947*, 250.

2166. Wassermann A.: Brit. 499958 (1939).

2167. Wassermann A.: Fr. 838379 (1938).

2168. Wassermann A.: US 2224071 (1940).
2168a. Watanabe T.. Nishizaki S., Kurita S.: Kogyo Kagaku Zasshi *73*, 461 (1970).
2169. Watkins C. H., de Rosset A. J.: Petroleum Ref. *36*, No. 3, 201 (1957).
2170. Watkins C. H.: *7th World Petrol. Congr.*, PD 20, Paper 5. Mexico City 1967.
2171. Watson F. D., Franse A. D., Pettefer R. L.: US 3247089 (1966).
2172. Watson W. B., Bentley H. R.: US 3232862 (1966).
2173. Watts R. N.: US 1904582 (1933).
2174. Wehner K., Kaufmann H., Welker J.: GDR 42228 (1965).
2174a. Wehner K., Welker J., Seidel G.: Ropa a uhlie *10*, 288 (1968).
2175. Weisser O.: *Thesis*. Institute of Chemical Technology, Prague 1959.
2176. Weisser O., Trdlička V.: *Scientific Papers of the Institute of Chemical Technology*, Prague 1961, Technology of Fuels *5*, 7.
2177. Weisser O.: Chem. listy *56*, 1222 (1962); Intern. Chem. Engng. *3*, 388 (1963).
2178. Weisser O., Landa S., Mostecký J.: Ropa a uhlie *5*, 201 (1963).
2179. Weisser O., Mostecký J., Landa S.: Brennstoff-Chem. *44*, 286 (1963).
2180. Weisser O., Landa S., Pecka K.: Chem. Technik *16*, 463 (1964).
2181. Weisser O., Popl M.: *Research Report*. Institute of Chemical Technology, Prague 1965.
2182. Weisser O.: *Thesis II. Institute of Chemical Technology*, Prague 1965.
2183. Weisser O.: Chem. průmysl *17*, 307 (1967).
2184. Weisser O., Landa S., Hála S., Petrov A. D., Kaplan E. P.: Neftekhimiya *8*, 421 (1968).
2185. Weisser O., Mostecký J., Eyem J., Kubelka V.: *Rozbor technické přípravy pinenmerkaptanu a analytické vyhodnocení produktu thiolace terpentýnu* (*Preparation and Analysis of the Product of Thiolation of Turpentine*). Research Report. Institute of Chemical Technology, Prague 1969.
2185a. Weisser O., Hála S.: Collection Czech. Chem. Commun. 37, No. 12 (at press).
2186. Weisser O., Šešulka V., Klvaňa D.: not yet published.
2187. Weisser O., Jaffar A.: not yet published.
2188. Weisser O., Mostecký J., Landa S.: Czech. 108423 (1963).
2189. Weisser O., Landa S., Mostecký J., Popl M.: Czech 122948 (1967).
2190. Weisz P. B., Prater G. D.: *Advances in Catalysis*, Vol. 13, 137 (1962).
2191. Welker J.: Freiberger Forschungs-H. *A 264*, 63 (1963).
2192. Welker J.: Freiberger Forschungs-H. *A 340*, 223 (1964).
2193. Welker J., Risse G.: Chem. Technik *16*, 549 (1964).
2194. Welker J.: *Proceedings, International Conference on Catalysts*, p. 441. Karlovy Vary 1964.
2195. Welker J., Wehner K.: *Proceedings, International Conference on Catalysts*, p. 459. Karlovy Vary 1964.
2196. Welker J.: Chem. Technik *17*, 213 (1965).
2197. Welker J., Wehner K.: Ropa a uhlie *7*, 163 (1965).
2198. Welker J., Onderka E.: Freiberger Forschungs-H. *A 387*, 93 (1966).
2199. Weller S., Pelipetz M. G., Friedman S.: Ind. Eng. Chem. *43*, 1572 (1951).
2200. Weller S., Pelipetz M. G., Friedman S.: Ind. Eng. Chem. *43*, 1575 (1951).
2201. Weller S., Pelipetz M. G.: *Proc. 3rd World Petrol. Congr.*, Sect. IV, 91. The Hague 1951.
2202. Weller S. W.: *Catalysis* (Ed. P. H. Emmett), Vol. IV, p. 513. Reinhold Publ. Corp., New York 1956.
2203. Werntz J. H.: US 2402698 (1946).
2204. Westgren A.: Z. anorg. Chem. *239*, 82 (1938).
2205. White P. T., Porter F. W. B.: Ger. 1119844 (1961).
2206. White R. J.: US 3153100 (1964).
2207. White R. J.: US 3248318 (1966).

2208. Wichterle O.: *Organická chemie (Organic Chemistry)*, p. 324, 377. Nakl. ČSAV, Prague 1955.
2209. Wichterle O., Petrů F.: *Anorganická chemie (Inorganic Chemistry)*, p. 251. Nakl. ČSAV, Prague 1956.
2210. Wilderwanck J. C., Jellinek F.: Z. anorg. Chem. *328*, 309 (1964).
2210a. Wilkinson H. F.: US 3468788 (1969).
2211. Wille H. C.: Fr. 1401376 (1965).
2212. Wille H. C.: *Die Chemie der Braunkohle* (Ed. A. Lissner, A. Thau), Vol. 2, p. 294. W. Knapp-Verlag, Halle 1953.
2213. Williams E. C., Williams C. Ch.: Brit. 464952 (1937).
2214. Williams E. C., Allen C. C.: Canad. 385013 (1939).
2215. Williams E. C., Allen C. C.: US 2052268 (1936).
2215a. Wilmot P. D.: Petroleum *21*, 225 (1958).
2216. Wilson R. L., Kemball C., Galwey A. K.: Trans. Farad. Soc. *58*, 583 (1962).
2216a. Wilson R. L.: *Thesis*. The Queen's University Belfast, 1962.
2217. Wilson R. L., Kemball C.: J. Catal. *3*, 426 (1964).
2218. Wilson W. A., Voreck W. E., Malo R. V.: Ind. Eng. Chem. *49*, 657 (1957).
2219. Wilson W. B.: US 3278418 (1966).
2220. Winsor J., Carruthers J., Peet W. A.: Erdöl u. Kohle *20*, 272 (1967).
2221. Winstrom L. O., Harris W. B.: US 2671763 (1954).
2222. Winstrom L. O.: US 2716135 (1955).
2223. Winstrom L. O.: US 2822397 (1958).
2224. Winstrom L. O.: US 2875158 (1959).
2225. Winter H., Free G., Mönnig H.: Brennstoff-Chem. *18*, 320 (1937).
2225a. Wiser W. H., Singh S., Qader S. A., Hill G. R.: Ind. Eng. Chem., Prod. Res. Dev. *9*, 350 (1970).
2226. Wolf F., Morgner M.: GDR 12749 (1957).
2227. Wood F. C., Bradley W. E., Peralta B.: US 3186936 (1965).
2228. Wood F. C.: US 3338819 (1967).
2228a. Wood F. C., Dhondt R. O., Cheadle G. D.: US 3458433 (1969).
2228b. Wood F. S., Menzel R. L., Marks M. E., Garst R. G.: US 3382168 (1968).
2229. Worrel C. J.: US 2951874 (1960).
2229a. Wu W. R. K., Storch H. H.: US Bureau of Mines, *Bull. 1968*, No. 633, 195 pp.
2229b. Wunderlich D. K.: US 3441500 (1969).
2230. Wustrow W., Mädrich O., Macura H.: Ger. 1085287 (1960).
2231. Wyk van J. W.: US 2859167 (1958).
2232. Yamada M.: Koru Taru *12*, 8 (1960); C.A. *60*, 11978 (1964).
2233. Yamada M.: Koru Taru *12*, 14 (1960); C.A. *60*, 11978 (1964).
2234. Yamada M.: Kogyo Kagaku Zasshi *64*, 1071 (1961); C.A. *57*, 3371 (1962).
2235. Yamamoto S.: US 3128243 (1964).
2236. Yamamoto S.: US 3211668 (1965).
2237. Yamauchi T., Matsuda S.: Bull. Japan Petrol. Inst. *2*, 76 (1960); C.A. *55*, 955 (1961).
2237a. Yan Tsoung-Yuan: US 3444074 (1969); C. A. 71, 32103 (1969).
2238. Yen T. F., Erdman J. G., Pollack S. S.: Anal. Chem. *33*, 1587 (1961).
2239. Yen T. F., Erdman J. G., Saraceno A. J.: Anal. Chem. *34*, 694 (1962).
2239a. Young B. J., Duir J. H.: US 3364133 (1968).
2240. Young D. A.: US 3239451 (1966).
2241. Young D. A.: US 3287256 (1966).
2242. Young D. A.: US 3342725 (1967).
2242a. Young P. A.: Brit. J. Appl. Phys. *1968* [2], 1 (7), 936.

2243. Yukhnovskii G. L.: Ukrain. Khim. Zhur. *3*, No 2, 65 (1928); C. A. 23, 377 (1929).
2244. Yura S., Hara H.: Jap. 174870 (1948).
2245. Yuriev J. K., Borisov A. E.: Ber. *69*, 1395 (1936).
2246. Yuriev Yu. B.: Uch. Zapiski Mosk. Gos. Univ. *79*, 1 (1945).
2247. Zakharenko V. A.: Khim. i Tekhnol. Topliv i Masel *3*, No. 8, 64 (1958).
2248. Zakharenko V. A., Lozovoi A. V.: *Trudy Inst. Gor. Iskop.* Tom. IX, p. 96, Moscow 1959.
2249. Zakharenko V. A.: *Trudy Inst. Gor. Iskop.*, Tom IX, p. 107, Moscow 1959.
2250. Zakharenko V. A.: *Trudy Inst. Gor. Iskop.* Tom IX, p. 154, Moscow 1959.
2251. Zakharenko V. A.: *Khim. i Tekhnol. Smol Termicheskoi Pererabotki Tverdykh Topliv* (*Chemistry and Technology of Tars from Thermal Processing of Solid Fuels*), p. 57. Izd. Nauka, Moscow 1965.
2252. Zakharenko V. A.: *Khim. i Tekhnol. Smol Termicheskoi Pererabotki Tverdykh Topliv* (*Chemistry and Technology of Tars from Thermal Processing of Solid Fuels*), p. 71, Izd. Nauka, Moscow 1965.
2253. Zaidman N. M., Orechkin D. B., Gladovskaya M. F., Martynova E. N.: Khim. i Tekhnol. Topliv i Masel *6*, No. 1, 25 (1961).
2254. Zalai A., Birthler R.: Acta Chim. Acad. Sci. Hung. *31*, 301 (1962).
2254a. Zalevskii B. K., Opalovskii A. A., Sobolev V. V., Fedorov V. E.: Izv. Sibir. Otd. AN SSSR, Ser. Khim. Nauk *1967* (3), 14; C. A. 68, 109783 (1968).
2254b. Zavelskii D. Z., Vylegzhanina E. P., Akimova G. S., Kasatkina E. I., Grachev A. M., Bitepazh Yu. A., Osmolovskii G. M.: Brit. 1135915 (1968).
2255. Zbirovský M., Šilhánek J.: Chem. průmysl *16*, 528 (1966).
2256. Zeise H.: *Thermodynamik*, B III/1, pp. 41; 93; 138. Hirzel Verlag, Leipzig 1954.
2257. Zeldovich Ya. B.: Zhur. Fiz. Khim. *10*, 583 (1939).
2258. Zelikman A. N., Krein O. E.: Zhur. Fiz. Khim. *29*, 2081 (1955).
2259. Zelikman A. N., Khristyakov Yu. D., Indenbaum G. V., Krein O. E.: Kristallografiya *6*, 389 (1961); C. A. 57, 5391 (1962).
2260. Zelvenskii Ya. D., Nedumova E. S., Prokopets V. E.: Khim. Prom. *1961*, 77.
2261. Zerbe I. C., Falkens K.: Petroleum Refiner *26*, No. 3, 144 (1948).
2262. Zerbe K., Grosskopf K.: Brennstoff-Chem. *19*, 61 (1938).
2263. Zhorov Yu. M., Panchenkov G. M., Kuznetsov O. I., Bazilevich V. V.: Tr. Mosk. Inst. Neftekhim. i Gaz. Prom. No. 51, 148 (1964).
2264. Zhorov Yu. M., Kuznetsov O. I., Panchenkov G. M.: Neftekhimiya *6*, 396 (1966).
2264a. Zhorov Yu. M., Kuznetsov O. I., Panchenkov G. M.: Tr. Mosk. Inst. Neftekhim. Gazov. Prom. *1967*, No. 69, 82.
2265. Zhorov Yu. M., Kuznetsov O. I., Panchenkov G. M.: USSR 186391 (1966).
2266. Zhuk N. P.: Zhur. Fiz. Khim. *28*, 1523 (1954).
2267. Zimmermann K. T.: Angew. Chem. *74*, 151 (1962).
2268. Zogala J.: Freiberger Forschungs-H. *A 36*, 31 (1955).
2269. Zogala J., Cír F.: Czech. 91408 (1959).
2270. Zogala J., Cír F., Kaška R., Frank J.: Czech. 96085 (1960).
2271. Zyryanov B. F., Kalashnikova N. I., Orechkin D. B.: Tr. Vost.-Sibir. Filiala AN SSSR, Ser. Khim. *26*, 141 (1959).

Author Index

A

B

C

D

E

F

G

I

J

K

L

M

N

O

P

Q

R

S

T

W

Y

Z

Subject Index

A

B

C

D

F

G

H

N

O

Numerical Patent Index

Canadian

Czechoslovakian

French

German